Breeding Vegetable Crops

Edited by

MARK J. BASSETT

Vegetable Crops Department
University of Florida
Gainesville, Florida

AVI PUBLISHING COMPANY, INC.
Westport, Connecticut

Breeding Vegetable Crops

Book design and illustrations for cover and title page
by Alma Orenstein

Copyright 1986 by
THE AVI PUBLISHING COMPANY, INC.
P.O. Box 831
250 Post Road East
Westport, Connecticut 06881

Library of Congress Cataloging-in-Publication Data

Main entry under title:

Breeding vegetable crops.

   Includes bibliographies and index.
   1. Vegetables—Breeding.   I. Bassett, Mark J.
SB324.7.B74   1986   631.5'3   86—19954
ISBN 0—87055—499—9

Printed in the United States of America
A B C D 5432109876

Contents

4   Tomato Breeding
EDWARD C. TIGCHELAAR

5   Cucumber Breeding
R. L. LOWER and M. D. EDWARDS

6   Squash Breeding
THOMAS W. WHITAKER and R. W. ROBINSON

Contributors

David W. Davis, Department of Horticultural Science and Landscape Architecture, University of Minnesota, St. Paul, MN 55108

Michael H. Dickson, Department of Horticultural Sciences, New York State Agricultural Experiment Station, Geneva, NY 14456

P. D. Dukes, U.S. Vegetable Laboratory, Agricultural Research Service, U.S. Department of Agriculture, Charleston, SC 29407

M. D. Edwards,[1] Department of Horticulture, University of Wisconsin, Madison, WI 53706

J. Howard Ellison, Department of Horticulture and Forestry, Cook College, Rutgers University, New Brunswick, NJ 08903

Walter H. Greenleaf,[2] Department of Horticulture, Auburn University, Auburn, AL 36849

Earl T. Gritton, Department of Agronomy, University of Wisconsin, Madison, WI 53706

Alfred Jones, U.S. Vegetable Laboratory, Agricultural Research Service, U.S. Department of Agriculture, Charleston, SC 29407

Karl Kaukis, Senior Agronomist Emeritus, Agricultural Research Department, Pillsbury Co., LeSueur, MN 56058

R. L. Lower,[3] Department of Horticulture, University of Wisconsin, Madison, WI 53706

Hubert C. Mohr,[4] Department of Horticulture and Landscape Architecture, University of Kentucky, Lexington, KY 40506

C. E. Peterson, Agricultural Research Service, U.S. Department of Agriculture, Department of Horticulture, University of Wisconsin, Madison, WI 53706

---

[1]Current address: Agricultural Research Service, U.S. Department of Agriculture, Department of Genetics, North Carolina State University, Raleigh, NC 27650.

[2]Current address: 1007 N. College St., Auburn, AL 36830.

[3]Current address: 1450 Linden Dr., University of Wisconsin, Madison, WI 53706.

[4]Current address: 1649 Linstead Dr., Lexington, KY 40504.

Leonard M. Pike, Department of Horticultural Sciences, Texas A&M University, College Station, TX 77843

R. W. Robinson, Department of Horticultural Sciences, New York State Agricultural Experiment Station, Geneva, NY 14456

Edward J. Ryder, U.S. Agricultural Research Station, P.O. Box 5098, Salinas, CA 93915

J. M. Schalk, U.S. Vegetable Laboratory, Agricultural Research Service, U.S. Department of Agriculture, Charleston, SC 29407

M. J. Silbernagel, Agricultural Research Service, U.S. Department of Agriculture, Washington State University, Irrigated Agriculture Research and Extension Center, P.O. Box 30, Prosser, WA 99350

P. W. Simon, Agricultural Research Service, U.S. Department of Agriculture, Department of Horticulture, University of Wisconsin, Madison, WI 53706

Edward C. Tigchelaar, Department of Horticulture, Purdue University, West Lafayette, IN 47907

D. H. Wallace, Departments of Plant Breeding and Biometry, and Vegetable Crops, Cornell University, Ithaca, NY 14853

Thomas W. Whitaker, Agricultural Research Service, U.S. Department of Agriculture, and Department of Biology, University of California, San Diego, P.O. Box 150, La Jolla, CA 92038

Preface

Students and teachers interested in plant breeding have many good textbooks from which to choose. All of these books present the established procedures of plant breeding, and a few of these textbooks show how plant breeding methods are applied to specific agronomic crops. There are relatively few books dealing with the genetic improvement of horticultural crops. *Advances in Fruit Breeding,* edited by Jules Janick and James Moore, has provided ample treatment of genetic improvement of many fruit crops, emphasizing the woody perennials of the temperate zone. *Breeding and Genetics in Horticulture* by C. North includes vegetable improvement, but the treatment given to each vegetable crop is very brief, often only a page or less. *Flower & Vegetable Plant Breeding* by L. Watts also addresses the improvement of several vegetable crops, but the maximum length of presentation for each crop is usually three pages. The last extensive description of the "state of the art" was the *Yearbook of Agriculture 1937,* which devoted 223 pages to the genetic improvement of vegetables.

The purpose of writing *Breeding Vegetable Crops* is to give extensive, up-to-date treatment to the genetic improvement of 14 vegetable crops. Each crop has its own unique requirements, opportunities, and challenges. Emphasis has been placed on the practical aspects of applying breeding techniques and current genetic knowledge to vegetable improvement. This book is intended for advanced students who already have had training in genetics and plant breeding; therefore, there are no chapters that present the fundamentals of these subjects. Great contributions to knowledge have been made during the last few decades in the various disciplines supporting genetic improvement of vegetables, but the reports of these researchers are widely scattered in the journal literature.

Each author contributing to *Breeding Vegetable Crops* has had long experience with his chosen crop and is able to "fill in the gaps" left by the brevity and highly specialized nature of journal reports. The task of each contributor involved evaluating the merits of these research reports, choosing only those aspects of value to vegetable breeders, and describing how to exploit this knowledge in a breeding program designed to meet various industry needs.

Both English and metric units of measure are used in this book, each where appropriate. The primary readership addressed by this book is the English-speaking agricultural science community. Because agricultural production is usually performed and reported in English units (at least in the United States) and much field research is conducted in English units, the clarity and easy accessibility of information are best served by being flexible on this issue. The rule is simple: describe and present the work in the same units originally used for the work and data collection.

M. J. Bassett

Breeding Vegetable Crops

1
Sweet Potato Breeding

ALFRED JONES
P. D. DUKES
J. M. SCHALK

The sweet potato *Ipomoea batatas* (L.) Lam. is an asexually propagated vegetable grown in commercial quantities along the East Coast, in New Jersey, Maryland, Virginia, North Carolina, South Carolina, Georgia, and Florida; in the Gulf states of Alabama, Mississippi, Louisiana, and Texas; in other southern and western states including Tennessee, Oklahoma, Arkansas, New Mexico, and California; and on the islands of Hawaii and Puerto Rico. Currently about 35% of the U.S. crop is produced in North Carolina, followed by 19% in Louisiana, 10% in California, and 8% in Texas (*89*). Thus, over half of the U.S. production is concentrated in two states, North Carolina and Louisiana. The area devoted to sweet potato production declined drastically from 1949 to 1970 but appears to have stabilized during the 1970s at about 119,000 acres (48,000 ha) (Table 1.1). Total production and per capita consumption followed the same trend, although the declines were not as severe due to increasing yields per acre.

The sweet potato is grown in most of the tropical and subtropical regions of the earth, where it is an important staple of subsistence farmers. The vines as well as the roots are eaten or fed to livestock (*92*). It is an important source of industrial starch in Japan, and its potential for fuel alcohol conversion is under study in many countries. In the United States, its cultivation is extended to its temperate limits. Sweet potato can be planted about 4 weeks after the average date of the last killing frost and generally requires a frost-free growing season of 4 or 5 months. Some cultivars will produce edible-sized roots in as

1

TABLE 1.1. **Trends in Area under Production, Yields, and Per Capita Consumption of Sweet Potatoes in the United States, 1940–1979**[a]

Years	Production Area (×1000 acres)	Av. yield (tons/acre)	Total (×1000 tons)	Per capita consumption (lb)
1940–1944	731	2.4	1754	19.4
1945–1949	551	2.6	1429	15.2
1950–1954	361	2.6	954	9.5
1955–1959	282	3.3	934	8.6
1960–1964	180	4.1	737	7.1
1965–1969	146	4.8	701	6.0
1970–1974	116	5.4	623	5.3
1975–1979	119	5.7	678	5.5

[a]Adapted from Walsh and Johnson (89).

little as 100 days, but if night temperatures are too cool they may not grow normally even though they survive. Although U.S. yields have increased in recent years to 5.4 or 5.7 tons/acre (12–13 MT/ha) (Table 1.1), yields in more tropical countries frequently reach 15.6–17.8 tons/acre (35–40 MT/ha). The lower U.S. yields are primarily due to a cooler climate and to the fact that markets require rather small roots, which results in harvesting prior to maximum yield.

ORIGIN AND GENERAL BOTANY

The exact origin of the sweet potato is not known, but an American origin is generally accepted. Available evidence suggests southern Mexico through Central America and northern South America as the probable area of origin. Until more extensive collections are made, especially in Central America and northern South America, the exact center cannot be specified. The sweet potato is obviously better adapted to seasonal variations characterized by wet and dry seasons than by warm and cold seasons. It is a perennial but is handled as an annual in the temperate United States, where its hardiness and drought tolerance are well recognized.

A member of the Convolvulaceae, the sweet potato has 90 chromosomes (Fig. 1.1) and is the only known natural hexaploid morning glory (3,32). Most of the wild species collected have proven to be diploid (2n = 30) (35), although some have been tetraploid (62,63) and occasional collections have been triploid (77). There is considerable uncertainty regarding the phylogenetic relationships of the wild species. Austin's (4) recent taxonomic study provides a good start toward clarification of the sweet potato complex, but clear understanding awaits more complete collections and cytotaxonomic studies (93).

Although hybridization with other species of the genus has been demonstrated by Japanese workers (53,78), the technique is very difficult. Generally hybridization is restricted to crosses within the species. Because it is hexaploid, there is extensive variability within the species available for exploitation by plant breeders (34,92). Each seedling is genetically different from all others and is potentially a new cultivar (Fig. 1.2).

In the United States, the sweet potato is sometimes referred to commercially as a "yam." The designation is technically incorrect and is quite confusing to those who have

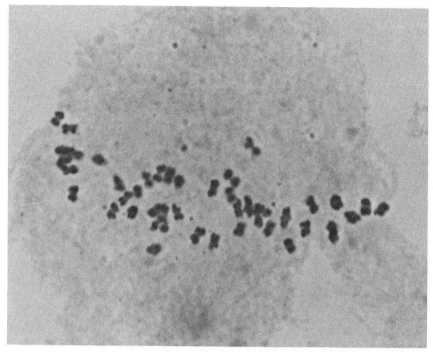

FIGURE 1.1. Chromosome pairing at metaphase I of a sweet potato pollen mother cell, illustrating that bivalent pairing is the rule and involves an average of 87.6 of the 90 chromosomes.
After Jones (32).

lived in tropical areas where yams are an important food. The 50 or more species of edible yams in the tropics belonging to the genus *Dioscorea* (59) are not even distantly related to the sweet potato. Frequently the edible sweet potato is wrongly referred to as a "tuber." Since a tuber is a fleshy subterranean stem or shoot, the sweet potato, which is a root, cannot be a tuber. Reference to the sweet potato as a tuber leads to another common error, which is to refer to a "mature" tuber or root. The sweet potato is perennial, and the storage root is capable of continued enlargement and does not mature in the sense of reaching some final size or stage of development. From the standpoint of function, the sweet potato root system consists of absorbing roots and fleshy or storage roots (20).

FLORAL BIOLOGY AND CONTROLLED POLLINATION

Sweet potato flowers are similar to those of other morning glories and occur in axillary inflorescences of 1–22 buds (Figs. 1.3 and 1.4) (34). They open in groups of two or more soon after daybreak and generally fade by noon. Colors of the floral parts vary from white through degrees of lavender to complete lavender. The tube depth varies from 28 to 63 mm and the limb width (corolla diameter) from 26 to 56 mm. The five petals are fused and have stamens attached at their bases with anthers that are normally white but may be light or dark lavender. Filaments vary in length from 5 to 21 mm, and this affects the position of anthers in relation to the stigma. Any number of anthers may be below, equal to, or above the stigma position as a result, and this pattern differs with cultivars. The stigma is

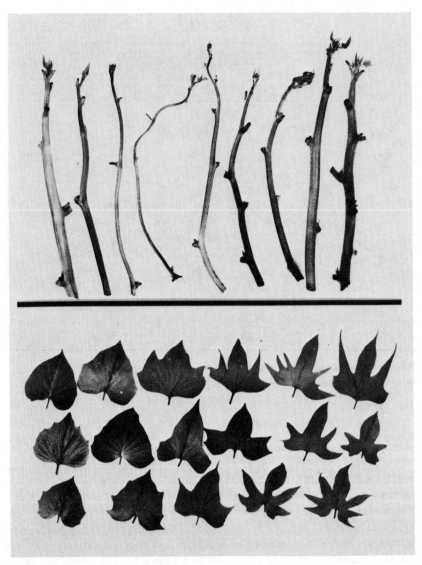

FIGURE 1.2. Every sweet potato seedling is different from all the others; thus the diversity of types is vast as illustrated by leaf and stem variations.
After Jones (*33*).

generally white and bilobed, but may be light or dark lavender. Styles may be from 8 to 29 mm long. There are two ovaries in the pistil, each containing two ovules. Sepals are leaflike and persistent and may be glabrous or pubescent. At the base of the corolla there are conspicuous yellow glands that contain insect-attracting nectar (*4,62*). Capsules contain one to four seeds and may be glabrous or pubescent (Fig. 1.5). Mature seeds are flat on two sides and round on the other with diameters of 3–5 mm (*60*). At maturity they are about half or less of their maximum green size, are usually dark brown or black (although some may be tan and others speckled), and weigh about 2 g per 100 seeds, which varies with parental type from about 1.3 to 3.0 g (*22,42*).

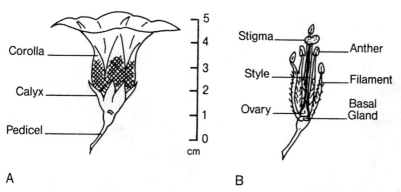

FIGURE 1.3. Parts of the sweet potato flower: (A) side view; (B) with calyx and corolla removed.

The seeds are hard and may retain viability for 20 years or more. Germination is consequently very irregular unless some means of seed scarification is used. The seed can be soaked in concentrated sulfuric acid for 20–60 min, washed in water or neutralized with a solution containing bicarbonate of soda, and rinsed in clear water (85). Also, seeds can be scarified by hand with a sharp needle or with a mechanical scarifier before sowing. Seeds are subject to infestation by the seed weevil *Megacerus impiger* Horn. in the field or in storage (41). A small (5 × 5 cm) segment of a household pest strip (20% 2,2-dichlorovinyl dimethyl phosphate) enclosed in a plastic bag with the seed has given satisfactory control.

A rather extensive literature cites the difficulty of obtaining flowering in sweet potato and includes studies of the causes of that condition (20,57,58). More recent work indicates that this condition was probably due to chance associations in the original plants used in U.S. breeding programs interested in dark orange flesh types. There are wide

FIGURE 1.4. (A) Sweet potato flowers and the subsequent seed capsules occur in axillary inflorescences called cymes. (B) A cyme with an unusually large number of seed capsules.

FIGURE 1.5. Seed capsules may be (A) glabrous or pubescent and (B) contain from 1 to 4 seeds, which are usually dark brown to black, sometimes speckled or tan. After Jones (*40*).

genetic differences in flowering incidence, as well as strong environmental influences (Table 1.2) (*32, 92*). Many breeding programs routinely induced non-flowering plants to flower by grafting or by various physiological shocks, perpetuating the non-flowering trait in subsequent generations. The best solution to non-flowering has been genetic: selection for good flowering. After about three generations of selection virtually every plant will flower sometime during the season (*61*). Further, no undesirable traits have been associated with good flowering (*42,45,84*). It may be necessary or desirable to use poorly flowering plants as parents in special situations, but in general it is best to use profusely flowering types if they can be found with the desired traits. In cases where poorly

TABLE 1.2.   Genetic Flowering Differences[a]

Flowering stage score[b]	Flowering stage percentages, 1965				
	6/24	7/16	8/6	9/16	11/1
0	26.2	15.7	12.3	20.5	6.4
1	6.0	7.2	3.3	1.9	3.9
2	20.0	6.6	4.5	11.1	3.9
3	7.4	3.3	4.1	4.3	9.1
4	14.2	3.3	1.6	4.5	16.9
5	26.2	63.9	74.2	57.7	59.8

[a]After three cycles of mass selection most plants will flower sometime during the season, but flowering incidence will change as the season progresses, as demonstrated in this population of 485 plants, where 92% flowered sometime during the season (*34*).

[b]0, no evidence of flower buds; 1, buds initiating; 2, 3 internodes have buds; 3, few buds, most mature; 4, abundance of buds, but no flowers; 5, flowering.

flowering parental types are used, the complex compatibility and sterility systems may be an important consideration. Techniques of making controlled crosses have been outlined recently by one of the authors (*40*). Since the use of polycross and mass selection techniques appears to be the most efficient way to breed sweet potatoes, techniques for controlled crossing will not be repeated here.

Trellises to facilitate open pollination by naturally occurring insects can be constructed of metal reinforcing rods, stakes, or bamboo canes placed alongside each plant and tied together near the top by a single cord anchored at the ends of the row (Fig. 1.6). A string tied at the base of the support is then twined around the vines and secured near the top of the support. During wet periods good air movement is important to control capsule rots and assure high seed quality. Ground mulches and/or herbicides may be used to hold weed growth to a minimum, and the trellises may be spaced sufficiently far apart to allow cultivation between rows. Insects and diseases can reduce seed set, and pesticide treatments can help alleviate the problem (*31,46,50*). Infestations of corn earworm [*Heliothis zea* (Boddie)] or fall armyworm [*Spodoptera frugiperda* (Smith)] can be very severe in the flowers. Chemical treatments should be made in the evenings when bee activity is reduced, and contact sprays or baits should be favored over those with residual activity to avoid killing insect pollinators.

The field layout is not especially critical except that long, narrow nurseries should be avoided to better assure random crossing. Seed mature in about 1 month, and seed harvest occurs over an extended time since flowering occurs over an extended period. Some plants will drop their capsules or shatter the seed if not harvested promptly after maturing. Care must be taken not to overfertilize, and especially high nitrogen should be avoided in the nursery because luxuriant growth is not necessary for good seed set. In fact, best seed

FIGURE 1.6. Elaborate trellises are not necessary for good sweet potato seed production, but they facilitate open pollination by insects.

set occurs on fairly small plants, and excessive foliage contributes to increased disease problems and lower seed quality.

Natural pollination is accomplished by insects during the morning hours when many species, chiefly Hymenoptera, can be observed visiting the flowers. Undoubtedly pollen transfers are made by many of these visitors, but honey bees and bumble bees are thought

TABLE 1.3. Some Diseases of Sweet Potato in the United States

Scientific name	Common name
Fungal diseases	
Fusarium oxysporum f. sp. *batatas* (Wr). Snyd. & Hans.	Fusarium wilt or stem rot
Fusarium oxysporum Schlect.	Surface rot
Fusarium solani (Mart.) Appel & Wr.	Fusarium root rot
Sclerotium rolfsii Sacc.	Southern blight: sclerotial blight and circular spot
Ceratocystis fimbriata Ell. & Halst	Black rot
Monilochaetes infuscans Ell. & Halst ex. Harter	Scurf
Rhizopus stolonifer (Ehr. ex. Fr.) Lind., and other *Rhizopus* spp.	Soft rot; ring rot
Diplodia tubericola (Ell. & Ev.) Taub.	Java black rot
Diaporthe batatatis Harter & Field	Diaporthe dry rot; stem rot
Phyllosticta batatas (Thuem.) Cbe.	Phyllosticta leaf blight
Cercospora batatae Zimm., and other *Cercospora* spp.	Cercospora leaf spot
Albugo ipomoeae-panduratae (Schw.) Swing.	White rust
Plenodomus destruens Harter	Foot rot
Macrophomina phaseoli (Maubl.) Ashby	Charcoal rot
Septoria bataticola Taub.	Septoria leaf spot
Bacterial diseases	
Streptomyces ipomoea (Person & W. J. Martin) Waks. & Henrici	Pox or soil rot
Erwinia chrysanathemi Dupes	Bacterial stem and root rot
Viral diseases	Feathery mottle: common strain, russet crack strain, internal cork strain
	Mild mottle
	Vein mottle
	Sweet potato mosaic viral complex
Diseases caused by nematodes	
Meloidogyne incognita (Kofoid & White) Chitwood	Southern root-knot nematode
Meloidogyne javanica (Treub.) Chitwood	Javanese (tropical) root-knot nematode
Meloidogyne hapla Chitwood	Northern root-knot nematode
Rotylenchulus reniformis Linford & Oliveira	Reniform nematode
Belonolaimus longicaudatus Rau	Sting nematode
Belonolaimus gracilis Steiner	Sting nematode
Ditylenchus dipsaci (Kuhn) Filipjev	Brown ring rot
Pratylenchus coffee (Zimmermann) Goodey	Root lesion nematode

to be the most important. Beehives placed near trellis areas can assure adequate numbers of pollinators.

DISEASES AND INSECT PESTS OF SWEET POTATO IN THE UNITED STATES

A number of diseases occur on sweet potatoes in the plant bed, during field production, or in storage (Table 1.3) (7,27,71,86,87). These diseases are a major concern of the plant breeder as resistances to some are now or soon will be necessary in any new cultivar (67). The only practical control of fusarium wilt is resistance. Resistance to root knot is readily available and new cultivars should contain at least an intermediate level (72). The release of a cultivar susceptible to internal cork would cause a serious problem to the trade; therefore, breeders must evaluate for resistance or tolerance to it (64,76). Soil rot (pox) is an increasingly important problem, and resistance to it has been found (75). There are obvious differences in degrees of susceptibility or levels of resistances to most of the other diseases, and breeders should take advantage of these.

At least 19 species of insects feed on sweet potato roots (Table 1.4) (12). Injury by a number of these is difficult to distinguish at harvest after the insects have left and damage has been altered by subsequent growth of the roots. Resistances to four types of insect injury have been identified. For convenience, damage by seven species that cause similar injury can be considered together as the WDS complex (wireworm–*Diabrotica*–*Systena*)

TABLE 1.4. Soil Insects of Sweet Potato in the United States

Scientific name	Common name
WDS complex (selection for resistance on basis of similar injury)	
Conoderus falli Lane	Southern potato wireworm
Conoderus vespertinus Fabricius	Tobacco wireworm
Diabrotica balteata LeConte	Banded cucumber beetle
Diabrotica undecimpunctata howardi Barber	Spotted cucumber beetle
Systena blanda Melsheimer	Pale-striped flea beetle
Systena elongata Fabricius	Elongate flea beetle
Systena frontalis Fabricius	A flea beetle
Grubs (selection for resistance on basis of similar injury)	
Phyllophaga ephilida Say	A white grub
Plectris aliena Chaplin	A white grub
Sweet potato flea beetle (selection for resistance on basis of typical injury)	
Chaetoncnema confinis Crotch	Sweet potato flea beetle
Sweet potato weevil (selection for resistance in controlled tests)	
Cylas formicarius elegantulus Summers	Sweet potato weevil
Others (no selection for resistance except in association with above groups)	
Conoderus amplicollis Gyllenhal	Gulf wireworm
Euzophers semifuneralis Walker	American plum borer
Melanotus communis Gyllenhal	A wireworm
Metriona spp.	Tortoise beetles
Noxtoxus calcaratus Horn	A flower beetle
Peridroma saucia Hubner	Variegated cutworm
Scolytid	Ambrosia beetle
Typophorus nigritus viridicyaneus Crotch	Sweet potato leaf beetle

as developed by Cuthbert and Davis (Fig. 1.7) (*13*). Grub injury is easily recognized as they gouge broad, usually shallow, areas in the roots (Fig. 1.8). Feeding by sweet potato flea beetle larvae leaves narrow channels or grooves just under the root skin (Fig. 1.9). Sweet potato weevils can be found in the root and identified directly (Fig. 1.10). No matter which insect species may be causing injury problems, genetic resistance should be considered as a possible solution. Even intermediate levels of resistance can be of significant economic importance (*16,81*). Little is known regarding the physiological mechanisms involved in soil insect resistances except that some factor seems associated with the root skin (*14*).

Some cultivars are more susceptible to flooding damage than others, a condition that is generally worsened by cool temperatures (10°C). Jewel is quite susceptible to this kind of damage (*1,88*). Chilling of roots during storage, shipping, or marketing below 5°C can cause a physiological disorder known as hardcore (*6*). Some cultivars are more tolerant to low-temperature exposure than others (*25*).

MAJOR BREEDING ACHIEVEMENTS OF THE RECENT PAST

Cultivars released since 1970 demonstrate the success breeders have had in development of multiple-pest resistance in combination with high yield and good culinary qualities

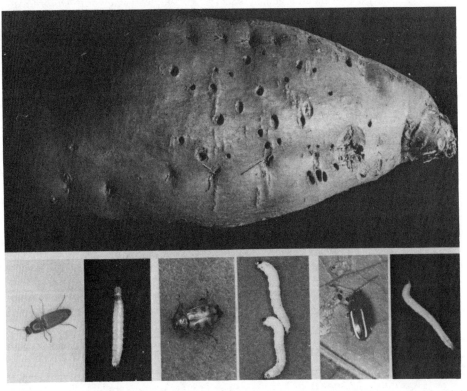

FIGURE 1.7. Soil insects of the WDS complex cause similar kinds of injury to sweet potato roots. Pictured here from left to right are adults and larvae representative of the wireworm, *Diabrotica,* and *Systena.*
After Cuthbert (*12*).

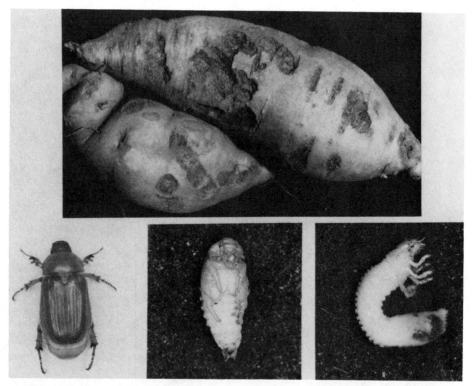

FIGURE 1.8. Grubs tend to feed on the underside of roots, causing this type of injury.
Pictured are adult, pupae, and larva of the white grub *Plectris aliena* Chapin.
After Cuthbert (*12*).

(Table 1.5). Jewel, released by the North Carolina Agricultural Experiment Station (AES)
in 1971 (*79*), is currently the most popular and widely grown cultivar in the United States
and accounts for more than 75% of the commercial plantings. It has a very wide adaptability, high quality, consistently dependable yields, and good storage characteristics and
sprouts well. One reason for its success is its high levels of resistances to fusarium wilt
and southern root knot. Certainly this cultivar is one of the main reasons for the increased
yields recorded for 1970–1979 (Table 1.1). All of the other releases since 1970 except
Georgia Jet (*26*) have sufficient resistance to fusarium wilt for effective control under field
conditions, and most have at least a moderate level of root-knot resistance. Those
cultivars developed by the Louisiana AES in recent years generally have at least some
resistance to soil rot, a disease of increasing importance. Most new cultivars have resistance to the internal cork virus, a potentially serious storage disease.

There has been a trend toward higher yielding types with excellent processing and
baking traits. Travis (*28*) has averaged 35–40% higher marketable yields than Jewel in
most recent regional trials. Eureka (*83*) and Vardaman (*2*) have been rated higher than
Jewel in baking and canning trials. The other lines (Table 1.5) were developed to meet
special needs; and although their general adaptability may not be as good as that of Jewel,
they do represent major accomplishments in meeting those special needs. Jasper (*29*) was
released for use in soil rot infested areas; Rojo Blanco (*90*), to meet the limited but
important white flesh market; Painter (*23*), for its potential in the Maryland–Virginia area;

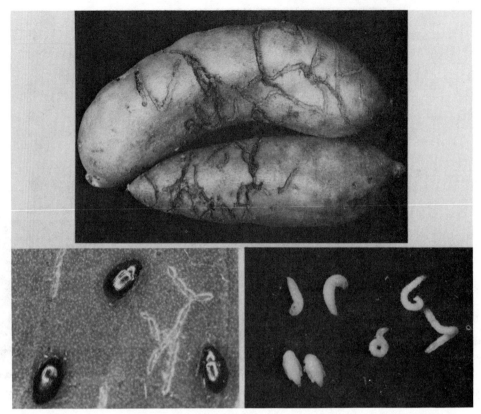

FIGURE 1.9. Narrow channels of the sweet potato flea beetle larvae can ruin the appearance of the susceptible cultivars. Pictured are adults, pupae, and larvae (from left to right).
After Cuthbert (*12*).

Georgia Jet, as an early sizing cultivar for use in South Georgia; Oklamex Red (*30*) and Caromex (*11*), to meet the unique environmental requirements for production in New Mexico; Pope (*10*), for use in areas subject to flooding; Eureka, because it appears to do well in California soil rot infested areas; and Vardaman, for its adaptation to Mississippi conditions.

New sources of resistances to many of the disease and insect pests have resulted through use of mass selection techniques (*15,43,45*). The U.S. Vegetable Laboratory in Charleston, South Carolina, has released nine breeding lines in recent years with unique combinations of high-level resistances (Table 1.6). All have orange flesh and the five most recent releases (W-115, W-119, W-125, W-149, and W-154) have generally acceptable baking and canning quality. As a group they tend to be more susceptible than current cultivars to sclerotial blight in plant beds. One line, W-51, was released because of its resistance to a resistance-breaking race of southern root-knot nematode first reported in 1973 (*65*).

The success in finding resistance to the resistance-breaking race of southern root knot demonstrates the value of a wider gene base, in this case provided through use of mass selection techniques. The degree of control provided by host plant resistance to insects has been demonstrated in sweet potato to be as good or better than that obtained with chemical

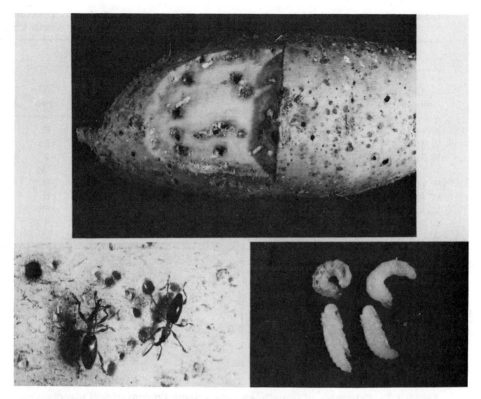

FIGURE 1.10. Worldwide, the most serious insect pest of sweet potato is the sweet potato weevil, which damages the roots during field production and continues to do so during storage. Pictured here are adults, pupae, and larvae of the weevil that occurs in the United States.
After Cuthbert (*12*).

TABLE 1.5. Disease and Insect Reactions of U.S. Sweet Potato Cultivar Releases, 1971–1981[a]

Cultivar	Previous designations	Disease reactions				Insect reactions		
		Internal cork	Fusarium wilt	Southern root knot	Soil rot	WDS	Flea beetle	White grub
Jewel (*79*)	NC-240	R	R	R	S	S	R	S
Redmar (*68*) (Goldmar) (*69*)	Md-2416		I–S	I–R		I–S	S	S
Georgia Jet (*26*)	GA-41		S	S		S	S	S
Jasper (*29*)	L9-190	R	R	I–R	R	I–S	I–R	S
Painter (*23*)	VP9-51	S	R	I–R		I–S	S	S
Carver (*91*)	TI-1885		R	I		I	R	S
Rojo Blanco (*90*)	CL-22-72		I	S		I	I	S
Oklamex Red (*30*)	OK-64-59		R	I				
Caromex (*11*)	NC-320		R	I		I–S	I	S
Pope (*10*)	NC-345	I–R	R	R		I–S	I	I
Travis (*28*)	L4-62	R	I–R	I	I–R	S	R	S
Eureka (*83*)	L4-131	R	R	I	I–R	S	R	I
Vardaman (*2*)	M3-702	R	R	S	I–S	S	S	S

[a]R, resistant; I, intermediately resistant; S, susceptible.

TABLE 1.6. Disease and Insect Reactions of Recent Sweet Potato Breeding Line Releases by the U.S. Vegetable Laboratory, Charleston, South Carolina[a]

	Disease			Insect			
Line number	Internal cork	Fusarium wilt	Southern root knot	WDS	Flea beetle	White grub	Sweet potato weevil
W-13 (44)	R	R	R	R	R	R	I–R
W-51 (18)	R	HR	HR	S	S		
W-71 (48)	R	HR	R	HR	R	I	R
W-115 (48)	R	R	R	R	HR	I–R	R
W-119 (48)	R	HR	HR	HR	R	R	R
W-125 (48)	R	R	R	R	HR	R	R
W-149 (48)	I	HR	HR	R	R	I–R	I–R
W-154 (48)	R	R	HR	R	R	R	I–R
W-178 (44)	R	R	R	R	S	R	

[a]HR, highly resistant; R, resistant; I, intermediately resistant; S, susceptible.

TABLE 1.7. Estimates of Genetic and Insecticide Control of Soil Insects[a]

Selection or cultivar	Roots damaged (%)			Control by source (%)[b]		
	Untreated	Treated	Mean	Genetic	Insecticide	Both
WDS at Charleston, South Carolina, 1975						
W-13	35	19	27	61	18	79
W-3	65	34	50	28	34	62
Goldrush	90	57	74	—	37	—
Sweet potato flea beetle at Charleston, South Carolina, 1975						
W-13	3	2	3	89	4	93
W-3	4	3	4	85	4	89
Goldrush	27	17	22	—	37	—
All insects including white grub (P. aliena) at Charleston, south Carolina, 1975						
W-13	27	20	24	70	8	78
W-3	65	39	52	29	29	58
Goldrush	91	66	79	—	27	—
White grub (P. ephilida) at Sunset, Louisiana, 1978						
L3-64	8	0	4	87	12	100
W-94	12	3	7	81	14	96
W-99	16	5	11	75	17	92
L4-89	31	10	20	52	32	84
SC 1149-19	46	23	34	28	36	64
Centennial	64	46	55	—	28	—

[a]Adapted from Cuthbert and Jones (16) and from Rolston et al. (81).
[b]Control estimates based on injury to the untreated susceptible cultivar.

treatment of susceptible cultivars (Table 1.7) (*16,81*). These recent breeding achievements suggest promise for even greater progress in the near future.

CURRENT GOALS OF BREEDING PROGRAM

The primary breeding goals for sweet potato are much as they always have been: to develop cultivars with the highest possible quality that can be produced at the lowest possible cost (Fig. 1.11). This implies high yield, good culinary qualities, and the many necessary traits for efficient production and marketing. These include good storage characteristics and resistances to storage diseases, high sprout production and resistances to plant bed diseases, and good field performance and resistances to field pests in types suitable for continued improvement efforts (plants that, therefore, flower and set seed without special treatments).

Breeders continue to seek higher levels of resistances to fusarium wilt and root knot, which are important diseases throughout the sweet potato production areas of the country.

FIGURE 1.11. Selection for a large number of well-shaped roots is one way to achieve high marketable yields.

Resistances to soil insects are receiving more emphasis especially for the lower parishes of Louisiana, where severe infestations in recent years have contributed to significant reductions in sweet potato production (*81*). Because sweet potatoes are increasingly grown in rotation with other crops requiring liming of soil and consequent increases in soil pH, soil rot has become a more severe problem especially in the major production areas of California, Louisiana, and North Carolina (*66*). Resistance offers the only practical solution to the soil rot problem and is therefore receiving major attention in the three states mentioned. Studies suggesting the possibility of resistance to sweet potato weevil (*24,48,73,74,81*), and the fact that it is of primary importance throughout the tropical growing areas of the world, have spurred greater interest and increased effort in breeding for weevil resistance. Virus diseases are a special concern, but there is little current

FIGURE 1.12. Suitability for planting with cut root pieces is under genetic control. The planted piece (arrow) either decays or fails to enlarge and the new hill consists of marketable shaped roots. By the elimination of beds for plant production this trait may contribute to reduced labor requirements.

TABLE 1.8. Correlation of Flesh Color with Root Specific Gravity[a]

Root specific gravity[b]	Number of plants, flesh color:[c]						
	W	CR	VLO	LO	O	DO	Total
1.00	0	1	1	3	4	3	12
1.02	2	3	2	3	9	9	28
1.04	2	9	3	10	15	13	52
1.06	6	19	10	7	10	2	54
1.08	3	11	7	7	4	4	36
Total	13	43	23	30	42	31	182

[a]Although a number of studies have demonstrated significant correlations of white or light flesh and high dry matter, orange and dark orange types do occur with high specific gravity or dry matter as demonstrated in the fifth generation of mass selection population I. [After Jones et al. (45).]
[b]Determined by flotation, Cherokee and Goldrush = 1.00.
[c]Subjective scores: W, white; CR, cream or yellow; VLO, very light orange; O, orange; LO, light orange; DO, dark orange.

breeding activity for resistances except to discard those types obviously susceptible to internal cork or russet crack. Only modest efforts are being made to find and incorporate resistances to reniform nematodes (8), sclerotial blight (19), scurf, and the various storage diseases.

Secondary breeding goals involve special-purpose types such as direct planting cultivars that can be planted with cut root pieces (Fig. 1.12), short vining or dwarf types for small-farm or garden cultivars, white-flesh types for cooking, and high-yielding cultivars for liquid-fuel production (biomass types). It remains a possibility to breed other special-purpose kinds if new developments provide sufficient demand. For instance, a high dry matter, orange-flesh sweet potato especially suited for making chips and french fries is possible (39,45) (Table 1.8, Fig. 1.13). Vine tips can be eaten and selection for high quality in that respect is possible as well as selection for forage value.

The sweet potato offers considerable promise for biomass production because of its adaptation to dry environments and its perennial growth. Many crops mature and cease growing after a relatively short growth period. In contrast, sweet potatoes continue growth as long as temperatures are not too cool, thus making use of available sunlight until harvested. The statistics on U.S. sweet potato yields (Table 1.1) are not applicable when considering biomass production because harvests are begun when prices are favorable for U.S. #1 size roots. Therefore, relatively small-sized roots are harvested prior to maximum root development, which would result in maximum bulk yields. Good data for biomass yield estimates are not yet available, but 18–22 tons/acre (40–50 MT/ha) is a reasonable expectation and as much as 40 tons/acre (90 MT/ha) may be possible.

SELECTION TECHNIQUES FOR SPECIFIC CHARACTERS

In sweet potato breeding one must remember while making selections for specific characters that the ultimate goal is to find unique combinations of favorable characteristics. One must resist the temptation to approach breeding as a gamble, to keep a plant for further testing on the slim chance that it will be outstanding. For instance, it does little good to select for root-knot resistance and then to keep those plants with deep orange flesh

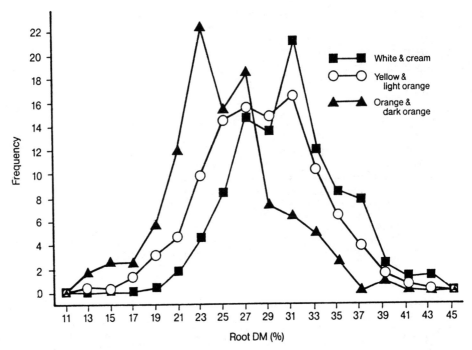

FIGURE 1.13. Frequency polygons for dry matter (DM) of three flesh-color groups of sweet potato, suggesting two genetic systems for DM, one associated with orange flesh and the other with white flesh. Although the lighter flesh colors are correlated with high DM (.61, significant at 1% level), dark orange selections with high DM are available in sufficient frequency for effective selection progress. From mass selection population H/3: white and cream ($n = 222$), yellow and light orange ($n = 567$), and orange and dark orange ($n = 127$).
After Jones (39).

whether they have resistance or not. For selection to be effective, plants that otherwise would have been saved must be discarded because of a particular unfavorable trait. Efficient breeding procedures and firmly set selection priorities remove most of the gamble and assure positive progress. The following procedures are taken from those of the authors and may not necessarily conform to procedures used in other breeding programs.

Preliminary Selections

The cost involved in testing each plant increases during the selection process. Efficiency demands that undesirable plants be discarded as early in the program as possible. Therefore, efficient systems follow sequential screening procedures that evaluate as many plants for as many traits as possible in the seedling stage.

Seedlings can be started in seed trays filled with horticultural-grade vermiculite or Jiffy mix (Fig. 1.14). We use commercially made plastic trays (11 × 22 × 2 in.) with drainage holes placed on black plastic to prevent contamination by microflora of the greenhouse benches and to help control moisture. Scarified seed are planted in holes made in the medium with a plywood template containing 20 × 10 rows of ¼-in. round pegs that are 1 in. long (200 per tray). When the majority of the seedlings have about four leaves they are transplanted to steam-sterilized greenhouse bench medium on 3- to 4-in. centers in holes

FIGURE 1.14. Seedlings started in trays filled with horticultural-grade vermiculite.

made with a long template (length determined by width of the bench) containing ¾-in. round pegs that are 2½ in. long. We have rather high sides (10 in.) on our benches to allow a deep medium for good root formation. Through the years we have found that many soil mixtures work well but that high proportions of coarse sand allow better moisture control. At the time the seedlings are transplanted, they are inoculated with a water suspension of composite inoculum containing about 2000 root-knot nematode eggs and 50,000 propagules of fusarium wilt organism per plant (*17*). The inoculum is poured around the seedlings before the holes are closed and then the soil mixture is firmed around the seedlings with two fingers. Temperatures must be carefully maintained at 24°C or above (30°C optimum) and the soil kept moist (near field capacity) to favor good nematode development. Plants are fertilized periodically with nutrient solution approximating a 1–2–3 (N–P–K) nutrient ratio. After 35–45 days of growth in the inoculated soil, soil moisture can be reduced until plants begin wilting, which favors expression of fusarium wilt and promotes storage root formation. Many of the plants suceptible to fusarium wilt will die and others will show typical symptoms such as yellowing, stunting, wilting, discoloration of the vascular system, and/or stem splitting. About 60 days after transplanting, the seedlings should be carefully lifted from the soil, their roots rinsed in water and examined with 2.75× magnifiers. At that time plants with fusarium wilt symptoms or evidence of root-knot susceptibility are discarded. Most plants should have small storage roots forming that can be cut to evaluate flesh color, and only those with dark orange flesh are saved for further testing. The authors save 10–15% of the seedlings at this stage. The total number of seedlings evaluated depends on availability of greenhouse space and plot land for subsequent field evaluations.

The selected seedlings are transplanted to black-plastic-covered field beds previously prepared with a complete fertilizer relatively high in nitrogen to force rapid vine growth. The black plastic helps maintain a warm soil, which encourages fusarium wilt development as well as good plant growth and provides effective identification of escapes from the greenhouse wilt screening. Each spring is slightly different; but as soon as weather permits and plant growth is sufficient, five vine cuttings about 12 in. long from each seedling are planted in five-plant field plots conforming as nearly as possible to commercial practices (Fig. 1.15). Plants are spaced 12 in. apart in the row with skips of 5 ft between plots to reduce root mixtures during harvest. In general, sweet potatoes are grown on ridges 8–10 in. high, depending on soil type, soil drainage, row width, and farming equipment in use. Row widths usually are 42–48 in. but may be less (36 in.) on sandy soils where ridges can be low. Vine growth of sweet potatoes is rapid providing good weed competition and some farmers use no herbicides. In experimental tests where vine types vary and in most farming operations, locally recommended herbicides are generally used.

After about 110 days of growth the seedlings are dug and judgments made whether or not to test them further. Those with obviously poor yield, poor shape, lobing, veining, cracking, or otherwise unacceptable root appearance are discarded immediately. The use of frequent plots of check cultivars, such as Jewel, assists greatly in these judgments. Roots from promising-looking plants are examined more closely for signs of insect or disease injury and lines discarded if they appear susceptible. Roots from selected plants are picked up in wet-strength paper bags, labeled with appropriate identity, cured, and stored. Curing roots immediately after harvest for 4–7 days at 27°–30°C and from 85 to 90% relative humidity promotes rapid healing of injuries through the formation of new cork layers and reduces decay. After the curing period, temperatures are lowered to 13°–16°C, but a high relative humidity is maintained. During the curing and storage periods,

FIGURE 1.15. Orange fleshed seedlings with resistance to fusarium wilt and root-knot nematode are tested in five-plant plots during the first year. The five plants in each plot are propagated from vine cuttings from resistant seedlings.

starch is converted to sugars and dextrins, a process that occurs at different rates and to different degrees in each selection. These differences are recognized in the trade and cultivars are classed as moist-flesh (''yam'') or dry-flesh types. Soon after harvest, sample roots of the seedling selections are cut and acceptable flesh color confirmed. Those that appear mottled or very light orange are discarded, as are those that are pithy or float in water.

At a convenient time after harvests, roots of each line are baked at 176°C for 2 hr and rated for culinary acceptability. We use two replicates of two roots each in these first baking tests. Since there will be more lines than can be baked at one time, an acceptable cultivar, such as Jewel, should be included as a control with each lot of samples baked. We rate five traits subjectively after cutting the roots from end to end. A scale of 1 to 5 is used for each trait, where 1 is excellent, 2 is good, 3 is questionable but worth retesting, 4 is probably not acceptable, and 5 is definitely unacceptable. The five traits rated are (1) color, (2) fiber, (3) discoloration or darkening, (4) general appearance or eye appeal, and (5) taste. Separate notes are made about any unusual quality or specific trait such as bland, too dry, very moist, grainy, very sweet, or pithy. When the average rating of the four roots is 3.0 or more for any one of the five baking traits, that line is discarded. This is a critical stage in the selection process and very vulnerable to individual bias, which must be carefully avoided. Judges must be objective and highly experienced. Sometimes it may be best to skip the taste test at this point and to use only the other four traits. Fiber is a particularly difficult trait to rate (47). One must remember that sweet potatoes are roots and must have some fiber even in those with excellent quality. It helps to think in terms of rating degrees of ''objectionable'' fiber. We rate fiber visually and by feel using a standard table fork in these early tests.

Sweet potatoes are propagated vegetatively with sprouts from roots placed in various kinds of plant beds (86). In the colder regions of production, plant beds must be heated to prevent chilling of the roots or sprouts; in more moderate temperature zones, a simple plastic cover is all that is required; while in much of the south a soil cover of 2–3 in. is sufficient. Vine cuttings from early plantings are also used for later plantings.

As roots are taken out of storage for bedding in early spring those selections with excessive shrivelling or storage rots can be discarded. Those that sprout poorly or have severe sclerotial blight when grown in plant beds covered with black plastic need not be tested further (Fig. 1.16). Those selections remaining make up the second-year seedlings and are subjected to more critical evaluations.

Second-Year Seedlings

In the second year, seedlings are yield tested in 25-plant plots replicated four times. Two replications are harvested after about 100 days to identify short-season types and the other two replications after about 120 days. We use four controls in these tests: Jewel and Centennial for yield comparisons, W-13 as an insect-resistant check, and SC 1149-19 as an insect-susceptible check. At harvest all entries are rated for insect injury, shape, freedom from cracking, veining or lobing, and good general appearance. The weights of each of four grades of roots are recorded: U.S. #1, roots with 2- to 3.5-in. diameters and 3- to 9-in. lengths, free of defects, and well shaped; Canner, roots with 1- to 2-in. diameters and 2- to 7-in. lengths; Jumbo or Oversize, roots that exceed the diameter and length requirements of U.S. #1 but are of marketable quality; and Cull, roots with 1-in. or larger diameters and so misshappen, cracked, or unattractive that they do not fit as

FIGURE 1.16. The use of black plastic on plant beds increases sclerotial blight incidence and aids in selection for resistance in seedlings entering the second year of testing.

marketable in any of the other grades. Grading boards with three holes of 1-, 2-, and 3.5-in. diameters assure uniform grading standards. After curing, baking quality is tested as outlined earlier.

Insect damage ratings are obtained from 10-plant plots replicated four times with the same four control cultivars as above and grown in an area where natural insect infestations are generally high. The numbers of holes or scars caused by each of the three groups of insects outlined in Table 1.4 under WDS complex, grubs, and sweet potato flea beetle are recorded for each root of all entries. With high natural infestation levels, the percentages of roots injured by larvae of the WDS complex or the sweet potato flea beetle may be adequate and are much less tedious to obtain (*51*). Where insect-rearing facilities are available, artificial infestations may be used to supplement natural insect populations (*82*).

The root-knot and fusarium wilt resistances of selections should be confirmed during this second year of testing in carefully controlled greenhouse tests. Uniform vine cuttings of the selections can be obtained from plant beds or field plots. This work is best done during the warm growing season because a supply of healthy, vigorous terminal cuttings are available and conditions are generally best for disease evaluations.

The fusarial inoculum should consist of a composite of virulent isolates (six or more isolates, if possible) of the wilt pathogen (*Fusarium oxysporum* f. sp. *batatas*) selected from widely separated production areas in the United States. The inoculum is produced and standardized to approximately 50,000 propagules per milliliter (*17*). This inoculum density should result in complete mortality of the standard susceptible cultivars Porto Rico and Nemagold within 2 weeks after inoculation. The fresh terminal cuttings with their expanded leaves removed are dip-inoculated and immediately placed in a greenhouse bench-bed containing a soil–sand mixture. Standard control cultivars with intermediate and high resistance are always included for comparative purposes. Optimum conditions for disease development are maintained. Disease readings are started at the first indication of wilt (stem rot) symptoms, which usually occur 5–7 days after inoculation. These readings are continued every other day for a total of eight times. After 21 days the remaining live plants are removed and their stems sliced to determine the number and extent of vascular infection. A disease index is computed from the disease data, and each selection is compared to the resistant standard cultivars. Only those selections with wilt resistance equal to or better than the intermediate-resistant cultivar Centennial are retained in the program.

Highly virulent isolates of the root-knot nematode species are maintained on susceptible host plants until eggs are needed for inoculations. Eggs of the nematodes are extracted from the fresh roots using the sodium hypochlorite procedure (43). About 2000 eggs are used to inoculate each cutting of each line. Standard resistant and susceptible cultivars are always used in these tests. Inoculation is accomplished by simply pouring an aqueous suspension of eggs into and around the holes in the soil of the bench in which the cuttings have been placed. The plants are grown under optimum conditions for growth and disease development for 50 to 60 days. They are then gently uprooted and their roots washed carefully and examined under magnification to obtain gall and egg mass indices for each plant. Plants are evaluated using a scale of 1 to 5 where 1 equals high resistance (no galls or egg masses) and 5 equals high susceptibility (severely galled or massive egg masses). In some special cases root samples are taken from some outstanding lines and the eggs extracted to determine the actual amount of reproduction. This is perhaps the ultimate method for evaluating for resistance, since this will show the amount of reproduction that has occurred. This is, however, a very time-consuming and tedious procedure. The root-knot indices are compared to the standard cultivars in the tests, and only those lines with a high level of resistance are retained. Since all of our seedlings are routinely evaluated in the preliminary testing procedure, a very high percentage of these lines (>90%) have high levels of resistance to both diseases. Promising selections still retained after these second-year trials are tested in the advanced trials the third year.

Trials of Advanced Lines

The advanced lines represent the best selections from two or more years of evaluation. These are handled in very much the same way as the second-year seedlings. As data are collected from increasing numbers of trials, better judgments regarding the cultivar potential of the selections can be made. Some idea of yield stability can be obtained from results of several years, and more data are available on sprouting, storage, and baking quality. During this period, canned roots are subjected to quality ratings, and root-knot nematode resistances are tested under field conditions in a uniformly infested nursery.

There has been an increasing interest in sweet potato weevil resistance and techniques for screening have been developed (24,74,80). Reliance on natural infestations has not been successful for us. Rearing of weevils is relatively easy, but quarantine regulations restrict doing so in certain areas of the country. The specific techniques are still in a period of refinement but some general comments are in order. By using laboratory screening techniques during winter months the frequencies of resistant lines in field tests can be increased, and the numbers of lines decreased, allowing more replications to be grown in a test of the same size. We find 10-plant plots to be sufficiently large and try to have at least eight replications. The mechanisms of resistance are not known at this time, but we suspect that effective field resistance may involve more than one mechanism. The levels of resistance attained (Table 1.6) do not approach immunity but probably are sufficient to be economically important.

Techniques for evaluation of soil rot resistance in the laboratory or greenhouse have not been developed to the extent that it can be done routinely in a large breeding program. However, field screening in infested soils can be very effective as demonstrated by recent releases from Louisiana (Table 1.5).

Parental Selection

Probably the most important aspect of selection from a breeding standpoint is that of parent plants because it determines the direction of the breeding program and, ultimately, its success or failure. The selection of a parent plant is based on two things: its phenotypic expression and that of its offspring, i.e., its ability to transmit desired traits to the next generation. Heritability estimates give us a general idea of what can be expected in regard to transmission of a desirable trait, but each parent will perform differently because of the complex hexaploid inheritance of sweet potato. Thus, progeny testing is a valuable tool in determining the breeding value of a particular selection.

Quantitative Genetic Considerations

Because inheritance in sweet potato is best described in quantitative genetic terms, it is very important for sweet potato breeders to develop a working concept of what heritability estimates are and how to interpret them. Heritability is the degree of correspondence between phenotypic values and breeding values (21). Heritabilities are not measures of desirability. They have nothing directly to do with frequency of good versus poor types, and they say nothing about means or distributions. A high estimate says nothing about how good our materials are, only that the better parents should tend to give the best progeny and, on the other hand, the poor parents can be counted on to have poor progeny on average. They merely estimate how well we can evaluate the parents and predict what the offspring will be like with the particular breeding materials and techniques of evaluation being used. If those breeding materials are from a wide gene base, the estimates will probably have wider application than if they were from a narrow gene base.

Heritability estimates can be obtained through a number of statistical procedures (21). For purposes of this discussion it is sufficient to consider two general narrow-sense heritability (h^2) estimation methods, one from variance–covariance techniques (36,52) and the other from twice the regression of offspring means on parent means (39,43,51). A considerable number of h^2 estimates has accumulated in recent years, and this body of knowledge can assist us in the interpretation of new estimates and can provide guidance in planning breeding programs (Table 1.9). By comparing estimates obtained by regression with those obtained by variance–covariance, we can see that the regression results are more conservative. It is apparent that the variance–covariance technique tends to yield overestimates. This illustrates the importance of knowing how an estimate was obtained in order to interpret it properly.

The size of the h^2 estimate provides guidance for planning selection sequences and

TABLE 1.9. Partial List of Narrow-Sense Heritability (h^2) Estimates

Trait	h^2 estimate (%)	Statistical technique	Reference
Root weight (yield)	41	Variance–covariance	52
	41 ± 4	Regression	49
	44	Variance–covariance	55
	25 ± 13	Regression (after two cycles of selection)	39
Growth cracks	51	Variance–covariance	52
	37 ± 4	Regression	49
Flesh color	66	Variance–covariance	52
	53 ± 14	Regression	39
Flesh oxidation	64	Variance–covariance	52
Dry matter	65 ± 12	Regression	39
Crude protein	57	Variance–covariance	56
Fiber	47 ± 4	Regression	49
Skin color	81	Variance–covariance	52
Sprouting	39 ± 14	Regression	39
	37 ± 2	Regression	49
Vine length	60	Variance–covariance	37
Leaf type	59	Variance–covariance	37
Flower buds/cyme	50	Variance–covariance	37
Fusarium wilt reaction	86	Variance–covariance	36
	89	Variance	9
	50	Regression	9
Nematode egg mass index			
(*M. incognita*)	75 ± 23	Regression	43
	57 ± 37	Regression	43
(*M. javanica*)	69 ± 18	Regression	43
WDS root injury (%)	45 ± 12	Regression	51
Sweet potato flea beetle			
root injury (%)	40 ± 7	Regression	51
Weevil resistance			
(*C. puncticollis* Boh)	84	Variance–covariance	24

techniques. When the estimate is high we can rely on the parental phenotype and not worry so much about progeny testing. We should be able to increase frequencies of desirable types rather quickly with the selection methods used to make the estimate. If the estimate is low, we may want to evaluate parental candidates by progeny tests. Since the h^2 estimate is basically a ratio of the additive genetic variance over the phenotypic variance, a low estimate may indicate a need for greater precision in techniques. Maybe one should consider more replications, more locations, artificial infestations, greenhouse tests, or laboratory techniques in order to reduce the phenotypic variance. Of course, if the estimate is low because of a small genetic component, reducing the phenotypic component may not be of much help.

What size h^2 estimate can be considered favorable in sweet potato breeding? An exact answer is difficult, but comparison of some actual selection advances with h^2 estimates from materials of similar genetic makeup does give some clues. Both frequencies and levels of resistance to the WDS complex of soil insects were increased considerably by recurrent selection (*15*) (Fig. 1.17). The h^2 of WDS resistance was later estimated as 0.37 ± 0.11 by regression of offspring means on parent means in somewhat similar materials

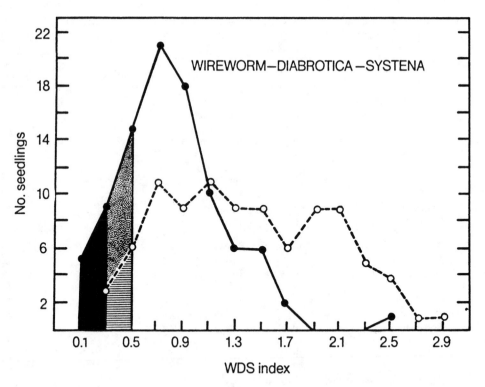

FIGURE 1.17. Frequency polygon for WDS injury to selected (solid line) and unselected (broken line) sweet potato populations. The cross-hatched area shows the portion of the unselected population with an acceptable level of resistance; the lightly shaded area shows the increased frequency of the same level of resistance in the selected population; and the heavily shaded area shows the portion of the selected population with higher levels of resistance than found in the unselected population. Thus, through four cycles of mass selection, both frequency and level of resistance were increased.
After Cuthbert and Jones (*15*).

(*51*). The h^2 of root flesh oxidation after treatment with catechol was estimated as 0.64 by variance–covariance methods (*52*), and subsequent selection experiments for low oxidation demonstrated rapid improvement (*38*) (Table 1.10). Obviously estimates over 0.60 are quite adequate for good selection advance. Probably estimates as low as 0.30 by regression and 0.40 by variance–covariance could be considered favorable provided that the selection techniques have enough precision.

Mass selection in sweet potato has been shown to be an effective way of combining favorable characters in parental types (*45*). In general, traits are sufficiently independent to allow effective selection with independent culling levels. It is important to note that mass selection not only provides for improvement in the mean performance of successive generations but also for higher levels of performance than occurred in the base population (Fig. 1.17). It is quite likely that we can develop higher levels of sweet potato weevil resistance than we now have.

DESIGN OF THE COMPLETE BREEDING PROGRAM

The philosophy of sweet potato breeding is slowly changing, with a gradual acceptance of mass selection methods based on quantitative genetic principles (*33*). Rapid generation

TABLE 1.10.   Selection Advance for Low Root Flesh Oxidation Following Two Selection Schemes[a]

Population/generation	Root flesh oxidation classes (%)[b]					Mean[c] score	No. plants scored
	1	2	3	4	5		
C/4[d]	3	38	23	14	22	3.1a	153
A/6[e]	17	53	22	6	2	2.2c	157
D/7[f]	7	46	24	13	10	2.7b	150

[a]After Jones (38).

[b]Increasing darkening of flesh 10 min after dipping in 0.25 M catechol solution.

[c]Means not followed by same letter are different at .01 significance level.

[d]C/4: initiating mass selection population with plants in classes 1 and 2 selected (30%) to begin the selection advance study. These results are from remnant seed grown and evaluated at the same time as A/6 and D/7.

[e]A/6: the selected clones from C/4 were polycrossed in isolation and seedlings of A/5 were evaluated as they were removed from the greenhouse bench. The selected plants were polycrossed to produce seed for A/6. Therefore, this population represents results from two cycles of selection in the seedling or first year of growth followed by crossing of the selected plants.

[f]D/7: seed from the same selections as used in population A were used to start a mass selection scheme with evaluations made at harvest and seed from low-oxidizing plants used to start the next cycle of selection. Therefore, this population represents results of three cycles of mass selection.

advance can be achieved by high selection pressures in open-pollinated populations by using flowering plant types. The concept that profuse flowering would be negatively associated with other favorable traits, such as high yield, has been shown to be unfounded (45,84). These techniques provide a sound basis for sweet potato improvement quite different from the classic pedigree breeding procedures, which were based on qualitative genetic principles.

As with other crops, sweet potato breeders are faced with long- and short-range breeding goals (Table 1.11). Long-range goals might encompass the search for new sources of resistances and their introduction into plants acceptable or almost acceptable to the trade, usually in combination with many other traits already considered essential. The new sources may come from plant exploration, genetic engineering, related species, increases or decreases in ploidy levels, somatic hybridization, or use of some other novel genetic technique. This may require looking ahead 15 to 20 years or more. Short-term goals encompass the development of new cultivars with as many of the known desirable traits as possible or some combination of them suitable for special environmental uses or purposes. Often, meeting short-term goals requires looking ahead 5 to 10 years. More breeding programs concentrate on short-term goals than long-term goals, which is proba-

TABLE 1.11.   Short and Long-Term Breeding Goals

Short-term goals	Long-term goals
Cultivars	Parents
Individual plant value	Average line value
Small numbers of plants	Large numbers of plants
Precise evaluation	Less precise evaluations
Quick improvement	Gradual continuous improvement
Adaptive research	Fundamental research
Expression	Transmission
Backcross procedures	Mass selection procedures

bly as it should be. Both objectives are important to maintain long-term stability in crop production and to meet increasing global food needs.

Mass Selection Populations

A comprehensive sweet potato breeding program addressing both long- and short-term goals should contain one or more mass selection populations to provide new parental types and to assure the necessary wide gene base for continuous selection advance in future years (Fig. 1.18). In early cycles of mass selection one complete sexual generation can be

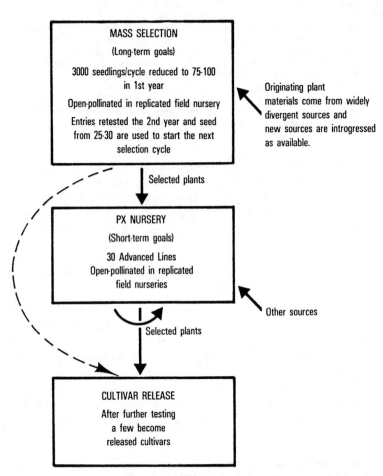

FIGURE 1.18. A comprehensive breeding program addressing both long- and short-term goals should contain mass-selected populations to provide new parental types and to assure the wide gene base necessary for continued selection advances in future years. The authors use two mass selection populations to provide new selections for the PX nursery, each with 2 years per cycle and grown on seed increase trellises in alternate years. Seedling selections from the PX nursery also provide materials for the next PX nursery. Selections from the mass selection population could become cultivars, but the probability is lower than for those from the PX.

completed each year by advancing from true seed to true seed each year. In the mass selection procedure, selected plants are not used in the next cycle, only seed from them. In the beginning cycles it is important to keep records on sources of each plant in order to be sure that the various sources remain represented for about three generations. After that time there is no need to keep records of origins, and seed from selected plants can be bulked to begin subsequent cycles of selection. Our experience indicates that light selection can begin in these early generations. For instance, progeny rows of seedlings from individual selections can be inoculated with root-knot nematodes in the greenhouse and the best resistance within each progeny selected. In some sources it may be necessary to keep low levels of resistances or even some susceptible plants to maintain that source in the population. If orange flesh occurs in low frequency, one can begin some light selection for that trait to assure its presence in later stages of the program. Field tests can be limited to nonreplicated five-plant plots. Controls consisting of the best cultivars should be included for yield comparison.

An exact number of cultivars necessary to start a mass selection program cannot be defined, but there are some general guidelines. For long-term objectives with flexibility to include selection for as yet unknown traits (such as resistance to future disease or insect pests), one should strive for as wide a gene base as possible. For instance, our most recent mass selection population (population J) was started with 350 collections from Taiwan, Japan, New Guinea, Hawaii, Nigeria, Thailand, Peru, Cuba, Philippines, Puerto Rico, New Zealand, Guatemala, Uruguay, Cook islands, Marquesas islands, Spain, Canada, and one of our previous wide gene base populations (population F). These were open-pollinated and about 3000 seedlings started in the greenhouse with 700 representing as near as possible all the various sources moved to a trellis area. Seed from about 200 were used to start the next cycle. In the third cycle the number selected as seed parents was reduced to about 100 among 700 trellised plants. For more limited short-term goals, one might start with as few as six plants, with the realization that a narrow gene base might provide better chances for rapid advance to meet some pressing need.

After about three cycles of intercrossing with light selection, it probably will be necessary to change to 2-year cycles in order better to evaluate yield, sprouting, and storage traits (45). Seedlings can be evaluated for flesh color and resistances to fusarium wilt and root knot in the greenhouse; and vine cuttings can be transplanted to the field for evaluation of yield, insect resistances (by including insect-resistant and -susceptible controls), lack of defects (cracking, veining, and lobing), acceptable shape, and general appearance using five-plant nonreplicated plots. About 150 of the best selections can be stored and evaluated for keeping quality in storage and for culinary qualities. In the spring the remaining selections can be rated for sprouting and other bedding traits, such as resistance to sclerotial blight. The best 75–100 selections are then planted on trellises in four or five replications. These are intercrossed by naturally occurring insects and seed collected with appropriate source labels. During the same season, vine cuttings or sprouts from the plant bed are used to plant replicated field trials. Data are collected on yield, insect resistances, and other traits of interest, and the best 25 or 30 selections identified. Seed from these are used to start the next cycle of selection.

This procedure should provide for rather rapid selection advance and concomitant narrowing of the gene base. Thus, it may be necessary to begin a second mass selection population that can be timed to alternate with the original mass selection population. In this way the breeder can have two mass selection populations with only one trellised seed increase nursery each year. New plant collections can be introgressed into the mass selection populations by inclusion in the seed increase nursery. In some cases it will be necessary to induce flowering by grafting or some other means (20,40,70). The identity of seedlings from these sources probably should be kept for about three generations, and it may be necessary to make some compromises and reduce selection pressure when evaluating them.

Selections from the mass selection populations can be tested directly for cultivar potential but are more likely to provide a source of parental material for a more restricted polycross nursery. Samples of seed from each cycle can be stored for later use in comparative studies of actual progress attained (15,45).

The Polycross Nursery

Technically, "polycross" refers to natural intercrossing of a group of plants in an isolated crossing block and would be applicable to the mass selection nursery, but for our purposes we define the polycross nursery in a more restricted manner. We use the term to refer to a limited number of parents (30 or less) randomly crossed in isolation by naturally occurring insects for the purpose of deriving new cultivars or advanced breeding lines. In our breeding strategy this refers to that group of plants intercrossed to meet our most pressing immediate or short-term goals (Fig. 1.18).

The authors grow a polycross nursery of about 30 advanced lines replicated four times in each of two locations each year. Generally sufficient seeds are produced in one location, but the second provides assurance against adverse environmental conditions that might lower seed set in one location. These nurseries are only about 200 ft each way when 30 plants are set 4–6 ft apart on four rows spaced sufficiently wide to allow for tractor cultivation between rows.

It is important to place some inflexible limits on the number of entries in the polycross to avoid the temptation to add just a few more parental selections. When too many parents are used, the average effect of each parent is diluted; and the hard decisions of which lines to drop are avoided, resulting in reduced selection progress. Each time a new line is added to the polycross, one of the previous entries must be dropped. If one assumes effective selection, the limitation on the number of parents in the polycross assures gradual and continuous improvement. In 5 or 10 years, the breeder will not still be crossing the same set of parents and their immediate progeny, but instead will be crossing an entirely new set of parents, each selected because of some superiority over previous entries. Some of the new entries should come from the mass selection populations and from other sources to prevent too much inbreeding and narrowing of the gene base.

By determining the average values of the important traits of parents in the polycross each year the breeder can visualize how much progress has been made and where selection pressures need increasing to obtain the favorable combinations required. At the same time, progeny tests can and should be used further to evaluate each parent. Where time or facilities prevent formal progeny tests, good notes taken during the selection process can be very valuable. For instance, the percentages of progeny discarded because of poor flesh color or susceptibilities to fusarium wilt and root knot will give a good idea of whether the seedlings from a particular parent will survive even the initial greenhouse screening procedures. If the proportion of seedlings kept is very small, more seed may need to be started from that parent, or it should be considered a candidate for replacement. If the seed set of some plant is very low, it can be entered twice in each replication or grafted to induce better flowering and eventually replaced.

Seedlings from the polycross provide a major source of potential cultivars and are screened for the many essential traits as detailed in previous sections. Selections made in the first year are tested in the second-year seedling trials and in the advanced line trials for 2 years or more. The most promising selections may then be vegetatively increased for submission to the national trials. Although the breeding program makes use of seed-propagated plants, the final hexaploid genotypes must be multiplied exclusively and indefinitely by vegetative means. A cultivar such as Jewel is actually one seedling multiplied clonally.

PERFORMANCE TRIALS OF ADVANCED LINES

National sweet potato trials are conducted annually by members of the National Sweet Potato Collaborators Group, which includes personnel of AES in Alabama, Arizona, Arkansas, California, Georgia, Indiana, Iowa, Kansas, Kentucky, Louisiana, Maryland, Mississippi, Missouri, New Jersey, New Mexico, North Carolina, Oklahoma, South Carolina, Tennessee, Texas, and Virginia and of the USDA, Agricultural Research Service. This group, organized in 1939 (5), has contributed greatly to sweet potato improvement through the cooperative efforts of its members. Current data on all phases of sweet potato research are assembled annually, distributed to the members, and discussed at meetings held each year in early February.

Breeders are free to enter any new selection they consider worthy of consideration to the group for testing at 20–24 locations. Not all locations provide yield tests, but they may supply information on other aspects such as disease and insect resistances or baking and canning qualities. Generally yield trials are conducted at about 18 locations. New entries are grown in the Observational Trials the first year of submission. These trials consist of 25-plant plots that are not replicated because of the limited number of bedding roots usually available. The Observational Trials typically contain about 6–10 selections plus Jewel and Centennial as controls. At the annual meeting the members vote whether to retain an entry in the Observational Trials, test it in the Advanced Trials, or drop it from further consideration. An entry may remain in the Observational Trials more than 1 year, but usually a recommendation either to advance it or drop it is made after 2 years.

Advanced Trials are usually conducted at the same locations as the Observational Trials but have four replications of 25 plants. Typically these trials contain only three to six selections and Jewel and Centennial. Harvested roots are weighed by grades, and the percentage of U.S. #1 is calculated by dividing the weight of U.S. #1 by the total marketable weight (culls and cracks not included). Comments made at harvest are made available to other members to assist in line evaluations. Also, any unusual conditions encountered are noted, which provides valuable information on performance under such stress conditions as prolonged periods of drought, cool wet weather during harvest, or severe disease or insect infestations. Notes are also provided on storage qualities, sprouting characteristics, culinary qualities, and production factors such as disease and insect reactions. About six or seven stations present results from taste panels for both baked and canned samples. Baked roots are rated on a scale of 1 to 10 for eye appeal, color intensity, color uniformity, freedom from discoloration, smoothness, moistness, lack of fiber, and flavor. Weighted baking and canning scores based on all subsidiary traits using a scale of 1 to 100 are then computed for each entry and the controls. Data from each location and summaries of all locations are included in the annual progress reports.

The members of the national group vote whether to drop a selection from further consideration by the group, to retain it for testing the next year, or to recommend that the originator name and release it for commercial production. Recommendations from the group are not binding on the breeder, who may release a selection the group has dropped from further consideration without a recommendation for release. There are many instances where such deviations from the national group evaluation are well justified. There may be a pressing need for a new cultivar with resistance to some specific pest in a limited production area, or a selection may appear very well suited to special environmental conditions or market needs of a particular area.

The national trials provide much more information on potential cultivars than an individual AES could obtain otherwise. A reasonably good estimate of yield stability can be made because of the many environments sampled in the 2 or 3 years of trials. Comments by other scientists about performances of the selections under various growing conditions are of immense value in decisions about release of new cultivars. This is especially true when extreme stress conditions have been encountered. The trials also give workers in cooperating states an opportunity to observe the performances of selections prior to release. Thus grower questions about new cultivars can be answered and informed recommendations made.

REFERENCES

1. Ahn, J. K., Collins, W. W., and Pharr, D. M. 1980. Influence of preharvest temperature and flooding on sweet potato roots in storage. HortScience *15*, 261–263.
2. Allison, M. L., Ammerman, G. R., Garner, J. W., Hammett, H. L., Singleton, C., Withers, F. T., and Palmertree, H. D. 1981. Vardaman, a new early maturing sweet potato variety for Mississippi. Miss. Agric. For. Exp. Stn., Inf. Sheet *1305*.
3. Austin, D. F. 1977. Hybrid polyploids in *Ipomoea* section *batatas*. J. Hered. *68*, 259–260.
4. Austin, D. F. 1978. The *Ipomoea batatas* complex. I. Taxonomy. Bull. Torrey Bot. Club *105*, 114–129.
5. Bowers, J. L., Harmon, S. A., and Dempsey, A. H. 1970. History of group and early research. South. Coop. Ser. Bull. *159*, 1–7.
6. Ceponis, M. J., Daines, R. H., and Hammond, D. F. 1974. Hardcore in Centennial sweet potatoes. N.J., Agric. Exp. Stn., Circ. *610*.
7. Chupp, C., and Sherf, A. F. 1960. Vegetable Diseases and Their Control. Ronald Press, New York.
8. Clark, C. A., Wright, V. L., and Miller, R. L. 1980. Reaction of some sweet potato selections to the reniform nematode, *Rotylenchulus reniformis*. J. Nematol. *12*, 218.
9. Collins, W. W. 1977. Diallel analysis of sweet potatoes for resistance to fusarium wilt. J. Am. Soc. Hortic. Sci. *102*, 109–111.
10. Collins, W. W. 1982. 'Pope' sweet potato. HortScience *17*, 265.
11. Collins, W. W., Pope, D. T., and Hsi, D. C. H. 1979. 'Caromex' sweet potato. HortScience *14*, 646.
12. Cuthbert, F. P., Jr. 1967. Insects affecting sweet potatoes. U.S., Dep. Agric., Agric. Handb. *329*.
13. Cuthbert, F. P., Jr., and Davis, B. W. 1970. Resistance in sweet potatoes to damage by soil insects. J. Econ. Entomol. *63*, 360–363.
14. Cuthbert, F. P., Jr., and Davis, B. W. 1971. Factors associated with insect resistance in sweet potatoes. J. Econ. Entomol. *64*, 713–717.
15. Cuthbert, F. P., Jr., and Jones, A. 1972. Resistance in sweet potatoes to *Coleoptera* increased by recurrent selection. J. Econ. Entomol. *65*, 1655–1658.
16. Cuthbert, F. P., Jr., and Jones, A. 1978. Insect resistance as an adjunct or alternative to insecticides for control of sweet potato soil insects. J. Am. Soc. Hortic. Sci. *103*, 443–445.

17. Dukes, P. D., Jones, A., and Cuthbert, F. P., Jr. 1975. Mass evaluation of sweet potato seedlings for reaction to fusarium wilt and root knot nematodes. Proc. Am. Phytopathol. Soc. 2, 133.

18. Dukes, P. D., Jones, A., Cuthbert, F. P., Jr., and Hamilton, M. G. 1978. W-51 root knot resistant sweet potato germplasm. HortScience 13, 201–202.

19. Dukes, P. D., Jones, A., and Schalk, J. M. 1979. Evaluating sweet potato for reaction to sclerotial blight caused by Sclerotium rolfsii. HortScience 14, 123.

20. Edmond, J. B., and Ammerman, G. R. 1971. Sweet Potatoes: Production, Processing, Marketing. AVI Publishing Co., Westport, CT.

21. Falconer, D. S. 1960. Introduction to Quantitative Genetics. Ronald Press, New York.

22. Fujise, K., Yunoue, T., and Chishiki, T. 1955. Studies on the habits of flowering and seed-setting in varieties of sweet potatoes (Ipomoea batatas Lam.). Bull. Kyushu Agric. Exp. Stn. 3, 109–142 (in Japanese, English summary).

23. Graves, B. 1976. Painter—a new sweet potato variety. Veg. Grow. News 30 (11), 1.

24. Hahn, S. K., and Leuschner, K. 1982. Breeding sweet potato for weevil resistance. In Sweet Potato. R. L. Villareal and T. D. Griggs (Editors), pp. 331–336 AVRDC, Tarnan, Taiwan.

25. Hammett, H. L., Alberton, R. C., and Withers, F. T. 1978 Hardcore in some sweet potato cultivars and lines. J. Am. Soc. Hortic. Sci. 103, 239–241.

26. Harmon, S. A. 1974. 'Georgia Jet,' an early, high yielding high quality sweet potato cultivar for Georgia. Ga., Agric. Exp. Stn., Res. Rep. 193.

27. Harter, L. L., and Weimer, J. L. 1929. A monographic study of sweet potato diseases and their control. U.S., Dep. Agric., Tech. Bull. 99.

28. Hernandez, T. P., Constantin, R. J., Hammett, H., Martin, W. J., Clark, C., and Rolston, L. H. 1981. 'Travis' sweet potato. HortScience 16, 574.

29. Hernandez, T. P., Hernandez, Travis P., Constantin, R. J., and Martin, W. J. 1974. Jasper, a soil rot-resistant sweet potato variety. La., Agric. Exp. Stn., Circ. 100.

30. Hsi, D. C. H., and Corgan, J. N. 1979. 'Oklamex Red' sweet potato. HortScience 14, 79–80.

31. Jones, A. 1964. A disease of pollen mother cells of sweet potato associated with Fusarium moniliforme. Phytopathology 54, 1494–1495.

32. Jones, A. 1965. Cytological observations and fertility measurements of sweet potato (Ipomoea batatas (L.) Lam.). Proc. Am. Soc. Hortic. Sci. 86, 527–537.

33. Jones, A. 1965. A proposed breeding procedure for sweet potato. Crop Sci. 5, 191–192.

34. Jones, A. 1966. Morphological variability in early generations of a randomly intermating population of sweet potatoes, Ipomoea batatas (L.) Lam. Ga., Agric. Exp. Stn., Tech. Bull. [N.S.] 56.

35. Jones, A. 1968. Chromosome numbers in Ipomoea and related genera. J. Hered. 59, 99–102.

36. Jones, A. 1969. Quantitative inheritance of fusarium wilt resistance in sweet potatoes. J. Am. Soc. Hortic. Sci. 94, 207–209.

37. Jones, A. 1969. Quantitative inheritance of ten vine traits in sweet potatoes. J. Am. Soc. Hortic. Sci. 94, 408–411.

38. Jones, A. 1972. Mass selection for low oxidation in sweet potato. J. Am. Soc. Hortic. Sci. 97, 714–718.

39. Jones, A. 1977. Heritabilities of seven sweet potato root traits. J. Am. Soc. Hortic. Sci. 102, 440–442.

40. Jones, A. 1980. Sweet potato. In Hybridization of Crop Plants. W. R. Fehr and H. H. Hadley (Editors), pp. 645–655. Am. Soc. Agron. Inc., Madison, WI.

41. Jones, A., and Cuthbert, F. P., Jr. 1971. Unpublished data. U.S. Vegetable Laboratory, U.S. Department of Agriculture, Charleston, SC.

42. Jones, A., and Dukes, P. D. 1976. Some seed, seedling and maternal characters as estimates of commercial performance in sweet potato breeding. J. Am. Soc. Hortic. Sci. 101, 385–388.

43. Jones, A., and Dukes, P. D. 1980. Heritabilities of sweet potato resistances to root knot caused by Meloidogyne incognita and M. javanica. J. Am. Soc. Hortic. Sci. 105, 154–156.

44. Jones, A., Dukes, P. D., and Cuthbert, F. P., Jr. 1975. W-13 and W-178 sweet potato germplasm. HortScience *10*, 533.

45. Jones, A., Dukes, P. D., and Cuthbert, F. P., Jr. 1976. Mass selection in sweet potato: Breeding for resistance to insects and diseases and for horticultural characteristics. J. Am. Soc. Hortic. Sci. *101*, 701–704.

46. Jones, A., Dukes, P. D., and Cuthbert, F. P., Jr. 1977. Pesticides increase true seed production of sweet potato. Hort Science *12*, 165–167.

47. Jones, A., Dukes, P. D., Hamilton, M. G., and Baumgardner, R. A. 1980. Selection for low fiber content in sweet potato. HortScience *15*, 797–798.

48. Jones, A., Dukes, P. D., Schalk, J. M., Mullen, M. A., Hamilton, M. G., Patterson, D. R., and Boswell, T. E. 1980. W-71, W-115, W-119, W-125, W-149, and W-154 sweet potato germplasm with multiple insect and disease resistances. HortScience *15*, 835–836.

49. Jones, A., Hamilton, M. G., and Dukes, P. D. 1978. Heritability estimates for fiber content, root weight, shape, cracking and sprouting in sweet potato. J. Am. Soc. Hortic. Sci. *103*, 374–376.

50. Jones, A., and Jackson, C. R. 1968. Fungi from floral parts of sweet potato. HortScience *3*, 76–77.

51. Jones, A., Schalk, J. M., and Dukes, P. D. 1979. Heritability estimates for resistances in sweet potato to soil insects. J. Am. Soc. Hortic. Sci. *104*, 424–426.

52. Jones, A., Steinbauer, C. E., and Pope, D. T. 1969. Quantitative inheritance of ten root traits in sweet potatoes. J. Am. Soc. Hortic. Sci. *94*, 271–275.

53. Kobayashi, M., and Miyazaki, T. 1977. Sweet potato breeding using wild related species. *In* Tropical Root Crops. J. Cock, R. MacIntyre, and M. Graham (Editors), No. 1, pp. 53–57. Int. Development Res. Center, Ottawa, Canada.

54. Li, L. 1967. Study on the combining ability in crosses among five varieties of sweet potato. J. Agric. Assoc. China [N.S.] *58*, 33–45.

55. Li, L. 1975. The inheritance of quantitative characters in a randomly intermating population of sweet potatoes (*Ipomoea batatas* (L.) Lam.). J. Taiwan Agric. Res. *24*, 32–42 (in Chinese, English summary).

56. Li, L. 1977. The inheritance of crude protein content and its correlation with root yield in sweet potatoes. J. Agric. Assoc. China *100*, 78–87 (in Chinese, English summary).

57. Martin, F. W. 1965. Incompatibility in the sweet potato. A review. Econ. Bot. *19*, 406–415.

58. Martin, F. W. 1970. Sterility in some species related to the sweet potato. Euphytica *19*, 459–464.

59. Martin, F. W. 1977. Selected yam varieties for the tropics. *In* Tropical Root Crops. J. Cock, R. MacIntyre, and M. Graham (Editors), No. 1, pp. 44–49. Int. Development Res. Center, Ottawa, Canada.

60. Martin, F. W., and Cabanillas, E. 1966. Post-pollen-germination barriers to seed set in sweet potato. Euphytica *15*, 404–411.

61. Martin, F. W., and Jones, A. 1971. Flowering and fertility changes in six generations of open pollinated sweet potatoes. J. Am. Soc. Hortic. Sci. *96*, 493–495.

62. Martin, F. W., and Jones, A. 1973. The species of *Ipomoea* closely related to the sweet potato. Econ. Bot. *26*, 201–215.

63. Martin, F. W., Jones, A., and Ruberte, R. M. 1974. A wild *Ipomoea* species closely related to the sweet potato. Econ. Bot. *28*, 287–292.

64. Martin, W. J. 1970. Virus diseases. South. Coop. Ser. Bull. *159*.

65. Martin, W. J., and Birchfield, W. 1973. Further observations of variability in *Meloidogyne incognita* on sweet potato. Plant Dis. Rep. *57*, 199.

66. Martin, W. J., Jones, L. G., and Hernandez, T. P. 1967. Sweet potato soil rot development in olivier silt loam soil as affected by annual applications of lime or sulfur over a seven year period. Plant Dis. Rep. *51*, 271–275.

67. Martin, W. J., Nielsen, L. W., and Morrison, L. S. 1970. Diseases. South Coop. Bull. *159*.

68. Maryland AES 1971. Mimeo Rel. Notice (unnumbered). Jan. 19. Redmar.
69. Maryland AES 1972. Mimeo Rel. Notice (unnumbered). Jan. 17. Goldmar.
70. Miller, J. C. 1939. Further studies and technic used in sweet potato breeding in Louisiana. J. Hered. *30,* 485–492.
71. Miller, P. R. 1960. Plant pests of importance to North American agriculture—index of plant diseases in the United States, U.S. Dep. Agric., Agric. Handb. *165.*
72. Morrison, L. S. 1970. Nematode diseases. South. Coop. Ser. Bull. *159.*
73. Mullen, M. A., Jones, A., Davis, R., and Pearman, G. C. 1980. Rapid selection of sweet potato lines resistant to the sweet potato weevil. HortScience *15,* 70–71.
74. Mullen, M. A., Jones, A., Arbogast, R. T., Schalk, J. M., Patterson, D. R., Boswell, T. E., and Earhart, D. R. 1980. Field selection of sweet potato lines and cultivars for resistance to the sweet potato weevil. J. Econ. Entomol. *73,* 288–290.
75. Nielson, L. W. 1970. Fungus and miscellaneous diseases. South. Coop. Ser. Bull. *159.*
76. Nielson, L. W., and Terry, E. R. 1977. Sweet potato (*Ipomoea batatas* L.). *In* Plant Health and Quarantine in International Transfer of Genetic Resources. W. B. Hewiet and L. Chiarappa (Editors), pp. 271–276. CRC Press, Inc., Cleveland, OH.
77. Nishiyama, I. 1971. Evolution and domestication of the sweet potato. Bot. Mag. *84,* 377–387.
78. Nishiyama, I., Miyazaki, T., and Sakamoto, S. 1975. Evolutionary autoploidy in the sweet potato (*Ipomoea batatas* (L.) Lam.) and its progenitors. Euphytica *24,* 197–208.
79. Pope, D. T., Nielsen, L. W., and Miller, N. C. 1971. Jewel, a new sweet potato variety for North Carolina. N.C., Agric. Exp. Stn., Bull. *442.*
80. Rolston, L. H., Barlow, T., Hernandez, T. P., Nilakhe, S. S., and Jones, A. 1979. Field evaluation of breeding lines and cultivars of sweet potato for resistance to the sweet potato weevil. HortScience *14,* 634–635.
81. Rolston, L. H., Barlow, T., Jones, A., and Hernandez, T. 1981. Potential of host plant resistance in sweet potato for control of a white grub, *Phyllophaga ephilida* Say (Coleoptera:Scarabaeidae). J. Kans. Entomol. Soc. *54,* 378–380.
82. Schalk, J. M., and Jones, A. 1982. Methods to evaluate sweet potato for resistance to the banded cucumber beetle in the field. J. Econ. Entomol. *75,* 76–79.
83. Scheuerman, R. W., Hernandez, T. P., Martin, W. J., Clark, C., Constantin, R. J., and Hammett, H. 1981. 'Eureka' sweet potato. HortScience *16,* 689.
84. Shikata, S. 1980. Utilization of random mating population in sweet potato breeding. Bull. Chugoku Natl. Agric. Exp. Stn., Ser. A *26* (in Japanese, English summary).
85. Steinbauer, C. E. 1937. Methods of scarifying sweet potato seed. Proc. Am. Soc. Hortic. Sci. *35,* 606–608.
86. Steinbauer, C. E., and Kushman, L. J. 1971. Sweet potato culture and diseases. U.S. Dep. Agric., Agric. Handb. *388.*
87. Thornberry, H. H. 1966. Plant pest of importance to north american agriculture-index of plant virus diseases. U.S. Dep. Agric., Agric. Handb. *307.*
88. Ton, C.-S., and Hernandez, T. P. 1978. Wet soil stress effects on sweet potatoes. J. Am. Soc. Hortic. Sci. *103,* 600–603.
89. Walsh, P. A., and Johnson, E. 1980. Agricultural Statistics. U.S. Government Printing Office. Washington, D.C.
90. Whatley, B. T., and Phills, B. R. 1977. 'Rojo Blanco' sweet potato. HortScience *12,* 265.
91. Whatley, B. T., and Phills, B. R. 1977. 'Carver' sweet potato. HortScience *12,* 266.
92. Yen, D. E. 1974. The sweet potato and Oceania. B. P. Bishop Mus. Bull. *236,* 1–389.
93. Yen, D. E. 1976. Sweet potato. *In* Evolution of Crop Plants. N. W. Simmonds (Editor), pp. 42–45. Longmans, Green, New York.

2
Watermelon Breeding

HUBERT C. MOHR

INTRODUCTION

In the 1964–1978 period the acreage of watermelons in the United States varied from a high of 286,320 in 1965 to a low of 210,700 in 1975. The trend toward reduced acreage is noticeable during this period. Yield per acre has tended to improve (from a low of 95 cwt per acre in 1969 to 116 cwt per acre in 1978), but total annual production has tended to decline from a high of 29,602,000 cwt in 1965 and to a low of 23,034,000 cwt in 1974. Thus, there appears to be an overall downward trend in production.

The principal acreages of watermelons are in Texas, with approximately 25,000 acres

BREEDING VEGETABLE CROPS

in spring production and another 25,000 acres in summer production; Florida, with 50,000 acres, all in spring production; and Georgia, with approximately 10,000 acres in spring production and 20,000 acres in summer production. It is noteworthy that Florida's production per acre is the highest, with approximately 9,000,000 cwt from 50,000 acres (*1*).

Commercial production continues to be dominated by the cv. Charleston Gray, which was introduced in 1954. Several other cultivars have been introduced since that time, notably Crimson Sweet and Jubilee (both in 1963), and these have achieved considerable popularity as high-quality melons. The consumption of watermelons in the United States has fallen from 18.2 lb per capita in the 10-year period 1946–1955 to approximately 13.5 lb in the 1967–1976 period. This represents a 25% decrease in per capita consumption, obviously a matter of concern to producers (*8*). The answer to this problem seems to lie in the development of superior quality cultivars that also enable growers to produce more efficiently. Thus, watermelons could be more competitively priced in relation to other dessert fruits. New cultivars that would permit production closer to the markets could help reverse the downward per capita trend by reducing the transportation costs from distant production areas of the rather bulky product. Smaller-fruited new cultivars could be priced lower, also helping to overcome buyer resistance to the high price per melon demanded for larger fruit.

ORIGIN AND GENERAL BOTANY

Climatic Requirements

The watermelon is tender to frost and most cultivars require a relatively long growing season (*21*). Plant growth and fruit development are favored by high temperatures and abundant sunlight. Atmospheric humidity appears to be a factor principally as it may affect the occurrence and severity of fruit and foliage diseases.

Taxonomy and Origin

Until recently the watermelon was classified as *Citrullus vulgaris* Schrad., but in 1963 Thieret called attention to the correct name, *Citrullus lanatus* (Thunb.) Mansf. (*54*). In 1857 David Livingstone, the famous missionary–explorer, found both bitter and sweet melons growing wild together in Africa and noted that the natives used them as a source of water in the dry season. It has thus been generally concluded that Africa is the center of origin of the genus (*6*).

Culture of the watermelon goes back to prehistoric times, as revealed by pictures made in ancient Egypt. Names for the watermelon appear in the old literature of the Arabic, Berber, Sanskrit, Spanish, and Sardinian languages. Apparently early cultivation of the crop occurred in the Mediterranean area and as far east as India. Now the watermelon is also important in the warmer parts of the Soviet Union as well as in other parts of Asia Minor, the Near East, China, and Japan. It is believed to have been brought to the United States by the earliest European colonists and is recorded as being grown in Massachusetts as early as 1629.

Uses

Although the watermelon is used as a fresh dessert fruit in America, it has some other uses elsewhere in the world. In the Soviet Union, it is reported that watermelon has been used in making a beer, and sometimes the juice is boiled down to make a heavy sweet syrup. In the Mediterranean area, watermelon may be a staple food for both human and livestock consumption. In Asia, the seeds are roasted, with or without salting, and eaten.

Species Descriptions and Cross Compatibility

Cogniaux and Harms [1924, cited by Shimotsuma (*50*)] recognized four *Citrullus* species, viz. *C. vulgaris* Schrad. [now called *C. lanatus* (Thunb.) Mansf.], *Citrullus colocynthis* (L.) Schrad., *Citrullus ecirrhosus* Cogn., and *Citrullus naudinianus* (Sond.) Hook., all originating in Africa (*50*). Descriptions of these species, as explored by Shimotsuma (*50*), are as follows:

C. vulgaris is an annual watermelon originating in Southern (perhaps Central) Africa. It is widely distributed in Egypt and in southern, western, and central Asia. It has large, broad green leaves, which are orbicular to triangular–ovate in shape and deeply three to five lobed or sometimes simple. Medium-sized flowers are monoecious and have short pedicels. Fruits are of medium to large size, with thick rind and solid flesh with high water content. Flesh color may be red, yellow, or white. Seeds are ovate to oblong, are strongly compressed, and have white or brown seed coats.

C. colocynthis is a perennial, indigenous to northern Africa. It differs from *C. vulgaris* chiefly in the size of plant organs. Leaves are small with narrow lobes, and are hairy and grayish in color. Flowers are monoecious and small. Bloom is profuse in autumn, when fresh vegetative growth also occurs. Seeds are small and brown. Fruits are small, not exceeding 3 in. in diameter, with rind and spongy flesh that is always bitter.

C. naudinianus and *C. ecirrhosus* are both perennials indigenous to the desert areas of South-West Africa. The vegetative characters of *C. naudinianus* differ from those of the other species. Leaves are deeply palmatifid and covered with dense, fine hairs. Tendrils are simple, straight, elongated, or slightly curved at the apex. Flowering is dioecious, and flowers are not formed until the second year of growth. Fruits are ellipsoid in shape, of medium to large size, with thin rind and soft juicy flesh. Seeds are white and would not germinate under natural conditions. *C. ecirrhosus* closely resembles *C. colocynthis* in vegetative characteristics, but its leaves are more divided, are covered with dense fine hairs, and have strongly recurved margins. Tendrils are lacking. Fruits are subglobose with white flesh and bitter like *C. colocynthis*. No flowers are produced until the second year of growth.

A fifth species, *Citrullus fistulosus*, was investigated and was found more closely to resemble *Cucumis* species. It was suggested that it be reclassified, since it would not hybridize with the above four *Citrullus* species and has a different chromosome number. All four *Citrullus* species could be crossed with each other successfully, and F_1 seeds from almost all crosses germinated well. F_1 seedlings grew normally and set fruit with good seeds.

It was concluded by Shimotsuma that a bitter-fruited form of *C. vulgaris* is the ancestor of the cultivated watermelon. The four species *lanatus, colocynthus, naudinianus,* and *ecirrhosus* all have 22 chromosomes ($2n$) according to results of Shimotsuma's cytological studies (*50*) (see Table 2.1).

General Botany

The watermelon plant resembles plants of *Cucumis,* but the stems are angular in cross section and the leaves, which are cordate at the base, are pinnately divided into three or

TABLE 2.1.   List of *Citrullus* Species Investigated and Their Major Characteristics[a]

Species	Origin	Periodicity	Chromosome number (2n)
C. vulgaris Schrad. (cultivated watermelon)	Japan	Annual	22
C. vulgaris Schrad. (bitter wild race)	Cape Province (South Africa)	Annual	22
C. vulgaris Schrad. (non-bitter wild race)	Cape Province (South Africa)	Annual	22
C. colocynthis (L.) Schrad.	Rabat (Morocco)	Perennial	22,
C. ecirrhosus Cogn.	South-West Africa	Perennial	22
C. naudinianus (Sond.) Hook.	South-West Africa	Perennial	22
C. fistulosus Stocks	India	Annual	24

[a]Adapted from Shimotsuma (50).

four pairs of lobes. A non-lobed (entire) leaved mutant was found and may be used as a genetic marker whenever it is unique (31).

Roots

The root system is very extensive but shallow, consisting of a tap root and many lateral roots growing within the top 2 ft of soil. Early destruction of the tap root (as in transplanting) may be advantageous in getting superior yields (13).

Growth Habit

The plant typically is a trailing (vine) annual, but recently dwarf forms, which may be described as "bush," have been located and are being used to develop cultivars suitable for growth in a limited space. Dwarfs may have a total diameter of as little as 2 ft, whereas the more rampant vine cultivars may be 30 ft or more in diameter. Dwarfing is primarily related to shortened internodes (32).

In the typical vine, there is a delay in the development of lateral branches, with a dominant single "runner" reaching several feet in length before branching is initiated, and maintaining dominance for some time thereafter. In some of the new dwarf types, multiple branching occurs simultaneously from the crown of the plant when it is quite young, which provides more potential bearing areas for concentrated fruit set (34).

Pollination

While watermelon is classed as a naturally cross-pollinated crop, there is a large amount of self-pollination and sibbing that occurs normally. This was demonstrated by interplanting two parental lines, one of which carried a marker gene, and then examining the progeny from the marker parent (31).

Fruit

The range of fruit size in existing cultivars is great, from the 3-lb melons typical of the New Hampshire Midget cultivar to the large-fruited Cobb Gem and Tom Watson cultivars, which may occasionally produce melons in excess of 100 lb. Most present-day cultivars produce melons of approximately 25 lb, but the future market trend appears to be toward smaller-fruited cultivars.

Fruit shape varies, ranging from long, cylindrical fruits to spherical fruits, with the melons of most cultivars intermediate in shape. Rind colors vary from white to various shades of green, with some cultivars having striped rind and some mottled rind. Rind ranges from brittle to tough and flexible, from very thin to excessively thick, and from very soft to very hard.

Flesh color may be white, yellow, orange, pink, or red; flesh texture varies from fine grained and "melting" to firm, coarse, and fibrous. Seeds range in size from very small (size of tomato seed) to large (size of pumpkin seed). Seeds usually range in color from white to black, but red, green, brown, and mottled colors also occur.

The greater part of the fruit is the edible flesh, which is mostly derived from the placentae, whereas the cantaloupe and squash are largely pericarp. The pericarp of the watermelon is its hard outer "rind."

FLORAL BIOLOGY AND CONTROLLED POLLINATION

Flowers

The flowers of watermelon are smaller and less showy than those of most other species of cultivated Cucurbitaceae (46). They are located in leaf axils, most commonly singly. Most cultivars are monoecious (producing pistillate and staminate flowers separately), but a few older cultivars are andromonoecious (producing perfect flowers and staminate flowers). The pistillate or hermaphroditic flowers normally occur in every seventh-leaf axil, the intervening axils being occupied by staminate flowers. The corolla is greenish yellow, united in a tube, deeply five lobed. Three stamens are inserted at the base of the corolla.

Pollination

Watermelons are generally pollinated by honey bees (41). In andromonoecious cultivars the hermaphroditic flowers must be visited by insects to effect pollination; therefore, the hermaphroditic flowers of watermelon do not confer the advantage of self-pollination one might expect, and andromonoecy has no advantage over monoecy in maintaining pure lines. The method of pollination naturally favors considerable cross-pollination, and consequently considerable genetic variability within a cultivar is maintained.

Flowers open shortly after sunrise and normally remain open only one day. The pistillate flower, and the staminate flower in the axil just below it, open on the same day. Usually the anthers have dehisced when the corolla expands, and the pollen is visually evident in sticky masses adhering to the anther. The stigma is receptive throughout the day, but it has been established that fruit-setting frequency following artificial self-pollination is much higher when pollination occurs between 6 and 9 A.M. than when it occurs later in the day. High atmospheric humidity favors fruit setting. The ovary, which is inferior, is readily apparent, and its relative size is an important factor in fruit setting. Small, poorly developed ovaries rarely set, whereas pollination of flowers that have large ovaries results in a high percentage of success. The largest ovaries are usually found on flowers near the tips of the most vigorous branches of a plant.

Fruit Setting

Unlike some other members of the cucurbit family, the watermelon does not have flowering peaks or fruiting cycles (41). However, there is an inhibitory influence produced by fruit already set that reduces further fruit setting. Thus, if the controlled pollination desired on a plant has not been obtained, it is necessary to continue to remove all of the open-pollinated fruit until success in setting the controlled pollination is attained.

Controlled Pollination

Controlled pollinations are made by bending back the petals of the staminate flower until they break, leaving the stamens protruding prominently. Holding this staminate flower by the pedicel, the anthers, with their masses of sticky pollen, are lightly brushed against the stigmatic surface of the pistillate flower.

Prior to pollination, the unopened buds should be protected from insect visitation. This may be done with small screen cages placed over the selected flowers (56) or by preventing the buds from opening with paper clips (41) or Scotch Tape. After pollination, the protection must be replaced and continued for at least one day. A helpful procedure for locating flowers to be used in controlled pollinations is the placement of plastic flags at each flower to be used. Different colors can be used to designate the location of developing fruits, flowers to provide pollen, and pistillate buds not yet opened. Tags should be attached to the pedicel of the pistillate flower identifying the pollen parent and indicating the date of pollination.

MAJOR BREEDING ACHIEVEMENTS OF THE PAST

Fusarium Wilt Resistance

A major objective of watermelon breeding has been to develop cultivars with disease resistance, particularly resistance to fusarium wilt, *Fusarium oxysporum* f. sp. *niveum* (E. F. Smith) Snyder and Hansen. One can understand this emphasis when considering that the alternative to effective resistance is a crop rotation that would permit watermelon production on the same land only once every 8–10 years. The organism responsible for fusarium wilt is a soil-inhabiting fungus that increases rapidly when watermelons occupy the land. It invades the roots of the plant and enters the vascular system where it plugs the xylem vessels, resulting in the wilting of the plant and eventual death. Plants suspected of being under attack can be checked by cutting through the stem of a branch showing wilting. If the interior shows a brown discoloration, it verifies that it is caused by the *Fusarium* organism.

The acute need for a resistant cultivar was recognized by W. A. Orton, an early scientist in the USDA, who initiated one of the first organized plant breeding programs with disease resistance as a goal (35). In 1911 he introduced the cv. Conqueror, which was the first vegetable cultivar developed by hybridization for resistance to a plant disease (36). The source of resistance was the inedible stock "citron." While Conqueror was acceptable, it lacked quality and consequently failed to achieve popularity. In addition, its resistance to fusarium wilt was not consistently high, a complaint that has plagued most so-called wilt-resistant cultivars of watermelon up to today.

Efforts to develop a cultivar that would resist the ravages of fusarium wilt under all conditions prevalent in the various watermelon-growing areas have occupied the efforts of numerous plant breeders and/or plant pathologists for many years. Up to the present time this goal has not been fully achieved, but progress has been such that several recent cultivars, e.g., Calhoun Gray, Summit, Smokylee, and Dixielee, have resistance sufficient for most situations. It has been determined that inheritance of resistance may involve several genes. Crall (*8*, p. 8), who has been involved in breeding for resistance to fusarium wilt for a number of years, states that ''wilt resistance in watermelons is governed by multiple factors, mostly recessive, so that the development of resistant varieties has been a rather slow process. . . .''

Parris (*37*) has reviewed the inheritance of resistance to fusarium wilt and indicates that there may be recessive or dominant genes involved depending on the source of resistance being used by the breeder. No clear definition of the mode of inheritance of resistance has been published to date. Henderson *et al.* (*17*) established that there is at least one major dominant gene for resistance in the cv. Summit. All cultivars previously named have what Crall and Elmstrom (*9*) refer to as ''high-type resistance.'' Summit was determined to inherit resistance as a dominant factor (*17*). Many of the early cultivars with lower levels of resistance obviously carried recessive genes for resistance since crosses with suscepti- ble cultivars produced only susceptible F_1s (*37*). It may be concluded that resistance to wilt in watermelon is governed by multiple factors, most of them recessive. This has complicated the development of dependable cultivars possessing high-type resistance. The likelihood of cross-pollination between individuals heterozygous for wilt susceptibili- ty in seed fields may account for the so-called running out of resistance experienced with some of the cultivars released as ''resistant.'' Establishment of lines homozygous for the dominant gene for resistance, reported by Henderson *et al.* (*17*) in the cv. Summit, should give added impetus to the trend toward development of commercial F_1 hybrids with resistance. Other than the identification of the dominant resistance gene of Summit, no specific designation of individual resistance genes is available.

Barnes (*4*) ran screening tests of resistant cultivars using isolates of the *Fusarium* organism from various watermelon production areas of the United States. His results seem to indicate that these isolates vary in virulence, which could partially account for the differences in degree of resistance reported by different workers.

Anthracnose Resistance

Anthracnose, caused by the pathogen *Glomerella cingulata* var. *orbiculare* [formerly *Colletotrichum lagenarium* (Pass.) Ell. and Halst.] is a very destructive disease of water- melon. Anthracnose is found almost everywhere that watermelons are grown with the possible exception of some parts of California. It becomes particularly severe in periods of high humidity. It is caused by a fungus that is spread by airborne spores. It attacks both plant and fruit and may cause serious damage to the fruit during post-harvest handling. While field applications of fungicides can reduce the problem, resistance is clearly desirable.

Resistance to the disease was first located in some melons obtained from Africa. The Reverend Rush F. Wagner, Old Umtali Mission, Umtali, South Africa, sent seed of five cultivars of watermelon (not taxonomically classified) to D. V. Layton (*25*). Their seed and leaf characteristics were similar to those of commercial cultivars although two were described as inedible. All were susceptible to fusarium wilt, but were resistant to anthrac- nose. The three edible lines were crossed to U.S. cultivars and the F_1 was found to be resistant. It was determined that resistance was dominant over susceptibility, and segrega- tion in the F_2 (and backcross F_1) suggested a single factor pair. The cvs. Congo, Fairfax, and Charleston Gray were the first introduced as anthracnose resistant; and Charleston Gray has been the dominant watermelon cultivar for over 20 years, with much of its popularity attributable to its disease resistance.

It has now been established that there are several different races of the pathogen causing anthracnose. The resistance derived from the original African watermelon introductions is not effective against some of these races. Fortunately, these new races are not as yet of widespread distribution, and this is permitting breeders and plant pathologists time in which to locate sources of resistance and to breed cultivars that can replace those susceptible to the new races.

Resistance to Gummy Stem Blight

A fungus *Didymella bryoniae* (formerly *Mycosphaerella citrullina*) causes a serious disease of watermelons, particularly in warm humid climates. It is called gummy stem blight and may attack fruit, seedling, cotyledons, leaves, and stems. If plants escape early damage, the organism may continue to grow slowly and cause severe injury later. A reddish-brown exudate appears on stems (near the crown) of infected plants, frequently as melons approach maturity, and wilting is followed by death of the plant. Resistance has been located in some plant introductions of the wild citron type, and breeders are working toward incorporating it into acceptable cultivars (53). The disease is controllable with fungicidal sprays; but, as is the case with anthracnose, resistance would provide additional protection and could reduce the costs of the spray program by reducing the frequency of application required for control.

Durability

Much of the early breeding emphasized durability of the melon to the rough handling in which it was often subjected. Melons were selected for thick, tough and flexible rinds and for firm flesh. Cultivars not capable of withstanding rough handling were failures, no matter how excellent in other respects. Now the emergence of a market for smaller melons opens the possibility for packaging and an associated lessening of emphasis on durability factors. Still, some attention should continue to be given in this area since extreme brittleness will not be acceptable. Such melons crack open in the field, sometimes even before maturity. Excessive supplies of moisture during development of the fruit seem to favor occurrence of this defect. Selecting for more flexible rind will reduce its prevalence. A technique for evaluation of flexibility has been proposed by Ivanoff (20). Andrus (2) demonstrated that flesh firmness is an important factor in durability. Firmness is apparently heritable, but its mode of inheritance has not been determined.

Rind may be extremely hard, as in the cv. Peacock, or very soft as in Calhoun Sweet. Most cultivars fall between these extremes, indicating that multiple factors are involved.

Flavor

The flavor of watermelon is not a clearly definable trait, but some evaluators recognize an off-flavor, which they call "caramel." To them, this flavor is objectionable, but others do not find it so. It seems to be associated with some of the "intense-red" breeding lines and appears to be heritable since it can be eliminated by selection.

Sugar content, as indicated by refractometer readings, is the most widely recognized criterion for flavor preference. The higher the reading, the greater the preference. Progress in breeding for soluble solids content has been good, with recently introduced cultivars running above 12%, whereas older cultivars generally are less than 9%.

Bitterness is an undesirable characteristic that is inherited from *C. colocynthis*. It is dominant to non-bitter flavor and inheritance is controlled by one gene. It is mentioned because interspecific hybridization may be desirable in watermelon breeding if desired genes are present only in *C. colocynthis*.

Seeds

Seeds enter into an evaluation of watermelon quality. The quantity of seeds is heritable, and consumers obviously find large numbers of seeds objectionable. They may prefer relatively small numbers, but seedsmen do not like to handle cultivars with few seeds per fruit. The seed price of a cultivar producing few seeds is relatively high.

Seed size has been investigated and heritability reported (24,39). The author has found seed sizes below those reported, with some actually as small as tomato seed! Consumers apparently are not concerned with seed size, except that in some of the very small melons, 3–5 lb, a large number of large seeds results in a relatively low percentage of edible flesh, which is furthermore difficult to obtain.

Seed colors include white, tan, brown, black, red, green, and mottled black and brown. There is a rather strong prejudice against white seed color, which the consumer associates with melon immaturity. Black seed seems to intensify red flesh color and permits a very attractive color combination in the cut melon. Occasionally the breeder may have difficulty in attaining uniformity of seed characteristics when all other important characteristics are sufficiently uniform to justify naming and release.

Seedless Triploid Hybrids

No cases of naturally occurring parthenocarpy have been reported in watermelon. Much attention was attracted, therefore, to the research reported by Kihara (23) concerning a technique of creating triploid F_1 hybrids that produced seedless fruit. Production of seedless fruit by triploid plants in the grower's field is dependent upon the flowers of the triploids being pollinated with pollen from diploid plants. The diploid pollen stimulates parthenocarpy, but the ovules fail to develop because of the sterility accompanying the triploid condition (27).

To produce triploid seeds a tetraploid line must first be established. The use of colchicine to induce polyploidy is discussed by Eigsti (12). The cross of a diploid pollinator with a tetraploid female will yield triploid seed, but the reciprocal cross (tetraploid pollinator by diploid female) is unsuccessful. This fact partially accounts for the high price of triploid seed, which in 1982 was approximately 20 times that of open-pollinated cultivars. Tetraploid watermelons characteristically produce much smaller numbers of seeds than diploids. This also makes it expensive to maintain the tetraploid lines (Fig. 2.1).

An additional disadvantage to growers of seedless melons is the difficulty in germinating the triploid seed. The Japanese recommended removal of part of the seed coat to facilitate germination. They also recommended that seed be germinated at a temperature of 86°F, which would entail starting them in hotbeds or greenhouses and transplanting to the field in many areas. These additional cultural operations are obviously expensive and, coupled with the high cost of seed, apparently raised production costs so high that the prices that consumers would pay did not leave acceptable profits for the producer.

Thus, while seedless triploids are a notable plant breeding achievement, producing a triploid watermelon crop has not been considered economically feasible by most U.S. watermelon growers.

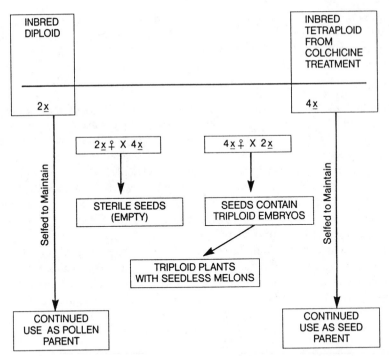

FIGURE 2.1. Procedure for producing seed of seedless triploid hybrid watermelons. The tetraploid line must be used as the female parent, because the reciprocal cross using the diploid as female results in empty seed. Seedless melons are the result of stimulative parthenocarpy (pollination with normal diploid pollen). Thus a diploid cultivar should be interplanted with the seedless hybrid.

Fruit Size

Older watermelon cultivars typically produced rather large fruit, and there was keen competition among growers to see who could produce the largest. Partially this related to the considerable market for "slicing watermelons" for restaurants and cafeterias. Also, as previously noted, melons were shipped in bulk and it was less time-consuming to harvest and load larger fruits.

The need for an early watermelon to mature in New Hampshire led Yeager (*61*) to import small-fruited cultivars from Japan and the Soviet Union that were early. From these he developed White Mountain and New Hampshire Midget, which required only 65 days to mature fruits weighing 2–4 lb. More recently the cultivar Petite Sweet (5–10 lb) was developed (*15*), and other small-fruited cultivars are under development, particularly at the Florida Agricultural Research Center at Leesburg and at the Kentucky Agricultural Experiment Station at Lexington.

Most of the present-day cultivars produce fruits in the range of 20–30 lb, but the trend seems to be toward fruits of 10–15 lb, and perhaps even smaller. The mode of inheritance of fruit size has been reported as polygenic (with 25 genes involved) by Poole and Grimball (*40*). A continuous array of sizes is obtained from the F_2 of large-fruited parents crossed with small-fruited parents.

Fruit Shape

Melon shape was determined to be qualitatively inherited, with one pair of alleles (round vs. elongated) involved (*40,42,59*). The F_1 was intermediate and the F_2 segregated in the ratio 1 round : 2 intermediate : 1 elongated. However, there appear to be modifier genes influencing shape. Some "elongate" cultivars are decidedly long–cylindrical in shape, others are tapered toward the ends. Some "round" cultivars produce melons almost perfectly spherical, whereas others are "blocky" in shape. Most elongated cultivars have a tendency to produce so-called gourd-neck fruit, which are classed as culls; whereas round-fruited cultivars tend to be quite susceptible to "hollow-heart", also a serious defect. The intermediate shape appears to be less subject to either of these defects, and this advantage is obtained in F_1 hybrids between long- and round-fruited parent lines.

Rind Color

A single gene determines the intensity of green color in the rind. Dark green is dominant to light green (called "gray") (*59*). Striped green is recessive to solid dark green but dominant to solid light green (*59*). Greenish white mottling of the rind is determined by a single recessive gene (*59*). Rind colors or color patterns are of little practical importance except that dark green melons seem to be more prone to sunburning, which results in collapse of rind tissues and may be a serious defect in regions of high light intensity. Barham (*3*) determined that a rind color change from green to orange is a good maturity indicator and is controlled by a single recessive gene pair. Unfortunately, the plant develops a chlorotic condition that accompanies the melon rind color change.

In addition to color there is a heritable characteristic known as "furrowing" or "grooving" of rind. It is recessive to smooth surface and apparently is of no practical importance.

Plant Habit

Taxonomists describe the plant of watermelon as a long, trailing vine. Considerable variation has been observed with regard to the vigor and extent of the vegetative growth of the various vine cultivars. Very limited vine growth occurs with cultivars such as New Hampshire Midget and Petite Sweet; but Charleston Gray, the presently dominant cultivar, produces vines that extend to a diameter of 20 ft or more, and there are cultivars with more rampant growth than this. No investigation of the genetics of vine size and vigor has been reported.

A mutant with much shortened internodes was discovered by the author (*32*). The resultant dwarf plant habit (Fig. 2.2) is recessive and is determined by a single gene (*dw1/dw1*). As with vines, there is heritable variability in the mature plant size, but this generally falls within the range of 1.5–6 ft for plant diameter. Such plants make feasible the culture of watermelons in home gardens where the area available is restricted. Another dwarfing gene (*dw2/dw2*), which also controls multiple branching from the crown of the plant (Fig. 2.3), was more recently discovered (*34*). When *dw1/dw1* was crossed with the dwarf *dw2/dw2*, the F_1 plants produced normal vines. The progeny segregated in the F_2 ratio of 9 vines : 3 short internode dwarfs : 3 multiple branched dwarfs : 1 "double-recessive dwarf," the latter being a totally new plant type (Fig. 2.4). A cultivar with the double-recessive genotype for dwarf is Kengarden, which is covered under the Plant

FIGURE 2.2. Branching typical of cv. Bush Desert King (*dw1/dw1*). Note short internodes and tendency for stems to be twisted (leaves removed). Stems are thickened and tend to be brittle.

FIGURE 2.3. Branching typical of Japanese Dwarf (*dw2/dw2*). Note multiple branching from the crown of the plant, which is unique. Stems are normal except for short internodes (leaves removed).

FIGURE 2.4. Very short internodes (leaves removed) typical of double-recessive dwarf ($dw1/dw1, dw2/dw2$). Crown branching derived from cv. Japanese Dwarf.

Variety Protection Act. In addition to suitability for garden culture, dwarf plant habits may have potential for adaptation to machine harvesting.

Leaf Shape

Typically, watermelon leaves are deeply indented and prominently lobed. A mutant without indentation and lobing on the leaves was discovered by the author (*30*). Lobed leaf shape is incompletely dominant to non-lobed. A single gene determines the non-lobed leaf character. It is an excellent seedling marker since the mutants are vigorous, fertile, and readily identified. When a line homozygous for this marker is used for a seed parent, it serves to identify F_1 hybrid progeny (normal leaf) and distinguishes them from progeny resulting from self-pollination (non-lobed type). The latter may be removed in the thinning of surplus seedlings, leaving only hybrids.

Sex Expression

A single pair of genes determines monoeciousness versus andromonoeciousness (46). Most cultivars are monoecious (A/−). Andromonoecious plants (a/a) have both staminate and perfect flowers. However, this does not appear to be advantageous since insect visitation of the perfect flowers is necessary to effect pollination.

Both monoecious and andromonoecious plants usually have a fixed ratio of six staminate flowers to one pistillate (or perfect) flower. The author has located lines that have a ratio of three to four staminate flowers to one pistillate flower, but the mode of inheritance has not yet been determined. The greater frequency of pistillate flowers when combined with multiple branching may enhance the concentration of fruit set near the crown of the plant, an important consideration in developing a plant suitable for mechanized harvesting.

Male sterility was present in an induced mutation discovered by Watts (57). Normal watermelon leaves are pubescent, but the mutant is glabrous. This distinctive pleiotropic marker associated with male sterility makes it easier to identify male-sterile plants in a segregating population. Male sterility is inherited as a single recessive gene (gms/gms).

Flesh Color

Yellow flesh is recessive (y/y) to red and is determined by a single gene. Another shade of yellow, called "canary," is dominant to pink flesh. The red flesh color of C. lanatus is recessive (wf/wf) to the white flesh color (Wf/−) of C. colocynthis (51). The intense-red flesh color now being incorporated in many of the recently introduced cultivars is derived from the old cv. Peacock. The mode of inheritance has not been reported.

CURRENT GOALS OF WATERMELON BREEDING PROGRAMS

Yield and Quality

High yield per acre, involving both numbers of melons and total weight, has been a major goal of past programs and is still one of the top priorities in breeding watermelons. While melon prices to the grower are most commonly quoted in terms of price per pound, retailers often prefer to sell on a price per melon basis; hence the need for the two kinds of yield data.

It appears that earlier preferences for large-sized melons are shifting toward the smaller-sized fruit and may eventually result in packaging of watermelons such as is practiced with cantaloupes and honeydew melons. This trend may also favor eventual mechanized harvest of the commercial crop.

With preference toward smaller melons, there is the need for thinner (but tougher) rind, firm flesh, and smaller seed size (to avoid being out of proportion with edible flesh).

Color of the melon flesh is being given more attention. Intense red seems to be preferred to paler red or pink. Surprisingly, a yellow-fleshed cultivar recently introduced has achieved considerable popularity, and so flexibility appears to exist in this goal.

Soluble solids, reflecting "sweetness" of the melon, continue to be given considerable emphasis in breeding, and recently introduced cultivars show marked improvement in this characteristic.

Flesh texture and firmness are factors of obvious concern in the evaluation of quality, but they have not been clearly defined. While observation indicates that these characteristics are heritable, no information concerning mode of inheritance is available. Two types of texture are roughly described either as "melting" (or fine-grained) flesh or as "fibrous" (and/or coarse-grained) flesh. More recently there has appeared "firm flesh," a trait associated with shipping durability (2). Investigations into consumer preferences for different types of flesh is needed as well as inheritance studies.

Melon shape may determine the suitability of the fruit for packaging if this becomes the preferred method of shipping. Also, it is a factor influencing adaptability to machine harvesting.

Earliness

Time required to reach maturity is an important heritable characteristic. It is important to commercial growers because the price paid for the first melons harvested in an area is usually higher than that received later.

Early maturity for regions with a short growing season is the determining factor in whether watermelons will be planted. Generally the earliest maturing cultivars of the past have been small fruited. Obviously, there are physiological limitations on the amount of photosynthate that can be produced in a short growing period, which limits fruit size and number. Since recent trends favor smaller fruit, there may be no disadvantage to this.

The length of time required for fruit to mature following pollination has been reported in a number of different cultivar release publications. In early descriptive work, watermelons generally were reported to require approximately 30–35 days from setting to maturity. Data presented in later cultivar release publications indicate that fruit of early-maturing cultivars may require as little as 26 days to mature (Petite Sweet) and late-maturing ones as much as 45 days (Super Sweet) (15). Inheritance studies are needed so intelligent breeding can be undertaken. The number of days from seeding to maturity is given in many cultivar release publications and in most seed catalog descriptions of cultivars. Environmental conditions will play an important role in the expression of earliness, but genetically controlled plant characteristics also play an important role. A more thorough investigation of those characteristics contributing to early maturity needs to be undertaken in order that a more intelligent approach to breeding for this important characteristic can be taken.

Exploiting the Dwarf Habit

Cooperation between the plant breeder and the agricultural engineer usually precedes the development of machinery to harvest a crop mechanically. Excessively vegetative plants tend to clog up machinery and make engineering the equipment a difficult task. Until the discovery of dwarfing genes in watermelon, there was little hope that the typically rampant vine growth of the species would permit mechanized harvest. Now, mechanized harvest appears to be a difficult but feasible goal for watermelon breeders.

The economics of inflation have resulted in a home gardening boom. Since space is a limiting factor for the majority of gardeners, the extensive vine growth of watermelon eliminated it from the list of crops most gardeners plant. The dwarf forms, however, have been acceptable. Plant breeders need to combine dwarfness with early maturity, high quality, increased productivity, as well as other desirable characteristics to produce new cultivars suitable for home gardens and for mechanized harvest as well. Also, development of inbred lines for producing F_1 hybrid dwarfs should not be overlooked.

Disease Resistance

Incorporating the genes for "high-type" resistance to fusarium wilt into new cultivars is very important since there is no practical substitute for resistance to this soil-borne disease.

While anthracnose and gummy stem blight are controllable with fungicides, this is expensive and counter to the trend to avoid chemicals for control if possible. Since new races of the pathogens involved are likely to occur, it behooves the watermelon breeder constantly to seek alternative sources of resistance. These two diseases are problems principally in high-humidity areas, and resistance to them is not essential in cultivars developed for arid climates.

Development of F_1 Hybrid Cultivars

The popularity of F_1 hybrids of all vegetable crops has included watermelons in recent years. One of the most sensational F_1 hybrids among vegetables is the seedless triploid watermelon. Unfortunately, the technique of producing seedless triploids results in relatively high-priced seed, which when combined with difficult germination of the seed, has caused reluctance among commercial growers to produce seedless melons. Improvements in techniques that make triploid seed more economical and easier to germinate could result in a large-scale shift to this superior type of watermelon. Also, the development of dwarf seedless triploids could expand interest on the part of home gardeners.

It has been reported that inbreeding watermelons does not reduce vigor or productivity and tends to produce uniformity in many characteristics (41). Data to establish that there is yield superiority (and to what extent) in F_1 hybrid watermelons are lacking. This needs to be explored before setting a goal of establishing inbred lines with the intention of developing commercial hybrids.

Several techniques have been proposed for producing F_1 hybrid seed cheaply, including the use of the marked male-sterile mutant discovered by Watts (57) as a seed parent. Open-pollinated seed can be produced by using the genetic marker discovered by Mohr et al. (31).

From experiments concerning the cost of producing F_1 seed by hand pollination and by hand removal of male blossoms, it was determined that costs were not prohibitive by Singletary and Moore (52). This is not surprising considering that each fruit can produce a large number of seeds (approximately 225).

One obvious advantage that could relate to use of an F_1 hybrid would be that resistance to disease controlled by dominant genes in only one parent could be transmitted to the F_1 progeny without converting the other parent. Of course, this assumes that the two parents make a good F_1 in other respects. Also, the uniformity in size of fruit that is associated with F_1 hybrids would be advantageous to retailers who prefer to sell at a price per melon rather than per pound. F_1 hybrids provide the best way to take advantage of the intermediate fruit shape, where one parent is round fruited and the other long fruited. The intermediate fruit shape has less gourd necking and less hollow heart than the parental fruit shapes.

The characters of watermelon for which inheritance data have been obtained are listed in Table 2.2 along with suggested gene symbols. Insufficient genetic information exists for a mapping of the chromosomes, as has been possible with more intensely investigated crops such as the tomato. Such information as exists on linkage is presented in Table 2.3.

TABLE 2.2. Inherited Characters of Watermelon and Their Gene Symbols[a,b]

Preferred symbol	Synonym	Character description	Reference
Seed, color, shape			
w		*White* seed coat	29
t	(b^t)	*Tan* seed coat	29
r		*Red* seed coat	39
l		*Long* seed	39,40
s		*Short* seed	40
Fruit flesh color, flavor			
R	Y	*Red* flesh	38
y	(r)	*Yellow* flesh (recessive to red)	38,42
Wf*	(W)	*White flesh* (dominant to red)	51
C		*Canary* yellow flesh (dominant to pink)	38
su*	(su^{Bi})	*Suppressor* of bitterness	7
Fruit rind color, texture			
g	(D)	*Light green* skin (recessive to dark green but dominant to light green skin)	59
g^s	(d^s)	*Striped green* skin (dominant to light green)	38,59
m		*Mottled* skin	38,59
p		*Pencilled* lines on skin	38,59
go*	(c)	*Golden*	3
f		*Furrowed* fruit surface (recessive to smooth)	38
e	(t)	*Explosive* rind	38,42
Fruit shape			
O		*Elongate* fruit (incompletely dominant to spherical)	40,59
Dwarf habit			
dw-1		*Dwarf-1*	26,32
dw-2		*Dwarf-2*	26,34
Leaf shape, color, texture			
nl*		*Non-lobed* leaves	30
go*	(c)	*Golden*	3
gms*	(ms_g)	*Glabrous male sterile*	57,58
Male sterility			
gms*	ms_g	*Glabrous male sterile*	57,58
Sex inheritance			
a		*Andromonoecious* (recessive to monoecious)	38,46
Disease resistance			
Ar*	(B) Gc	*Anthracnose* resistance	16,25
pm		*Powdery mildew* susceptibility	45

[a]Adapted from Robinson *et al.* (*44*).
[b]Asterisks indicate a proposed new symbol.

TABLE 2.3. Seven Cases of Probable Linkage in Watermelon[a]

Genes	Crossover percentages or correlation coefficients indicated	Reference
1. M and G	39.9 ± 4.3%	59
M and g^s	33.8 ± 4.6%	59
2. M and fruit weight	$r = .38$ in F_2; $r = .65$ in backcross to light-weight parent	59
3. A and O	13.6 ± 2.9% in backcross; 20.7 ± 1.0% in F_3; 35.0 ± 3.5% in F_2	40
4. C and O	26.5 ± 8.2% in F_2	
5. O and fruit weight	$r = -.31$ total variance; $r = -.86$ Mendelian phenotypes	40
Same	$r = -.34$ F_2 total variance; $r = -.41$ F_2 total variance.	59
6. A and fruit weight	Linkage probable, but estimation impossible, difference between mean weight: $A = 5.08 \pm 0.24$, $a = 4.42 \pm 0.22$ kg	40
7. L and W	15.8 ± 1.2% in backcross; 19.3 ± 1.1% in coupling F_2; 21.5 ± 7.2% in F_3; 24.7 ± 4.9% in repulsion F_2	39

[a]Crossover percentages are shown for qualitative characters; correlation coefficients are shown where one of the phenotypes is a quantitative character not convertible. [Adapted from Poole (38).]

SELECTION TECHNIQUES FOR SPECIFIC CHARACTERS

Dwarfs

Hypocotyl length of double-recessive dwarfs is approximately one-half that of vines. This permits elimination of the vine genotype in segregating populations of seedlings grown in peat pots for transplanting. In this way only dwarf plants are transplanted in the field.

The compact growth of dwarf seedlings is a characteristic of much value in production of transplants. Double-recessive dwarf seedlings remain in a condition suitable for setting out for several weeks, whereas vine types become overgrown and impossible to handle efficiently within a few days. By planting four seeds per peat pot (3 in. square) of lines expected to segregate, and clipping all long hypocotyl seedlings with scissors, nearly 100% of the pots will have a dwarf plant. The seedling populations described above are produced by selfing or crossing plants with the following genotypes: $dw1/dw1$, $Dw2/dw2$ and $Dw1/dw1$, $dw2/dw2$.

Fusarium Wilt Resistance

Techniques for screening lines segregating for resistance vary considerably in the severity of the test. Some breeders use field testing on land with a history of severe wilt; others rely upon greenhouse screening of seedlings, sometimes in 3 in. square peat pots with survivors transplanted to the field. In any test, it is important to have a strain or race of the pathogen that is highly virulent. Failure to do this may result in later breakdown of resistance when the supposedly resistant line is exposed to a more virulent strain (4).

One greenhouse technique involves planting seeds in peat pots containing soil mixed with wheat grains on which mycelium of the wilt organism is growing. The inoculum is prepared by transferring mycelial fragments from a pure culture of a virulent strain of the organism to autoclaved flasks of whole wheat grains.

Most of the seedlings under this treatment die before reaching the transplanting stage, which saves time, effort, and space in the field. This technique (*33*) was used in selecting the resistant Texas W-5 Breeding Line described by Crall and Elstrom (*9*) as having "high-type" resistance and used by them in breeding the new wilt-resistant cvs. Smokylee, Dixielee, and Sugarlee.

Anthracnose Resistance

Resistance to anthracnose is dominant to susceptibility and is governed by a single pair of genes. The most prevalent races of the pathogen are designated as races 1 and 3, and these are the races to which most present cultivars are resistant. However, race 2 seems to be spreading and threatens to become a serious problem. So far, no cultivars are resistant to it, although resistance has been located in a plant introduction. Presumably the same inoculation and selection techniques will apply in developing cultivars resistant to race 2 that have been used successfully for resistance to races 1 and 3. These are described by Hall *et al.* (*16*).

Gummy Stem Blight Resistance

Resistance to gummy stem blight has been located in two plant introductions (*53*). One carries genes that promote a higher level of resistance than the other, but both are useful in developing cultivars with acceptable resistance. Greenhouse screening of seedlings, followed by transplanting of resistant segregates, is providing satisfactory progress. The screening technique is similar to that used in breeding for anthracnose resistance. The pathogen is grown on autoclaved frozen snap beans in 125 ml Erlenmeyer flasks for 7–10 days under fluorescent light at room temperature. Two cultures are blended in 500 ml of distilled water with a food blender for 2 min. Filtration of this inoculum through a Whatman No. 4 filter removes most of the snap bean substrate. It was found that this inoculum contained from 10^5 to 5×10^5 spores per milliliter. This was sprayed onto seedlings until run-off occurred. Plants were then incubated for 48 hr at $25° \pm 2°C$ in a moisture-saturated atmosphere. This technique was developed by Sowell and Pointer (*53*), and followed in screening to locate the resistant plant introductions PI 189225 and PI 271778.

Fruit Shape

Selections for fruit shape can be made in the seedling stage; long, narrow cotyledons indicate elongated fruit shape, whereas round cotyledons indicate that round fruit will be produced (*59*).

Earliness

Selections for early maturity are more reliable when based on data that include the date of first fruit set *and* the number of days from fruit set to maturity. Data on soluble solids can assist in verifying that fruit is fully mature. A hand refractometer is useful for this purpose.

Tough Rind

A simple method of selecting for rind toughness is to cut a thin strip or ribbon of rind $\frac{1}{16}$ to $\frac{1}{8}$ in. thick and 3 in. long and bend it into a circle. If it makes a complete circle, it is very tough; but if it breaks early in the attempt to make a circle, it is tender ("explosive"). The method is described in detail by Ivanoff (*20*).

Leaf Type

The non-lobed leaf type is readily selected in the young plant. Actually, the first two or three "true leaves" ("juvenile" leaves) are often non-lobed in plants that are genotypically "lobed leaf," but thereafter there is no question of identity of the homozygous non-lobed plant. With a little experience, the heterozygotes can be recognized by their intermediate lobing as described by Mohr *et al.* (*31*).

Seeds

Evaluation of seed characteristics must await the harvesting of fruit. During the extraction of seed from selected fruit an assessment may be made of the approximate numbers of seeds, using the subjective ratings very few, few, average or normal, above average, and excessive. These ratings will be related to size of the fruit. Thus, the *actual* number of seeds in a very small fruit (such as in New Hampshire Midget) may be no more than in a larger fruited cultivar, but because they are concentrated in a small space, they may be classified as excessive. Seed color and size must be uniform in a line ready for release because the seed trade demands this. The size of "ideal" seed has not been determined, but seedsman are reluctant to handle very small seed and/or cultivars with few seed per fruit. On the other hand, few seed per fruit may be preferred by watermelon consumers.

Poole and Grimball (*39*) studied inheritance of various characteristics of seed and reported in detail; most of the simply inherited traits of watermelon are presented in tabular form with their gene symbols and references in Table 2.2.

Miscellaneous

New and heritable biological variations can be discovered by careful observation while making selections in segregating populations. Photographs and objective measurements should be made, and the mode of inheritance studied and reported.

Plastic flags, such as used by civil engineers and foresters, are a cheap and effective means of marking selected plants for further observation and/or controlled pollinations.

DESIGN OF THE COMPLETE BREEDING PROGRAM

Sequence of Development

Breeding programs are usually designed to solve problems of the commercial grower, although in recent years some programs have placed emphasis on the needs of the home gardener. Actually breeders should consider both groups, as they may find segregants that are not suitable for their primary clientele, but represent notable advances for the other group.

Developing a Project

The first step in any breeding program is the development of a project proposal. In this the breeder states the principal objectives and the procedures to be followed in attaining them. A thorough review of the literature on the subject provides the background. Breeding projects are long term by reason of the amount of time required to attain homozygosity following hybridizing. Projects are generally approved for a period of approximately 5 years, subject to renewal after revision. This revision affords an opportunity for the breeder to modify the project to take advantage of changes in clients' needs and to utilize new information and new germplasm.

Familiarity with Parental Material

Choice of parents is a critical matter that can make or break the program. The breeder must become intimately acquainted with existing cultivars, not only with respect to the characteristics primarily sought, but also to all other characteristics of importance. If undesirable characteristics are present in a parent, they may be very difficult to eliminate from the ensuing generations.

Critical evaluation of potential parent material should continue each year with examination of new cultivars and/or plant introductions that appear to be promising. Unusual characteristics may be found and, if sufficiently promising, may cause a major change in project objectives.

Pollination Techniques

A literature review will reveal descriptions of techniques used in other projects. It is advisable constantly to seek to improve techniques, since controlled pollinations comprise a considerable portion of the physical work in breeding watermelons. Local conditions may permit labor-saving modifications. For example, honey bees do not appear in our fields until several hours after daybreak, which permits pollinating unprotected flowers during that time (if the breeder starts early).

Examples of Some Complete Programs

The Florida Watermelon Breeding Program

This program has been quite productive, with the naming and recent introduction of the three cvs. Smokylee, Dixielee, and Sugarlee.

Smokylee combined the high-type resistance of Texas W-5 to fusarium wilt with the anthracnose resistance of Charleston Gray. In yield and quality it appears to be comparable to Charleston Gray, but should prove superior to it where fusarium wilt is a serious problem.

Dixielee's pedigree (Fig. 2.5) illustrates how the breeder brought together the following desirable characteristics: high-type wilt resistance, anthracnose resistance, firm flesh, and intense red flesh color. In 1961 three crosses were made: W-5 × Wilt Resistant Peacock 132 (source of intense red color, firm flesh, and tough rind), W-5 × Fairfax (source of anthracnose resistance), and W-5 × Summit (high-type wilt resistance). Note that all three involved W-5, which indicates the importance that the breeder attached to the high-type of resistance to fusarium wilt. Further indication of this is evident in the 1962 pollinations in which the three F_1 lines were backcrossed to W-5. In 1964 the line carrying genes from Wilt Resistant Peacock 132 was outcrossed to Graybelle, which introduced resistance to anthracnose into this line. The line carrying anthracnose resistance from Fairfax was merged with the line carrying high-type wilt resistance from Summit. Selections were made until the F_4, when the line was combined with the F_3 of the line from Wilt Resistant Peacock 132, which had been selected for the important characteristics contributed by that cultivar. The next seven generations involved selection and selfing until the line was sufficiently homozygous to be released. In total, 14 years were required

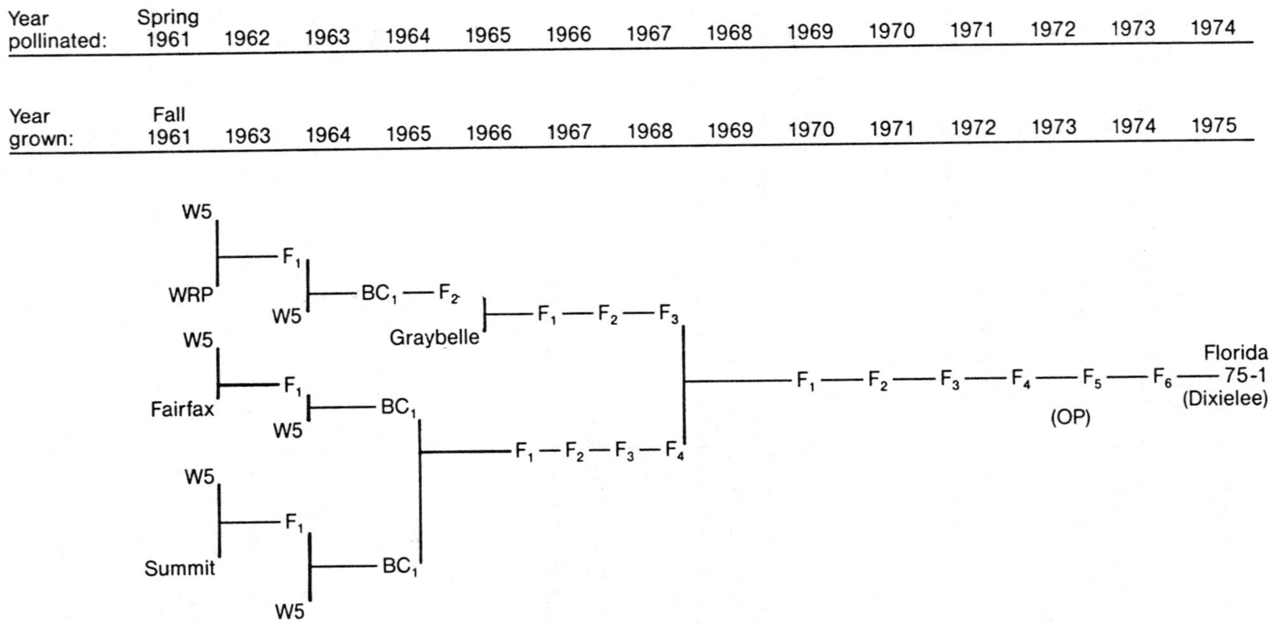

FIGURE 2.5. Pedigree of the cv. Dixielee developed at the Agricultural Research Center, Leesburg, Florida, illustrates the time span involved in combining disease resistance with superior quality factors. W-5, Texas W-5; WRP, Wilt Resistant Peacock 132; F_1–F_6, filial generation following a cross (from self-pollination except as noted); BC_1, first backcross generation; OP, open pollination.
After Crall and Elmstrom (9).

to attain the objectives, and some years of field testing in various trial locations still remained before naming and release.

Comparisons of the fruit characteristics of Dixielee with those of the dominant cv. Charleston Gray are given in Table 2.4. Notable superiority with respect to the percentage of soluble solids (flavor), flesh firmness (durability), and flesh color are indicated, and the rind firmness is equal to that of Charleston Gray although the rind is much thinner.

Results of consumer preference tests given in Table 2.5 indicate that the majority of consumers evaluating the two cultivars preferred Dixielee to Charleston Gray.

Tests such as presented in Tables 2.4 and 2.5 provide valuable information to other breeders who contemplate using the new cultivar as a parent in their programs. Unfortunately, evidence of yield superiority, as derived from replicated yield trials, has been until recently the principal concern of those concerned with authorizing the naming and release of new cultivars. Present trends, however, seem to be toward more complete evaluation of the breeding line, involving qualitative characteristics as well as quantitative.

The Florida program leading to the release of Dixielee has been presented as an example of a program that was well organized and carried through to the successful attainment of objectives. It is not meant to imply that Dixielee will necessarily attain the status of dominant cultivar because of its demonstrated merits. History reveals that many fine cultivars have failed to attain popularity with growers because of some defect that the breeders either considered to be of minor importance, or failed to recognize in their evaluations.

The pedigree of Dixielee clearly indicates the time span required in its development, but does not adequately portray the amount of work involved. In the early stages of development, extensive and time-consuming greenhouse screening for selections resistant to fusarium wilt and anthracnose were required.

It was also necessary to obtain numerous self-pollinations in the incorporation of the genes controlling flesh color and firmness because the segregants for these characteristics

TABLE 2.4. Comparison of Fruit Characteristics of Dixielee and Charleston Gray Watermelons, 1977[a]

Characteristics	Dixielee	Charleston Gray
Length (in.)	11.2	19.1
Diameter (in.)	10.8	9.6
Shape ratio	1:1	2:1
Rind firmness (lb)[b]		
Blossom end	17	17
Side	19	19
Flesh firmness (lb)[c]		
Heart	5.3	4.8
Outside seeds	6.3	5.6
Rind thickness (in.)	0.4	0.7
Soluble solids (%)	11.2	8.9
Flesh color	Dark red	Red

[a]Study conducted by A. K. Showalter, University of Florida, Gainesville, and reported by Crall and Elmstrom (9).
[b]Determined by pressure tester with 5/32-in. probe.
[c]Determined by pressure tester with 23/32-in. probe.

TABLE 2.5. **Consumer Preferences for Dixielee and Charleston Gray Watermelons**[a]

Factor evaluated	Test number	Number of tasters	Preference (%)		
			Dixielee	Charleston Gray	No choice
Sweetness	1	63	91	6	3
	2	287	78	16	6
	Mean		85	11	4
Flavor	1	63	76	11	13
	2	287	70	19	11
	Mean		73	15	12
Color	1	63	75	16	9
	2	287	78	10	12
	Mean		77	13	10
Texture	1	63	65	25	10
	2	287	67	20	13
	Mean		66	22	12
Off-flavor	1	63	0	0	—
	2	287	1	0	—
	3	332	4	0	—
	Mean		2	0	—
Overall	1	63	89	11	0
	2	287	77	18	5
	3	332	85	14	1
	Mean		84	14	2

[a]Studies conducted in 1977 by R. K. Showalter and Mary N. Harrison, University of Florida, Gainesville, and reported by Crall and Elmstrom (9).

could not be determined until the fruit was examined internally. Late-season pollinations (after fruit evaluations) are not dependable. When a number of selections have been made, as is usually the case, the breeder faces the problem of finding space to grow all of them and time to pollinate and evaluate them. Thus the task is not as simple as the pedigree might lead one to conclude.

The Kentucky Watermelon Breeding Program

In the development of Dixielee, emphasis was placed on improvement of disease resistance and quality. Other breeding programs in the United States have similar objectives. The Kentucky Program seeks to introduce new cultivars with dwarf plant habit (sometimes referred to as "bush"). Two different recessive genes for dwarfing are involved, with the double-recessive producing a third type of dwarf. Initiation of the program at Texas A&M University, College Station, followed the discovery of the first dwarf, apparently a mutant in a planting of the cv. Desert King. Following determination of the mode of inheritance, breeding lines of this type ($dw1/dw1$) of dwarf were established (33). The principal objective was to develop a commercially acceptable dwarf.

Undesirable characteristics of the Desert King bush mutant were yellow flesh, yellow rind, very late maturity, and disease susceptibility. Vine-type cultivars were selected as parents to overcome these disadvantages, and the desired genes were transmitted to several bush breeding lines. At this point in the program the author moved to the University of Kentucky, Lexington, leaving the materials that had been developed behind, except for a small packet of seed of one advanced line designated W-45.

In 1962, a dwarf mutant was reported in the Japanese cv. Asahi Yamoto by M. Shimotsuma. Seed of it was obtained for comparison with W-45. Crossing this Japanese dwarf with W-45 was the beginning of the Kentucky program. In the F_2 the new double-recessive dwarf appeared. It was superior in plant characteristics to the single-gene dwarfs (Figs. 2.2–2.4).

The principal objective then became to develop dwarf cultivars that combined the most desirable characteristics of vine cultivars with this new plant growth habit. Several years were required to work out of the mode of inheritance and to obtain seed from self-pollinations of the double-recessive dwarf segregants. Exceptionally late maturity was a serious problem. Plants had to be started in the greenhouse and transplanted to the field to have sufficient growing time to mature seed. Unless earlier maturity could be combined with the double-recessive dwarf characteristic, the new cultivar was virtually worthless.

In 1968, an outcross to the vine cultivar New Hampshire Midget was made to introduce genes for earlier maturity. New Hampshire Midget had demonstrated outstanding earliness in our search for a parent to transmit this characteristic. In 1967, the F_1 of this cross was selfed, and in 1968, some F_2 segregants appeared that matured seed in the normal growing season.

By 1970, maturity of some double-dwarf segregants in the F_4 was satisfactory for Kentucky. Twenty-one different F_5 lines were established by selfing. These were evaluated in 1971, with emphasis placed on melon quality, since plant type and maturity were acceptably uniform. One of these 21 lines showed a satisfactory level of uniformity of melon quality factors except for seed color and size. In 1972 and 1973, continued selection and selfing established the desired uniformity in respect to all characteristics.

Traditionally the next step would have been extensive testing to evaluate commercial acceptability, as was done prior to the release of Dixielee. However, Kengarden, as the new cultivar was named, was designated as a home garden cultivar, with its principal merit the dwarf habit of plant growth. The latter characteristic being unique, Kengarden qualified for coverage under the Plant Variety Protection Act. Assignment of the right to propagate Kengarden was made by the University of Kentucky and royalties from sale of seed are received by the university.

The breeding program at Kentucky continues to emphasize development of dwarf types, and numerous improvements can be made in the double-recessive dwarf type. The single-gene dwarfs offer promise also. Progress with them can be more rapid, since one-quarter of the F_2 progeny from crosses with vine cultivars are of the desired plant growth habit, whereas only one-sixteenth of the progeny of double dwarfs crossed with vine habit segregate as the desired plant type.

A Typical Year's Work

Greenhouse Activities

One year can be saved by making a cross in the greenhouse or by selfing an F_1 to obtain seed for the F_2. This advantage relates to the abundant seed per fruit (average approximately 225 seeds per fruit). Peat–lite (peat plus vermiculite) mixtures are commercially available and work nicely if occasional "feedings" of a complete nutrient solution are applied. Large-sized (1 gal. or more) plastic nursery pots make satisfactory containers. If vine-type plants are involved, cords for the vines to climb can be dropped from overhead wires. (Dwarf types do not require this, and some lines can be considered as potential "patio pot plant" cultivars.)

Screening for resistance to fusarium wilt can be undertaken either to evaluate the level of resistance in a breeding line by the percentage of seedling survival or, if done near planting season, to transplant survivors to the field where further selections can be made. In the former case, ordinary pots may be used and the screening may be conducted at any time. Soil-heating cable may be necessary to assure optimum soil temperature for the development of the *Fusarium* organism. In the second case seedlings should be started in 3-in. square peat pots, five to six seeds per pot (also with heat supplied by soil-heating cable). Usually the majority of seedlings will be killed if the lines are in an early generation, and the few resistant survivors can be transplanted to the field for selfing and/or sibbing.

Anthracnose resistance can be handled in the same fashion: either a test to confirm homozygosity of resistance, or a test followed by transplanting of resistant segregants. A humidity chamber, which is needed to maintain high humidity levels, can readily be constructed using plastic film. The inoculum is sprayed on the young seedlings, and this is followed by periodically spraying a fine mist of water into the chamber. When lesions appear on the leaves, the plants may be rated for resistance, and those free of lesions may be transplanted to the field if desired.

In lines segregating for dwarfness, seedlings may be started in 3-in. square peat pots (three to four seeds per pot), and the vine segregants can be eliminated by clipping all seedlings with long hypocotyls at the soil line. The dwarf seedlings that remain are ideal for transplants as they make slow-growing, compact plants that do not become entangled with each other in the seed flat. In fact, they make such ideal transplants that we recommend them to bedding-plant producers.

Field Activities

Seed to be planted is treated with a seed protectant such as Thiram. Seedlings are grown in peat pots for transplanting, and set out in beds covered with black plastic film (3 ft wide). Row spacings and in-row spacings are determined from notes on the size of plants from the previous generation of the breeding line. Holes are punched in the plastic and seedlings set with a trowel. They are watered-in to assure contact of roots with soil. The plastic provides almost complete weed control in the row, as well as a uniform moisture supply. Standard spray recommendations for control of insects and diseases are followed. Observations and note-taking start early and continue at regular intervals. Plants that are selected for meritorious characteristics are marked with colored plastic flags and are selfed when at the proper stage. Controlled pollinations are marked with distinctive flags and inspected periodically. When the pollination is unsuccessful, follow-up attempts are made.

As melons begin to mature, those which come from controlled pollinations are carefully marked with a felt marker pen (waterproof ink) and transferred to cold storage to await examination of melon characteristics. The remaining melons in the field, which resulted from open pollination, are cut to determine internal characteristics. Notes concerning the extent of variability of the line in this regard are useful.

Seed is extracted from fruits that have desirable melon characteristics, and notes describing these are taken. Seed from each melon is handled separately by placing it on a square rack of ¼-in. mesh "hardware cloth." It is washed with water and the rack is placed in a well-ventilated room to dry. When dry it is packaged and stored in a metal cabinet at room temperature until needed for planting. Seed has been found to have good germination for at least 5–6 years, which is longer than necessary in a breeding program.

During the winter, notes are studied and priorities are established concerning which selections will be planted in the next growing season. Usually seed has been saved from more selections than the scope of an average breeding program will permit growing.

TRIALS OF ADVANCED LINES

Evaluation of Commercial Potential

Since a major goal of past breeding efforts with watermelon has been that of a high level of resistance to fusarium wilt, most of the testing of advanced lines has emphasized performance on land infested with the fusarium wilt organism.

The Southern Cooperative Watermelon Trials include trial locations in Alabama, Arizona, Arkansas, Florida, Kentucky, Louisiana, Missouri, North Carolina, Oklahoma, South Carolina, Texas, Virginia, and even the non-southern states of Indiana and Iowa. The trials are under the sponsorship of the Southern Region of the American Society for Horticultural Science. Entries first must be in an observational trial; and if the majority of the cooperators vote favorably, they advance to a replicated yield trial. Results of both observational and replicated trials are summarized and distributed to all cooperators annually. Data taken include the following: marketable yield in numbers of fruit per acre and in weight of melons per acre, average weight per melon, soluble solids (percentage) by location, maturity class, fusarium wilt resistance rating, and anthracnose resistance rating. Cooperators vote for the entry they rate as "best in trial" (both replicated and observational). The number of locations provides assurance of exposure of the entry to a wide range of environmental conditions. Entries that have done well in these trials have proved to be successful as commercial cultivars when introduced. These trials do not profess to evaluate cultivars being developed for home gardens, nor have they made provision to adjust plant spacing to permit dwarf plants to be competitive.

Evaluation of Garden Value

The All-America Trials[1] place emphasis on unique developments in breeding, whereas the Southern Cooperative Watermelon Trials have concentrated on improvements in yield and disease resistance. Trials of cultivars for garden use should emphasize superior qualitative characteristics, whereas trials of lines developed for commercial use must emphasize yield and disease resistance, and such factors as contribute to durability under existing shipping conditions.

Data such as were taken for the cultivar Dixielee (see Tables 2.4 and 2.5) are very helpful in evaluating the future performance of a new introduction. Cooperation with persons expert in evaluation of consumer preferences is required.

The Southern Cooperative Watermelon Trials depend upon the participation of breeders who eventually expect to submit their own prospective introductions for evaluation.

---

[1]All-America Selections, 628 Executive Drive, Willowbrook, IL 60521.

REFERENCES

1. Agricultural Statistics 1964–1978. U.S. Department of Agriculture Statistical Reporting Service. U.S. Government Printing Office, Washington, DC.
2. Andrus, C. F. 1949. Factors influencing breakage resistance in watermelons. Proc. Am. Soc. Hortic. Sci. *67*, 487–489.
3. Barham, W. S. 1956. A study of the Royal Golden watermelon with emphasis on the inheritance of the chlorotic condition characteristic of this variety. Proc. Am. Soc. Hortic. Sci. *67*, 487–489.
4. Barnes, G. L. 1972. Differential pathogenicity of *Fusarium oxysporum* f. sp. *niveum* to certain wilt-resistant watermelon cultivars. Plant Dis. Rep. *56*, 1022–1026.
5. Bennett, L. S. 1936. Studies on the inheritance of resistance to wilt (*Fusarium niveum*) in watermelon. J. Agric. Res. *53*, 295–306.
6. Boswell, V. R. 1949. Our vegetable travelers. Natl. Geogr. Mag. *96*, 192–193.
7. Chambliss, O. L., Erickson, H. T., and Jones, C. M. 1968. Genetic control of bitterness in watermelon fruits. Proc. Am. Soc. Hortic. Sci. *93*, 539–546.
8. Crall, J. M. 1953. History and present status of watermelon improvement by breeding. Soil Sci. Soc. Fla., Proc. *13*, 71–74.
9. Crall, J. M., and Elmstrom, G. W. 1979. 'Dixielee.' Circ.—Fla., Agric. Exp. Stn. *S-263*.
10. Crall, J. M., and Elmstrom, G. W. 1981. 'Sugarlee.' Circ.—Fla., Agric. Exp. Stn. *S-277*.
11. Dutta, S. K., Hall, C. V., and Heyne, E. G. 1960. Observations on the physiological races of *Colletotrichum lagenarium*. Bot. Gaz. (Chicago) *121*, 163–166.
12. Eigsti, O. J. 1971. Seedless triploids. HortScience *6*, 1–2.
13. Elmstrom, G. W. 1973. Watermelon root development affected by direct seeding and transplanting. HortScience *8*, 134–136.
14. Furasota, K., and Miyazawa, A. 1957. Crossing experiments with *Citrullus vulgaris* Schrad., and *Citrullus colocynthis* Schrad. Natl. Inst. Genet. (Jpn.) Annu. Rep. *7*, 45–47.
15. Hall, C. V. 1970. Petite sweet and super sweet. Kans., Agric. Exp. Stn., Contrib. *460*.
16. Hall, C. V., Dutta, S. K., Kalia, H. R., and Rogerson, C. T. 1960. Inheritance of resistance to the fungus *Colletotrichum lagenarium* (Pass.) Ell. and Halst. in watermelons. Proc. Am. Soc. Hortic. Sci. *75*, 638–643.
17. Henderson, W. R., Jenkins, S. F., Jr., and Rawlings, J. O. 1970. The inheritance of fusarium wilt resistance in watermelon, *Citrullus lanatus* (Thunb.) Mansf. J. Am. Soc. Hortic. Sci. *95*, 276–282.
18. Henderson, W. R. 1977. Effect of cultivar, polyploidy and "reciprocal" hybridization on characters important in breeding seedless watermelon hybrids. J. Am. Soc. Hortic. Sci. *102*, 293–297.
19. Henderson, W. R., and Jenkins, S. F. 1977. Resistance to anthracnose in diploid and polyploid watermelons. J. Am. Soc. Hortic. Sci. *102*, 693–695.
20. Ivanoff, S. S. 1954. Measuring shipping qualities of watermelons. J. Hered. *45*, 155–158.
21. Jones, H. A., and Rosa, J. T. 1928. Truck Crop Plants, pp. 463–478. McGraw-Hill Book Co., New York.
22. Kanda, T. 1931. The inheritance of seed-coat colouring in the watermelon. Jpn. J. Genet. *7*, 30–48.
23. Kihara, H. 1951. Triploid watermelons. Proc. Am. Soc. Hortic. Sci. *58*, 217–230.
24. Konsler, T. R., and Barham, W. S. 1958. The inheritance of seed size in watermelon. Proc. Am. Soc. Hortic. Sci. *71*, 480–484.
25. Layton, D. V. 1937. The parasitism of *Colletotrichum* (Pass.) Ell. and Halst. Iowa, Agric. Exp. Stn., Bull. *223*.
26. Liu, P. B. W., and Loy, J. B. 1972. Inheritance and morphology of two dwarf mutants in watermelon. J. Am. Soc. Hortic. Sci. *97*, 745–748.

27. Lower, R. L., and Johnson, K. W. 1969. Observations on sterility of induced autotetraploid watermelons. J. Am. Soc. Hortic. Sci. 94, 367–369.

28. Loy, J. B., and Liu, P. B. W. 1975. Comparative development of dwarf and normal segregants in watermelon. J. Am. Soc. Hortic. Sci. 100, 78–80.

29. McKay, J. W. 1936. Factor interaction in Citrullus. J. Hered. 27, 110–112.

30. Mohr, H. C. 1953. A mutant leaf form in watermelon. Proc. Assoc. South. Agric. Work. 50, 129–130.

31. Mohr, H. C., Blackhurst, H. T., and Jensen, E. R. 1955. F₁ hybrid watermelons from open-pollinated seed by use of a genetic marker. Proc. Am. Soc. Hortic. Sci. 65, 399–404.

32. Mohr, H. C. 1956. Mode of inheritance of the bushy growth characteristic in watermelon. Proc. Assoc. South Agric. Work. 53, 174.

33. Mohr, H. C. 1963. Utilization of the genetic character for short internode in improvement of the watermelon. Proc. Am. Soc. Hortic. Sci. 82, 454–459.

34. Mohr, H. C., and Sandhu, M. S. 1975. Inheritance and morphological traits of a double recessive dwarf in watermelon, Citrullus lanatus (Thunb.) Mansf. J. Am. Soc. Hortic. Sci. 100, 135–137.

35. Orton, W. A. 1907. A study of disease resistance in watermelons. Science 25, 288.

36. Orton, W. A. 1911. The development of disease-resistant varieties of plants. Conf. Int. Genet. 4th, pp. 247–265.

37. Parris, G. K. 1949. Watermelon breeding. Econ. Bot. 3, 193–212.

38. Poole, C. F. 1944. Genetics of cultivated cucurbits. J. Hered. 35, 122–128.

39. Poole, C. F., Grimball, P. C., and Porter, D. R. 1941. Inheritance of seed characters in watermelon. J. Agric. Res. 63, 433–456.

40. Poole, C. F., and Grimball, P. C. 1945. Interaction of sex, shape and weight genes in watermelon. J. Agric. Res. 71, 533–552.

41. Porter, D. R. 1933. Watermelon breeding. Hilgardia 7, 533–552.

42. Porter, D. R. 1937. Inheritance of certain fruit and seed characters in watermelons. Hilgardia 10, 489–509.

43. Porter, D. R. 1937. Breeding high-quality wilt-resistant watermelons. Bull.—Calif. Agric. Exp. Stn. 614.

44. Robinson, R. W., Munger, H. M., Whitaker, T. W., and Bohn, G. W. 1976. Genes of the Cucurbitaceae. HortScience 11, 554–568.

45. Robinson, R. W., Provvidenti, R., and Shail, J. W. 1975. Inheritance of susceptibility to powdery mildew in the watermelon. J. Hered. 66, 310–311.

46. Rosa, J. T. 1928. The inheritance of flower types in Cucumis and Citrullus. Hilgardia 3, 233–250.

47. Sachs, M. 1977. Priming of watermelon seeds for low-temperature germination. J. Am. Soc. Hortic. Sci. 102, 175–178.

48. Schenck, N. C. 1961. Resistance of commercial watermelon varieties to Fusarium wilt. Proc. Fla. State Hortic. Soc. 74, 183–186.

49. Shimotsuma, M., and Ogawa, Y. 1960. Cytogenetical studies in the genus Citrullus. III. Bitter substances in fruits of C. colocynthis Schrad. Jpn. J. Genet. 35, 143–152.

50. Shimotsuma, M. 1963. Cytogenetic and evolutionary studies in the genus Citrullus. Seiken Jiho 15, 24–34.

51. Shimotsuma, M. 1963. Cytogenetical studies in the genus Citrullus. VII. Inheritance of several characters in watermelons. Jpn. J. Breed. 13, 235–240.

52. Singletary, C. C., and Moore, E. L. 1965. Hybrid watermelon seed production. Miss. Farm Res. 28, 5.

53. Sowell, G., Jr., and Pointer, G. R. 1962. Gummy stem blight resistance of introduced watermelons. Plant Dis. Rep. 46, 883–884.

54. Thieret, J. W. 1963. The correct name of the watermelon. Taxon 12, 36.

55. Tomes, M. L., and Johnson, K. W. 1963. The carotene pigment content of certain red-fleshed watermelons. Proc. Am. Soc. Hortic. Sci. *82,* 460–464.
56. Walker, M. N. 1943. A useful pollination method for watermelons. J. Hered. *34,* 11–13.
57. Watts, V. M. 1962. A marked male-sterile mutant in watermelon. Proc. Am. Soc. Hortic. Sci. *81,* 498–505.
58. Watts, V. M. 1967. Development of disease resistance and seed production in watermelon stocks carrying the *msg* gene. Proc. Am. Soc. Hortic. Sci. *91,* 579–583.
59. Weetman, L. M. 1937. Inheritance and correlation of shape, size and color in the watermelon, *Citrullus vulgaris* Schrad. Iowa, Agric. Exp. Stn., Bull. *228.*
60. Winstead, N. N., Goode, M. J., and Barham, W. S. 1959. Resistance in watermelon to *Colletotrichum lagenarium* races 1, 2, and 3, Plant Dis. Rep. *43,* 570–577.
61. Yeager, A. F. 1950. Breeding improved horticultural plants. N.H., Agric. Exp. Stn., Bull. *380.*

3
Pepper Breeding

WALTER H. GREENLEAF

67

BREEDING VEGETABLE CROPS

Although peppers (*Capsicum* spp.) are one of the lesser vegetable crops in dollar terms, they are an important constituent of many foods, adding flavor, color, vitamin C, and pungency, and are therefore indispensable to the United States and world food industries. Mexican foods (over 60 kinds) are the fastest growing line of ethnic foods in the United States. Grocery store sales have increased from $60 million in 1970 to over $300 million in 1980. The Economics and Statistics Service of the United States Department of Agriculture (USDA) reports that the United States produced an annual average of 261,000 tons of green peppers for the fresh market and for processing from 1978 to 1980 (Tables 3.1 and 3.2). These peppers were worth over $109 million yearly. The five major producing states are Florida, California, Texas, New Jersey, and North Carolina. Texas, New Mexico, Arizona, and California produce most of the hot chili peppers in the United States.

Production of bell peppers in the United States from 1978 to 1980 is shown in Table 3.1. According to Marshall (*91*), the United States produced a total of 686,000 tons of all kinds of peppers in 1976 on 111,000 acres (Table 3.2). This was second only to China's 1,310,000 tons produced on 319,000 acres.

TABLE 3.1. Green Pepper Production for Fresh Market and Processing in the United States, 3-Year Average 1978–1980[a]

	Acres harvested	Tons	Tons/ acre	Price $/ton	Price $/100 1b	Value (×$1000)
Seasons						
Winter (Florida only)	5,303	31,742	6.0	469	23.4	14,888
Spring	12,409	59,434	4.8	542	27.1	32,204
Summer	23,801	94,957	4.0	304	15.2	28,888
Fall	13,776	74,781	5.4	445	22.2	33,272
Total (or average)	55,289	260,932	4.7	419	20.9	109,252
Major producing states						
Florida	17,747	94,050	4.9	469	26.0	48,854
California	8,073	72,351	9.0	286	14.3	20,657
Texas	8,439	36,000	3.8	573	28.7	20,626
New Jersey	6,939	22,451	3.2	328	16.4	7,361
North Carolina	6,805	13,050	1.9	325	16.3	4,245
Total (or average)	49,003	237,902	4.9	428	21.4	101,743

[a]After USDA (*2*).

**TABLE 3.2. Estimated Acreage, Production, and Grower Value
of Pepper Cultivars Grown in the
United States, 1976[a]**

Pepper type	Harvested acres	Production (tons)	Value (×$1000)
Bell	68,167	415,215	109,409
Paprika	15,940	143,475	24,310
Pimiento	12,900	42,270	8,935
Jalapeno	2,880	32,525	7,357
Cayenne	1,786	6,685	2,850
Small Yellow Pickling	1,685	14,935	6,040
Cherry	1,640	7,750	2,855
Cubanelle	1,405	8,250	2,040
Hot Banana	1,255	7,440	1,710
Tabasco	1,200	2,740	1,715
Small Chili	760	1,515	440
Sweet Banana	565	2,315	505
Pepperoncini	235	535	365
Mexican	170	350	175
Serrano	95	570	305
Total	110,683	686,570	169,011

[a]After Marshall (*91*).

ORIGIN OF *Capsicum*

Authorities are generally agreed that *Capsicum* originated in the New World tropics and subtropics. Safford recovered dried pepper pods from 2000-year-old burials in Peru (*129*). De Candolle traced the origin of cultivated plants and concluded from a lack of reference to this genus in ancient languages that no *Capsicum* was indigenous to the Old World (*40*). Peppers were unknown in Europe until the sixteenth century, having been introduced into Spain by Columbus on his return trip in 1493. Cultivation spread from the Mediterranean region to England by 1548 and to Central Europe by the close of the sixteenth century (*14*). The Portuguese carried *Capsicum* from Brazil to India prior to 1885 and cultivation was reported in China during the late 1700s.

TAXONOMY AND CENTERS OF ORIGIN

The genus *Capsicum* of the nightshade family Solanaceae comprises some 20–30 species of the New World tropics and subtropics. Modern taxonomists recognize five major cultivated species (Fig. 3.1): *Capsicum annuum* L., *Capsicum frutescens* L., *Capsicum chinense* Jacquin, *Capsicum pendulum* Willdenow, and *Capsicum pubescens* Ruiz & Pavon. Eshbaugh has proposed that *C. pendulum* and the closely related species *Capsicum microcarpum* Cavanilles be reclassified as botanical varieties of *Capsicum baccatum*. The latter species was originally described by Linnaeus. In this classification, the larger fruited cultivar *C. pendulum* Willd. becomes *C. baccatum* L. var. *pendulum* (Willd.) Eshbaugh and the smaller fruited wild *C. microcarpum* Cav. becomes *C. baccatum* L. var. *baccatum* (*53*).

The five major cultivated species are derived from different ancestral stocks found in three distinct centers of origin. Mexico is the primary center for *C. annuum*, with Guate-

FIGURE 3.1. Representative varieties of five major *Capsicum* species. (A) *C. annuum* L. *cv.* Truhart Perfection Pimiento; (B) *C. chinense* Jacquin Acc. 1555. Scales are in inches.

mala a secondary center; Amazonia for *C. chinense* and *C. frutescens,* and Peru and Bolivia for *C. pendulum* and *C. pubescens. C. annuum* and C. frutescens are widely distributed from Mexico through Central America and throughout the Caribbean region. *C. chinense* is the most commonly cultivated species in South America. Wild forms exist of all but *C. pubescens,* which is known only in cultivation (*141*). The most comprehensive modern sourcebook on the genus *Capsicum* is by Jean Andrews (*1a*).

In the United States *C. annuum* is the major cultivated species. Only Tabasco and Greenleaf Tabasco, both *C. frutescens,* are grown to a limited extent in Louisiana for

FIGURE 3.1. (*Cont'd.*) (C) *C. frutescens* L. *cv.* Tabasco; (D) *C. baccatum* var. *pendulum* (Willd.) Eshbaugh. Scales are in inches.

FIGURE 3.1. (Cont'd.)   (E) *C. pubescens* Ruiz et Pavon. Scale is in centimeters.

making Tabasco sauce from the red ripe fruit and vinegar sauce from the yellow–green immature peppers pickled in vinegar. *C. pendulum* is the third species grown to a very limited extent in California for pickling in the mildly pungent immature yellow–green stage and sold under the brand name Mild Italian. Modern pepper breeders must acknowledge the achievements of the native Indian cultivator–breeders, probably women, who developed from the wild species a wide range of fruit shapes and sizes, from the small fiery Tabasco to the sweet giant bell type, by the time the Spaniards arrived. This development is thought by Pickersgill to have occurred during the last 4000 years (*113*). A probable contemporaneous development, corn (*Zea mays*) from its fully interfertile wild ancestral species *Euchlaena mexicana* (teosinte), is thought by Beadle (*8*) to have taken place between 8000 and 15,000 years ago, probably in Mexico. This time period is based on the earliest corn yet identified from archaeological finds in the southwestern United States and Mexico, which has been dated back 7000 years.

Species Characteristics

The five major cultivated species of *Capsicum* can usually be distinguished by a combination of flower and fruit characteristics (Table 3.3). *C. annuum*, the commonly cultivated species in the United States, has white flowers, blue to purple anthers, a toothed calyx, and typically single-fruited nodes with the possible exception of an occasional double-flowered axil in a lower main fork. *C. frutescens* has greenish flowers, a non-toothed, non-constricted calyx that encloses the fruit base, blue anthers, and mostly single-fruited nodes but with a few double-flowered nodes on each plant, as in Tabasco, unless the

TABLE 3.3. Taxonomic Characters Distinguishing *Capsicum* Species[a]

Species	Corolla color	Corolla throat spots	Corolla shape	Anther color	Calyx teeth	Seed color	Flowers/node
C. annuum	White	None	Rotate	Blue–purple	Present	Tan	1
C. frutescens	Greenish white	None	Rotate	Blue	None	Tan	1–3[b]
C. chinense	White to greenish white	None	Rotate	Blue	Present	Tan	1–5
C. galapogense	White	None	Rotate	Yellow	None	Tan	1
C. chacoense	White	None	Rotate	Yellow	Present	Tan	1
C. schottianum	White	Yellow	Rotate	Yellow	None	Black	5–7
C. baccatum	White	Green–yellow	Rotate	Yellow	Present	Tan	1–2
C. praetermissum	White to lavender	Yellow	Rotate	Yellow	Present	Tan	1
C. eximium	White to lavender	Yellow	Rotate	Yellow	Present	Tan	2–3
C. pubescens	Purple	None	Rotate	Purple	Present	Black	1
C. cardenasii	Blue	Greenish yellow	Campanulate	Pale blue	Present	Tan	1–2

[a]After Lippert *et al.* (*89*).
[b]Certain wild forms may produce up to five flowers.

plants are stunted. Certain wild forms of *C. frutescens* apparently produce up to five fruits per node. *C. chinense* has white or greenish white flowers, blue anthers, a constricted, toothed calyx, and typically from one to three fruits per node. *C. pendulum* is easily identified from the white flowers with the yellow corolla throat spots, yellow anthers, and the long, curved, characteristically pendant fruit pedicels and leaf petioles. *C. pubescens* with its larger, showy purple flowers, soft pubescent leaves, yellow–orange fruits, and black seeds is unique.

Species Crossability and Hybrid Fertility

When recording crosses the accepted convention is always to write the female parent first. *C. annuum* and *C. frutescens* will cross reciprocally with *C. chinense,* producing partially fertile hybrids. With *C. chinense* as the bridge, these three species can form a common gene pool. The crossability of *Capsicum* species and the fertility of their hybrids is shown in Fig. 3.2 and Table 3.4. Dumas de Vaulx and Pitrat reported the successful hybridization of two cultivars of *C. annuum*—Doux Long des Landes (DL), a sweet, long-fruited cayenne type, and Yolo Wonder (YW)—with two Acc. of *C. baccatum* var. *pendulum* by two new methods: (1) Double pollination, first with *C. baccatum* pollen followed with pollen of *C. annuum* cv. Nigrum. The best results with this method, two to three hybrids per pollinated flower of DL and 0.4 of YW, were obtained when the second pollination followed the first 3–4 days later. (2) The second method, nitrous oxide (N_2O) gas treatment of the female gametophyte in a pressure chamber at 6 atm for 4 hr prior to pollination with *C. baccatum* pollen, yielded seven hybrids per pollinated flower (*49*). It is a general rule that wide crosses are more successful when the smaller fruited wild type is used as the female.

All *Capsicum* species have $2n = 24$ chromosomes. The F_1 hybrid *C. chinense* PI 152225 × *C. frutescens* var. Tabasco will illustrate the chromosome pairing relationship and fertility level to be expected in F_1 hybrids between *C. annuum, C. chinense,* and *C. frutescens.* In this cross, chromosome pairing is surprisingly regular, with 12 bivalents usually forming at metaphase I (MI) of meiosis and with mostly regular disjunction at MI and MII, resulting in 12 chromosomes being distributed to most pollen grains in the pollen

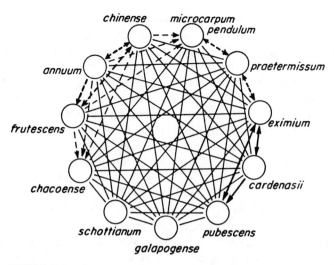

FIGURE 3.2. Cross compatibility of *Capsicum* species. —,
Highly fertile F$_1$; ---, partially fertile F$_2$; —, highly sterile F$_1$;
---, no viable F$_1$ seed.
After Lippert *et al.* (*89*).

tetrads. Pollen fertility, however, as determined by staining with iron acetocarmine, was
only about 21% as compared to 80–90% in the parents. Seed fertility is correspondingly
reduced, with the F$_1$ hybrid, Tabasco, and *C. chinense* averaging 6.3, 21.1, and 28.1
seeds per fruit, respectively. A number of F$_2$ seedlings emerged only partially from the
seed coat. Many of these, extricated manually, developed abnormally. Those that survived grew slowly into highly abnormal dwarf plants with thick and malformed leaves.
Such partially sterile interspecific F$_2$ populations would not be expected to produce
Mendelian ratios as do intraspecific crosses. For example, there was a deficiency of

TABLE 3.4. Crossability of *Capsicum* Species and Hybrid Fertility[a]

Cross	Initial cross	Viability		
		F$_1$ seed	F$_2$ seed	Backcross seed
C. annuum × *C. frutescens*	−			
C. annuum × *C. chinense*	++	++	++	++
C. annuum × *pendulum*[b]	E	E	+	−
C. annuum × *C. pubescens*	−			
C. frutescens × *C. annuum*	+	+	+	+
C. frutescens × *C. chinense*	+	+	+	+
C. frutescens × *C. pendulum*	++	++	+	+
C. chinense × *C.* × *frutescens*	+	+	+	+
C. chinense × *C. annuum*	+	+	+	+
C. chinense × *C. pendulum*	+	+	−	−
C. chinense × *C. pubescens*	E	E	−	−
C. pendulum × *C. pubescens*	−			

[a]After Smith and Heiser (*141*). E, seed germinated by embryo culture only; −, no viable seed; +, few
viable seed; ++, many viable seed.

tobacco etch virus (TEV) resistant plants in the F_2 from *C. chinense* PI 152225 (resistant) × Tabasco (susceptible) due to zygote abortion resulting from gene imbalance (genic sterility). During meiosis in the F_1 hybrid, pairing and crossing-over between homoeologous chromosomes having small structural rearrangements would produce gametes with chromosomes having duplications, deficiencies, or both. Whereas the union of such gametes would often result in nonviable F_2 zygotes, their gene imbalance is usually covered by the normal gametes from the backcross parent, thus producing viable zygotes and nearly normal Mendelian ratios in the backcross (BC). Stebbins ascribed the cause of this type of sterility, common in interspecific hybrids, to *cryptic structural hybridity* since it is not revealed by the seemingly normal chromosome pairing relationship in meiosis (*64,148*).

The F_1 hybrid Tabasco × *C. pendulum* is an example of a more highly sterile cross with only about 3% good pollen. Meiosis is again surprisingly regular, but less so than in the cross *C. chinense* × Tabasco. Some chromosome bridges are observed at MII and laggard groups of chromosomes give rise to micronuclei. Hence this hybrid exhibits both chromosomal and genic sterility, but gene imbalance appears to be the major cause of gamete abortion.

Karyotype and Species Crossability

Ohta presented the idiograms of six species (Fig. 3.3). Of these, *C. frutescens, C. pendulum,* and *C. microcarpum* have similar karyotypes, each with three similar, readily distinguishable chromosomes, namely, one large satellited (Sat-I), one smaller satellited (Sat-II), and one with a subterminal constriction and a heterochromatic (non-staining) distal region. The remaining nine chromosomes, all with median centromeres, are similar in size and could not be distinguished. The similarity in karyotype of the species corresponds with their crossability (*105*).

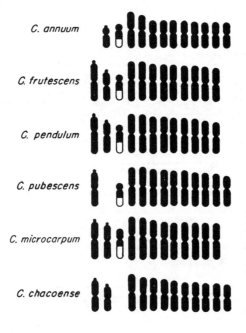

FIGURE 3.3. Karyotypes of *Capsicum* species. Top to bottom: *C. annuum* L., *C. frutescens* L., *C. baccatum* var. *pendulum* (Willd.) Eshbaugh, *C. pubescens* Ruiz & Pavon, *C. baccatum* var. *microcarpum* (Cav.) Eshbaugh, and *C. chacoense* Hunziger.
After Ohta (*105*).

FLOWER STRUCTURE AND POLLINATION

The flowers of wild *Capsicum* species are pentamerous, but large-fruited cultivars have five to seven corolla lobes. The stamens alternate with the petals and correspond with them in number (Fig. 3.4). While crosses can be made at any time during daylight hours, the best times are in the early morning or in the late afternoon when the flowers are in the mature bud stage and have not been disturbed by insects. The necessary tools are a bottle of 95% ethyl alcohol with a screw cap, a pair of sharp-pointed forceps, a spear needle, a 14-power hand lens, a plastic squeeze bottle filled with water and fitted with a fine-tipped nozzle, a roll of cheesecloth, a pair of scissors, some lightweight cotton string, and several balls of different colors of thread to mark the pollinated flowers.

Crossing and Selfing

The stigma of the flower chosen for crossing is first carefully examined with the hand lens and, if found contaminated with pollen, is pinched off. If the flower has already been visited by bees, the pollen will appear evenly distributed over the stigma and the flower must be discarded.

If some pollen is present on the style but not on the stigma, or if contamination occurred during manipulation of the flower, the flower may still be saved for crossing if the anthers are first removed with forceps, the stigma and style washed off, the excess water blown off by mouth, and any droplets remaining on the stigma gently blotted off with the thumb and fingers. The stigma is then again carefully examined with the hand lens and, if free from signs of earlier pollen penetration as evidenced by swelling and puffiness, it may be pollinated. Pollen is gently transferred to the stigma either from mature undehisced anthers by scooping it out through the lateral sutures with the spear needle, or by touching a freshly dehisced anther to the stigma with the forceps.

Several flowers are usually emasculated and prepared for pollination at one time to speed up the process. The hands and tools must be disinfected between different crosses or selfs, and care must be taken that the fingers are not damp with alcohol when the stigmas are blotted free of water droplets. Pollinated flowers are marked by loosely tying colored thread around the delicate pedicels, preferably enclosing a leaf petiole for protection. Different colors of string can be used for different crosses on the same plant, and white for

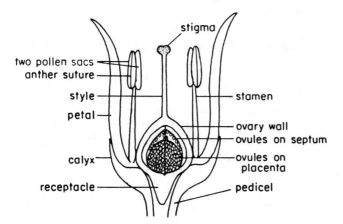

FIGURE 3.4. Diagram of *Capsicum* flower.

the selfs. Pollinated flowers must be protected from bees by a double layer of cheesecloth, loosely wrapped around the branch, enclosing leaves and flowers, and securely fastened. Appropriately marked plastic labels (insects will chafe paper) describing the cross, with the date, are attached to a bamboo stake marking the chosen plant. Pollinated flowers should be periodically checked and the cheesecloth removed in 4–6 days. Fruits should mature in about 45 days.

In an insect-screened greenhouse, several plants can be quickly selfed by touching the flowers of each with a different fingertip before again disinfecting the hands with alcohol. Pollen in mature, undehisced anthers can be stored at $-5°C$ and 97% relative humidity for about 10 days (*112*).

Natural Crossing

There is considerable natural crossing in peppers by bees, which probably accounts for some of the variability found in pepper cultivars. Odland and Porter reported from 7.6 to 36.8% crossing in the field, with a mean of 16.5% (*103*). They transplanted single plants of each of six cultivars possessing recessive fruit and plant characters into the middle of solid block plantings of Harris Early Giant, a cultivar with all dominant traits. Variance analysis showed that the cultivars differed significantly in the amount of natural crossing. The differences were probably due to differences in flower structure, such as proximity of the anthers to the stigma and to the kind and number of insect visitors. In an open field, the minimum safe isolation distance for peppers is estimated at 600 ft.

Time Scale of Female Gametogenesis

Dumas De Vaulx has provided a time scale of embryo sac development in *C. annuum* from meiosis to the first division of the zygote, based on his cytological studies on the origin of haploids (*47*). A knowledge of the time sequence of the reproductive stages is important for hybridization and cytological studies (Fig. 3.5).

HORTICULTURAL CLASSIFICATION OF PEPPER VARIETIES

This classification system for peppers was developed by Dr. P. G. Smith, Department of Vegetable Crops, University of California, Davis. The system is based on grouping cultivars that are horticulturally similar in major characteristics such as fruit shape, size, color, texture, flavor, and pungency, as well as in uses, and hence may provide alternate sources for various processed products of the pepper industry. This classification is useful because it makes order out of a seemingly unlimited diversity of cultivars. Although intended primarily for cultivars of *C. annuum*, this system can encompass cultivars of the other four cultivated species.

With Dr. Smith's approval, his text is reproduced with only minor modifications and additions, except for certain additional cultivars and of one new group, the Squash Group. Representative cultivars of the 13 groups are shown in Fig. 3.6. All fruit measurements are length (depth) × width in inches.

This classification is limited to the more important forms of *C. annuum* and to one cultivar of *C. frutescens* (Tabasco) grown in the United States, and to one exotic cultivar of *C. chinense* (Rocotillo). *C. annuum* is the most important cultivated species of the northern hemisphere, with its center of domestication in Mexico. Within this one species the fruit range from less than ¼ to over 10 in. long, and from very slender to more than 4

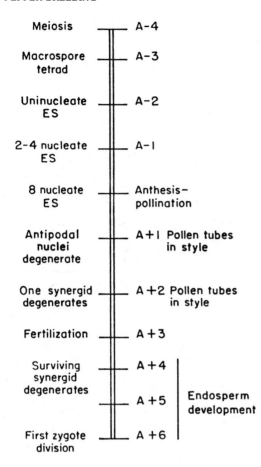

Meiosis	A-4
Macrospore tetrad	A-3
Uninucleate ES	A-2
2-4 nucleate ES	A-1
8 nucleate ES	Anthesis–pollination
Antipodal nuclei degenerate	A+1 Pollen tubes in style
One synergid degenerates	A+2 Pollen tubes in style
Fertilization	A+3
Surviving synergid degenerates	A+4
	A+5 Endosperm development
First zygote division	A+6

FIGURE 3.5. Time scale of female gametogenesis from meiosis to the first zygote division in *C. annuum*. ES, embryo sac; A − 4, 4 days before anthesis; A + 6, 6 days after anthesis.
After Dumas De Vaulx and Pitrat (49).

in. wide. Some are smooth, others are rough and irregular. Immature color may be varying shades of green to yellow, and the mature fruit, while most commonly red, may also be orange, yellow, brown, or even green.

Terminology may be confusing. The word *pimiento* is Spanish for any pepper, but in the United States it, or pimento, refers specifically to a large, sweet, thick-walled heart-shaped group of cultivars. Chili (chilli in European literature) is derived from the Mexican word *chile*, which in Mexico and Central America means any pepper. In Europe and the United States, chili refers to most pungent peppers, but not to non-pungent cultivars. The American Indian word *aji* (pronounced *ah-hee*), meaning hot pepper, was recorded by Columbus in his journal in 1493.

In recent years the Mexican influence has been felt by the increasing use of the term Long Green Chile for the Anaheim Chili cultivar group, and by the increased use of the name Ancho for the Mexican Chili. The use of Mexican names may be expected to increase.

The following classification is proposed:

Bell Group

Fruit large (3–5 in × 2–4 in.), smooth, thick-fleshed, blocky, blunt, three to four lobed, square to rectangular, or tapering in longitudinal sections. Color usually green when

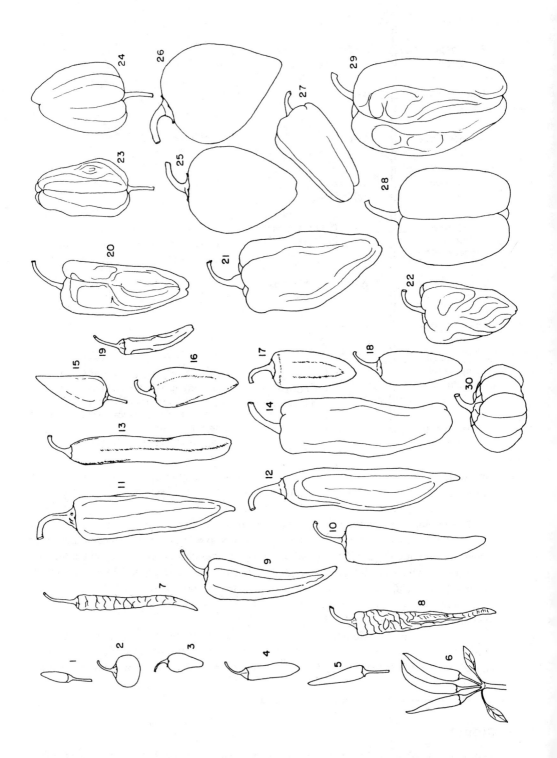

immature, turning red at maturity. A few minor cultivars are yellow when immature and others become orange–red at maturity. Mostly non-pungent, although a few pungent forms are known.

Common cultivars. Non-pungent; green, turning red when ripe: California Wonder (various strains), Yolo Wonder (various strains), Florida VR2, Florida VR2-34, Florida XVR3-25, Keystone Resistant Giant (various strains), Early Calwonder, Ruby King, World Beater, Miss Belle, Bell Boy, Big Bertha, Golden California Wonder (orange–yellow when ripe). Yellow, turning red when ripe: Golden Bell, Rumanian. Yellow-green turning red when ripe: Gypsy. Pungent: green, turning red when ripe: Bull Nose Hot. Yellow, turning red when ripe: Rumanian Hot. *Uses:* fresh market, salads, stuffed, sauteed, pizza, meat loaf, dehydrated (dried) processed meats, canning.

Pimiento Group

Fruit large, heart-shaped (2½–5 in. ✕ 2–3 in.), green, turning red, smooth, thick-walled, non-pungent.

Common cultivars. Pimiento, Pimiento Select, Pimiento Perfection, Pimiento Truhart, Pimiento Truhart-D, Pimsan, Pimiento L., Pimiento Bighart. *Uses:* fresh market, salads, soups, meat loaf, cheese, stuffed olives, processed meat, canning.

Squash or Cheese Group

Fruit small to large, wider than deep, scalloped or rounded, flat or semipointed, smooth or rough and involuted (1–2 in. × 2–4 in.), medium- to thick-walled, non-pungent. Medium green or yellow to mature red.

Common cultivars. Cheese, Yellow Cheese, Frommage, Sunnybrook, Tennessee Cheese, Gambo (Israel), Norron de Conserva (Spain), Antibois (France). *Uses:* processing, canning, pickling, stuffing, culinary. Mildly pungent: Rocotillo (*C. chinense*), from Peru is flavorful eaten raw in salads, or used as a garnish.

Ancho Group

Fruit large, heart-shaped (4–6 in. × 2–3 in.), smooth, thin-walled, stem indented into top, forming cup, mildly pungent.

FIGURE 3.6. Fruits of 30 representative pepper cultivars reduced to ¼ natural size. Country of origin follows the cultivar name. Synonyms are in parentheses. Fruits are borne erect or pendent as shown.

(1) Greenleaf Tabasco (Tabasco G), U.S. (2) Cherry Sweet, U.S. (3) Cascabella, Mexico. (4) Serrano Chili, Mexico. (5) Red Chili, U.S. (6) Santaka, Japan. (7) Cayenne Long Slim, U.S. (8) Cayenne Large Red Thick, U.S. (9) Hungarian Sweet Wax, Hungary. (10) Sweet Banana (Long Sweet Hungarian), Hungary. (11) College 64 L, U.S. (12) Anaheim TMR 23, U.S. (13) Pasilla, Mexico. (14) Aconcagua, Argentina. (15) Fresno Chili Grande, U.S. (16) Santa Fe Grande, U.S. (17) Caloro, U.S. (18) Jalapeno M., Mexico. (19) Golden Greek, Greece. (20) Cubanelle, U.S. (21) Long Spanish Bell (PI 164564), Spain. (22) Ancho 101, Mexico. (23) Rumanian Hot, Rumania. (24) Rumanian Sweet, Rumania. (25) Pimiento L., U.S. (26) Pimiento Bighart KL, U.S. (27) Gypsy (F_1 hybrid), U.S. (28) Emerald Giant (selection from Keystone Resistant Giant), U.S. (29) Big Bertha (F_1 hybrid), U.S. (30) Cheese, U.S.

I am indebted to Dr. K. W. Owens of PetoSeed Company for the inclusion and measurements of the cvs. Cheese and Golden Greek.

Reproduced by permission of PetoSeed Company, P.O. Box 4206, Saticoy, CA 93003, and of Horticultural Enterprises, P.O. Box 34082, Dallas, TX 75234. Fifth (1980) Pepper Conference, Las Cruces, New Mexico. Drawings by Teresa Rodriguez.

Common cultivars. Ancho, Mexican Chili, green, turning red to red-brown at maturity. Mulato, black-green, turning brown-black at maturity; imported from Mexico, little grown in the United States. *Uses (Ancho and Mexican Chili):* powder, whole pod (all). *Uses (Mulato):* dried whole or powdered or fresh roasted and peeled; distinctive flavor.

Anaheim Chili Group (Long Green Chile Group)

Pods long, slender (5–8 in. × ¾–1¾ in.), tapering to a point, smooth, flesh medium thick, medium to dark green, turning red. Moderately pungent to sweet.

Common cultivars. Moderately pungent: Sandia, Big Jim. Mildly pungent: Anaheim Chili, California Chili. Very slightly pungent: Mild California, New Mexican Chili. Nonpungent: Paprika (Bulgaria). In the United States paprika is a product, not a cultivar. Nonpungent (or nearly so) Anaheim types are used in the western United States to make this product. These cultivars are often called Paprika locally. *Uses:* dehydrated whole pods, powder (usually mixed with spices), color (especially paprika), canned (green pods), sauces (both green and red pods), many Mexican dishes.

Cayenne Group

Fruit long, slender (5–10 in. × ½–1 in.), thin-walled, medium green, turning red, characteristically wrinkled and irregular in shape, highly pungent.

Common cultivars. Mature fruit red: Cayenne Long Red, Cayenne Long Slim, Cayenne Long Thick. *Uses:* fresh market, dried, powder, hot sauce, pickling. Pasilla: long, glossy, dark blue-green, turning chocolate brown at maturity; imported from Mexico; used dried and toasted in sauces, distinctive, rich flavor. Non-pungent: Centinel and Doux Long des Landes (France). *Uses:* fresh, in salads, culinary.

Cuban Group

Fruit (3–6 in. × ½–2 in.), yellowish green, turning red, thin-walled, irregular, blunt, mildly pungent.

Common cultivars. Cuban, Cubanelle, Aconcagua, Golden Greek. Non-pungent, immature fruit green: Pepperoncini (Green Italian). *Uses:* fresh market, salads, pickled, frying. Golden Greek and Pepperoncini are harvested when fruits are 1½–2 in. long for use in salad bars; mature fruit are 3–4 in. long.

Jalapeno Group

Fruit elongated (2–3 in. × 1–2 in.), rounded cylindrical shape, smooth, thick-walled, dark green, turning red, with or without corky network on skin of mature fruits, variable in shape, highly pungent.

Common cultivars. Jalapeno (various strains). *Uses:* as green pepper, canned (in oil and spices), fresh market, mature-dried, sauces.

Small Hot Group

Fruit slender (1½–3 in. × ¼–1 in.), medium- to thin-walled, green, turning red, highly pungent.

Common cultivars. Fresno Chili (several strains) and Serrano (several strains) are harvested in green stage only; Red Chili, Chile de Arbol, Japanese Chili, Santaka, Hontaka. *Uses:* fresh green, dried powder for seasoning and for sauces.

Cherry Group

Fruit small, spherical (½–2 in.) to somewhat flattened, thick flesh, green, turning red, pungent.
Common cultivars. Red Cherry Large, Red Cherry Small. Non-pungent: Sweet Cherry. *Uses:* pickling.

Short Wax Group

Fruit (2–3 in. × 1–2 in.), yellow, turning orange–red, smooth, medium to thick-walled, tapered.
Common cultivars. Pungent: Floral Gem, Cascabella, Caloro, Santa Fe Grande. Non-pungent: Petite Yellow Sweet. *Uses:* fresh, pickling, processing, sauces, cooking.

Long Wax Group

Fruit (3–5 in. × ¾–1½ in.), yellow, turning red, pointed or blunt.
Common cultivars. Pungent: Hungarian Yellow Wax. Non-pungent: Sweet Banana, Hungarian Sweet Wax, Long Yellow Sweet. *Uses:* fresh, pickled, sauces, relishes, canning.

Tabasco Group

Fruit slender (1–2 in. × ¼ in.), yellow or yellow-green to red, highly pungent, of the species *C. frutescens.*
Common cultivars. Tabasco, Greenleaf Tabasco. *Uses:* pickled yellow, vinegar sauce; red, Tabasco sauce.

Some Leading Bell Pepper Cultivars

Most of the following cultivars were chosen from among the better entries in the 1979–1981 National Pepper Cultivar Evaluation Trials (*95*). Others are superior varieties of long standing. This list is not complete and is intended to serve only as a guide and as seed sources for growers, home gardeners, and breeders: Marengo, Shamrock, Skipper, Early Calwonder, California Wonder 300 (Asgrow); Grande Rio 66 (Baxter); Hybelle, Ladybelle (Harris); Ferry Morse 6 C-X57 (F_1), 6 C-236 (OP), Big Belle; Keystone Hybrid 6700, Keystone Resistant Giant; PetoSeed 9275, 10275, 21476, California Wonder, Yolo Wonder L; Miss Belle (MSU).
Seed sources. Walter Baxter Seed Co., Weslaco, TX 78596; Joseph Harris Seed Co., Inc., Moreton Farms, Rochester NY 14624; Ferry Morse Seed Co., Box 1010, San Juan Bautista, CA 95045; Keystone Seed Co., Box 1438, Hollister, CA 95023; PetoSeed Co., Inc., Rt. 4, Box 1255, Woodland, CA 95695; Mississippi State U. Truck Crops Branch Experiment Station, Crystal Springs, MS 39059.
In 1981 Dr. Pieter Vandenberg, plant breeder for Nickerson International Plant Breeders S.A., P.O. Box 1787, Gilroy, CA 95020, wrote me this note. "In Spain, Italy, France, Greece they grow long blocky bell peppers under plastic almost the year around.

These peppers are pruned and trained. They use hybrids like Lamuyo (developed by INRA). We are currently selling Victor, Challenger, Chieftain. Sizes average 15–18 cm long and diameter of 8 cm.''

CULTURE

Production of Transplants

Greenhouse Plant Production

Various commercial ready-to-use growing media (without sand,) such as Jiffy Mix, Pro-Mix A, Pro-Mix BX, and Speedling Mix are satisfactory for growing peppers, provided they are kept sterile. Jiffy Mix, Pro-Mix BX, and Speedling Mix contain a wetting agent which is helpful. All made-up mixes should be steam pasteurized for 30 min at 82°C with free-flowing steam and only sufficient pressure to assure circulation between the spaced layers of flats. If a suitable steam sterilizer is not available, a commercial mix is preferred, because toxicity resulting from oversteaming is a common problem. Seeds will germinate in the toxic medium, but the seedlings will turn yellow and stop growing. Standard (14 × 21 × 3 in.) wood flats that have been treated with 2% copper naphthanate (never with creosote, which is highly toxic) are packed with 54 Jiffy Strip 2¼ in. square peat pots and filled with pasteurized medium. The medium is firmed carefully by hand to avoid breaking the peat pots. From 5 to 10 seeds are sown per pot. The flats are then drenched with a solution of a completely soluble fertilizer, such as NutriLeaf-60 or Peter's 20 : 20 : 20, 1 cup/55 gal. of warm water plus ½ cup $Ca(NO_3)_2$, ½ cup $MgSO_4$, and ½ cup of tribasic copper sulfate for protection against damping-off fungi. The plants are watered with the nutrient solution as needed. Fungi and insects are controlled by spraying with one of the following fungicides: Bravo 500, Manzate 200, or Dithane M45, plus one of the insecticides Lannate, Malathion, Diazinone, Cygone, Vydate, or Pyrenone. Kelthane plus Tedion will control spider mites. Vydate and pyrethroids will control leaf miners. Wettable powders are preferred to avoid toxicity from organic solvent carriers. One teaspoon (5 cc) of liquid soap or of a spreader–sticker, e.g., Ortho 77 or Volk, is added per gallon of spray. The plants should be ready for hardening off outdoors in about 4 weeks and should be ready for transplanting to the field in 5–6 weeks. Ideal greenhouse growing temperatures would be 68°F at night and 80°–85°F during the day. The seeding date should be about 6 weeks before the last average killing frost.

Speedling Plants

Many more plants can be produced per square foot of greenhouse area by the Speedling system. The plants are grown in patented, reusable styrofoam trays (21 × 13 in.) with 200 cone-shaped cavities (1 × 1 × 3 in. deep). Such plants are much smaller than bare-rooted field-grown transplants and hence are more easily handled in quantity for machine transplanting. However, they are also more delicate and take longer to become established and are not as tolerant of environmental stress. Larger, stronger seedlings can be produced in Speedling trays with larger 1½-in and 2-in. cells, holding 128 and 72 plants, respectively.

Field-Grown Transplants

Most pepper growers in the Southeast use field-grown plants produced in central Florida in early spring. The land is fumigated with Vorlex the preceding fall to control nematodes and fungi. Two thousand pounds of 8 : 8 : 8 fertilizer with minor elements is broadcast per acre. Starting early in February the seed is drilled in rows spaced 1 ft apart, at the rate of 35 lb/acre. Seeding is customer timed for different sections of the country. The young plants are side-dressed twice with 400 lb each application of 12 : 12 : 12. They are sprayed weekly with Dithane M-45 plus Lannate, Cygone, or Vydate plus a spreader–sticker. Plants are certified free from diseases and nematodes by state inspectors. One million plants are pulled per acre, packed, and shipped 2000 per crate. A cover crop of beggarweed follows the peppers.

Field Culture

Transplanting

Freshly dug, state-inspected, Florida field-grown transplants are probably the cheapest and best plants if properly handled, shipped promptly, and not allowed to dry out or to heat in transit. The processor provides the plants. Plants should be field set as early as possible, but after the danger of frost and of cold winds is past. This is about April 20 in central Alabama. Greenhouse-grown plants must be acclimated for 1 week in the open before tranplanting, which is done either by hand or with a mechanical transplanter for larger plantings. Transplanting into dry soil is not advised, even with the use of a starter solution. The starter solution should be very dilute, containing no more than 1 lb of 20:20:20 all-soluble fertilizer per 100 gal. of water plus 2 lb of Terrachlor for southern blight control. Dry weather will concentrate the fertilizer in the soil solution and evaporation, especially following a light rain, will bring it to the soil surface, where it can kill the plants. Plants should be spaced 15 in. in rows spaced 3.5 ft.

A Treflan premerge treatment at ½ lb/acre on light soils to 1.0 lb on heavy soils should be followed with a diphenamide (Dymid) overspray at 6.0 lb/acre within 2 weeks after transplanting and before weed seedlings have emerged. A lay-by treatment of 40 lb/acre of 10% granular Amiben about 4 weeks after transplanting completes the weed control schedule. A no-tillage system of culture is recommended to avoid root pruning of the plants, save moisture, and reduce costs.

Plastic Mulch Systems

Black and Rolston reported on the repellent action on aphids of reflected light from aluminum-foil-covered pepper beds (11). The number of aphids trapped over aluminum foil was only 10% of the number trapped over black polyethylene or no-mulch plots during the first 3 weeks after planting; and at the first harvest, only 10% of the plants showed virus symptoms versus 85% on black plastic and 96% on no-mulch beds. Later tests were conducted with aluminum-painted black plastic, which proved to be equally effective, but was cheaper and more easily handled. A 6-year average yield from 1974 to 1979, based on 8–10 harvested acres each year was 469 bu (25 lb/bu) for aluminum-painted mulch plots, as compared to the state average of 200 bu. Another advantage of this system is that a second crop of peppers, tomatoes, squash, cucumbers, green beans, or cowpeas can be grown on the same plastic mulch with economy of cultivation, weed control, and fertilizer, and with excellent yields. Chiseling 12–16 in. deep in advance of mulching to break any hard pan and promote deeper root penetration improves yields.

Florida Gradient Mulch System

This is a plastic-mulch-covered, raised-bed system with subirrigation from shallow water tables. The fertilizer is incorporated into the bed or is variously banded for availability to the seedlings and to avoid salt injury (54,60). This establishes a fertilizer gradient by diffusion downward by rain and upward from the water table, which is periodically raised by flooding to within about 3 ft from the soil surface. The roots grow into the most favorable nutrient concentrations.

Plants spaced 12 × 12 in. In double rows on 30-in. wide beds on 54-in. centers (19,360 plants/acre) have yielded 1196 bu/acre. With three rows per bed the yield was 1436 bu. Seeding is done with a machine that cuts holes in the plastic and places a compressed plug of a scientifically compounded growing medium containing fertilizer and seed into each hole. This is known as the plug mix seeding method (73). Because this intensive system of culture requires a high fertility level under the plastic and because all of the fertilizer is applied before planting, salt injury to the seedlings at the soil surface during dry periods and leaching of nutrients from around the planting hole by excessive rain have been major problems that have been only partially overcome by the use of controlled-release fertilizers, e.g., Osmocote, which is too costly for field culture.

Field Seeding of Pregerminated Seed in Gel

Pepper seeds have an extended germination period of 7–14 days. This results in much variation in seedling and mature plant size, fruit maturity, and yield. The use of pregerminated seed would avoid these problems by promoting quicker emergence and stand establishment. Peppers require high temperatures for rapid germination, but once germinated will grow below their minimum germination temperature. Essentials of the method are (1) germination of the seed at optimum temperature (30°C) in aerated water columns or immersion in a meshed bag with aeration, (2) separation of the germinated seed in the early radicle emergence stage by specific gravity in a sugar solution (25/75 wt/wt, specific gravity 1.105), (3) suspension in a gel, and (4) drilling or clumping the germinated seed in the field. Germinated seed must be handled with care to avoid damage to the radicle, and if not planted immediately can only be held for 2–3 days at 5°C. Seedling emergence of California Wonder (CW) was 50% in 7.5 days, whereas it took 16 days to obtain this percentage from dry seed. Gibberellic acid (GA_3) at 400 ppm promoted the germination of several cultivars to 81–90% in 3 days (98,143).

Campbell reported on the performance of the Gel-O-Flex Vegetable Planter, which he designed and built for the Campbell Soup Company (20). This self-propelled, hydraulic-pressure-operated machine will continuously drill or clump in hills, at desired rates, seeds of peppers, tomatoes, and other vegetables in Laponite or Viterra 2 gel at speeds up to 4 mph.

Disease Control in the Field

Irrigation and Blossom End Rot

Irrigation is essential for production of peppers, as drought is always a potential threat to the crop in the Southeast. Drought periods are common in May and June, triggering the onset of blossom end rot and of much fruit drop. The major cause of this disease has been shown to be insufficient availability of calcium in the soil solution, particularly during the period of maximum fruit set when the demand is greatest and when the plants are under drought stress. Blossom end rot can, to a large extent, be controlled with a uniform water supply during dry periods, and by liming the soil in advance.

Bacterial Leaf Spot

Genetic resistance is the only sure method of control. Seed treatment of susceptible cultivars is an important practice to prevent seedling infection from this source but cannot prevent infection from infested field soils. Georgia pimiento canners treat freshly harvested seeds for 15 min in a 4.2% calcium hypochlorite or a 2.6% sodium hypochlorite solution (50% Clorox), used at the rate of 3 gal./gal. of seed. The seed is then rinsed in water for 15 min and dried rapidly under a fan (44). Spraying with tribasic copper sulfate gives some control in the field but is ineffective in hot rainy weather.

Southern Blight

Crop rotation with corn, sorghum, or rye will reduce losses from this disease, caused by *Sclerotium rolfsii*. A transplanting solution containing 2–4 lb of 75% PCNB (Terraclor) WP/100 gal. of water has been widely used for control at the rate of about 1 cup/plant. This chemical must be kept uniformly suspended to avoid injury to the seedlings. Terraclor is also available as an emulsion containing 2 lb a.i./gal. or as 10% granules. The application rate with granules is 5–10 lb a.i./acre, broadcast. Biological control of *S. rolfsii* with the antagonistic fungus *Trichoderma harzianum* grown on molasses absorbed on diatomaceous earth granules has given equivalent control with peanuts at 50–100 lb/acre (3).

Viruses

Seed Disinfection   Resistant cultivars are the best means of virus control. However, pepper seed disinfection is a valuable precautionary measure for preventing the spread of a surface-borne pepper strain of tobacco mosaic virus (TMV). We routinely treat freshly harvested, moist seed for 2–3 min in 1–2% H_2SO_4 in a plastic container with agitation, rinse under the tap, dip in a Captan solution, and dry rapidly with a fan without rinsing. Black uses a 2% solution of HC1 (9). Nakayama treats fresh seed in 10% Clorox for 10 min. Floaters are decanted in three rinses of tap water. Demski treated dry seed for 15 min in a 50% Clorox solution, followed by a 15-min tap water rinse, without any apparent harm to the seed (45).

Oil Sprays   Tests with oil sprays in Florida have shown much promise for controlling aphid vectors of viruses on peppers. Proper equipment and method of application are essential to success. Zitter and Ozaki used a specially formulated mineral oil (JMS Stylet-Oil®) developed by Simons, in a 0.75% emulsion (3 qt/100 gal.) applied with Teejet TX-4 nozzles at 400 psi. No foliage injury appeared in any of the oil plots in spite of a cold winter season. Unsprayed Early Calwonder plants averaged 75% infection versus 15% in the oil-sprayed plants 10 weeks after virus spread began. The oil plots significantly outyielded the nonsprayed plots in both number and weight of marketable fruit. Spraying need not start until aphids are found on close inspection (138,166).

PAST BREEDING ACHIEVEMENTS

Truhart Perfection Pimiento

This leading pimiento cultivar, developed by Cochran by the pure-line pedigree breeding method between 1935 and 1942, originated from a single plant selection of the Perfection

Pimiento in a farmer's field in Georgia. The Perfection Pimiento itself was also derived from a single plant selection made by S. D. Riegel and Sons of Experiment, Georgia, from seed introduced into the United States from Spain in 1911. Riegel introduced the Perfection Pimiento in 1913. By 1935, growers and canners described this cultivar as having "run out," because commercial seed produced fruits of many shapes and sizes.

From the open-pollinated seed of the single superior plant Cochran grew the F_2 progeny in 1936. Six plants were selected for selfing under muslin-covered cages and six F_3 lines grown in 1937. One line proved to be superior to the others in productivity, fruit shape, and fruit wall thickness. Selections from this line were inbred twice more to produce the F_5 generation. Thereafter mass selection was practiced. Truhart Perfection, released in 1942, is still the leading commercial pimiento in the Southeast (23).

Truhart Perfection D Pimiento

This cultivar, released in 1963, was developed by Dempsey from a cross of Truhart with the pungent, bacterial leaf-spot-resistant cv. Santaka. The breeding procedure included backcrossing to Truhart and sibcrossing and selfing for 12 generations. Selection was for a smooth, tightly closed stylar scar that would prevent fungus entry into the fruit cavity and for resistance to bacterial leaf spot. In trials Truhart D outyielded Truhart by about 1000 lb/acre over a 3-year period. Fruit loss from internal mold was reduced from 10 to 5%. However, the monogenic recessive resistance of Santaka to bacterial spot was lost (42).

Carolina Hot

This cultivar was developed by Martin and Crawford between 1942 and 1954 by selection within the Cayenne cultivar grown for powder production in South Carolina. Selection was for strong plants with a deciduous calyx and for long, slender pods (6 × ¾ in.) borne well up in the plant canopy to reduce the amount of fruit rot from soil contact. Carolina Hot was believed to be the first cultivar of its type with a deciduous calyx. It was also more disease and nematode resistant than the original cultivar. The authors recommended that the fruit be allowed to ripen and dry on the plants rather than be harvested when succulent. Carolina Hot produced over 1 ton of dried pepper per acre, with a moisture content of about 10% (92).

Mississippi Nemaheart

This root-knot resistant pimiento pepper was developed by Hare between 1951 and 1966. It is the only pimiento with resistance to the southern root-knot nematode *Meloidogyne incognita* and its race *acrita,* the cotton root-knot nematode (70). The parentage involved Truhart and a woolly leaved, root-knot-resistant, erect-fruited, pungent Mexican pepper with light green foilage labeled M152B. Ten BC were made to Truhart, followed by seven generations of inbreeding. Root-knot screening tests were conducted during the winter in the greenhouse, and selections made for plant and fruit type in the field during the summer. Nemaheart lacked sufficient fruit size for commercial acceptance.

Pimsan 1, 2, 3 and Bighart Pimientos

These cultivars were developed by myself and co-workers at the Georgia and Alabama Experiment Stations between 1945 and 1969. The pedigree of Bighart involved 36 generations. This was a backcross–intercross program having as objectives the transfer to

Truhart of the *L* gene from Holmes' Bell, bacterial spot resistance from Santaka, and resistance to tobacco etch virus (TEV) from Cayenne SC 46252. Pimsan 1 and 2 both have the *L* gene and were released in 1956 and 1957, respectively. Pimsan 3 has the genes *L* and *et*[a] and was released in 1958. Bighart, released in 1969, possesses the *L* gene. The gene for TEV and that for bacterial spot resistance were lost for lack of technical assistance and facilities needed to produce inocula and to screen large segregating populations from selfed BC to the susceptible cvs. YW, Truhart, and Keystone Resistant Giant. Natural epiphytotics of bacterial spot (*Xanthomonas campestris* pv. *vesicatoria*) cannot be depended upon for field screening.

Bighart Yield, Fruit Size, Recovery, and Seed Yield

Bighart yielded significantly more than Truhart and the fruit was larger. Recovery of cored raw fruit was increased 17% and canned recovery weight 11%.

About 25% of the raw pimiento fruit is waste (Fig. 3.7) (*24*). Recent laboratory tests at Auburn University have shown a 68.6% recovery of canned product from whole fresh Truhart fruit when steam peeled vs. 65.6% when lye peeled (*139*). In canning plants the usual lye peel process recovers only about 30%.

The seed yield of Bighart was 1.6% of the fresh fruit weight and air-dried seed was 0.8%. One ton of fruit yielded 16 lb of dry seed vs. 41 lb from Truhart. One pound of Bighart seed contained 63,000 seeds (*68*).

Greenleaf Tabasco (Tabasco G)

Tabasco, the cultivar of the Tabasco industry in Louisiana, was introduced into the United States from Mexico about 1848. This pepper responds with a lethal wilt disease when infected with TEV (*63*). I transferred the TEV resistance of *C. chinense* PI 152225 to Tabasco by the BC method (*64,69*). Progeny from this cross had golden yellow–green rather than green foliage, which prompted me to introduce a second TEV-resistant cultivar of *C. chinense* into the pedigree at the BC$_3$ level (Fig. 3.8). There was much sterility during the early BC generations. This was expressed in partially developed, short stubby fruits with few or no seeds. The sterility was gradually overcome by additional backcrossing, interline crossing, and selection. The recessive mode of inheritance of TEV resistance required that alternate selfed generations be screened for TEV resistance before making the next BC. Greenleaf Tabasco, introduced in 1970, has saved the Tabasco

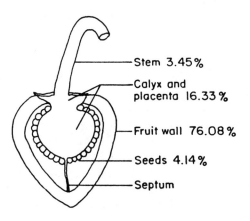

Stem 3.45%

Calyx and placenta 16.33%

Fruit wall 76.08%

Seeds 4.14%

Septum

FIGURE 3.7. Truhart Perfection Pimiento fruit showing mean percentages of constituent parts. Average fresh fruit weight 0.173 lb (78.25 g).
After Cochran (24).

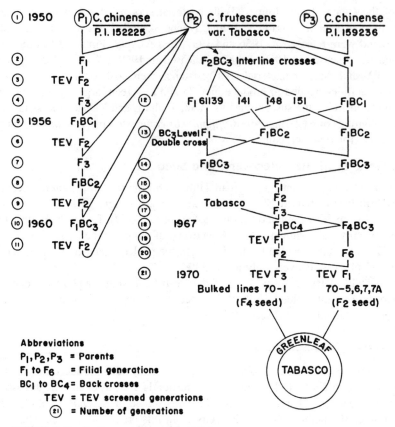

FIGURE 3.8. Pedigree of Greenleaf Tabasco.

industry in Louisiana from the aphid-transmitted etch wilt disease (65). Zitter reported that Tabasco G was also resistant to pepper mottle virus (PMV), an unexpected bonus (162).

Yolo Y, VR2, and Delray Bell

These three virus-resistant bell peppers were developed by Cook and co-workers in Florida (29,32,35). Yolo Y, released in 1966, originated from a single mutant plant of Yolo Wonder (YW) resistant to potato virus Y (PVY), discovered by Cook. This plant, the parent of Yolo Y, possessed a single recessive gene y^a that conferred resistance to mild strains of PVY. This cultivar also carries the gene L^1 (L^i) of YW which determines the imperfect localization response to infection with common TMV, described by Holmes (75). Yolo Y yielded slightly less than YW (690 vs. 668 bu/acre) in 17 trials in eight states.

Florida VR2, released in 1976, is resistant to TEV, PVY, and TMV. Sources of resistance were the small, pungent, fruited *C. annuum* PI 163192 and PI 264281, and Cayenne SC 46252. All carry the gene et^a, which confers resistance to mild strains of TEV and to mild and severe strains of PVY (26–28). F_1 hybrids of these cultivars with YW and California Wonder (CW) were variously backcrossed to YW, Yolo Y, CW, and Florida Giant. Four BC were made, the last one to Yolo Y, followed by two field and two greenhouse generations. Selection was for horticultural characters, for virus resistances, and for resistance to bacterial spot race 2 (*Bs2*) from PI 163192. Florida VR2 is homozygous for et^a, L^1, and *Bs2*.

Delray Bell, released in 1977, is resistant to TEV, PVY, and PMV. These viruses have been major problems in Florida's Delray Beach pepper production area. TEV and PVY resistance came from various breeding lines with the gene et^a, and that for PMV from Avelar (et^{av}). The gene L^1 for resistance to common TMV was lost in backcrossing to susceptible parents. Delray Bell will eventually develop a mild mottle from PMV but no fruit symptoms. Its yield was superior to that of commercial cultivars in virus-infested areas and was comparable in areas with no virus problems.

Yolo Y, VR2, and Delray Bell produced flat pods with few seeds in cold weather, probably from insufficient pollination. Early Calwonder was much less affected and was preferred by buyers.

Cook Breeder Line Releases

Cook released four advanced breeding lines in 1982: Florida VR2-34, XVR3-25, Florida VR4, and USAJ15 (*31*). Florida VR2-34 was developed from a bulk of several lines reselected from Florida VR2 for larger fruits with square nondepressed blossom ends. Preliminary trials indicate that VR2-34 has larger fruits and yields better than CW or VR2. It carries the same genes for disease resistances as VR2 (et^a, L^1, *Bs2*). XVR3-25 (et^a, L^1, *Bs1*, *Bs2*) is similar to VR2-34, but also carries resistance to race 1 of the spot bacterium (*X. campestris* pv. *vesicatoria*) from *C. chacoense* PI 260435. Florida VR4 (et^a, et^{av}, *Bs2*) is resistant to TEV, PVY, PMV, and race 2 of pv. *vesicatoria* but susceptible to common TMV. This line derives from the cross Florida VR2 × Delray Bell. All three of the new bell lines have large fruits, averaging 10 × 8 cm, with fruit walls 5–6 mm thick. USAJ15, a medium long, slightly curved, pointed hot pepper is resistant to TEV and PVY. It derives from a cross of PI 264281 (et^a) with a PVY-resistant pepper from Ecuador ECUAJI. Fruit size averages 9.6 × 1.8 cm. The fruit wall is 1.5 mm thick and the fruit weight ranges from 5 to 12 g. Maturity is 81 days from transplanting.

Cayenne 16 through Cayenne 20

These five cayenne peppers were selected by Black and Simmons for TEV and PMV resistance in a long-fruited cayenne strain grown in Louisiana for making hot sauce. This strain is longer and larger fruited than Carolina Hot. Cayenne 16, the largest, averages 7.5 in. (19 cm) long, weighs 26.5 g, and is slightly curved and pointed (*12*).

Texas Bell 76004, Long Green Chile 76042, and TAM Mild Jalapeno-1

These three peppers were developed by Villalon and released in 1979. They are the product of an intensive interdisciplinary breeding program initiated in 1971 by Texas A&M University to assist growers with pepper disease problems, primarily viruses, in the lower Rio Grande winter production area. All three cultivars, according to Villalon, are resistant to the south Texas strains of TEV, PMV, PVY and to Samsun latent TMV (SLTMV) (*157*). He conducted the first virus disease surveys in 1971, and by means of host range studies and diagnostic serological and electron microscopic techniques, found that TEV, PMV, PVY, cucumber mosaic virus (CMV), SLTMV, and tobacco ringspot virus (TRSV) were the most common viruses of peppers in south Texas. Over 100 commercial bell peppers were susceptible to all of these viruses. For sources of resistance, he screened 13 exotic germplasm stocks with TEV, using an artist airbrush inoculation technique at 125 psi. Resistant selections were crossed with commercial bell types and several large-fruited sweet lines developed by the BC method. Because genes for virus resistances are recessive, each F_2 generation was screened for virus resistance before making the next BC.

Villalon found that by intercrossing resistant lines of diverse origin higher levels of resistance could be obtained, indicating that modifier genes were additive in effect. Concurrently with the bell pepper BC, resistant pungent-fruited segregates were back-crossed to Long Green Chili, Red Chili, Serrano, Ancho, pimiento, cherry, yellow pickling, and paprika types, and multiple virus-resistant derivatives of these cultivars also developed. A significant feature of Villalon's breeding method was to screen first for a high level of TEV resistance. Once TEV resistance was fixed in F_3 he screened in F_4 for PMV, PVY, and SLTMV resistance. That he was able so quickly to obtain resistance to all four viruses in F_5 lines supports other evidence that multiple resistance is conferred by the stronger potyvirus resistance alleles. F_6 lines were tested in replicated statewide trials. Villalon recommends that growers test his new cultivars on a limited commercial scale for several seasons for adaptation to soils and climate and for processor and consumer acceptance before planting their entire acreage to them. He points out that the ultimate step, processor evaluation and consumer acceptance, must be taken by the industry itself.

Pepper Breeding in Brazil—Agronomico 8 and Agronomico 10

Nagai reported that strains of PVY were the most prevalent at Campinas, S.P., Brazil. By combining the PVY-resistant germplasms of several cultivars, such as Puerto Rico Wonder, Moura, Ikeda, Avelar, Casca Dura, and PI 264281 (P11), he developed the multiple-resistant cultivar Agronomico 8, released in 1967. This cultivar has shown a high level of resistance to a complex of potyviruses in the southeastern United States, in California, and in Europe. Following an outbreak of a new strain of PVY in Brazil, Nagai made additional crosses between selections within a pool of resistant germplasms and developed a new series of resistant lines designated Agronomico 10. This is now the leading cultivar in Brazil. He is currently incorporating a hypersensitive-type resistance to bacterial spot into Agronomico 10. This cultivar has shown resistance to PMV in Florida (99).

BREEDING FOR HORTICULTURAL CHARACTERS

Earliness

Pochard observed transgressive segregation for this trait (114). He determined earliness by two methods, first by ranking the cultivars by date of first bloom, starting with the earliest Antibois on June 12, as time zero. Vinedale was +2, YW +12, Quadrato Giallo +15, and Piment de Bayonne +22 days later. In the second method he used the percentage of ripe fruit harvested by August 30 as a measure of earliness. When the first method was used, 45% of the F_3 lines from the cross YW × Antibois (large, subspherical squash type) and 10% from the cross Quadrato Giallo (bell) × Antibois flowered earlier than Antibois. One of the F_3 lines of the latter cross flowered 14 days before Antibois. The second method yielded similar results, 13 of 78 F_3 lines producing 80–100% ripe fruit weight by August 30. This improvement resulted from the selection of the earliest 4% of the F_2 plants by either method. Inheritance of earliness in these crosses was oligogenic.

Fruit Shape and Size

In certain crosses of small oblate or round-fruited cultivars of *C. annuum* and of *C. chinense* with larger elongate-fruited cultivars, the F_1 is small fruited and oblate and the F_2 segregates 3 oblate : 1 elongate. All fruits with a length/width ratio above 1.0 are classed as elongate. Elongate fruit range from short and blunt to long and pointed. Elongate fruit is determined by polygenes that are recessive to the dominant gene O. This gene is present in *C. chinense* Acc. 1555 (Fig. 3.1B). Early, productive, smooth, nearly round deep-fruited segregates from BC_2 to pimiento give promise that a large, nearly round fruited cultivar can be developed from this cross.

Other crosses between oblate and elongate produce an intermediate F_1 and an F_2 with a continuous range of fruit shapes and sizes, typical of quantitative inheritance. Fruit size and shape in such crosses are determined by polygenes, with the genes for small fruit being dominant. Because about 30 genes determine fruit size, large fruits cannot be recovered in F_2 (*83*). Four or more BC are required to recover the fruit size of the larger parent.

According to Pochard, extra large fruit is undesirable because it is usually associated with lower productivity, irregular fruit shape, and poor quality. The French pepper breeding program stresses improvement in fruit appearance through more uniformity in shape and size. Plants are selected for large, glossy, firm, thick-fleshed fruit that will withstand shipping and that are resistant to blossom end rot. CW possesses these characteristics.

Selection for Fruit Quality

Selection, by taste, of plants with fruit having a strong, pleasing, high sugar/acid ratio and other desirable flavors is probably the most practical method for the breeder to incorporate quality into his breeding lines. Soluble-solids measurements with a refractometer could also prove helpful. Pochard reports that fruit acidity increases with maturity in all cultivars but particularly in the YW group. Both the fresh and canned product of F_4 and more advanced lines would need to be evaluated by carefully chosen taste panels. Such tests should be designed to provide statistically unbiased probability estimates. Laboratory tests for flavor compounds, pigment content, and vitamin C levels would be valuable for selecting the better advanced lines. High pigment, according to Pochard, is a partially dominant trait of polygenic inheritance.

The vitamin C content of YW fruit was only 150 mg/100 g fresh wt. vs. 190 mg in CW and 300 mg in Doux d'Alger (*114*). Rymal has developed a quick color spot test suitable for making quantitative estimates of vitamin C in pepper fruits in the field (*126*).

Fruit Flavor and Pungency

Flavor in peppers, according to Jones and Rosa, is due to several aromatic substances present in very small quantities but not connected with pungency (*81*). Buttery *et al.* first identified the important flavor component of bell peppers as 2-isobutyl-3-methoxyprazine (*19*). This compound has an extremely potent aroma, its odor being detectable in as little as 2 ppt in water solution or 1 drop in an olympic-sized swimming pool. Huffman *et al.*

defined and quantified the flavor components of fresh jalapeno and bell peppers by means of gas chromatography and mass spectrometry (77). They confirmed that the above compound is the major flavor component in both bell and jalapeno peppers. Although its concentration is very low, in the ppb range, jalapeno peppers, originally from Jalapa, Mexico, have a strong flavor. The highest concentration was found in the outer wall, with lesser amounts in the cross walls and in the placenta and with none in the seed. Thermal processing destroyed most of the flavor compound. The authors concluded that this odorant is synthesized in many cells in different parts of the fruit but is concentrated in the outer wall.

Haymon and Aurand identified 23 flavor compounds in Tabasco peppers, all of which were necessary to produce the characteristic Tabasco aroma. There was no single dominant flavor compound as in bell or jalapeno peppers (72).

Flavor of Pimiento vs. Bell Peppers

The flavor of canned pimientos has traditionally been considered superior to that of canned bell peppers, but until recently no data were available to decide this question. Also, pimientos that were flame peeled were not comparable with bell peppers that were either unpeeled, lye peeled, or steam peeled because the superior flavor of the pimientos could well have been due to the roasting process. Rymal *et al.* used the triangle test design and statistical probability to answer this question. Six taste panelists were presented with diced samples of three canned pimiento and three canned bell pepper cultivars in 15 different comparisons and asked to identify the odd sample. The test was replicated four times for a total of 360 decisions. The results showed that there was no significant difference in flavor, color, or texture between the pimiento and bell pepper samples when they had been similarly processed (128).

Pungency in peppers is due to capsaicin ($C_{18}H_{27}NO_3$), a fat-soluble, flavorless, odorless, and colorless compound, the structure of which was determined by Nelson in 1920 (100a). Capsaicin distribution in fresh jalapeno peppers was highest in the cross walls, followed by the placenta, seeds, and outer walls. Processing increased the capsaicin content many times throughout the fruit, indicating that heating had transformed a precursor into capsaicin (Table 3.5). Huffman *et al.* found no specialized structures that produced either the flavor compound or capsaicin, but Ohta observed receptacles high in capsaicin on the fruit cross walls (Fig. 3.9) (107).

Pepper cultivars differ greatly in capsaicin level, with those of *C. frutescens* among the highest (Table 3.6) (104). Pochard uses a quick chemical color test for pungency. A bit of placenta tissue is transferred to filter paper with a spear needle and a little of the oily secretion absorbed on the paper. When a drop of a 1% solution of vanadium oxytrichloride in carbon tetrachloride (Marquis reagent) is added, a bluish color develops if capsaicin is present. A negative color test could always be verified by taste (117). Van Blaricom and Martin have devised permanent color standards for this test (155).

Total Capsaicinoids and Scoville Units

Total capsaicinoids include capsaicin plus four structurally similar, pungent compounds. The pungency stimulus of peppers is predominantly piperine. The percentage of capsaicinoids in dried red capsicum powder and in oleoresins is an important index of quality for these products, which are widely used as spices by the world food industry to provide pungency and red color.

TABLE 3.5. Mean Values of Capsaicin Content in Various Parts of Fresh and Thermally Processed Jalapeno Pepper[a]

Pod portion	Capsaicin[b] (mg/100 g)	
	Fresh	Processed
Outer wall	0.12a	128.19ab
Cross wall	18.37c	345.96c
Placenta	8.2b	194.05b
Seed	0.45a	68.24a

[a]After Huffman et al. (77).
[b]Means followed by different letters are significantly different at the 1% level according to the Student–Newman–Keuls test. Each mean represents eight determinations.

The American Spice Trade Association uses the Scoville unit (SU) measure of pungency, which is the reciprocal of the highest dilution at which pungency can still be detected by a taste panel. The method is subjective, with results varying by up to 150% (152). It is nonetheless the preferred method of the U.S. industry because chemical tests are lengthy and laborious and do not relate directly to pungency. Chemical tests are, however, necessary to monitor the reliability of the SU method. Rajpoot and Govindarajan (122) found that paper chromatography used in conjunction with a spectrophotometer to measure absorbance by capsaicinoids at 615 nm gave reproducible results that correlated closely with SU values. They expressed this relationship by the least squares regression equation $y = -9.22 + 164.126x$ $(r \approx 1)$, where y is the SU (in 1000s), x the total capsaicinoids (%), and r the correlation coefficient, which is equal to or close to 1. y can therefore be calculated directly from x and vice versa. This method can also be used to estimate capsaicinoids in immature peppers that have been dried. In bell and paprika peppers that have very low concentrations of capsaicinoids (<0.1%), SU values may be overestimated due to the presence of nonpungent phenolic compounds, e.g., epigenol and lutein (Table 3.6).

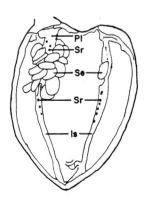

FIGURE 3.9. Secretory receptacles in the F_1 hybrid Takanotsume (pungent) × Large Bell (sweet). Pl, placenta; Sr, secretory receptacles of capsaicin; Se, seed; Is, interlocular septum.
After Ohta (107).

TABLE 3.6. Capsaicinoid Levels (as Percentage of Dry Matter and in Scoville Units, SU) in Capsicum Species, Cultivars, and Accessions[a]

Species	Variety or accession	Capsaicinoids (%)	SU (1000)
C. annuum	059-991	0.80	131
	Long Red Cayenne	0.55	90
	Fresno Chili	0.32	52
	AC 1874	0.16	26
	N58-248	0.67	110
	KUSE 132A	0.25	41
	KUSE 132B	0.26	43
	KUSE 132C	0.30	49
	KUSE 751A	0.32	52
	KUSE 751B	0.34	56
	KUSE 751C	0.35	57
	KUSE 751D	0.33	54
	Large Bell	0.06	10
	Bola	0.05	8
	Doux	0.04	7
C. frutescens	Tabasco	0.88	144
	AC 1443	0.96	157
	AC 1651	0.50	82
	AC 1585	0.46	75
	AC 1448	0.43	70
C. microcarpum	AC 1553	0.58	95
C. chacoense	AC 1256	0.35	57
	AC 1751	0.24	39
C. pendulum	AC 1941	0.41	67
	AC 1786	0.40	66
	AC 1233	0.30	49
C. pubescens	Roja	0.65	107

[a]Estimated SU pungency scale (1000): no pungency detectable (0.0–10); mildly pungent (11–30); moderately pungent (31–80); highly pungent (>80). KUSE: collected by Kyoto University Scientific Expedition to Afghanistan and Iran, 1955. [After Ohta (104).]

Rymal *et al.* have developed an injection–extraction method for the rapid determination of total capsaicinoids in single whole pepper fruits (127).

Inheritance of Pungency

Inheritance studies of pungency generally support the single dominant gene hypothesis. Webber reported a ratio of 25 hot- to 5 sweet-fruited plants in the F_2 of the cross Red Chili (hot) × Golden Dawn (sweet) (160). Deshpande observed an F_2 ratio of 202 pungent to 70 nonpungent plants, $\chi^2(3:1) = .078$, $P = .90$, in a cross of a cayenne with a sweet bell type. He assigned the symbol C for capsaicin to this gene (46).

However, Ohta, using a quantitative method for determining pungency, namely, a combination chromatography–taste threshold method, found varying degrees of pungency in F_1 and bimodal distributions in F_2 and in BC populations. His results clearly showed polygenic inheritance of pungency. A major gene determined pungency, but the polygenes acting in a cumulative manner, both positive and negative, determined various degrees of pungency (106). Ohta also found that high nighttime and daytime temperatures increased the capsaicin content.

Inheritance of Mature Fruit Color

Smith reported the mode of inheritance of red, brown, and green mature fruit colors in peppers (140). Chocolate brown fruit is common in Mexican cultivars of *C. annuum*. The cross Mexico Acc. 406 (red) × Mexico Acc. 401 (brown) produced a red F_1 and an F_2 ratio of 3 red : 1 brown. The BC to Acc. 401 produced 18 red : 19 brown, and the BC to Acc. 406 produced only red. The cross Oshkosh (yellow mature fruit) × Acc. 401 (brown) produced a new green mature fruit color in the F_2, the double-recessive genotype. The digenic ratio was 132 red : 46 brown : 44 yellow : 11 green, a good fit to a 9 : 3 : 3 : 1 ratio ($P = .50–.70$). Smith ascribed brown fruit color to the presence of a recessive chlorophyll retainer gene *cl*, which prevents the complete degradation of chlorophyll. The presence of chlorophyll with the red pigment lycopene (gene y^+) produces the brown fruit color. Hence, the mature fruit colors and their genotypes are Oshkosh (yellow) *y/y*, *cl+/cl+;* Acc. 406 (red) y^+/y^+, *cl+/cl+;* Acc. 401 (brown) y^+/y^+, *cl/cl;* and mature green *y/y, cl/cl*. The latter fruit color may prove valuable for breeding green peppers with a longer harvesting period and a longer fresh market life, since mature red bell peppers are not acceptable to the fresh market.

Inheritance of Immature Fruit Color

Odland and Porter studied the inheritance of immature fruit color in crosses between several cultivars of *C. annuum* (102). The cultivars, their immature fruit colors, and deduced genotypes were Ornamental, sulfury white, *sw1/sw1, sw2/sw2, . . . , swn/swn;* Hungarian Yellow Wax, yellow green or lettuce green, *sw1+/sw1+, sw2/sw2, . . . , swn;* Oshkosh and Red Cherry, both dark green or cedar green, *sw1+/sw1+, sw2+/sw2+, . . . , swn*. The cross Ornamental × Oshkosh produced a medium-green F_1 and an F_2 ratio of 15 green : 1 sulfury white. Because of incomplete dominance, the 15 green F_2 plants include four shades of green, corresponding to the number of *sw+* genes present in their genotype. The BC to Ornamental produced a ratio of 1 medium green : 2 yellow green : 1 sulfury white. Similarly, the cross Oshkosh × Hungarian Yellow Wax produced a medium-green F_1 and an F_2 ratio of 1 dark green : 2 medium green : 1 yellow green. Subsequently, Odland confirmed the duplicate-factor hypothesis with additional crosses of Harris Earliest, Harris Early Giant, and Cayenne, all cedar green, with Ornamental (sulfury white). However, certain F_2 families of the crosses with Harris Early Giant and with Cayenne segregated 63 green : 1 sulfury white. Other F_2 families from the same crosses produced 15 : 1 ratios, indicating that some plants of these two cultivars carried three *sw+* gene pairs (101).

Inheritance of Pedicel Length

A long, slender pedicel is desirable to allow for expansion of the developing fruit, especially of large-fruited bell peppers. A short pedicel on a determinate plant results in many deformed fruits. Subramanya and Ozaki reported that long pedicel was partially dominant over short and polygenic in inheritance. F_2 populations exhibited a continuous range of phenotypes. Estimated additive gene heritability was 88% and the genetic advance based on the selection of 5% of the F_2 plants with the longest pedicels was 42%. At least three loci determined this trait (*151*).

Inheritance of Multiple Fruitedness

C. annuum typically sets one fruit per axil, whereas *C. chinense* sets from one to four. The transfer of this trait to *C. annuum* promises potentially higher yields. Subramanya reported on the inheritance of multiple flowers in the cross Delray Bell × PI 159236. He concluded that probably three major dominant genes control double flowers but that more genes are required to produce three to four flowers per node (*150*). J. E. Watson and W. H. Greenleaf (unpublished) concluded that seven semirecessive additive genes determined multiple fruitedness in *C. chinense* Acc. 1555. Both studies indicated that the transfer of double fruitedness to *C. annuum* was possible by the BC method. Our results indicate that genes for this trait would need to be concentrated in selections made in alternate selfed generations before each BC.

Breeding New Ideotypes

Genetic restructuring of the plant habit of large fruited bell and pimento cultivars from indeterminate to determinate promises to increase the yield per plant.

Hoyle (*76a*) described four variations in the indeterminate habit of chili peppers, and concluded that the ideal chili plant is 18–24 in. tall, erect and compact, and produces up to 15 fruits. For this type plant to develop, the fruit must be set early so that all the branches on both sides of the main fork develop equally. Unbalanced plants produce fewer fruits.

Subramanya defined two promising determinate ideotypes (*148a*). The first has a single unbranched stem with one or two fruits per node and a single unbranched terminal cluster of four fruits for a total of 13 fruits. The second type, also with a single stem bears all of its fruits (up to 12) in a single, branched compact terminal cluster. The first plant type lacks foliage cover and may be subject to sunscald, whereas the second is bushy and compact and protects the fruit well. Both plant types are structured for resistance to breakage from a heavy fruit load and also to permit close spacing. No cultivars of these two new ideotypes have yet been released.

Among potential parents for breeding determinate plant types of *C. annuum* are Truhart Pimento (for fruit size and long fruit pedicel), *C. chinense* Acc. 1555, PI 159236, and Rocotillo (for multiple fruited nodes and disease resistance), and Santaka and Frommage (with the gene *fa* for compact determinate fruit clusters).

Selection for Seedling Emergence Free from the Seed Coat

Failure of pepper seedlings to emerge free from the seed coat (sticky seed) has been observed in otherwise promising lines. This major weakness prevents the separation of the cotyledons and retards the growth of affected seedlings. It can be avoided by rigorous selection.

Internal Fruit Proliferation

This undesirable genetic trait is common in a wide variety of large-fruited pepper cultivars, ranging from 0 to 25% of affected fruit (*89*). The fruitlike outgrowths from the base of the placenta or fruit wall should be avoided when making selections.

BREEDING FOR DISEASE RESISTANCE

Fruit Rots, Cercospora Leaf Spot, and Powdery Mildew

Most pepper cultivars are susceptible to fruit rots of several kinds. Morgan-Jones, mycologist at Auburn University, identified the following from infected field-grown fruit: (1) *Alternaria tenuissima*, (2) *Colletotrichum dematium*, (3) *Colletotrichum gloeosporium*, (4) *Curvularia lunata*, and (5) *Phoma destructiva*. I observed high field resistance to fruit rot, including soft rot caused by *Erwinia carotovora*, in *C. chinense* Accs. 1555, 1554, 906 (Uvilla Grande), and others. Resistance appears to be dominant and dependent on few genes. The mode of inheritance to specific organisms would need to be determined by controlled inoculation. Bartz and Stall found that Jalapeno was the most resistant to *E. carotovora* of some 26 cultivars tested by puncture inoculations of stem, calyx, and fruit wall (6).

Among other cultivars the following have been reported from India as resistant to anthracnose (*Colletotrichum capsici*) fruit rot: *C. annuum* cvs. Chinese Giant, Yolo Y, Hungarian Yellow Wax, Spartan Emerald, and Paprika; to cercospora leaf spot (*C. capsici*): California Wonder, Canape (F1), Merrimack Wonder, and *C. microcarpum;* to powdery mildew (*Leveillula taurica*): *C. microcarpum, C. pendulum,* and *C. pubescens;* moderately resistant cultivars of *C. annuum* were World Beater, Florida 1063-2, Bull Nose, Midway, Spanish Long, PI 159252, PI 288982, and Chilli Long (*154*). From the F_1, F_2, and the BC to Pimiento of the cross *C. annuum* Acc. 46101 (Brazil, res.) × Pimiento (susc.), Hare concluded that resistance was semidominant and its inheritance trigenic.

Bacterial Leaf Spot

This disease, caused by the bacterium *X. campestris* pv. *vesicatoria,* can quickly defoliate pepper plants in hot humid weather, resulting in sun-scalded fruit, heavy crop loss, and even crop failure. Fieldhouse and Sasser have developed a selective culture medium that will support growth of pv. *vesicatoria* but essentially prevent the growth of all other species of *Xanthomonas,* contaminant bacteria, and fungi (*57*). Two races of this bacterium occur in Florida, of which race 2 is the more common. PI 163192, a cayenne type from India, and *C. chacoense* PI 260435 possess single dominant gene resistance to race 2 and 1, respectively, when the plants are inoculated by infusion into the leaves (*33,34*).

Sowell, and Sowell and Dempsey reported that *C. annuum* PIs 163189, 163192, 271322, and 322719 were resistant to both races when inoculated by spraying the plants. Of these, PI 322719 was perhaps the more valuable to breeders because of its larger, non-pungent fruits (*144,145*). Dempsey transferred the resistance of PI 163192 to pimiento. This resistance has held up at Experiment, Georgia, for over 20 years. Adamson and Sowell confirmed the monogenic dominant resistance in PI 163192 and showed further that PI 322719 carried a different single dominant gene and that PI 163189 possessed two or more additive genes, at least one of which was linked with the dominant gene in PI 163192 (*1*). Cook incorporated resistance to race 2 into VR2 and to both races into XVR3-25 (*31*).

Stall, however, reported a breakdown of race 2 resistance from PI 163192 in Florida due to a high mutation rate in pv. *vesicatoria* for a change in race, namely, one cell in 40,000. This high rate of mutation according to Stall would prevent the use of this single dominant gene (vertical resistance). By the use of the leaf infusion technique with a sufficiently low concentration of bacteria (2.5×10^3 cfu ml^{-1}), whereby separate lesions were produced on the leaves rather than a confluent necrosis, Stall could differentiate resistant plants in the F_2 of the cross PI 271322 × Early Calwonder by their lower lesion counts. The F_2 ratio of 140 susceptible : 10 resistant plants indicated that two recessive genes determine this horizontal resistance, which the bacterium would presumably be less likely to overcome (147).

Phytophthora Root Rot

This soilborne disease, caused by the fungus *Phytophthora capsici,* is common in the furrow-irrigated Southwest and California. Kimble and Grogan reported that of 613 Acc. of *C. annuum* only 13 exhibited various degrees of resistance, determined by survival counts of seedlings 30 days after inoculation. PIs 188376, 201232, and especially 201234, a cayenne type from Central America, proved resistant with only 7–15% of dead plants (84). Smith *et al.* found resistance dominant in F_1 hybrids with YW. F_2 and F_3 progeny gave 3 : 1 and 15 : 1 ratios, with acceptable χ^2 probabilities, indicating that one or two independent dominant genes (duplicate factors) conferred resistance. However, there was much heterogeneity among the ratios. The authors nevertheless concluded that a high level of resistance could be transferred to desirable cultivars (142).

In Brazil, Matsuoka conducts tests with *P. capsici* in the greenhouse at 20°–30°C. He sows 20–30 seeds per row and inoculates seedlings 40 days later by pouring 30–40 ml per row of a suspension containing about 4.8×10^4 zoospores per milliliter. Young pathogenic cultures are grown on V-8 juice agar under continuous fluorescent light of 2000 lux. To induce sporulation, 20 ml of distilled water are added per petri dish culture and the surface is rubbed gently to collect the sporangia. Most of the young sporangia discharge zoospores within 2 hr at room temperature. Cold treatment was not necessary for the release of zoospores (93).

Pochard and Chambonnet devised a quantitative method, suitable for making resistant individual plant selections (120). Young potted plants in the early flower bud stage are decapitated above the seventh or eighth leaf with a razor blade. The cut stem surface is inoculated with a 4-mm-diameter disk from a mycelial culture in a petri dish and protected from desiccation with aluminum foil. The plants are usually transferred to a growth chamber maintained at 22°C and 12 hr of fluorescent light of 900 lux at plant level. Measurements of the length of the stem necrosis were made to ±1 mm every third day. The rate of fungus advance is usually negatively correlated with the survival rate of seedlings by the root inoculation method. The authors demonstrated transgressive segregation in a cross of the moderately resistant cv. Phyo 636 with a susceptible cultivar. The hybrid possessed both a higher level and a more stable resistance than Phyo 636. The latter is similar to YW; it derives from the cross PI 201234 × YW, followed by three BC to YW and five selfed generations. Resistance determined by the above method must be confirmed by the seedling root inoculation method.

Pochard concluded that PI 201234 remains the best source of resistance so far identified. It has been used extensively to incorporate resistance into several French cultivars. He described resistance in F_2 as plurimodal, determined by at least two complementary dominant genes, their expression depending on the genetic background. Significant variety–isolate interaction showed that resistance was variety–isolate specific, not general. Mulato proved highly resistant in Bulgaria, both in the field and when artificially inoculated.

Southern Blight

This important disease of peppers caused by *Sclerotium rolfsii* is common in the tropics and subtropics. This organism usually exists as a saprophyte in the soil and expresses its pathogenic potential only under favorable conditions of nutrition, moisture, and temperature (30°–35°C). Typically the stem is invaded at the soil surface, where the fungus can be identified by its cottony mycelium and by its white and later dark brown sclerotia. Success in breeding for resistance is dubious because there are no well-defined levels of genetic resistance and the fungus is an omnivorous, facultative parasite. However, significant differences in survival of pepper cultivars have been reported (*43*), but the genetic basis and nature of this resistance remains obscure. In the currant tomato (*Lycopersicon pimpinellifolium*) PI 126932 resistance is associated with a thick cork periderm that develops when the plants are about 6 weeks old (*96*).

Southern Bacterial Wilt

This is an important disease of peppers and tomatoes in the tropics and subtropics as well as in parts of the southern United States. It is caused by the bacterium *Pseudomonas solanacearum*. In field tests on Guadeloupe, Kaan and Anais screened 35 pepper cultivars of diverse types for resistance (*82*). Of these, 18 were highly susceptible, including Florida Giant, Florida VR2, Puerto Rico Wonder, Titan, Yolo Wonder, Allbig, World Beater, Aconcagua, Agronomico 8, Avelar, Truhart Perfection Pimiento, Vinedale, Sunnybrook, and Sweet Cherry. Moderately resistant were Bastidon, Doux d'Espagne, Doux de Valence, Chinese Giant, Saint-Remy, Narval, and Largo Valenciano B209. These cultivars have large, elongated, thick-walled fruits. Cubanelle, Hungarian Sweet Wax, and Sweet Banana were also moderately resistant. Highly resistant were the more primitive cvs. Antibois, Chay 3, Conic, and two Cook lines derived from a cross of the Philippine cv. Bontoc with Pimiento. Local cultivars of *C. chinense* and *C. frutescens* on Guadeloupe appear unaffected by *P. solanacearum* and unlike most cultivars of *C. annuum* will set fruit at high night temperatures.

Root Knot

Hare reported that Santaka and a pepper labeled 405 B Mexico (both *C. annuum*) each carry a single dominant gene that confers resistance to the southern root-knot nematode *Meloidogyne incognita* and to its race *acrita*, the cotton root-knot nematode. He assigned the symbol *N* to both genes. From Hare's data it appears probable that this resistance was also effective against *M. arenaria* and *M. javanica*. He selected six homozygous resistant lines in F_3, three from crosses of Santaka with Ruby King and California Wonder Special, and three from crosses of 405 B Mexico with Burlington and Truhart Pimiento. The six lines were also highly resistant to the above two species. Early California Wonder, used as the check, was susceptible to all four nematodes, but Ruby King, Burlington, and California Wonder Special were only slightly susceptible to *M. arenaria* and resistant to *M. javanica*. All of the above-mentioned cultivars are susceptible to *M. hapla,* the northern root-knot nematode (*70*).

Screening for Root-Knot Resistance

Chopped Galled Roots Method

A simple method is to chop galled roots, preferably of okra, tomatoes, or squash, into small pieces with a hatchet on a wood block. Mix thoroughly 1 vol chopped roots with 4

vol moist field soil. (This is important to prevent decomposition of the roots by fermentation.) Place a sheet of newspaper in the bottom of a standard wood flat (21 × 14 × 3 in.)
and cover with a 1-in. layer of moist sand or of a peat–sand mixture (unsterilized) for
drainage. Level and tamp. Cover with a 1-in. layer of the root-knot inoculum; level and
tamp.

Finish with a 1-in. layer of moist, sterile medium. Sow seeds into the sterile medium in
five rows and cover with more sterile medium. Drench flats with a tribasic copper sulfate
plus NutriLeaf-60 solution (1 tsp of each per gallon) to control damping-off. Thin seedlings in cotyledon stage to 20–24 per row. Duration of the test should be about 40 days
from emergence. The temperature of the root medium must not exceed 85°C at any time
because higher temperatures cause a breakdown of resistance (76). Seedlings are classified visually into five classes: 1, heavily knotted; 2, moderately knotted; 3, slightly
knotted; 4, very few knots; 5, no knots. Classes 4 and 5 seedlings are resistant. True
breeding resistant F_3 lines are easily recognized. With 16 F_3 seedlings all resistant, P =
0.99 that the population is homozygous (159).

Egg Inoculation Method

Hussey and Barker tested several methods of egg preparation of *Meloidogyne* species
(78). Probably the best method, which gave a 70% egg hatch and 58% larval penetration,
was to treat the chopped galled roots in ⅕ strength Clorox for 4 min with agitation.
Shepherd drains the Clorox solution through a 200-mesh screen into a 500-mesh screen to
catch the eggs and then rinses off the digested roots under the tap (131). The eggs are left
standing in 1–2 liters of water for about 1 hr to free them of Clorox, and then rescreened.
One-milliliter aliquots are counted under the binocular microscope, and the egg suspension is diluted to 1000 eggs/ml. Pepper seedlings are inoculated in the cotyledon stage
with 10 cc each of the inoculum. Duration of the test is about 40 days from emergence.

VIRUSES

Viruses are among the more serious agents of disease in peppers, sometimes causing
whole fields to be abandoned prior to harvest. Virus RNA in the nucleus interferes with
the normal synthesis of chlorophyll, causing chlorosis and a mottling of the foliage known
as mosaic. Symptoms vary greatly in severity from a mild mottle to severe mottling, leaf
puckering, leaf distortion, shoe stringing, and extreme plant stunting (rosetting). Some
viruses cause systemic necrosis, which may be lethal. Only rarely is a lethal wilt induced,
as by TEV in Tabasco.

Resistance to Common TMV

Holmes' research with viruses laid the foundation for breeding TMV-resistant peppers
(74,75). He discovered the necrotic local lesion reaction to TMV infection in *C. frutescens* cv. *Tabasco* and in *C. annuum* cv. Minimum Blanco. Localized necrotic lesions
develop on the inoculated leaves in 3–4 days (Fig. 3.10). The lesions are interpreted as a
hypersensitive reaction to infection. The invaded cells promptly die, thus localizing the
virus in the lesions and preventing systemic spread. In 5–7 days the inoculated leaves turn
yellow and abscise, leaving the plant free of virus. In contrast, most cultivars of *C.
annuum* produce a systemic mottle disease known as mosaic. Holmes found that the local
necrotic lesion reaction was dominant in crosses with mottling reaction plants and inher-

FIGURE 3.10. Virus symptoms on tobacco. (A) Larger lesions induced by common TMV vs. (B) smaller lesions produced by SLTMV on *Nicotiana glutinosa;* (C) TEV-infected leaf and (D) healthy leaf of *Nicotiana tabacum.*

ited as a monogenic dominant trait. He assigned the symbol L to this gene and l to the recessive allele and transferred it from Tabasco to bell peppers. This important gene prevents systemic infection in the greenhouse and field by common strains of TMV.

Holmes found an additional gene at the L locus in Long Red Cayenne, in Sunnybrook, and in a selection from cv. Sweet Meat Glory. Plants with this gene produced yellowish primary lesions that tended to become necrotic, the necrosis spreading along the leaf veins. Leaf abscission would usually occur with recovery and freedom from virus, but a few plants would develop scattered secondary (systemic) lesions in noninoculated leaves. Holmes designated this gene l^i to denote the imperfect localization of the virus. The gene L was completely dominant over l^i and l, but l^i/l genotypes would respond with systemic necrosis, especially at higher temperatures (approx. 30°C), and all of the plants would die. Pochard reports that l^i acts as a completely dominant gene with most European tomato strains of TMV (Aucuba type) at 20°–25°C. The inoculated cotyledons abscise, thus freeing the plants of virus (*119*).

Several important commercial pepper cultivars have the l^i/l^i genotype, among them Keystone Resistant Giant, YW, Yolo Y, Florida VR2, VR2-34, XVR3-25, and the Dutch greenhouse cv. Verbeterde Glas. These cultivars are not infected by common strains of TMV under normal greenhouse and field environments. The allelic series of Holmes $L>l^i>l$ has been redesignated $L^3 > L^2 > l^1 > L^+$ by Boukema (Table 3.7) to include recent data and to conform to present rules for gene symbols (*15*).

Pepper Strains of TMV

McKinney isolated a pepper strain of TMV from a TEV-resistant pungent cultivar from South Carolina, Cayenne SC 46252 (*94*). This strain, unlike common TMV, produced no local lesions on this cultivar, but a systemic mottle disease. A second distinguishing feature was its failure to infect tomato (*Lycopersicon esculentum*). A third was that it appeared to be latent (symptomless) in Turkish Samsun tobacco. This virus has since been reported from Alabama, Florida, Louisiana, and south Texas. Greenleaf *et al.* designated it the Samsun latent strain of TMV (SLTMV) (*67*). Several strains of this virus have been identified by Rast in Holland (*123*) where it poses a threat to the considerable greenhouse production of bell peppers, which was 22,000 tons from 435 acres in 1977. SLTMV continued to spread despite soil steaming, seed and tool disinfection, and hand dips in skim milk (*150*). SLTMV stunts the plants, roughens the fruit, and reduces yield.

Resistance to SLTMV

An intensive search has been made to discover local lesion-type resistance to SLTMV similar to that so effectively provided by the L^2 gene (*L* gene of Holmes) against strains of common TMV. Greenleaf *et al.* screened 125 PIs of several species (*67*), and Simmons tested 1105 Accs., mostly of *C. annuum,* but including a few of *C. chinense* and *C. pendulum.* No local lesion response was found, but differences in susceptibility were noted. Simmons listed 28 PIs, of which some plants remained apparently symptomless after repeated inoculation (*137*). The progeny of 10 selected plants from PIs 297486, 179870, 174809, and 174111 showed intermediate levels of resistance expressed in a lower virus titer and mild symptoms (*10*).

Boukema (*15*) screened 524 PIs of 10 species with Rast's strains P8 and P11 and discovered 10 PIs of *C. chinense* that produced local necrotic lesions with strain P8. These PIs possessed a third gene L^3, allelic with Holmes *L* series. Breeding for L^3 resistance appeared promising until Rast isolated strain P14 in 1979, which overcame L^3 resistance (Table 3.7). It was concluded that oligogenic or polygenic sources of resistance were needed to control this mutable virus.

Additional TMV Strains in Peppers

A pepper strain of TMV that produces local lesions on *L* gene peppers that are followed by systemic infection but that will not infect tomato (*L. esculentum*) has been reported from Argentina (*56*); and a tomato strain that reacts similarly on peppers has been reported from Hungary (*38*). Of some 20 cultivars of several species tested, only *C. chinense* PI 159236 proved resistant. This PI responded with small necrotic local lesions that eventually resulted in leaf abscission and freedom from virus.

Other Pepper Viruses

In surveys of viruses in pepper fields in Louisiana, Sciumbato and Whitam found that cucumber mosaic virus (CMV), potato virus Y (PVY) and tobacco etch virus (TEV) accounted for over 90% of infected plants. Tomato spotted wilt virus (TSWV) and the pepper strain of TMV (SLTMV) made up most of the remainder. Common TMV was rare, and surprisingly, pepper mottle virus (PMV) was absent from pepper fields in Louisiana (*130,161*).

Villalon identified the same viruses in peppers in south Texas, plus PMV and tobacco

TABLE 3.7. Relation between Genotypes for Resistance in *Capsicum* and Strains of TMV[a]

	Genotype symbols		TMV strains			
				Pepper strains		
Accession	Original (Lippert *et al.*)	Proposed (Boukema)	Tomato (P_0)	P_{11} (P_1)	P_8 ($P_{1.2}$)	P_{14} ($P_{1.2.3}$)
C. annuum cv. Early Calwonder	L^+L^+	L^+L^+	+	+	+	+
C. annuum cv. Verbeterde Glas	L^iL^i	L^1L^1	−	+	+	+
C. chinense Ru 72-292	L^iL^i	L^1L^1	−	+	+	+
C. frutescens cv. Tabasco	$L\ L$	L^2L^2	−	−	+	+
C. chinense PIs[b]		L^3L^3	−	−	−	+

[a]After Boukema (*15*).
[b]PIs 152225, 159233, 159236, 213917, 215024, 224424, 257117, 257284, 315008, 315023.

ring spot virus (TRSV) (*157*). PMV was first identified by Zitter in Florida (*162,163*). PVY, TEV, and PMV are long flexuous rods, classified as potyviruses, named after the type virus PVY. The various strains of TMV are long straight pods, classed as tobamoviruses. CMV, the type virus of the cucumovirus group, is isohedral (*25*).

Sources of Virus Infection

Weed Hosts

Mechanical transmission of viruses from weeds to peppers is generally difficult. Whitam transferred CMV, PVY, TEV, and TSWV to pepper from 18 species of weeds (*161*). The largest number of successful transmissions were from *Solanum nigrum* (black nightshade) 26/42; *Medicago anabica* (spotted bur clover) 15/85; *Rudbeckia amplexicaulis* (Blackeyed Susan) 12/54; *Melilotus officinalis* (yellow sweet clover) 10/51; *Geranium carolinianum* (cranesbill) 8/40; and *Senecio glabellus* (butterweed) 5/69. Of 615 attempted inoculations, 47 yielded CMV, 45 PVY, 21 TEV, and 7 TSWV.

Crop Plant Hosts

Sciumbato assayed cultivated crop plants growing near pepper fields and mechanically transferred the following viruses to peppers: CMV, TEV, TMV, and potato virus X (PVX) from tomato; CMV from cantaloupe; CMV and TEV from eggplant; TMV, TEV, and PVX from tobacco and PVX from mustard.

Virus Transmission by Aphids

CMV, PVY, TEV, and PMV are transmitted from diseased weeds or crop plants to peppers by aphids, in a nonpersistent stylet-borne manner. Common vectors of CMV in Louisiana were *Myzus persicae* (green peach aphid) and *Aphis gossypii,* and of PVY and TEV, *M. persicae* and *Aphis craccivora*. CMV is reported to be transmitted by over 60 species of aphids (*25*). TSWV is transmitted by thrips.

Seed Transmission of Viruses

Transmission of common TMV in tomatoes and of SLTMV in peppers through the endosperm of the seed is rare, and the embryos of both species appear to be immune.

Seedling infection may occur from surface-contaminated seed during seedling emergence (*36,45,153*). However, Sciumbato found that only 1–2% of Tabasco seedlings became infected by SLTMV during germination when left undisturbed, but that handling during transplanting and especially pruning the plants to make them bushy, resulted in nearly 100% infection. Seed treatment removes this source of infection. CMV, PVY, TEV, and PMV are not seed transmitted.

Yield Reduction from Viruses

Sciumbato reported yield reductions from SLTMV-inoculated Tabasco of 38 and 22% in two tests, and from TEV in bell peppers and Cayenne of 23 and 21%, respectively. CMV reduced the yield of bell pepper by 97% and of Cayenne 61% (*130*).

Villalon reported similar yield reductions in bell peppers inoculated with four viruses prior to field planting (Table 3.8). Field losses were similar for jalapeno and chili peppers (*157*).

TABLE 3.8.   Yield Reduction in Six Bell Pepper Cultivars Caused by Four Viruses[a]

Variety	Virus	Tons/acre[b]	Yield reduction (%)
Tamu Bell 7506	Check	9.6a	—
	TEV	8.1b	15.6
	SLTMV	7.0c	27.1
	PMV	5.7d	40.6
	PVY	4.8d	50.0
Lucky Green Giant	Check	8.4a	—
	TEV	5.2a	38.1
	SLTMV	4.2bc	50.0
	PVY	3.7c	56.0
	PMV	2.0d	76.2
VR-2	Check	6.4a	—
	TEV	5.8a	9.4
	PMV	3.3b	48.4
	PVY	3.2b	50.0
	SLTMV	2.9b	54.7
Keystone	Check	7.7a	—
Resistant	TEV	6.5ab	15.6
Giant #3	PMV	5.8bc	24.7
	PVY	5.1bc	33.8
	SLTMV	4.5c	41.6
Delray Bell	PVY	6.7a	+1.5
	Check	6.6a	—
	TEV	6.3a	4.5
	PMV	6.0a	9.1
	SLTMV	3.5b	47.0
Pip	Check	4.9a	—
	PVY	3.6b	26.5
	SLTMV	3.3b	32.7
	TEV	3.3b	32.7

[a]After Villalon (*157*).
[b]Figures not followed by the same letter differ significantly at $P = .05$ according to Duncan's multiple-range test.

Virus Inoculation Methods

Carborundum Leaf-Wiping Method

This is the standard inoculation method for mechanically transmissible viruses, e.g., TMV, TEV, PVY, PMV, and CMV. The procedure outlined is to serve only as a guide. Virus experiments require careful planning in advance. Young, vigorously growing test plants are needed, e.g., peppers 5–6 weeks and tobaccos 7–8 weeks from seed. The plants are spaced out on the greenhouse bench and labeled in advance. The leaves to be inoculated are marked and dusted with 400- to 600-mesh carborundum. Avoidance of chance virus contamination by contact or by aphids is critical, and the hands need to be washed with soap and water between operations.

Water checks are always inoculated first. Inoculum is prepared by grinding leaves or whole shoots of infected source plants in a mortar or blender. A weighed sample is ground in a measured volume of water or phosphate buffer (KH_2PO_4, 0.01–0.1 M, pH 7.0–7.5) to facilitate grinding and to determine the dilution factor. The crude extract is filtered through cheesecloth. TMV extracts are used at $\frac{1}{100}$ and TEV, PVY, PMV, and CMV at $\frac{1}{10}$ to $\frac{1}{20}$ dilution w/v.

Transmission difficulties with CMV due to virus inhibitors in pepper sap were overcome by Pochard by the use of an extraction solution containing 0.025 M phosphate buffer, pH 7.0, 5% sodium bisulfite, 1.7 g/liter sodium diethyldithiocarbamate, 0.5% caffeine, and 100 mg of activated vegetable charcoal per cubic centimeter of inoculum.

Test plants are each inoculated in three to four leaves. A wad of cotton or a cloth pad saturated in the inoculum serves as a wiper. The leaf is supported with one hand and is wiped gently three or four times. The plants are rinsed off promptly with tap water when the inoculations are completed. Symptoms develop in 3–10 days, depending on the virus, and maximum virus titer is reached in about 14 days.

Spray Gun Method of Inoculation

This method works well with highly infectious viruses like TMV or SLTMV, but is not as reliable with the less infectious viruses TEV, PVY, PMV, and CMV. Carborundum must be added to the inoculum, about 5% by volume. The plants are sprayed forcefully from a distance of 3–4 in. at 60–100 psi. With this method hundreds of 5- to 6-week-old seedlings in flats can be inoculated in a few minutes. Villalon uses an artist's airbrush at 125 psi. He considers this method only 90% as reliable as the leaf-wiping method but satisfactory for large-scale screening (*158*).

Frozen Storage of Viruses

Cook maintains CMV at a high titer in tobacco by making transfers every 5–7 days. For prolonged storage he freezes finely cut leaves in small test tubes loosely plugged with cotton and placed over $CaCl_2$ within larger rubber-stoppered tubes. In such storage he estimates the longevity of CMV to be 6 months, the potyviruses 1 year, and TMV 10 years (*30*). Crude extracts of TMV diluted $\frac{1}{10}$ to $\frac{1}{50}$ can also be stored frozen for several years.

Virus Identification

Identification and maintenance of a virus on a suitable host is essential before screening tests for resistance can begin. Identification can be difficult and may involve the cooperative efforts of the breeder, a virologist, an electron microscopist, an entomologist, and an immunologist (to produce antiserum), as well as the help of technicians. For such a team to function properly would require the development and formal approval of a cooperative research project, so that all the scientists can officially justify their involvement.

Differential Host Reactions

This is an important method of virus identification. The unknown virus is inoculated into several tester hosts. Symptoms are compared with those induced by known viruses on the same hosts. Characteristic symptoms on one or more hosts matching those of a known virus can be diagnostic for the unknown virus, e.g., Tabasco wilt for TEV. To identify a new virus requires additional tests.

The identification by Zitter of the new potyvirus PMV in south Florida by this method, supplemented by immunodiffusion tests, is shown in Table 3.9. He separated the virus isolates from peppers into four basic types: TEV-C, a common mild strain; TEV-S, a severe strain; PVY-C, a common mild strain; and PMV (*9,162*).

Cross Protection

This is a sensitive biological test of relationship between strains of the same virus. One strain must be of a known virus that produces local lesions on a tester host in which the unknown virus is systemic. If the two viruses are closely related, the presence of the one systemic in the host will *protect* against entry by the challenge virus, and no local lesions will appear. Examples are common TMV vs. SLTMV on Pimiento Bighart (*69*), and

TABLE 3.9. Reaction of Pepper Cultivars to Florida Potyviruses[a]

Cultivars	*Datura stramonium* positive		*Datura stramonium* negative	
	TEV-C	TEV-S	PVY-C	PMV
Early Calwonder	M	M	M	M
PI 264281	R	M	I	M
SC 46252	R	M	I	M
23-1-7	R	M	I	M
Yolo Y	M	M	I	M
23-1-7 × Yolo Y	R	M	I	M
Avelar	R	MM	I	MM
Agronomico 9	R	MM	I	MM
Ambato Immune	R	MM	I	MM
PI 342947	R	M	I	M
PI 152225	R	R	I	I
PI 159236	R	MM	I	I
Serrano Acc. 2207	R	M	I	M
PI 281367	R	M	I	M
Tabasco	MW	MW	MM	LM
Greenleaf Tabasco	R	R	I	I

[a]Susceptible reaction as determined visually, by indexing on California Wonder or serological tests. TEV-C and TEV-S are, respectively, mild and severe strains of TEV. R, Resistant; I, probably immune; M, chlorotic mottle; MM, mild mottle; MW, mottle and wilt; LM, local lesions on inoculated leaves and occasionally on systemically infected leaves; systemic mottle, with leaf and stem necrosis on young plants, followed by death or regrowth from below the necrotic stem region. [Adapted from Zitter (*162, 164*) and Black (*9*).]

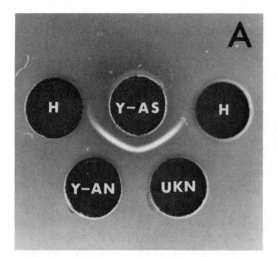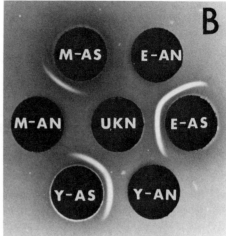

FIGURE 3.11. Immunodiffusion patterns and techniques for the identification of tobacco viruses. (A) Identification of a known virus as PVY. Y-AS, Antiserum to PVY; H, juice from a healthy plant; Y-AN, juice from a PVY-infected plant; and UKN, juice from a plant infected with the unknown virus: (B) Identification of viruses in a doubly infected plant. UKN, Juice from a plant suspected to be infected with PVY, TEV, TVMV, or a combination of these; Y-AS, antiserum to PVY; Y-AN, juice from a plant infected with PVY; E-AS, antiserum to TEV; E-AN, juice from a plant infected with TEV; M-AS, antiserum to tobacco vein mottling virus (TVMV), M-AN = juice from a plant infected with TVMV. AN = antigen (virus); AS = antiserum.
After Gooding (*61*).

Fulton's local lesion *Vinca* strain of CMV vs. virulent systemic strains of CMV on peppers (*59*).

Serology

Definitive identification of a virus may require serological methods as well as electron microscopy. For serological tests, antisera to known viruses are prepared against which the unknown virus (antigen) is tested. Antisera are produced by injecting partially purified virus into rabbits, taking blood samples after a suitable time interval and separating the serum as described by Ball (*4*). (Virus antisera can be obtained from American Type Culture, Rockville, Maryland.) Some viruses are more strongly immunogenic than others, meaning that they produce stronger precipitin reactions and hence a more reliable test. Among the available methods of identification, namely, particle morphology, physical–chemical properties, inclusion bodies, host range, symptomatology, cross protection, and antigenic homology, Gooding considers the last mentioned *the most reliable parameter currently known that can be used alone for identification* (*61*). The main advantage of the agar-gel double-diffusion technique is that *the known virus, the unknown, and control extracts from healthy plants are tested in the same system* (Fig. 3.11).

Enzyme-Linked Immunosorbent Assay (ELISA)

This sensitive immunological technique for plant virus identification was originally developed for human and animal virus diseases, but has also been shown to have wide applicability to plant viruses. In this test, antigen (virus), antiserum, and a serum or antigen-specific enzyme are successively adsorbed to a special plastic microplate. When the enzyme substrate is added, a color develops, the intensity of which is proportional to the degree of homology of virus and antigen (52).

Inheritance of Virus Resistance

Potyvirus Resistance Alleles

TEV resistance in *C. annuum* Cayenne SC 46252 is recessive and monogenic. A few plants remained symptomless after repeated inoculation and gave negative recovery tests on Tabasco, indicating that modifier genes increased the resistance level to near immunity. Resistance in *C. chinense* PI 152225 (identical with PIs 152233 and 159241) was likewise recessive and monogenic. Resistance in both species was expressed in a reduced rate of virus multiplication in plant tissues as compared with susceptible hosts, rather than immunity. The symbols et^a and et^c were assigned to the respective resistance genes to denote their species origin (64). Cook subsequently demonstrated monogenic recessive resistance in *C. annuum* PI 264281 (P11) and in SC 46252 (P34) to a common strain of etch (TEV-C) and apparent immunity of both cultivars to PVY-N^{YR}, a more virulent mutant of PVY-N.

The genes in P11 and P34 were allelic and apparently identical with et^a (Cook's invalid gene ey^a) (26). Cook later discovered a single PVY-N immune plant in YW, which was the progenitor of Yolo Y. This plant possessed a single recessive gene, which he designated y^a. This gene proved to be allelic with et^a (27). Strain PVY-N^{YR} subsequently appeared in Yolo Y (28). Cook had thus shown that et^a was allelic with y^a, but did not distinguish clearly between et^a, ey^a, and y^a.

Zitter and Cook reported a third gene that conferred resistance to PMV in *C. annuum* cv. Avelar from Brazil. This gene was allelic with and dominant over et^a. Because Avelar progeny that were tolerant to PMV were also resistant to TEV and PVY, a single gene in Avelar must confer resistance to all three viruses (165). This allele, here designated et^{av}, has a higher potency than et^a, which protects only against TEV-C (common strain) and PVY-N^{YR}. Zitter reported similar reactions to TEV-C and PMV in PI 159236 and Avelar but observed a higher level of resistance to PMV in this PI (162,163). Subramanya elucidated the genetics of this difference in the cross Delray Bell (et^{av}) × PI. Surprisingly, the F_1 was susceptible to PMV and the F_2 and both BC segregated 1 resistant : 1 susceptible (149). The interaction of the resistance alleles of these two cultivars resulted in susceptibility, similar to that observed in the heterozygotes L^1/L^+ (l^i/l of Holmes) in peppers and $TM2^a/TM2^+$ in tomatoes when inoculated with TMV. The published data therefore support the existence of an allelic series of resistance genes to the Florida potyviruses. Only the dominance relationship and allelism of the genes in PI 159236 and PI 152225 remain to be determined (Table 3.10). The symbols et^{c1} and et^{c2} (formerly et^c of Greenleaf) are tentatively assigned to these two genes. The reason for the choice of the etch symbol et for these alleles (except for Cook's gene y^a) is that the inheritance data show that genes for resistance to TEV also confer resistance to PVY. An earlier report by Bawden and Kassanis that TEV replaces PVY in mixed infections supports this conclusion (7). Resistance conferred by the gene et^a was effective against the Guadeloupe strain but not the Puerto Rico strain of PVY (28,82).

Barrios *et al.* (5) reported a dominant gene with resistance to TEV in LP-1, a *C. frutescens* cv. LP-1 remained symptomless when inoculated with TEV. The F_1 hybrid LP-1 × Tabasco also remained symptomless when inoculated, showing neither mottle nor wilt symptoms. The F_2 segregated 98 mosaic and wilt resistant to 42 wilted plants ($\chi^2 =$

TABLE 3.10. Resistance Alleles to Florida Potyviruses[a]

Virus strains of increasing virulence (left to right)			
PVY-C	TEV-C	PMV	TEV-S
PVY-N	PVY-N[YR]		

Sources of resistance			
Yolo Y	PI 264281	Avelar	PI 152225
	SC 46252	Delray Bell	Tabasco G
	VR2	PI 159236	

Increasing potency and dominance of resistance alleles (left to right)

$$y^a \quad < \quad et^a \quad < \quad et^{aw} \nless et^{c1} \quad < \quad et^{c2}$$

[a]PVY-N, mild strain from tomato, Yolo Y is immune; PVY-N[YR], mutant strain of PVY-N that infects Yolo Y; PVY-C, mild strain from Early Calwonder, Yolo Y is immune; TEV-C, common mild strain from Early Calwonder, Avelar is immune; TEV-S, severe strain from Avelar that induces a mild mottle in this variety as does PMV. PI 159236 is rated resistant to PMV, whereas Avelar is only tolerant. The symbol $\nless$ means "not dominant," the heterozygote et^{av}/et^{c1} being susceptible to PMV. Allelism and dominance relationship of et^{c1} and et^{c2} with respect to TEV-S are not established.

1.86, $P = .10-.20$ for a 3:1 ratio). The F_1 hybrid LP-1 × Almeda (*C. frutescens*, exhibits a TEV susceptible mottling response) also remained symptomless. The F_2 segregated 160 mosaic mottle resistant to 65 mottled plants ($\chi^2 = 1.81, P = .10-.20$ for a 3:1 ratio). The nature of LP-1 resistance, immunity vs. a low titer symptomless tolerance, was not investigated.

Cucumber Mosaic Virus Resistance

Rusko and Csillery report that CMV is the most common and destructive virus in Hungary and commented that, despite serious losses worldwide, few breeding programs have focused primarily on this virus (*125*). The reason is that the task is difficult because inheritance of resistance is polygenic and varies with plant age and strains of the virus. For example, young Tabasco plants up to 8 weeks old are invaded systemically and develop leaf and stem necrosis and, if they survive, produce distorted, mottled new growth. In contrast, plants 10 weeks or older are able to localize the virus in the older inoculated leaves or in young lateral shoots. In mature plants the virus is not a serious problem.

According to Pochard, no complete resistance to CMV has been discovered in peppers, not even to particular strains of the virus (*116*). He believes that a higher, more durable resistance will require a combination in a single genotype of three kinds of resistance, which he designates Ra, Rb, and Rc. Ra confers ability to escape infection if the inoculum dose is low. Its detection employs Fulton's avirulent *Vinca* strain of CMV, which produces only local necrotic lesions on peppers in numbers presumably proportional to their susceptibility to more virulent systemic strains (Table 3.11). In this, the NF (necrotic Fulton strain) test, young pepper plants are decapitated above the seventh leaf, inoculated in the sixth or seventh leaf, and the number of local lesions produced counted.

Rb is a hypersensitive-type resistance that localizes the virus through necrosis of the invaded tissue. It is detected in plants decapitated above the fourth leaf and inoculated in the third leaf with a virulent local strain of CMV. In susceptible lines the third inoculated leaf becomes necrotic and abscises. Both the third and fourth axillary shoots that develop within a month become necrotic, whereas in hypersensitive lines the fourth axillary shoot

TABLE 3.11. **Resistance Levels to CMV in *Capsicum*
                Species and Cultivars Determined by the
                Local Lesion Test with Fulton's
                Vinca rosea Strain**[a]

Accession	Average number of lesions/leaf
C. chacoense	85
C. chinense Acc. 2	71
C. frutescens Acc. 10	50
C. pubescens Acc. 2	43
C. annuum cv. Bighart Pimiento	42
C. baccatum Acc. 5	31
C. annuum cv. Hatvani	31
C. annuum cv. Yolo Y	31
C. praetermissum Acc. 1	27
C. annuum cv. Javitott Cecei	6.3
C. frutescens Acc. 11	1.7
C. baccatum Acc. 2	1.1
C. frutescens Acc. LP1	0.8
C. annuum Acc. Perennial (Singh)	0.03

[a]After Rusko and Csillery (*125*).

usually remains symptomless and virus free. The absence of virus can be checked by necrotic local lesion tests on *Vigna sinensis* (NV tests).

Rc resistance is non-necrotic and is expressed in a slow rate of virus multiplication in the inoculated sixth or seventh leaf. The degree of resistance is indicated by the number of local lesions produced in periodic subinoculations to *Vigna sinensis*. Decapitation above the seventh leaf is advantageous for this test, as abscission of the inoculated leaf is then delayed for over a month as compared with 7–8 days when decapitated above the fourth leaf.

Field selection for CMV resistance is not reliable, because the amount of inoculum is variable and other viruses may also be present. Cultivars classified by type of resistance were *C. baccatum* (Rb), Moura (Ra, Rc), Ikeda (Rc), Val (Rb), and LP1 (Ra, Rb). The field tolerance of the Brazilian cvs. Moura, Ikeda, and Avelar and of other small-fruited introductions is difficult to assess by the above tests. Rb test symptoms on them were severe. One hope resides in the possible association of resistance in these peppers with the unique, large, unpigmented lesions they produce with the Fulton strain. Pochard observed transgressive segregation for CMV resistance in the cross Val × Babas, both of Rb type. Val derives from Antibois and Babas from the cross *C. baccatum* × Bastidon. The breeding objective is to combine the genes that determine the different identifiable mechanisms of resistance.

SOURCES OF GERMPLASM FOR BREEDING

The largest collection of *Capsicum* species in the United States is maintained at the Southern Region Plant Introduction Station, Experiment, Georgia. Professional breeders around the world have access to this collection of about 2500 PIs. Many have been screened for disease resistances and horticultural characteristics (*109*).

BREEDING METHODS

A survey of the methods by which pepper cultivars have been developed reveals that the following were used: (1) pedigree breeding with selections from superior cultivars; (2) pedigree breeding following hybridization between superior cultivars; (3) transfer of single genes from primitive cultivars or wild forms to leading cultivars by the BC pedigree method; (4) intercrosses between different BC families with different recurrent parents and with different target genes from diverse germplasms to combine several disease resistances and new horticultural traits.

Backcross–Intercross Scheme for Breeding PMV- and CMV-Resistant Bell Peppers

Assume that *C. chinense* Acc. 1555 proved immune to PMV in preliminary screening tests and carries either the gene et^{c1} or et^{c2} that we wish to transfer to Cook's large-fruited, multiple-disease-resistant bell pepper XVR3-25. Additional desirable traits that could be transferred from Acc. 1555 are a round fruit shape (gene O), earliness, ripe-fruit rot resistance, and multiple fruitedness. We also wish to transfer the resistance to CMV from Perennial. If preliminary Ra, Rb, and Rc tests were to show that Acc. 1555 also has a high level of CMV resistance, the cross with Perennial would not be needed. A breeding procedure designed to combine the resistance genes for PMV and CMV is shown in Fig. 3.12.

Diallel Crosses and Pepper Improvement[1]

Conventional plant breeding may involve selection of superior individuals from a variable population, hybridization among selected parents followed by pedigree selection, or character introduction into a population by BC breeding. Quantitative plant breeding generally involves crosses among a number of selected parents to initiate a program of general population improvement. Pepper traits that are inherited quantitatively, or whose expression depends upon the accumulation of many genes each contributing small increments to the total expression, include fruit size, fruit yield, carotenoid content, and adaptation to environmental conditions.

Problems to be considered by the quantitative plant breeder are (1) which parents and how many parents to include in matings to provide desirable genetic input, and (2) how to handle the combined population throughout the breeding program to increase the expression of the desired traits.

The diallel cross offers one approach to the evaluation and selection of parents to be combined to form a genetically variable population. It also provides information on genetic control of quantitative traits, which is useful in choosing the breeding procedures that will accomplish the desired improvements. Additionally, the diallel mating scheme provides valuable information on heritability and heterosis, and the results may be used in predicting performance of synthetic populations that may be formed from various combinations of parents.

Simply stated, a diallel cross involves the mating of selected parents (p) in all possible two-parent combinations, with evaluations made on the resulting hybrids. A complete diallel square consists of p^2 crosses, including selfs and reciprocals. A more commonly

---

[1]This section was written by Dr. L. F. Lippert, University of California, Riverside, and is presented with his permission.

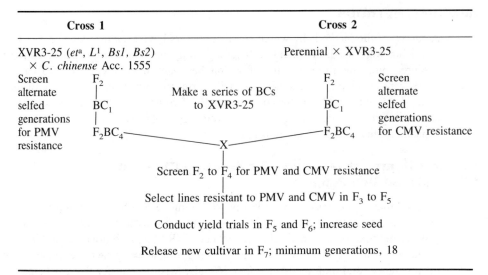

Cross 1	Cross 2

XVR3-25 (et^a, L^1, $Bs1$, $Bs2$)   Perennial × XVR3-25
  × *C. chinense* Acc. 1555

Screen F_2 F_2 Screen
alternate | Make a series of BCs | alternate
selfed BC_1 to XVR3-25 BC_1 selfed
generations | | generations
for PMV F_2BC_4 F_2BC_4 for CMV resistance
resistance   ─X─

Screen F_2 to F_4 for PMV and CMV resistance

Select lines resistant to PMV and CMV in F_3 to F_5

Conduct yield trials in F_5 and F_6; increase seed

Release new cultivar in F_7; minimum generations, 18

FIGURE 3.12. Backcross–intercross scheme for breeding PMV- and CMV-resistant bell peppers.

used diallel scheme eliminates selfs and reciprocals and evaluates $p(p-1)/2$ hybrid combinations.

Results from a nine-parent diallel cross in chili pepper were reported by Lippert (*87*) and Marin and Lippert (*90*). Parents were selected pepper cultivars and breeding lines representing mainly Anaheim chili types, but including one high-color Mexican chili type (60M4) and one small-fruited chili (Red Chili) with high fruit number per plant. The purpose of the diallel was to determine types of gene action controlling characters of importance in chili production and to evaluate the nine parents for combining ability, with eventual selection of parents for incorporation into a synthetic population with broad genetic base.

Biparental hybrids and parental selfs were made in the greenhouse and the diallels of 36 hybrids and nine parents were tested during two seasons in southern California. Records included fruit number per plant, dry weight yield per fruit and per plant, percentage of mature fruit at harvest, fruit length and width, and total carotenoid content of the dry fruits. Fruits from this study were also analyzed for pod component percentages by separation into endocarp, exocarp, stem, septa, and seed. The goal of a breeding program would be to maximize the percentage of endocarp, which contains the valuable carotenoid pigments, without reduction of total dry-fruit yields.

The evaluation of parents was accomplished by comparing the performance of each parent in hybrid combination with all other parents. This evaluation is termed general combining ability (GCA) and relates to the additive gene effects within the population. The performance of individual F_1 hybrids compared to average performance of the parent lines involved is specific combining ability (SCA) and relates to nonadditive gene effects. The combining ability effects have the following association in the performance of a particular hybrid:

$$x_{ij} = \mu + g_i + g_j + s_{ij},$$

where x_{ij} is the hybrid performance for a given trait, μ the overall population mean, g_i and g_j the GCA effects for parents i and j, and s_{ij} the SCA effect for hybrid x_{ij}.

Analyses of variance of the diallel data provided significant differences among hybrids for fruit number per plant, dry weight per fruit, fruit length and width, and total carotenoids, as well as for all components for the total dry pod. Separation of the among-crosses variation into GCA and SCA indicated that additive gene effects (GCA) were more important than nonadditive effects (SCA), both in magnitude of variances and significance levels. GCA effects of each parent for four characteristics evaluated in this study are presented in Table 3.12. High positive values for dry-fruit weight and carotenoid levels relate to desirable parental performance. High positive values for endocarp component coupled with negative values for seed indicate desirable parents for those pod components.

Selection of parents for inclusion in a mating program to improve characteristics can be assisted by predicting the performance in synthetic populations from the combination of various parents. The formula for these predictions is

$$\text{performance of synthetic} = \overline{F}_1 - (\overline{F}_1 - \overline{P})/n,$$

where $\overline{P}$ and $\overline{F}_1$ are mean values of selected parents and hybrids between these parents, and n is the number of parents incorporated into the synthetic.

Performance was calculated for a synthetic population based on the mating of three parents, Sweet Pickles, Gentry 456, and 60M4, each selected for high carotenoid content in the dry fruits and for high GCA for this trait. Performance of a four-parent synthetic was predicted with the addition of 57M75, a fairly large-fruited type, but with low GCA for carotenoids, to estimate increase in fruit size and dry-fruit yields. Values of seven characters predicted for these two synthetics are shown in Table 3.13.

Several mating designs and selection schemes may be used to improve the expression of quantitative traits in genetically variable populations. Recurrent selection is effective for traits under additive gene control. Superior individuals identified from an original population are intermated to form a population for the next cycle of selection. Reciprocal

TABLE 3.12.　General Combining Ability Effects for Parental Entries for Characters Measured on the Two-Season Performance of F_1 Hybrids in a Nine-Parent Chili Pepper Diallel Experiment[a]

Parent	Dry-fruit wt./plant	Carotenoid content	Pod components	
			Endocarp	Seed
Sweet Pickles	0.53	749	−0.70	0.75
College 6-4	−5.41	− 41	1.30	−2.57
Oxnard Chili	−4.37	−315	0.21	−0.19
Red Chili	5.27	26	−7.14	7.83
Lark	−1.60	−827	−0.71	−0.05
Gentry 456	3.41	711	0.77	1.11
60M4	2.96	567	2.22	0.05
57M30	−2.65	−488	1.64	−3.99
57M75	1.86	−383	2.39	−2.95

[a]Adapted from Lippert (87) and Marin and Lippert (90).

TABLE 3.13. Predicted Performance for Synthetic
Populations Formed from Three and Four
Parents Selected for Total Carotenoids[a]

	Predicted synthetic performance	
Character	Three-parent[b]	Four-parent
Total fruit/plant	31.8	31.0
Mature fruit (%)	76.5	79.6
Dry wt/plant (g)	96.2	99.6
Dry wt/fruit (g)	3.95	4.04
Fruit length (mm)	102	110
Fruit width (mm)	29.5	28.6
Total carotenoids (μg/g)	6415	5945

[a]After Lippert (87).
[b]Sweet Pickles, Gentry 456, and 60M4 comprise the three-plant synthetic,
with 57M75 added into the four-parent synthetic.

recurrent selection or reciprocal full-sib selection provides simultaneous evaluation and selection within two populations, with one population tested against the other in each cycle of selection. These schemes have the advantage of improving populations for traits responding to both additive and nonadditive gene action, that is, improvement for both GCA and SCA. Each of these breeding programs has application to a self-pollinated crop such as pepper.

Genic Male Sterility

Genic male sterility is useful for making F_1 hybrids because tedious and costly hand emasculation of individual flowers is avoided. Also, F_1 hybrid peppers are generally more vigorous and more uniformly productive than open-pollinated cultivars. Male-sterile plants are found as mutants in about 0.01% of the plants in large fields. Shifriss found one completely pollen-sterile plant among 24 unfruitful mutants of various types selected from a field of 10,000 plants of the squash pepper cv. Gambo. Earlier, Shifriss and co-workers found one male-sterile plant each in the bell peppers Allbig and California Wonder. Meiosis in the male steriles appeared to be normal, but the microspores degenerated soon after the tetrad stage, and no fertile pollen was present at anthesis when examined under the microscope in 1% acetocarmine or acetoorcein (132).

The three *ms* genes mentioned proved to be nonallelic. The practical value of male-sterile mutants depends on the economic importance of the parent cultivar, on the degree of sterility and its stability, and on the combining ability with other cultivars in F_1 hybrids.

Because male steriles constitute only 50% of the plants (*ms/ms* $\times$ *ms$^+$/ms*), one half must be eliminated in the seedling stage, before transplanting to the field. This has not been feasible because closely linked marker genes have not been found and the *ms* genes have no obvious phenotypic effect prior to flowering. However, male-sterile plants are easily identified at anthesis and hand pollinated in an insect-screened greenhouse.

Use of Genic Male Sterility

Pochard reports only limited use of genic male sterility to produce F_1 hybrids in Bulgaria, France, and Yugoslavia. The limited use is due to abnormal anther development in ms/ms^+ hybrids at low temperatures and consequently poor pollination (118). On the positive side, however, Breuils and Pochard produced the popular F_1 hybrid Lamuyo–INRA by the use of the gene ms 509. Further, Shifriss writes that the Dutch seed company Bruinsma released a true ms/ms^+ hybrid in 1980, which carried an ms gene allelic with one of his ms mutants ($17,133$). Bruinsma Wonder, an F_1 hybrid greenhouse cultivar, is a cross of CW (ms/ms) × Sweet Westland, released earlier. Shifriss did not observe the anther abnormality in Israel and stated that they would use this technique in the future.

Cytoplasmic–Genic Male Sterility

Cytoplasmic male sterility (CMS) in peppers was first discovered by Peterson in *C. annuum* PI 164835 from India (110). Its modus operandi proved to be similar to that found in onions by Jones and Clarke (80). This type of male sterility is due to the interaction of sterility inducing S-type cytoplasm with a recessive nuclear male sterility inducing ms gene. The ms gene is only expressed in S cytoplasm. The only plasmon–genome combination that induces male sterility is Sms/ms. The other combinations Sms^+/ms, Sms^+/ms^+, Nms/ms, Nms^+/ms, and Nms^+/ms^+ all produce fertile pollen. The CMS system for producing F_1 hybrids has an advantage over the genic system for mass hybridization, because all of the female parent plants are male sterile as compared with only 50% with the ms-gene method. Three parent lines are required to produce F_1 hybrid seed by the CMS system: the male-sterile or A line Sms/ms, the maintainer line Nms/ms, and a fertility restorer or C line Nms^+/ms^+ or Sms^+/ms^+. The male-sterile line is maintained by hybridization, in isolation, with a maintainer line of the same cultivar. In the field, crossing would need to be by bees, supplemented by hand pollination during periods of maximum flowering. The commercial hybrid would be produced by crossing the Sms/ms line with a good combiner line or cultivar of similar type to generate heterosis. Unfortunately, Peterson's Sms/ms lines were unstable in fluctuating environments, producing pollen at lower temperatures, and hence could not be relied upon to produce F_1 hybrids.

Synthesis of Additional Cytoplasmic Male Steriles

Peterson grouped pepper cultivars according to the presence of restorer genes (ms^+) or of non-restorer genes (ms) (110). Shifriss and Frankel expanded this list (Table 3.14) (134). Duvick suggested that CMS could probably be found or induced in most crop species (51). On this premise and with a need for more stable CMS genotypes, Shifriss and Frankel searched for and discovered two additional sources of S-type cytoplasm in two hot peppers from India, PI 154-1 and PI 164682 (134). When crossed onto the bell pepper Yolo Y (Nms/ms), both F_1 hybrids were pollen fertile, but two F_2 populations, one from each cross, segregated ¼ male-sterile plants. Proof of CMS vs. genic sterility were the 1:1 and 1:0 ratios of male-fertile to male-sterile plants obtained from reciprocal crosses of the two types of cytoplasm (Sms^+/ms × Nms/ms). Peterson's S cytoplasm appears to be identical with the two new S cytoplasms because cultivars known to be restorers or non-restorers reacted similarly in crosses with all three cytoplasms. Ohta obtained similar results with the S cytoplasms of Fresno Chili, Delaware Bell, and Liberty Bell (108).

Shifriss and Guri incorporated the two new cytoplasms into the sweet pepper cvs Bikura, Yellow Yolo Y, and Zohar with six to eight BC and tested the stability of their pollen sterility in the field under conditions of natural cross-pollination. Recessive marker genes in the female parents permitted determination of the percentage of plants produced

TABLE 3.14. Male-Sterile and Restorer Genes in Pepper Cultivars[a]

Male-sterile gene (*ms*)		Restorer gene (*ms*+)	
Yolo Wonder	Fresno Chili	Floral Gem	PI A-J
Yolo Y	Long Red Cayenne	Anaheim Chili	PI A-8
California Wonder	PI 206421	California Chili	PI 154-1
Delaware Bell	PI 164835	Mexican Chili	PI 164682
Liberty Bell		Fushimiamanaga	PI 164738
Naharia		Takanotsume	PI 164847
Vinedale		Jalapeno	PI 195557
Ojishi		Puri Red	PI 201228
Pimiento		Serrano	PI 201231

[a]Adapted from Peterson (*110*), Shifriss and Frankel (*134*), and Ohta (*108*).

by self-pollination. The crosses and the percentages of selfed plants were Bikura (S*ms*/*ms*) × Yolo Y (S*ms*+/*ms*+), <1%; Zohar × Yolo Y, 28%; Zohar × Maor, 28%; Yellow Yolo Y × Maor, 18.5%. These results show that the degree of pollen sterility and stability of different CMS lines depends not only on the S*ms*/*ms* component but also on the rest of the genotype. The authors concluded that cultivars with the high pollen sterility and stability of Bikura, with <1% selfing, can serve as A lines in the production of F$_1$ hybrid seed by natural cross-pollination. Further improvement of CMS lines, now variable in pollen sterility, should be sought by (1) the use of genetically different maintainer lines, (2) the incorporation of recessive marker genes to permit elimination of selfed plants in the seedling stage, and (3) the production of F$_1$ hybrid seed under more constant high-temperature environments (*135*).

Breeding Success with the CMS System

This breeding method has only been occasionally successful. Shifriss produced YW-type S*ms*+/*ms* hybrids that exhibited incomplete pollen fertility restoration and set a considerable amount of flat, seedless fruit. In contrast, he found the method promising for the production of hot-pepper hybrids (*133*):

> From my experience one can plant the A line (S*msms*) towards anthesis under temperatures optimal for male sterility (July in Israel). In my cytoplasmic male sterile lines there is a lag period of 1–2 hr between flower and anther opening. If pollen from C-line (N*ms*+*ms*+) is available and bees are active, such lag period is advantageous if the S*msms* plants are partially pollen fertile. In our studies we obtained 100% hybrids when A and C lines were exposed to natural crossing. This technique is now being tried in Israel by the Hazera Seed Company. Since most of the hot cultivars contain restorer genes this technique is more promising with this group [*136*].

Pochard reports similar problems with the CMS system: viz. that "the sterility in the Peterson system is not always complete," and that "cytoplasmic systems are linked to deleterious effects on growth and fruit setting" (*118*).

HAPLOIDY

It has long been thought that haploidy offered the breeder a short-cut method for obtaining homozygous diploid lines in one step, by doubling the chromosomes of haploids with colchicine. This would save several years of inbreeding hybrids to the desired uniformity of the F_7 generation (22). All haploids originate by parthenogenesis from haploid female or male nuclei in the embryo sac (Fig. 3.13) (21,97,121). Experimentally they have been produced as a result of interspecific hybridization, by irradiation of buds and pollen, chemical treatments of pollen, N_2O gas treatment of the embryo sac, and by in vitro pollen culture (22,47,48).

Most *Capsicum* haploids have occurred naturally from $n-2n$ twin seedlings of poly-embryonic seeds. Haploidy, according to Pochard, is the most frequent mutation in *Capsicum*, occurring in 1 per 1000 to 1 per 10,000 plants. This frequency has been increased 6–10 times by selection in crosses between haploid-derived autodiploid lines. It was also shown that field-grown seed produced significantly more haploids than green-house-grown seed, namely, 2.0 vs. 0.4 per 1000 plants (121).

Cultivars differed significantly in the frequency of polyembryony, Goliath producing 0.65%, CW 0.28%, and Perfection Pimiento 0.06% vs. 2.8% in a haploid-derived auto-diploid of Goliath. Although polyembryony is only occasionally higher in haploid-derived autodiploids than in the original population, it can be increased by selection (121).

Cytological evidence indicates that $n-2n$ twins originate from a synergid and a fertil-ized egg nucleus. The $n-n$ type of twins originate from a synergid and an unfertilized egg nucleus, or from a synergid nucleus by division. The other synergid disintegrates after passage through it of the pollen tube (47). The antipodal nuclei in peppers disintegrate before fertilization. The chromosome numbers of multiple seedlings were, in ascending order of frequency, $2n-4n(1)$, $n-n(2)$, $2n-2n$ conjoined (19), $n-2n$ (41), $2n-2n$ unat-tached (76), three sets of triplets $2n-2n-2n$, and one set of quadruplets, probably also diploids. Rarely $n-n-2n$ triplets occur with the haploids conjoined. These are interpreted as arising by cleavage from a haploid proembryo developed from a synergid plus a zygote nucleus.

Several androgenic haploids have been documented. One appeared among 652 F_1 plants from the cross of an autodiploid of Perfection Pimiento with Floral Gem and had the yellow fruit color of Floral Gem (21). A pair of conjoined androgenic haploids of identical genotype has also been reported (121).

Diploidization of Haploids

Pochard decapitates young plants in the 8–10 leaf stage above the eighth leaf, to eliminate apical dominance, removes the axillary buds with a scalpel, and applies a drop of 0.5%

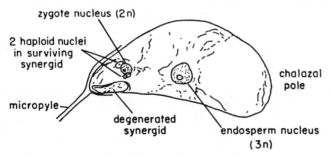

FIGURE 3.13. Postulated synergid origin of haploids in *Cap-sicum*. Embryo sac in a high haploid producer line of Doux Long des Landes, showing two haploid nuclei near the zygote nucleus in the surviving synergid, 4 days after pollination. After Dumas De Vaulx (47).

colchicine containing 1% of a pest oil (e.g., DuPont No. 7) to the wound for better penetration. New buds grow out within a month. The diploid sectors, generally fairly large, can be recognized by their darker green and smoother leaves and by the production of normal flowers with abundant pollen (*118*).

Haploidy as a Breeding Method for Peppers

Extensive research with haploid-derived autodiploids by Pochard and co-workers has cast serious doubt on the value of haploidy as a pepper breeding method. Autodiploid lines have proved inferior in fertility and stability as compared with standard inbred lines. From 20 to 30% of the plants of the first selfed generation (H_1) of an autodiploid from the F_1 hybrid YW $\times$ L107 were partially sterile, and in H_2 the mean seed yield per fruit of the best 14 lines was less than that of the standard F_9 parent lines. The sterility was transmitted through successive generations. Instability in plant height of a new cultivar candidate line, i.e., an increase in height between H_1 and H_5, rendered it useless. There was also excessive variation in plant and fruit characters of sib lines from the same autodiploid parent during successive selfed generations. Repeated haploid–diploid cycling increased the degree of instability. Such disturbing phenomena had not been observed previously in conventional inbred lines (*121*). Absolute homozygosity is obviously undesirable in *Capsicum*. In contrast, Thevenin produced superior F_1 asparagus hybrids of great uniformity by the use of haploids (*22*).

TRISOMICS AND CHROMOSOME MAPPING

Pochard identified 65 primary trisomics among 3500 F_2 progeny of the unusually self-fertile DL haploid, which averaged 2.3 seeds per fruit, more than five times the frequency of haploids from other pepper cultivars (Table 3.15). The primary trisomics could be distinguished from secondary and tertiary trisomics by their phenotype, frequency, and fertility and by their chromosome configurations in meiosis. Eleven of the 12 possible trisomics could be identified by plant, flower, and fruit characteristics and were given code names of flower colors (Table 3.16).

Pochard determined the mean lengths (μm) of the 12 chromosomes of DL and YW from six mitotic root tip plates of each and calculated the standard error of the mean chromosome length of each cultivar. Only three of the 12 chromosomes differed significantly in length. The same three chromosomes were also distinguished by Ohta (Fig. 3.3). The other nine, being metacentric and with their lengths intergrading too closely, did not permit identification.

Eleven trisomic phenotypes could be distinguished and associated with the genes they

TABLE 3.15. Chromosome Numbers of the Selfed Progeny from a Haploid Plant of Doux Long des Landes Propagated by Cuttings and Isolated in a Greenhouse from Other Pepper Genotypes[a]

Chromosome number	12	24	25	26–28	35–37	Total
Frequency of plants	3[b]	634	13	11	19	680
Percentage of population	0.44	93.2	1.91	1.62	2.79	100

[a]After Pochard and Dumas De Vaulx (*121*).
[b]Haploids from twin seedlings.

TABLE 3.16. Eleven Primary Trisomics of Doux Long des Landes[a]

Trisomic name	Trisomic[b] symbol	Chromosome no.	Description[c]	No. of trisomic plants[d]	Transmission (%)[e]	
					Selfed	× cv. Nigrum
Violet	VI	I	Anthers violet	11	4	none observed
Indigo	IN		Anthers bluish	9	47	47
Blue	BL		Foliage bluish green	13	24	17
Green	VE		Foliage intense green	9	32	32
Yellow	JA	XI	Foliage yellowish	6	21	28
Orange	OR		Mature fruit orange–red	5	14	12
Red	RO		Mature fruit dull dark red	4	43	51
Purple	PO	XII	Arbitrary name (Pourpre)	1	6	10
Black	NO		Immature fruit dull dark green	3	21	33
Brown	BR		Stamens yellowish brown	3	34	22
Dusky	BI		Arbitrary name (Bistre)	1	4	15

[a]After Pochard (*115*).
[b]First two letters in capitals of French color names.
[c]The diploid check DL has yellow anthers bordered with blue, medium-green leaves, and bright red fruits.
[d]Number of primary trisomics obtained from 3500 progeny of the selfed haploid of DL.
[e]Transmission percentage of each of the 11 trisomics when selfed (including some confounded tetrasomics), and when crossed with the cv. Nigrum.

carry. The procedure is to cross them with diploid cultivars with marker genes. From three to six F_1 trisomic plants are selected from each cross, selfed, and the F_2 segregation ratios determined. If a gene is located on a trisome, a 2:1 rather than a 3:1 ratio is obtained. The ratios will vary with the transmission frequency of the extra chromosome and the distance of the gene from the centromere (*117*). The 3:1 ratio from the diploid check serves as the control (Table 3.17).

FIELD TRIALS

Field trials are an essential part of the work of plant breeders, who wish to compare the yields of their best lines with the leading commercial cultivars. Field trials involve the concept of randomization and replication in an experimental design that permits statistical analysis of the data. If properly conducted, such trials will provide an unbiased, objective evaluation based on probability theory.

REGIONAL TRIALS

Regional trials, such as the National Pepper Cultivar Evaluation Trials, are intended to test potential new cultivars for adaptation over a wider area and also to speed up the evaluation (*95*). Superior performance in regional trials enhances the chances of commercial acceptance of a new cultivar.

TABLE 3.17. Locating the Gene L^1 for TMV Resistance by Analysis of Trisomics of Doux Long des Landes (L^+) × PM 165 (L^1 Cook Bell Line)[a]

F_2	L^+	Total	$\chi^2 D$	P	$\chi^2 H$	$P\chi^2 H$	K
Normal							
Diploid	33	151	.80	.30–.50	.10	.30–.50	2
VI I	—	—	—	—	—	—	—
IN	46	187	.016	>.50	.01	>.50	2
BL	20	82	.016	>.50	—	—	1
VE	28	109	.027	>.50	—	—	1
JA XI	58	206	1.09	.10–.20	1.76	.10–.20	2
OR	31	151	1.61	.20–.30	.03	>.50	2
RO	35	145	.058	.30–.50	1.97	.10–.20	2
PO XII	45	190	.175	>.50	.46	.50	2
NO	33	151	.80	.30–.50	.10	>.50	2
BR	S	S	S	S	3.57	.30–.50	4
BI	26	76	3.44	.05–.10	—	—	1
Total or average	355	1448	.181	>.50	7.85	>.50	10
BR							
a	39	75	29.2	<.001			
b	20	39	14.4	<.001			
c	21	40	16.1	<.001			
d	29	44	39.3	<.001			
Total or average	109	198	95.4	<.001	3.57	.30–.50	4

[a]L^+, Number of TMV susceptible plants; $\chi^2 D$, χ^2 deviations from 3:1 ratios; $\chi^2 H$, χ^2 for homogeneity of K ratios; P, probability of χ^2 for 3:1 ratios; $P\chi^2 H$, probability of homogeneity χ^2; S, significant deviation from a 3:1 ratio (note separate analysis); K, number of progenies originating from the same source; VI, IN, etc., are abbreviations of the French color names assigned to the 11 trisomics. [After Pochard (117).]

Observational Trials

Breeding lines in F_4 or F_5 are compared in single- or double-row plots of 12–24 plants per plot row. Inferior lines are discarded and the evaluation effort concentrated on the better lines. Numerical scores from 1 to 5 (5 the best) are assigned for plant vigor, plant uniformity, foliage cover (for protection of fruit from sunscald), earliness, productivity, fruit size, fruit shape, stylar closure (to prevent entry of molds and bacteria), fruit rot, and plant disease resistances. Measurements of the best lines for plant height × width, fruit length (depth) × width, and fruit wall thickness (mm) are desirable. Yield and average fruit weight estimates (lb) of the best lines are obtained by harvesting single-row plots and weighing and counting the fruits.

Yield Trials and Procedures

Yield trials are intended to compare the better breeding lines in F_5 or later generations, with leading commercial cultivars as checks.

There are many possible causes of variation in yield that can affect the accuracy and reliability of yield trial data, e.g., herbicide or spray damage, incomplete harvesting, or inconsistent fruit grading. These can be kept to a minimum, but others such as variation in soil fertility or unknown causes must be controlled by randomization and replication in standard experimental designs, which permit a variance analysis of the data.

One or two days prior to the first harvest, the yield trial entries are visually scored for plant and fruit characteristics and overall desirability. Stand counts are made prior to each harvest so that plot yields can be corrected separately for each harvest and added to obtain the total yield per plot per block at the end of the season. Complete and uniform harvesting for fruit size and maturity and standardized fruit grading supervised by the breeder are important to reduce variance in the data. Harvesting must not exceed the capacity to grade, count, and weigh completed blocks within 1–2 days of harvest. Harvesting must not stop partway in a block because of rain or time of day, if results are to be reliable. Fruits are graded as marketable or culls. Cull fruit are further classified by causes such as fruit rots, blossom end rot, sunscald, malformation, virus puckering, fruit worms, stink bugs, harvest injury, soil abrasion, and smallness. All fruit categories are counted and weighed quickly on direct-reading (net weight) spring scales to the nearest 0.1 lb. Greater accuracy is not warranted for such data. Data are recorded on a form with the various fruit classes shown. When completed, the marketable fruit of each entry, four plots from four blocks, are bulked and the entries ranked in order of yield. The entries are then scored visually on a scale from 1 to 5 for yield, appearance, earliness, fruit size, shape, color, stylar closure, fruit firmness, wall thickness (mm), and overall desirability.

Fruit Quality Determinations

Fruit quality determinations such as objective color measurements, pH, soluble solids, total solids, vitamin C, pigment content, flavor, pungency, yield, and appearance of canned product are best done by food scientists in the laboratory. Their cooperation and assistance should be sought in advance. Valuable assistance of this nature warrants coauthorship with the breeder in any cultivar release.

EXPERIMENTAL DESIGNS

The Randomized Complete Block

This is the simplest and most commonly used design with up to 12 entries and usually four replications. The entries are first numbered consecutively and then randomized in each block by the use of a table of random numbers. The pencil is pointed randomly at the page and two-digit numbers are drawn from the table in any direction. These are mentally divided by the number of entries and the remainders written as they occur. If a number has occurred before, it is ignored. Exact multiples of the number of entries, say 10, would be entry number 10. All four blocks are thus randomized and a field plan is drawn up. The individual plot rows should be long and narrow with 12–24 plants each. Individual blocks would ideally be square to minimize soil heterogeneity. The blocks can be separate or laterally contiguous, thus requiring only two border rows.

Usually no border rows are planted at the open ends of the blocks on the assumption that all plots there are equally favored. If spray alleys are needed, they should run between the blocks. Additional border rows will then be required. In the field, the entries in each block are identified by entry and block number. A duplicate record and plot of the yield trial layout is kept for insurance against loss.

The Latin Square

The Latin square is a more efficient design than the randomized block. The reason is that in the Latin square variance due to rows, columns, and entries is subtracted from the total variance, leaving a smaller residual error variance for testing mean differences between entries; whereas in the randomized block only two variances for blocks and entries can be accounted for. This design accommodates up to eight entries. For more complex designs, it is advisable to consult a statistician when planning the experiment.

Variance Analysis

For a detailed calculation of variance analysis for the randomized block and the Latin square, the reader is referred to Briggs and Knowles (*18*).

Duncan's Multiple Range Test

This is a test for significance between the means of all entries in a yield trial, in all possible combinations. First, a standard variance analysis is made of the data. To make the test, the variety means, the standard error with its degrees of freedom, and a table of Duncan's *significant studentized ranges* (r_p) at the 5% or 1% level of significance are required. The appropriate values from the table are multiplied by the standard error of the mean to give Duncan's *shortest significant ranges* (r_p). Duncan, Harter, and LeClerg *et al.* present the method of calculation in detail (*50,71,86*).

VARIETY RELEASE PROCEDURES

A new cultivar developed at the Alabama Agricultural Experiment Station is officially released and named by a variety release committee in consultation with the breeder. The breeder presents 2 years' yield trial data with leading commercial cultivars as checks. To be released, the new cultivar has to be superior in yield or have some compensatory advantage not present in other cultivars. A foundation seed increase requires 10–20 lb of seed.

Seed Increase and Marketing

It may prove difficult to find a reliable seed company that will take the financial risk to increase and market the seed of a new pepper cultivar. The market for the seed may be limited because canners save their own seed and guard it from their competitors. As an inducement to take the risk it may be necessary to make an exclusive release to a seed company. This involves a written agreement between the director of the agricultural experiment station and the company, with the responsibilities of both parties spelled out. Another possibility is for the station or the seed company to patent the cultivar. In this case, a royalty would be paid to the station per pound of seed sold. This money is needed to support additional plant breeding research. In the University of Florida system, seed of new cultivars is produced by the Florida Foundation Seed Producers Association or contracted for increase with private seed companies and paid for by Florida Foundation Seed (*58*).

PLANT VARIETY PROTECTION ACT

In 1930, plant patent legislation was enacted by the U.S. Congress for clonally propagated plants such as fruit trees and potatoes. In 1970 the *Plant Variety Protection Act* was passed to cover seed-reproduced cultivars of crop plants, but the following six vegetables were not included: peppers, okra, tomatoes, carrots, celery, and cucumbers. In 1980, the Congress passed an amendment that removed these exemptions. Today all crops are covered and the developer has the exclusive right to sell the seed of a cultivar for 17 years. A cultivar that has once been released cannot later be patented. Under the *Plant Variety Protection Act,* a seed sample is deposited with the National Seed Storage Laboratory at Fort Collins, Colorado. This seed is periodically renewed. Many nonpatented cultivars are also stored there.

The Netherlands Plant Breeder's Decree enacted in 1941 championed plant breeder's rights and set the example for other countries to follow. In 1962 DeHaan wrote, "The protection of the plant breeder's ownership and the allocation of bonuses have had a favorable influence on the improvement of crop plants in the Netherlands. More efficiently organized establishments, additional personnel, more technical collaborators, better equipment and increase of the number of plant breeders have resulted" (*41*,p. 4).

Capsicum GENES[2]

The preservation of genes is the key to future crop improvement. The availability and usefulness of the national and regional germplasm banks is currently being improved by the establishment of a series of descriptors for each crop that will permit a search for and identification by computer of specific germplasms needed by breeders, from among the thousands of seed stocks of crops in the gene banks. The needed information on which PIs carry genes for specific disease resistances or horticultural traits must first be determined by a host of researchers, including geneticists, plant pathologists, and plant breeders. Several European nations have also established national gene banks. Germplasm preservation is a worldwide concern and requires international cooperation. The same is true for gene identification and gene mapping.

Lippert *et al.* have proposed that the University of California at Davis and Riverside serve as depositories of *Capsicum* germplasm in the United States.

Rules of Gene Nomenclature

Lippert *et al.* and Robinson *et al.* have established rules for designating genes (*88,124*). A gene is named for the main diagnostic feature of its phenotype. Gene symbols are formed from the first letter or not more than the first three letters of the name. Dominant genes are assigned capital letters and recessive genes lowercase letters. Normal or wild-type genes are identified with the superscript $+$. Thus, the normal allele of the dominant gene A is A^+ and of the recessive gene y is y^+. Multiple-recessive alleles are distinguished by lowercase superscripts and dominant alleles by capital letter or arabic numeral superscripts. Polymeric or nonallelic mimic genes are distinguished thus: $sw1$, $sw2$, . . . , swn and their dominant alleles by $sw1^+$, $sw2^+$, . . . , swn^+, with the numerals printed on line with the symbol. Allelism tests are required before a new symbol is assigned to a mimic, e.g., to an ms gene, and "a gene symbol shall not be assigned to a character unless supported by statistically valid segregation data for the gene" (*124*). Genes from different species affecting the same trait may or may not be allelic. They can be distinguished by a superscript indicating species origin, e.g., et^a and et^c from *C. annuum* and *C. chinense*, respectively. Lippert *et al.* chose California Wonder as the standard on which to base dominance or recessiveness of genetic traits. European breeders prefer the sweet cayenne pepper Doux Long des Landes as the standard because it is early, prolific, non-pungent, and better tolerates low light intensity, plus the availability of trisomics for chromosome mapping.

[2]Table 3.18 gives a list of the *Capsicum* genes considered important to breeders.

TABLE 3.18. *Capsicum* Genes[a]

Symbol	Character	Reference
Color genes		
A	*Anthocyanin*: basic gene for purple color in foliage, stem, flower, and fruit; incompletely dominant; fully effective only in presence of *MoA*; located on chromosome of trisomic RO; *A*-6.5–*O*-18.8–*sw1* are linked in *C. annuum*	*88,89,110, 117*
al-1 to *al-5*	*Anthocyaninless*: nodes green, anthers yellow; *al-5* shows slight purple along line of dehiscence; epistatic to *A*, *As*, and *Asf*; nonallelic	*37,88,89*
As(P)	*Style anthocyanin*: purple in absence of *A* or *Asf*	*88,89*
Asf(W)	*Style* and *filament anthocyanin*; purple in absence of *A*	*88,89*
B	β-*Carotene*: high in mature fruit; interacts with *t* for higher level in mature fruit	*16,88,89*
c1(c)	*Carotenoid* pigment inhibitor: 1/10 reduction in red color of mature fruit	*88,89*
c2(c1)	*Carotenoid* pigment inhibitor: red color reduction much stronger than *c1*; these two independent gene pairs interact with *y* and *y*+ (red) to produce a range of mature fruit colors from red to ivory	*88,89*
cl(g)	*Chlorophyll* retainer in mature fruit: combines with *y*+ or *y* to produce brown or olive green mature fruit color, respectively	*16,88,89, 140*
im	*Intermediate maturity* of purple in originally nonpurple unripe fruit	*88,89*
MoA(B)	*Modifier of A*: intensifies purple color in presence of *A*; incompletely dominant; located on chromosome of trisomic BR	*88,89,117*
sw1–swn	*Sulfury white* immature fruit color: dominant alleles control various green shades; duplicate or cumulative in action; in linkage group *A*–6.5—*O*—18.8—*sw1*	*79,88,89 101,102,111*
t	High β-carotene content: complementary with *B*	*16,89*
y(r)	*Yellow* or orange mature fruit color: located on chromosome of trisomic IN	*88,89,117*
Ys	*Yellow spot*: on corolla of *C. pendulum*, monogenic dominant in interspecific cross with *C. annuum*, *C. chacoense*, *C. chinense*, and *C. frutescens*	*88,89*
yt1, yt2	*Yellow top*: young expanding leaves are yellow and gradually turn green	*37*
Disease and nematode resistance genes		
*Bs1	*Bacterial spot* resistance to *X. campestris* pv. *vesicatoria* race 1 in *C. chacoense* PI 260435	*30,33,34*
Bs2(Bs)	*Bacterial spot* resistance to race 2 of pv. *vesicatoria* in *C. annuum* PI 163192	
et[a]	*Tobacco etch* virus (TEV) resistance in *C. annuum* Cayenne SC 46252 and PI 264281: allelic with and dominant over *y*[a]	*26,27,64*
*et[av]	Resistance in Avelar to TEV and PVY, tolerance to PMV and to TEV-S; allelic with and dominant over *et*[a] and *y*[a]	*157,158*

TABLE 3.18. (*cont.*)

Symbol	Character	Reference
*et^{c1}	Resistance to PMV and TEV-C in *C. chinense* PI 159236: allelic with et^{av}; interaction of et^{av}/et^{c1} results in susceptibility to PMV	149
et^{c2} (et^{c},et^{f})	Resistance in *C. chinense* PI 152225 to PVY, PMV, and TEV-S: probable top allele of the series $et^{c2}>et^{c1}>et^{av}>et^{a}>y^{a}$, but allelism and dominance relationship with et^{c1} not established	64
L^{3}	*Localization* of pepper strains of TMV in *C. chinense*: member of the allelic series $L^{3}>L^{2}>L^{1}>L^{+}$	15
$L^{2}(L)$	*Localization* of common strains of TMV in *C. frutescens* (cv. Tabasco) and in certain cvs. of *C. annuum*: similar reaction found in *C. baccatum* vars. *pendulum* and *microcarpum*, but allelism with the latter not established	74,75
L^{1} (L^{i},l^{i})	Imperfect *localization* of TMV: reported only in *C. annuum*; incomplete dominance of L^{1}/L^{+} with some strains of TMV results in systemic necrosis especially at high temperature; located on chromosome of trisomic BR	75,117
N	Root-knot *nematode* resistance to *M. incognita* and *M. incognita* race *acrita*	70
v1, v2	*Veinbanding* virus resistance: combination of these two genes determines four reactions to virus infection of which only v1/v1,v2/v2 confers resistance; similar expression of resistance observed in some accessions of *C. chinense, C. frutescens, C. pendulum,* and *C. pubescens*	89
y^{a}	Resistance in Yolo Y to a mild strain of PVY: bottom allele of series $et^{c1}>et^{av}>et^{a}>y^{a}$	27
Plant morphology genes		
ca	*Canoe*: margins of cotyledons and leaves are rolled up, exposing only the abaxial surface	37
*dw1	*Dwarf* plant about 6 in. tall reported by Dale in Floral Gem: reduced female fertility	88,89
dw2(dw)	*Dwarf* plant 4–6 in. tall: normal number of nodes (8–10); from cross of *C. baccatum* var. *pendulum* × *C. annuum*	37
fa	*Fasciculate*: flowers and fruits borne in clusters on bunched, compounded nodes; bushy plants with determinate tendency	88,89
fi1 (fi mutant 1)	*Filiform* threadlike leaves: blossom irregularities; female sterile	88,89
fi2	Similar to *fi1*: narrow cotyledons and leaves; threadlike petals; carpels usually not fused to pistil; incompetely female sterile; allelism with *fi1* not tested	37
fr	*Frilly*: leaf margins undulated	37
Hl(H)	*Hairless* (smooth) stem in *C. annuum* var. *minimum* (Blanco): dominant over hairy stem in cv. Golden	74,89

(continued)

TABLE 3.18. (*cont.*)

Symbol	Character	Reference
	Dawn; Ikeno reported dominant digenic inheri-tance (15:1) in *C. annuum* of hairy over smooth, with stems, petioles, and leaves exhibiting various degrees of pubescence	
O	*Oblate* or round fruit shape in *C. annuum* and *C. chinense*: dominant over elongate; allelism not tested; in linkage group A–6.5—*O*—18.8—*swl*	*83,111*
P(D)	*Pointed* fruit apex: incompletely dominant over blunt	*89*
pc1,pc2,pc3	*Polycotyledon*: seedlings with three to four cotyledons and fasciated stem; pseudodichotomous branching with unequally developing shoots; non-allelic	*37*
rl	*Roundleaf*: length but not width of leaves is reduced, changing the length/width ratio from 1.50 to 1.24	*66*
ru	*Rugose* or savoyed mature leaves: appeared in *C. chinense* × *C. annuum* derivative; mature leaves darker green than normal; readily classified by fleshy cotyledons with down-curved margins; good seedling vigor and survival; homogeneous 3:1 F_2 and 1 : 1 BC ratios (W. H. Greenleaf, unpublished)	
tu	Cotyledons and leaves are rolled up like a *tube*, with only the abaxial surface exposed: relationship to *ca* not tested	*37*
up (*p, u*)	*Upright* or erect pedicel and fruit orientation: inter-mediate expression in some genetic backgrounds; located on the chromosome of trisomic NO	*89,100,117*
wl	Willow leaf: leaves narrowed, similar to *fi*, but wider; nonallelic; highly female sterile; male fer-tile; sets parthenocarpic, seedless fruits	*88,89*
Sterility genes		
*fs1	*Female-sterile* mutant in *C. annuum*: mature plants larger, otherwise normal; male fertile; sets par-thenocarpic, seedless fruits	*39*
fs2(fs)	*Female-sterile* mutant in *C. annuum* PI 159276 reported by Bergh and Lippert: similar to *fs1*; male fertile; allelism with *fs1* not tested	*88,89*
ms	*Male-sterile* nuclear gene induces male sterility in S cytoplasm in genotype *Sms/ms*: three different sources of S cytoplasm produced similar reactions with different *ms* and restorer genes, indicating identity	*108,110,134*
ms1,ms2,ms3	Spontaneous *male-sterile* mutants from *C. annuum* cvs. Allbig, California Wonder, and Gambo, re-spectively: nonallelic; many additional male-sterile mutants have been reported	*132*
ms9, ms509, ms705	γ-Irradiation-(*ms9*) and chemical (EMS)-induced (*ms509, ms705*) *male-sterile* mutants: *ms509* is used to produce the French F_1 hybrid bell pepper Lamuyo-INRA	*17*
Genes determining physiological traits		
C	*Capsaicin* or pungent fruit: modifiers increase or decrease capsaicin to produce a bimodal distribu-	*46,88,89* *104,117*

TABLE 3.18. (*cont.*)

Symbol	Character	Reference
	tion of pungency in F_2; located on chromosome 11 of trisomic JA	
Gi	*Graft incompatible*: reaction of *C. annuum* vegetatively grafted with other Solanaceae	*88,89*
Ps(S)	*Pod separates* easily from calyx: distinct from soft flesh *S*; expression improves with fruit maturity; subject to modifiers and background genotype	*62,79,146*
S(Ps)	*Soft* juicy fruit in Paprika, distinct from *Ps*: observed also in a *C. chinense* PI 152225 × Tabasco (*Ps, S*) derivative (W. H. Greenleaf, unpublished)	*85*

[a]Adapted from the gene lists of Lippert *et al.* (*88,89*). Reproduced by permission. Only genes considered important to breeders are listed. Redesignated symbols are given in parentheses. Trisomic tests are with Doux Long des Landes. Asterisks indicate proposed new symbols.

REFERENCES

1. Adamson, W. C., and Sowell, G. 1982. The inheritance of three sources of resistance to bacterial spot of pepper. Phytopathology *72*, 999 (Abstr. No. 552).

1a. Andrews, J. 1984. Peppers: The Domesticated Capsicums. University of Texas Press, Box 7819, Austin, TX 78713.

2. Annual Vegetables 1980. Crop Reporting Board, Economics and Statistics. U.S. Department of Agriculture, Washington, DC.

3. Backman, P. A., and Rodriguez-Kabana, R. 1975. A system for the growth and delivery of biological control agents to the soil. Phytopathology *65*, 819–821.

4. Ball, E. M. 1974. Serological Tests for the Identification of Plant Viruses. Phytopathol. Soc., St. Paul, MN.

5. Barrios, E. P., Mosokar, H. I., and Black, L. L. 1971. Inheritance of resistance to tobacco etch and cucumber mosaic virus in *Capsicum frutescens*. Phytopathology *61*, 1318.

6. Bartz, J. A., and Stall, W. M. 1974. Tolerance of fruit from different pepper lines to *Erwinia carotovora*. Phytopathology *64*, 1290–1293.

7. Bawden, F. C., and Kassanis, B. 1945. The suppression of one plant virus by another. Ann. Appl. Biol. *32*, 52–57.

8. Beadle, G. W. 1980. The ancestry of corn. Sci. Am. *245*, 112–119.

9. Black, L. L. 1982. Personal communication. Dep. Plant Pathol., Louisiana State Univ., Baton Rouge.

10. Black, L. L., and Price, M. A. 1980. Observations on the Samsun latent strain of tobacco mosaic virus in peppers. Natl. Pepper Conf., Louisiana State Univ., Baton Rouge, 5th, 1980. Abstr. No. 11.

11. Black, L. L., and Rolston, L. H. 1972. Aphids repelled and virus diseases reduced in peppers planted on aluminum foil mulch. Phytopathology *62*, 747.

12. Black, L. L., and Simmons, L. 1983. Cayenne pepper lines with multiple virus resistance. Natl. Pepper Conf., San Miguel Allende, Mexico, 6th, 1983. Abstr. No. 8.

13. Black, L. L., and Zitter, T. A. 1982. Personal communication. Louisiana State Univ., Baton Rouge, and Cornell Univ., Ithaca, NY.

14. Boswell, V. R. 1949. Garden pepper, both a vegetable and a condiment. Natl. Geogr. Mag. *96*, 166–167.

15. Boukema, I. W. 1980. Allelism of genes controlling resistance to TMV in *Capsicum* L. Euphytica *29*, 433–439.

16. Brauer, O. 1962. Studies on quality characteristics in F1 pepper hybrids *Capsicum annuum* L. Z. Pflanzenzucht. *48*, 259–276 (in German).

17. Breuils, G., and Pochard, E. 1975. Development of the pepper hybrid Lamuyo-INRA by the use of the male sterile gene *ms* 509. Ann. Amelior. Plant. *25*, 399–409 (in French).

18. Briggs, F. N., and Knowles, P. F. 1967. Introduction to Plant Breeding. Reinhold Publishing Corp., New York.

19. Buttery, R. G., Seifer, R. M., Lundin, R. G., Guadagni, D. G., and Ling, L. C. 1969. Characteristics of an important aroma component of bell peppers. Chem. Ind. (London) *15*, 490.

20. Campbell, G. M. 1980. GEL-O-FLEX vegetable planter. Natl. Pepper Conf., New Mexico State Univ., Las Cruces, 5th, 1980. Abstr. No. 22 (Campbell Inst. Agric. Res., Cairo, GA.).

21. Campos, F. F., and Morgan, D. T. 1958. Haploid pepper from a sperm. J. Hered. *49*, 134–137.

22. Chase, S. S. 1974. Utilization of haploids in plant breeding—breeding diploid species. *In* Haploids in Higher Plants, Advances and Potential. Proc. First Int. Symp., Univ. of Guelph. K. J. Kasha (Editor), pp. 211–230. Univ. of Guelph, Guelph, Ontario, Canada.

23. Cochran, H. L. 1943. The *Truhart Perfection* pimiento. Ga., Agric. Exp. Stn., Bull. *224*.

24. Cochran, H. L. 1963. A quantitative study of some anatomical constituents of the raw pimiento fruit. Proc. Am. Soc. Hortic. Sci. *83*, 613–617.

25. Commonwealth Mycological Institute and Association of Applied Biologists, Ferry Lane, Kew, Surrey, England. 1981. Descriptions of Plant Viruses Holywell Press, Ltd., Oxford.

26. Cook, A. A. 1960. Genetics of resistance in *Capsicum annuum* to two virus diseases. Phytopathology *50*, 364–367.

27. Cook, A. A. 1961. A mutation for resistance to potato virus Y in pepper. Phytopathology *51*, 550–552.

28. Cook, A. A. 1963. Genetics of response in pepper to three strains of potato virus Y. Phytopathology *53*, 720–722.

29. Cook, A. A. 1966. Yolo Y, a bell pepper with resistance to potato Y virus and tobacco mosaic virus. Circ.—Fla., Agric. Exp. Stn. *S-175*.

30. Cook, A. A. 1982. Personal communication. Dep. Plant Pathol., Univ. of Florida, Gainesville.

31. Cook, A. A. 1983. Pepper breeding line releases Florida VR2-34, Florida XVR3-25, Florida VR4 and USAJ15. Dep. Plant Pathol., Univ. of Florida, Gainesville.

32. Cook, A. A., Ozaki, H. Y., Zitter, T. A., and Blazquez, C. H. 1976. Florida VR-2, a bell pepper with resistances to three virus diseases. Circ.—Fla., Agric. Exp. Sta. *S-242*.

33. Cook, A. A., and Stall, R. E. 1963. Inheritance of resistance in pepper to bacterial spot. Phytopathology *53*, 1060–1062.

34. Cook, A. A., and Stall, R. E. 1982. Distribution of races of *Xanthomonas vesicatoria* pathogenic on pepper. Plant Dis. *66*, 388.

35. Cook, A. A., Zitter, T. A., and Ozaki, H. Y. 1977. Delray Bell, a virus resistant pepper for Florida. Circ.—Fla., Agric. Exp. Stn. *S-251*.

36. Crowley, N. C. 1957. Studies on the seed transmission of plant virus diseases. Aust. J. Biol. Sci. *10*, 449–464.

37. Csillery, G. 1980. Gene mapping of the pepper needs more initiatives. Contribution to the gene list. CAPSICUM *80*, 5–9. Fourth Eucarpia Congr., Wageningen, The Netherlands. (Inst. Veg. Crops, Dept. Budateteny. Pf. 95. Budapest H-1775, Hungary.)

38. Csillery, G., and Rusko, J. 1980. The control of a new tobamovirus strain by a resistance linked to anthocyanin deficiency in pepper (*Capsicum annuum*). CAPSICUM *80*, 40–43.

Fourth Eucarpia Congr., Wageningen, The Netherlands. (Inst. Veg. Crops, Dept. Budateteny. Pf. 95. Budapest H-1775, Hungary.)

39. Curtis, L. C., and Scarchuck, J. 1948. Seedless peppers. J. Hered. *39*, 159–160.

40. De Candolle, A. 1886. Origin of Cultivated Plants (Reprint of 2nd Edition, Hafner Publishing Co., New York, NY, 1967).

41. DeHaan, H. 1962. The Netherlands Plant Breeder's Decree. Euphytica *11*, 1–4.

42. Dempsey, A. H. 1963. Truhart Perfection D Pimiento. Ga., Agric. Exp. Stn., Leafl. [N.S.] *42*, 1–6.

43. Dempsey, A. H. 1976. Field resistance in peppers to southern blight (*Sclerotium rolfsii*). Natl. Pepper Conf., Univ. California, Davis, 3rd, 1976. Abstr. No. 9. (Agric. Exp. Stn., Experiment, GA.).

44. Dempsey, A. H. 1978. Calcium and sodium hypochlorite for pimiento pepper seed treatment. Natl. Pepper Conf., Louisiana State Univ., Baton Rouge, 4th, 1978. Abst. No. 2. (Agric. Exp. Stn., Experiment, GA.).

45. Demski, J. W. 1981. Tobacco mosaic virus is seedborne in pimiento peppers. Plant Dis. *65*, 723–724.

46. Deshpande, R. B. 1935. Studies in Indian chillis. Inheritance of pungency in *Capsicum annuum* L. Indian J. Agric. Sci. *5*, 513–516.

47. Dumas De Vaulx, R. 1977. Embryogenesis of haploids in the pepper (*Capsicum annuum* L.) (in French, English summary). CAPSICUM *77*, 67–73. Third Eucarpia Congr., Avignon-Montfavet, France, July 5–8, 1977. E. Pochard (Editor). Station Amelioration des Plantes Maraicheres, INRA. Domaine St. Maurice. 84140.

48. Dumas De Vaulx, R., and Chambonnet, D. 1980. Influence of 35°C treatments and growth substances concentrations on haploid plant production through anther culture in *Capsicum annuum*. CAPSICUM *80*, 16–20. Fourth Eucarpia Congr., Wageningen, The Netherlands.

49. Dumas De Vaulx, R., and Pitrat, M. 1977. Interspecific hybridization between *Capsicum annuum* and *Capsicum baccatum* (in French). CAPSICUM *77*, 75–81. Third Eucarpia Congr., Avignon-Montfavet, July 5–8, 1977. E. Pochard (Editor), Station Amelioration des Plantes Maraicheres, INRA.

50. Duncan, D. B. 1955. Multiple range and multiple F tests. Biometrics *11*, 1–42.

51. Duvick, D. N. 1959. The use of cytoplasmic male-sterility in hybrid seed production. Econ. Bot. *13*, 167–195.

52. ELISA 1977. Catalog No. 1-223-01. Dynatech Laboratories, 900 Slaters Lane, Alexandria, VA 22314.

53. Eshbaugh, W. H. 1970. A biosystematic and evolutionary study of *Capsicum baccatum* (Solanaceae). Brittonia *22*, 31–43.

54. Everett, P. H. 1978. Controlled release fertilizers for bell peppers. Natl. Pepper Conf., Louisiana State Univ., Baton Rouge, 4th, 1978. Abstr. No. 17. (Univ. Florida, AREC, Immokalee).

55. FAO Production Yearbook 1978. World production of peppers. Vol. 32, p. 155.

56. Feldman, J. M., and Oremianer, S. 1972. An unusual strain of tobacco mosaic virus from pepper. Phytopathol. Zeitschrift *75*, 250–267.

57. Fieldhouse, D. J., and Sasser, M. 1980. Detection of *Xanthomonas vesicatoria* (bacterial spot of pepper) on seeds, plants and soil residue. Natl. Pepper Conf., New Mexico State Univ., Las Cruces, 5th, 1980. Abstr. No. 10. (Dep. Plant Sci., Univ. Delaware, Newark).

58. Florida Foundation Seed Producers Association, Inc. Box 14006, Gainesville, FL 32604.

59. Fulton, R. W. 1958. Resistance in tobacco to cucumber mosaic virus. Virology *6*, 303–316.

60. Geraldson, C. M. 1975. The gradient mulch system for pepper production in Florida. Natl. Pepper Conf., Lake Worth, Florida, 2nd, 1975. Abstr. No. 16 (Univ. Florida, AREC, Bradenton).

61. Gooding, G. V. 1975. Serological identification of tobacco viruses. Tob. Sci. *19*, 135–139.

62. Greenleaf, W. H. 1952. Inheritance of pungency and of the deciduous character in peppers (*Capsicum annuum*). Proc. Assoc. South Agric. Work. *49*, 110–111 (Abstr.).

63. Greenleaf, W. H. 1953. Effects of tobacco etch virus on peppers (*Capsicum* sp.). Phytopathology *43*, 564–570.

64. Greenleaf, W. H. 1956. Inheritance of resistance to tobacco-etch virus in *Capsicum frutescens* and in *Capsicum annuum*. Phytopathology *46*, 371–375.

65. Greenleaf, W. H. 1975. The Tabasco story. HortScience *10*, 98.

66. Greenleaf, W. H. 1976. A roundleaf mutant in Bighart Pimiento pepper (*Capsicum annuum* L.). HortScience *11*, 463–464.

67. Greenleaf, W. H., Cook, A. A., and Heyn, A. N. J. 1964. Resistance to tobacco mosaic virus in *Capsicum*, with reference to the Samsun latent strain. Phytopathology *54*, 1367–1371.

68. Greenleaf, W. H., Hollingsworth, M. H., Harris, H., and Rymal, K. S. 1969. Bighart, an improved pimiento pepper (*Capsicum annuum* L.) variety. HortScience *4*, 334–338.

69. Greenleaf, W. H., Martin, J. A., Lease, J. G., Sims, E. T., and Van Blaricom, L. O. 1970. Greenleaf Tabasco, a new tobacco etch virus resistant Tabasco pepper variety (*Capsicum frutescens* L.). Leafl.—Ala., Agric. Exp. Stn. *81*, 1–10.

70. Hare, W. W. 1957. Inheritance of resistance to rootknot nematodes in pepper. Phytopathology *47*, 455–459.

71. Harter, H. L. 1960. Critical values for Duncan's new multiple range test. Biometrics *16*, 671–685.

72. Haymon, L. W., and Aurand, L. W. 1971. Volatile constituents of Tabasco peppers. J. Agric. Food Chem. *19*, 1131–1134.

73. Hayslip, N. C. 1974. A plug mix seeding method for field planting tomatoes and other small seeded hill crops. Univ. Fla., Agric. Res. Cent. Res. Rep. *RL 1974-3*.

74. Holmes, F. O. 1934. Inheritance of ability to localize tobacco mosaic virus. Phytopathology *24*, 984–1002.

75. Holmes, F. O. 1937. Inheritance of resistance to tobacco mosaic disease in the pepper. Phytopathology *27*, 637–642.

76. Holzmann, O. V. 1965. Effect of soil temperature on resistance of tomato to root-knot nematode (*Meloidogyne incognita*). Phytopathology *55*, 990–992.

76a. Hoyle, B. J. 1976. Yield of the chili pepper. Third Nat. Pepper Conf., Univ. of California, Davis. Sept. 1976. Abstr. No. 22.

77. Huffman, V. L., Schadle, E. R., Villalon, B., and Burns, E. E. 1978. Volatile components and pungency in fresh and processed Jalapeno peppers. J. Food Sci. *43*, 1809–1811.

78. Hussey, R. S., and Barker, K. R. 1973. A comparison of methods of collecting inocula of *Meloidogyne* spp., including a new technique. Plant Dis. Rep. *57*, 1025–1028.

79. Jeswani, L. M., Deshpande, R. B., and Joshi, A. B. 1956. Inheritance of some fruit characters in chilli. Indian J. Genet. Plant Breed. *16*, 138–143.

80. Jones, H. A., and Clarke, A. E. 1943. Inheritance of male sterility in the onion and the production of hybrid seed. Proc. Am. Soc. Hortic. Sci. *43*, 189–194.

81. Jones, H. A., and Rosa, J. T. 1928. Truck Crop Plants. McGraw-Hill Book Co., New York.

82. Kaan, F., and Anais, G. 1978. Breeding large fruited red peppers (*C. annuum*) in the French West Indies for climatic adaptation and resistance to bacterial (*Pseudomonas solanacearum, Xanthomonas vesicatoria*) and viral diseases (Potato Virus Y), Annu. Rep., pp. 265–273. Station Amelioration des Plantes, INRA, Domaine Duclos, Petit-Bourg, Guadeloupe (in French), in Plant Breed. Abstr. 1978. *48*, Abstr. No. 8867. Reprinted in CAPSICUM 77. Third Eucarpia Congr., Avignon-Montfavet, France, July 5–8, 1977. E. Pochard (Editor). Station Amelioration des Plantes Maraicheres, INRA. Montfavet 84140, Vaucluse, France.

83. Khambanonda, I. 1950. Quantitative inheritance of fruit size in red pepper (*Capsicum frutescens* L.). Genetics *35*, 322–343.

84. Kimble, K. A., and Grogan, R. G. 1960. Resistance to *Phytophthora* root rot in pepper. Plant Dis. Rep. *44*, 872–873.

85. Kormos, K., and Kormos, J. 1957. A soft paprika. Novenytermeles *6*, 33–44 (in Hungarian, English summary).

86. LeClerg, E. L., Leonard, W. H., and Clark, A. G. 1962. Field Plot Technique. Burgess Publishing Co., Minneapolis, MN.

87. Lippert, L. F. 1975. Heterosis and combining ability in chili peppers by diallel analysis. Crop Sci. *15*, 323–325.

88. Lippert, L. F., Bergh, B. O., and Smith, P. G. 1965. Gene list for the pepper. J. Hered. *56*, 30–34.

89. Lippert, L. F., Smith, P. G., and Bergh, B. O. 1966. Cytogenetics of the vegetable crops. Garden pepper, *Capsicum* sp. Bot. Rev. *32*, 24–55.

90. Marin, V. O., and Lippert, L. F. 1975. Combining ability analysis of anatomical components of the dry fruit in chili pepper. Crop Sci. *15*, 326–329.

91. Marshall, D. E. 1976. Estimates of harvested acreage, production and grower value for peppers grown in the United States. Available from W. R. Moore, Pickle Packers International, Inc. One Pickle and Pepper Plaza, Box 31, St. Charles, IL 60174.

92. Martin, J. A., and Crawford, J. H. 1958. Carolina hot pepper. Circ.—S. C., Agric. Exp. Stn. *117*.

93. Matsuoka, K. 1979. Personal communication. Federal Univ. of Vicosa, 36.570-Vicosa-MG-Brasil.

94. McKinney, H. H. 1952. Two strains of tobacco mosaic virus, one of which is seed borne in an etch-immune pungent pepper. Plant Dis. Rep. *36*, 184–187.

95. Miller, C. H. 1982. National pepper cultivar evaluation trials, 1977–1981. Dep. Hortic., N.C. State Univ., Raleigh.

96. Mohr, H. C., and Watkins, G. M. 1959. The nature of resistance to southern blight in tomato and the influence of nutrition on its expression. Proc. Am. Soc. Hortic. Sci. *74*, 484–493.

97. Morgan, D. T., and Rappley, R. D. 1954. A cytogenetic study on the origin of multiple seedlings of *Capsicum frutescens*. Am. J. Bot. *41*, 576–585.

98. Muehmer, J. K. 1980. Producing pepper transplants for the North in quantity by the pregermination technique. Natl. Pepper Conf., New Mexico State Univ., Las Cruces. 5th, 1980. Abstr. No. 19 (Ridgetown Coll. Agric. Technol., Ridgetown, Ontario, Canada. NOP2CO).

99. Nagai, H. 1980. Pepper breeding in Brazil. Natl. Pepper Conf., New Mexico State Univ., Las Cruces, 5th, 1980. Abstr. No. 1 (Instituto Agronomico, Campinas, SP, Brazil).

100. National Pepper Conferences Abstracts. 1984. Available from W. R. Moore, Clerk, Pickle Packers International, Inc., One Pickle and Pepper Plaza, Box 31, St. Charles, IL 60174.

100a. Nelson, E. K. 1920. The constitution of capsaicin, the pungent principle of *Capsicum* III. J. Am. Chem. Soc. *42*, 597–599.

101. Odland, M. O. 1948. Inheritance studies in the pepper, *Capsicum frutescens*. Minn., Agric. Exp. Stn., Tech. Bull. *179*.

102. Odland, M. L., and Porter, A. M. 1938. Inheritance of the immature fruit color of peppers. Proc. Am. Soc. Hortic. Sci. *36*, 647–657.

103. Odland, M. L., and Porter, A. M. 1941. A study of natural crossing in peppers, *Capsicum frutescens*. Proc. Am. Soc. Hortic. Sci. *38*, 585–588.

104. Ohta, Y. 1960. Physiological and genetical studies on the pungency of *Capsicum*. Capsaicin content of several varieties of *C. annuum* and related species. Seiken Jiho *11*, 63–72 (in Japanese, English summary).

105. Ohta, Y. 1962. Genetical Studies in the Genus *Capsicum*. 94 pp. Kihara Inst. Biol. Res., Yokohama, Japan (in Japanese, English summary).

106. Ohta, Y. 1962. Physiological and genetical studies on the pungency of *Capsicum*. Inheritance of pungency. Jpn. J. Genet. *37*, 169–175 (in Japanese, English summary).

107. Ohta, Y. 1962. Physiological and genetical studies on the pungency of *Capsicum*. IV. Secretory organs, receptacles and distribution of capsaicin in the *Capsicum* fruit. Jpn. J. Breed. *12*, 182–183 (in Japanese, English summary).

108. Ohta, Y. 1973. Identification of cytoplasms of independent origin causing male sterility in red peppers (*Capsicum annuum*). Seiken Jiho *24*, 105–106.

109. Peppers (*Capsicum* spp.) 1977. Catalog of seed available at the Southern Regional Plant Introduction Station, Experiment, GA. 30212.

110. Peterson, P. A. 1958. Cytoplasmically inherited male sterility in *Capsicum*. Am. Nat. *92*, 111–119.

111. Peterson, P. A. 1959. Linkage of fruit shape and color genes in *Capsicum*. Genetics *44*, 407–419.

112. Pety, C., and Nakayama, R. M. 1980. Effect of temperature and relative humidity on viability of stored chile (*Capsicum annuum*) pollen. Natl. Pepper Conf., New Mexico State Univ., Las Cruces, 5th, 1980. Abstr. No. 32 (Dep. Hortic., N.M. State Univ.).

113. Pickersgill, B. 1969. The archeological record of chili peppers (*Capsicum* spp.) and the sequence of plant domestication in Peru. Am. Antiq. *34*, 53–61.

114. Pochard, E. 1966. Experimental results of selection with peppers (*Capsicum annuum* L.). Ann. Amelior. Plant. *16*, 185–197 (in French).

115. Pochard, E. 1970. Description of pepper trisomics (*Capsicum annuum* L.) in the progeny of a haploid plant. Ann. Amelior. Plant. *20*, 233–256 (in French, English summary).

116. Pochard, E. (Editor) 1977. Methods for Studying Partial Resistance to Cucumber Mosaic Virus. CAPSICUM *77*, 93–104. Third Eucarpia Congr., Avignon-Montfavet, France, July 5–7, 1977. E. Pochard (Editor). Station Amelioration des Plantes Maraicheres, INRA. Montfavet 84140, Vaucluse, France.

117. Pochard, E. 1977. Locating genes in *Capsicum annuum* L. by trisomic analysis. Ann. Amelior. Plant. *27*, 255–266.

118. Pochard, E. 1982. Personal communication. Station Amelioration des Plantes Maraicheres, Montfavet 84140, Vaucluse, France.

119. Pochard, E., and Breuils, G. 1965. Resistance to tobacco mosaic and cucumber mosaic virus in peppers (*Capsicum*). Characteristics and mode of inheritance. 1éres J. Phytiat. Phytopharm. pp. 189–193 (in French).

120. Pochard, E., and Chambonnet, D. 1971. Methods of selection with pepper for resistance to *Phytophthora capsici* and to cucumber mosaic virus. Eucarpia *Capsicum* Conf., Torino. Ann. Fac. Sci. Agrar. Univ. Torino *7*, 270–281.

121. Pochard, E., and Dumas De Vaulx, R. 1979. Haploid parthenogenesis in *Capsicum annuum* L. Reprinted from The Biology and Taxonomy of the Solanaceae, No. 36. Linn. Soc. Symp. Ser. *7*, 455–472.

122. Rajpoot, N. C., and Govindarajan, V. S. 1981. Paper chromatographic determination of total capsaicinoids in capsicums and their oleoresins with precision, reproducibility and validation through correlation with pungency in Scoville units. J. Assoc. Off. Anal. Chem. *64*, 311–318.

123. Rast, A. T. B. 1977. Introductory remarks on strains of TMV infecting peppers in The Netherlands. CAPSICUM *77*, 83–84. Third Eucarpia Congr., Avignon-Montfavet, France.

124. Robinson, R. W., Munger, H. M., Whitaker, T. W., and Bohn, G. W. 1976. Genes of the Cucurbitaceae. HortScience *11*, 554–568.

125. Rusko, J., and Csillery, G. 1980. Selection for CMV resistance in pepper by the method developed by Pochard. CAPSICUM *80*, 37–39. Fourth Eucarpia Conf., Wageningen, The Netherlands. (Inst. Veg. Crops, Dept. Budateteny, Pf. 95, Budapest H-1775, Hungary.)

126. Rymal, K. S. 1983. Portable micromethod for quantitative determination of vitamin C in fruit and vegetable juices. J. Assoc. Off. Anal. Chem. *66*, 810–813.

127. Rymal, K. S., Cosper, R. D., and Smith, D. A. 1984. Injection-extraction procedure for the

rapid determination of capsaicinoids in fresh Jalapeno peppers. J. Assoc. Off. Anal. Chem. *47*, 658–659.

128. Rymal, K. S., Greenleaf, W. H., and Smith, D. A. 1980. Taste panel testing proves no difference in flavor between canned pimiento or canned red bell pepper. Highlights Agric. Res. *27*, 5. Ala. Agric. Exp. Stn.

129. Safford, W. E. 1926. Our heritage from the American Indians. Smithson. Inst., Annu. Rep. pp. 405–410.

130. Sciumbato, G. L. 1973. Studies on the viruses infecting pepper (*Capsicum* sp.) in Louisiana. Ph.D. Thesis, Louisiana State University, Baton Rouge.

131. Shepherd, R. L. 1979. A quantitative technique for evaluating cotton for root-knot nematode resistance. Phytopathology *69*, 427–430.

132. Shifriss, C. 1973. Additional spontaneous male sterile mutants in *Capsicum annuum* L. Euphytica *22*, 527–529.

133. Shifriss, C. 1982. Personal communication. Agricultural Research Organization, Volcani-Center, Bet Dagan, Israel.

134. Shifriss, C., and Frankel, R. 1971. New sources of cytoplasmic male sterility in cultivated peppers. J. Hered. *62*, 254–256.

135. Shifriss, C., and Guri, A. 1979. Variation in stability of cytoplasmic-genic male sterility in *Capsicum annuum*. J. Am. Soc. Hortic. Sci. *104*, 94–96.

136. Shifriss, C., and Sacks, J. M. 1980. The effect of distance between parents on the yield of sweet pepper × hot pepper hybrids, *Capsicum annuum* L. in a single harvest. Theor. Appl. Genet. *58*, 253–256.

137. Simmons, L. B. 1979. Virus resistance and symptom expression in *Capsicum* species. M.A. Thesis. Louisiana State Univ., Baton Rouge.

138. Simons, J. N. 1980. Use of mineral oil sprays to control aphid transmitted viruses. Natl. Pepper Conf., New Mexico State Univ., Las Cruces, 5th, 1980. Abstr. No. 9 (JMS Flower Farms, Inc. 1105 25th Ave., Vero Beach, Florida 32960).

139. Smith, D. A. 1985. Jar-roll: An improved process for glass packed pimientos. J. Food. Sci. (In Press).

140. Smith, P. G. 1950. Inheritance of brown and green mature fruit colors in peppers. J. Hered. *41*, 138–140.

141. Smith, P. G., and Heiser, C. B. 1957. Taxonomy of *Capsicum sinense* Jacq. and the geographic distribution of the cultivated *Capsicum* species. Bull. Torrey Bot. Club *84*, 413–420.

142. Smith, P. G., Kimble, K. A., Grogan, R. G., and Millet, A. H. 1967. Inheritance of resistance in peppers to *Phytophthora* root rot. Phytopathology *57*, 377–379.

143. Sosa-Coronel, J., and Motes, J. E. 1983. Effect of gibberellic acid and seed rates on pepper seed germination in aerated water columns. J. Am. Soc. Hortic. Sci. *107*, 290–295.

144. Sowell, G. 1976. Summary of reports on the resistance of plant introductions to diseases, insects and nematodes in *Capsicum*. U.S. Regional Plant Introduction Station, Experiment, GA.

145. Sowell, G., and Dempsey, A. H. 1977. Additional sources of resistance to bacterial spot of pepper. Plant Dis. Rep. *61*, 684–686.

146. Spasojevic, V., and Webb, R. E. 1972. Inheritance of abscission of ripe pepper fruit from its calyx. Arh. Biol. Nauka *23*, 115–119. Plant Breed. Abstr. 1974. *44*, Abstr. No. 7110.

147. Stall, R. E. 1982. Selection for components of horizontal resistance to bacterial spot of pepper. *In* Proceedings of the Fifth International Conference on Plant Pathogenic Bacteria, Cali, Columbia, 1981, J. C. Logano (Editor), pp. 511–517. Centro International de Agricultura Tropical.

148. Stebbins, G. L. 1950. Variation and Evolution in Plants. Columbia Univ. Press, New York.

148a. Subramanya, R. 1982. New pepper plant types and their potential in Florida. Proc. Fla. State Hort. Soc. *95*, 317–319.

149. Subramanya, R. 1982. Relationship between tolerance and resistance to pepper mottle virus in a cross between *Capsicum annuum* L. × *Capsicum chinense* Jacq. Euphytica *31*, 461–464.

150. Subramanya, R. 1983. Transfer of genes for increased flower number in pepper. HortScience *18*, 747–749.

151. Subramanya, R., and Ozaki, H. Y. 1980. Inheritance of pedicel length in pepper (*Capsicum annuum*). Natl. Pepper Conf., Univ. California, Davis, 5th, 1980. Abstr. No. 4 (Univ. Florida, AREC, Belle Glade).

152. Suzuki, J. I., Tausig, F., and Morse, R. E. 1957. Some observations on red pepper. I. A new method for the determination of pungency in red pepper. Food Technol. *11*, 100–104.

153. Taylor, R. H., Grogan, R. G., and Kimble, K. A. 1961. Transmission of tobacco mosaic virus in tomato seed. Phytopathology *51*, 837–842.

154. Ullasa, B. A., Rawal, R. D., Sohi, H. S., Singh, D. P., and Joshi, M. C. 1981. Reaction of sweet pepper genotypes to anthracnose, *Cercospora* leaf spot and powdery mildew. Plant Dis. *65*, 600–601.

155. Van Blaricom, L. O., and Martin, J. A. 1947. Permanent standards for chemical test for pungency in peppers. Proc. Am. Soc. Hortic. Sci. *50*, 297.

156. Van den Berkmortel, L. G. 1978. Sweet pepper cultivation and breeding in The Netherlands. Natl. Pepper Conf., Louisiana State Univ., Baton Rouge, 4th, 1978. Abstr. No. 10 (Bruinsma Seed Co., Naaldwijk).

157. Villalon, B. 1981. Breeding peppers to resist virus diseases. Plant Dis. *65*, 557–562.

158. Villalon, B. 1982. Personal communication. Tex., Agric. Exp. Stn., Weslaco.

159. Warwick, B. L. 1932. Probability tables for mendelian ratios with small numbers. Tex., Agric. Exp. Stn., College Station. [Bull.] *463*.

160. Webber, H. J. 1912. Preliminary notes on pepper hybrids. Am. Breed. Assoc., Annu. Rep. *7*, 188–199.

161. Whitam, H. K. 1974. The epidemiology of virus diseases of bell peppers (*Capsicum annuum* L.) in Louisiana. Ph.D. Thesis. Louisiana State Univ., Baton Rouge.

162. Zitter, T. A. 1972. Naturally occurring pepper virus strains in south Florida. Plant Dis. Rep. *56*, 586–590.

163. Zitter, T. A. 1973. Further pepper virus identification and distribution studies in Florida. Plant Dis. Rep. *57*, 991–994.

164. Zitter, T. A. 1983. Personal communication. Dep. Plant Pathol., Cornell Univ., Ithaca, NY.

165. Zitter, T. A., and Cook, A. A. 1973. Inheritance of tolerance to a pepper virus in Florida. Phytopathology *63*, 1211–1212.

166. Zitter, T. A., and Ozaki, H. Y. 1978. Use of oil sprays to delay spread of nonpersistent aphid borne viruses. Natl. Pepper Conf., Louisiana State Univ., Baton Rouge. 4th, 1978. Abstr. No. 5 (Univ. of Florida, AREC, Belle Glade).

4

Tomato Breeding

EDWARD C. TIGCHELAAR

INTRODUCTION

The cultivated tomato (*Lycopersicon esculentum* Mill.) is a relatively recent addition to the world's important food crops. Within the past century, it has become one of the most popular and widely consumed vegetable crops, with an annual world production approaching 50 million MT (Table 4.1). It is also America's most popular and pampered home garden vegetable, occupying space in more than 90% of the home gardens planted in the United States (6). Per capita consumption in the United States has more than tripled in the past 50 years to approximately 56 lb per person. Its versatility in fresh or processed form has played a major role in its rapid and widespread adoption as an important food commodity.

The tomato is a tender perennial that is almost universally cultivated as an annual. Despite its susceptibility to frost, the tomato can be grown outdoors successfully from the equator to as far north as Alaska. Cultivars have been developed for a variety of different environments, methods of production, and food uses. Its adaptation to fit many diverse

BREEDING VEGETABLE CROPS

TABLE 4.1. World Production of Tomatoes[a]

	Area (×10³ ha)	Production (×10³ MT)	Yield (MT/ha)
World	2,404	49,201	20.5
Continent			
Africa	354	4,769	13.5
North and Central America	314	9,847	31.3
South America	135	2,854	21.2
Asia	706	11,251	15.9
Europe	490	13,871	28.3
Oceania	10	208	21.3
Leading countries			
1. United States	182	7,663	42.1
2. USSR	395	6,400	16.2
3. Italy	126	4,294	34.1
4. China	284	3,930	13.8
5. Turkey	108	3,136	29.2
6. Egypt	139	2,421	17.5
7. Spain	64	2,050	32.0
8. Greece	40	1,669	41.8
9. Brazil	56	1,500	26.6
10. Rumania	72	1,393	19.3

[a]After Yamaguchi (122).

uses and environments is a reflection of the great wealth of genetic variability existent in the genus *Lycopersicon* and the relative ease with which this diversity can be exploited in applied breeding programs.

Tomatoes are readily available in the U.S. marketplace in fresh and processed form throughout the year. The fresh and processed tomato industries are distinct entities, and cultivars and systems of crop management are generally unique for each.

Fresh Tomatoes

Fresh tomatoes are available in the United States in greatest supply during the summer months (June–September). Although almost all states grow fresh tomatoes to some extent, Florida, California, and Mexico provide the bulk of annual supplies, largely as a result of their long cropping seasons and/or mild winters (54). Fresh tomatoes also rank as the leading greenhouse-produced vegetable; however, less than 5% of the annual fresh production in the United States is derived from this source. In Western Europe climatic limitations require more extensive use of greenhouse facilities to extend availability of fresh tomatoes throughout the year.

Processed Tomatoes

Processed tomato products reach the consumer in a variety of forms or as ingredients in a wide array of processed commodities. A major portion of the per capita increase in tomato consumption in the United States during the past four decades is attributable to increased use of processed tomato products (Table 4.2). The dramatic growth of the fast-food industry and the rapid rise in popularity of food items containing tomato, such as pizza, have fostered a steady rise in consumption of processed tomato products.

TABLE 4.2. Fresh and Processing Tomato Production and Yields in the United States (1940–1983)[a]

Period	Fresh market		Processing	
	Volume (cwt × 10⁶)	Yield (cwt/acre)	Volume (tons × 10⁶)	Yield (tons/acre)
1940–1943	14.6	66	3.4	5.4
1944–1947	17.1	64	3.5	5.8
1948–1951	17.6	75	3.2	8.2
1952–1955	19.3	84	2.8	10.1
1956–1959	19.5	93	3.0	12.1
1960–1963	20.2	127	4.4	15.3
1964–1967	20.6	135	4.7	16.3
1968–1971	18.8	130	5.6	19.8
1972–1975	19.7	152	6.8	21.2
1976–1979	21.8	172	6.9	21.6
1980–1983	26.2	209	6.6	23.7

[a]Adapted from Brandt *et al.* (*17*) and Sullivan (*104, 105*).

California is by far the leading producer of tomatoes for processing, supplying more than 85% of the processed tomato products manufactured in the United States each year. Italy, Spain, and Greece are major suppliers of processed tomato products to the world marketplace. Production in these areas is favored by long, dry growing seasons that facilitate crop management, improve predictability of supplies, and provide a consistent quality of raw product for processing purposes.

The tomato does not rank high in nutritional value. By virtue of volume consumed, however, it contributes significantly to dietary intake of vitamins A and C as well as essential minerals. Its popularity is due in large measure to its versatility and the variety it lends to our diet. As an ingredient in numerous foods, from lasagna to a well-spiced Bloody Mary, the tomato has truly become a dietary staple.

ORIGIN AND EARLY HISTORY

Numerous wild and cultivated relatives of the tomato can still be found in a narrow, elongated mountainous region of the Andes in Peru, Equador and Bolivia as well as in the Galapagos Islands. These primitive relatives of the edible tomato occupy diverse environments based on latitude as well as altitude and represent an almost inexhaustible gene pool for improvement of the species (*2*).

Domestication and cultivation of the tomato appears to have first occurred outside its center of origin by early Indian civilizations of Mexico. The cultivated tomato is common in Peru today; however, it is used as a food primarily by the non-Indian population. Where it is cultivated by the native Indians of Peru, it appears to be a recent addition to their diet. Quite the contrary is true in Mexico, where the tomato is widely used by Indians and great diversity is evident in cultivars being grown. Furthermore, the name "tomato" comes from the Nahuatl language of Mexico, and variants of this name have followed the tomato in its distribution throughout the world (*40*).

The first written account documenting the arrival of the tomato in the Old World appeared in 1554 by the Italian herbalist Pier Andrea Mattioli. The first cultivars introduced to Europe probably originated from Mexico rather than South America. These early introductions were presumably yellow, rather than red in color, since the plant was first known in Italy as *pomi d'oro* or golden apple. It was also known as the love apple, *pomme d'amour,* in France. This appealing name did little, however, to hasten its acceptance as a food crop. In most places, the tomato was remarkably slow to gain acceptance, except as an ornamental curiosity. Apparently the tomato's similarity to familiar poisonous members of the nightshade family such as mandrake and belladonna caused concern over its safety as a food. Such unfounded superstitions persisted widely, even into the twentieth century and undoubtedly had a major impact in slowing its adoption as a useful and nutritious food crop.

The first recorded mention of the tomato in North America was made in 1710. It was apparently brought from the Old World by early colonists but did not gain widespread acceptance, presumably because of the persisting view that its fruits were unhealthy and poisonous. Thomas Jefferson wrote of tomato plantings in Virginia in 1782 and makes frequent reference to its planting and culinary uses in later writings. However, it was not until 1830 that the tomato began to acquire the popularity that has made it the indispensable food commodity it has become today. Its history as a commercially processed commodity began at Lafayette College at Easton, Pennsylvania, in 1847, where the commercial "canning" possibilities of the "love apple" were first demonstrated. From this humble beginning, the tomato has become the leading processed vegetable crop in the United States today.

The increasing popularity of the tomato resulted in a rapid proliferation of new cultivars. In 1863, 23 cultivars were known; however, within two decades the number of cultivars available to growers had increased to several hundred (*61*). Liberty Hyde Bailey of the Michigan Agricultural College initiated a testing program in 1886 to clarify the classification of tomato cultivars and reported that much of the confusion was a result of indiscriminate renaming by seed suppliers (*61*).

Livingston was probably the first American to recognize the need for constructive breeding. Between 1870 and 1893, he introduced 13 cultivars developed by single plant selection to meet specific requirements of tomato producers and consumers. The worth of useful new cultivars was clearly appreciated during this early history of the tomato as evidenced by the fact that seed of the cv. Trophy was sold for 5 dollars per packet of 20 seeds when it was introduced in 1870.

The early progress of tomato breeding in the United States is poorly documented and is best illustrated by the length of time cultivars remained in demand and listed by seed suppliers (*61*). On this basis, the cvs. Red Cherry, Red Pear Shaped, and Trophy represented the most important, popular, or persistent cultivars during the early history of the tomato in the United States. Few, if any, of these cultivars are of commercial significance today; however, they represent the foundation upon which modern-day cultivars were developed.

Extensive efforts are under way to maintain old cultivars and the invaluable wild relatives that have served as the progenitors of the present-day tomato. The U.S. Seed Storage Laboratory in Fort Collins, Colorado, has the responsibility of maintaining seed of old cultivars. Tomato introductions from foreign countries and from germplasm collection expeditions are maintained by the North Central Regional Plant Introduction Station at Ames, Iowa, and are available to private and public breeders working on tomato improvement.

TABLE 4.3. The Species of the Genus *Lycopersicon*[a]

Species	Common name	Somatic chromosome number	Reproductive features[b]
L. esculentum	Common tomato	24	SP
L. pimpinellifolium	Currant tomato	24	SP + CP
L. cheesmanii	Wild species	24	SP
L. parviflorum	Wild species	24	SP
L. chmielewskii	Wild species	24	CP
L. pennellii	Wild species	24	SI
L. hirsutum	Wild species	24	SF, SI
L. chilense	Wild species	24	SI
L. peruvianum	Wild species	24	SI

[a]Adapted from Rick (*81*).
[b]SP, self-pollinated; CP, cross-pollinated; SF, self-fertile; and SI, self-incompatible.

BOTANICAL CLASSIFICATION

The tomato is a member of the nightshade family (Solanaceae) and the genus *Lycopersicon*, which contains several species commonly divided into two subgenera. The subgenus *Eulycopersicon* includes red-fruited species and *Eriopersicon* mostly green-fruited types. At present, nine species are recognized as distinctive entities within the genus (Table 4.3). Controversy still exists, however, regarding taxonomic classification of the wide variability found within the genus *Lycopersicon*. All members of the genus are annual or short-lived perennial herbaceous diploids with a somatic chromosome number of 24. Essentially all cultivated forms of the tomato belong to the species *esculentum*.

The relatives of the cultivated tomato have proven to be an invaluable source of germplasm for plant improvement (*76*). Interspecific crosses between *L. esculentum* and *Lycopersicon pimpinellifolium* are easily made and show few, if any, barriers to gene exchange. Both members of the subgenus *Eulycopersicon* are also compatible with members of the subgenus *Eriopersicon,* but in some cases only when the latter functions as the pollen parent (Table 4.4). Where such unilateral incompatibility exists, the F_1 interspecies hybrid can be crossed to *L. esculentum* only when the F_1 functions as the male, and to the wild parent only when the F_1 is used as female. The self-incompatibility common to many of the wild species is also transmitted to interspecies hybrids, and aberrant genetic segregation is common in such wide crosses (*58*). Embryo abortion may occur in crosses of *L. esculentum* with *Lycopersicon peruvianum,* but this barrier can be overcome by use of

TABLE 4.4. Survey of Intra- and Interspecific Breeding Barriers in *Lycopersicon*[a,b]

♀ \ ♂	*L. esculentum*	*L. pimpinellifolium*	*L. hirsutum*	*L. chilense*	*L. peruvianum*
L. esculentum	+	+	+	EA	EA
L. pimpinellifolium	+	+	+	EA	EA
L. hirsutum	+, UI	+, UI	+, SI, UI	?	EA
L. chilense	UI	UI	?	SI	EA
L. peruvianum	UI	UI	UI	EA	SI

[a]Adapted from Hogenboom (*45*).
[b]+, No serious barrier; SI, self-incompatibility; UI, unilateral incompatibility; EA, embryo abortion; ?, no research results known.

embryo culture (*92*). Thomas and Pratt (*106*) have recently shown that hybridization between *L. esculentum* and *L. peruvianum* can be enhanced by regenerating plants from embryo callus rather than direct embryo culture. Backcrosses of the embryo callus hybrids to the *L. esculentum* parent were also produced in this fashion to facilitate introgression between these two species. Successful intergeneric crosses have also been made between *L. esculentum* and the closely related *Solanum* species *S. lycopersicoides* (*75,118*). A relatively free (although sometimes difficult) exchange of genes between species is thus possible, although it may require the use of special techniques.

REPRODUCTIVE BIOLOGY

The cultivated tomato has been a favored crop for genetic studies because of the wealth of variability within the species and the ease with which it can be manipulated. It is normally highly self-pollinated; flowers are easily emasculated and pollinated; and individual crosses may yield as many as several hundred seed. Rates of natural cross-pollination in temperate zones vary from 0.5 to 4% (*80*); however, much higher rates occur in Peru, presumably as a result of native insect vectors that can transfer the pollen. Rick suggests that the change from moderate cross-fertilization to almost exclusive self-fertilization occurred following introduction to Europe, and was accompanied by a change in stigma position from outside to within the anther cone.

The tomato flower is normally perfect, having functional male (anthers) and female (pistil) parts (Fig. 4.1). Several (usually four to eight) flowers are borne in each compound inflorescence and a single plant may produce as many as 20 or more successive inflorescences during its life cycle. This feature facilitates crosses between cultivars that represent extremes in variation for maturity since flowering occurs over a long period of time.

Present cultivated varieties form a tight protective anther cone surrounding the stigma, which greatly reduces the possibility for natural cross-fertilization. Outdoors, flower movement aided by wind is sufficient to release pollen, but under greenhouse conditions, manual vibration of open flowers is required to effect pollination and fruit set. Genetic or environmental modification of stigma position can affect both fruit set and degree of cross-fertilization.

Emasculation for the purpose of controlled pollination must be done approximately 1 day prior to anthesis or flower opening to avoid accidental self-pollination. At this time, the sepals have begun to separate and the anthers and corolla are beginning to change from light to dark yellow, characteristic of fully opened flowers. The stigma appears to be fully receptive at this stage, thus allowing for pollination immediately after emasculation. With favorable environmental conditions, 200 or more seed may be obtained from a single pollination. Generally, under greenhouse conditions no protection is required following emasculation to prevent uncontrolled crossing. Making controlled pollinations under field conditions may be less efficient than under greenhouse environments because hot, drying winds may cause rapid desiccation of the exposed pistil before fertilization is achieved. Cool, dry, and relatively wind-free weather is preferred for high success rates with outdoor crossing, and protection of flowers with glassine bags may be necessary to avoid chance crosses.

Under optimal temperature and growth conditions, the tomato will complete its reproductive cycle in 95–115 days, depending upon cultivar (*18*). The first flowers open 7–8 weeks after seeding, and an additional 6–8 weeks elapse from first flower to ripe fruit.

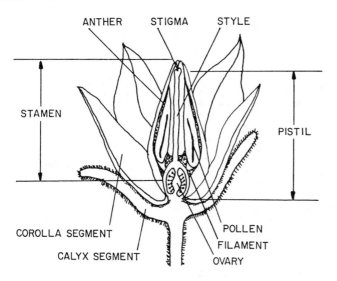

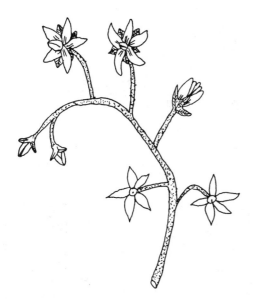

FIGURE 4.1. Tomato flower and inflorescence.

Seed is physiologically mature when the fruit reaches full ripeness. This makes it theoretically possible to complete three reproductive cycles a year using greenhouse facilities for off-season plantings.

Self-incompatibility is a common feature of the wild relatives of the tomato, and is transmitted to hybrids with *L. esculentum*. Self-incompatibility is of the *Nicotiana* type conditioned by a single locus. Genetic male sterility has also been reported frequently within the genus and many loci producing male sterility have been identified and described (*74*).

The tomato has proved to be an ideal plant for genetic studies because of its relatively simple reproductive biology, its ease of culture, and the wealth of genetic variation in

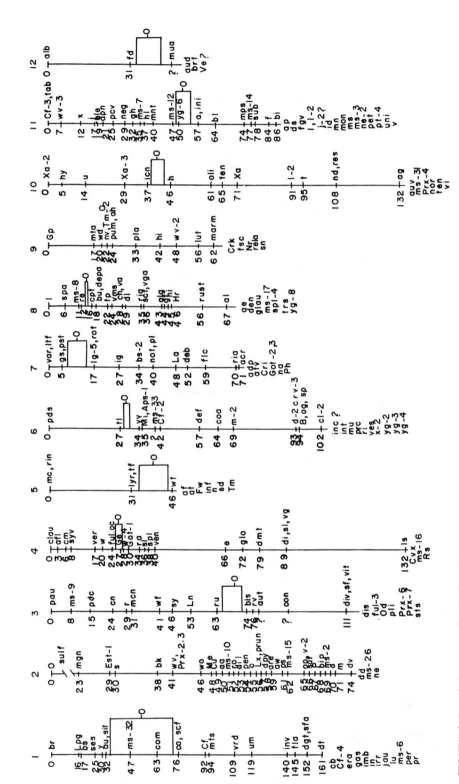

FIGURE 4.2. Linkage map for the tomato.

142

cultivated and wild forms. It has been more extensively studied genetically than any other major food crop (excepting possibly maize), and more than 970 genes had been reported by 1979. The Tomato Genetics Cooperative, established in 1951 by Dr. C. M. Rick, University of California at Davis, has provided an invaluable service to the many workers in tomato genetics by coordinating gene nomenclature and mapping efforts. The annual report of this organization[1] also provides brief research reports on tomato genetics as well as new cultivar pedigrees. In addition, descriptive lists of new genes are published periodically to update workers on the rapidly accumulating information on the genetics of the genus *Lycopersicon*.

The extensive genetic information that has accumulated from many years of research has permitted the development of genetic maps showing the relative location of many genes controlling a wide variety of traits (Fig. 4.2). Such maps have proven useful in the design and planning of breeding programs since linkage distances can be used to predict the probability of recombination between linked genes.

BREEDING HISTORY

The earliest written record of attempts to select and develop improved tomato cultivars date to the mid-nineteenth century in Europe. Livingston, in 1870, is generally recognized as the first tomato breeder in North America. However, as is the case with most cultivated crops, much of the very early improvement can be credited to those who first domesticated, cultivated, and consumed the crop. In the case of the tomato, the Indians of Mexico must certainly be credited for improvements that fostered its adoption as a new food crop in Europe and later in North America. As its popularity grew in the twentieth century, intensity of improvement efforts increased accordingly. During the past four decades, efforts by both public and private plant breeders have resulted in spectacular improvements in yield and other characters (see Table 4.2). New cultivars grown under improved methods of crop culture and management have resulted in fourfold increases in yield per hectare for California processing tomatoes since 1940. Without this progress, processed tomato products would undoubtedly be a commodity that only a select few could afford.

Many hundreds of new cultivars have been developed within the past 40 years to meet the diverse needs and varied situations and climates under which the tomato crop is grown. The recent trend has been toward development of cultivars to meet specific uses rather than multipurpose cultivars to meet several needs. For example, fresh-market and processing cultivars are distinct today, largely as a result of the different quality requirements for intended use. Likewise, cultivars for greenhouse culture generally differ from their outdoor counterparts because of the vastly different cultural systems used in production. Many breeding situations encountered today in tomato (as well as other crops) involve fitting the crop to its intended environment, cultural system, method of harvest and/or handling, and proposed food use.

The research and development in the 1950s and 1960s that converted processing tomatoes from a hand-harvested to an almost exclusively machine-harvested crop (Fig. 4.3) are an example of cooperative efforts among several disciplines to achieve a successful system. The development of the first machine-harvestable cultivar by G. C. Hanna at the University of California at Davis involved a major redesign of standard cultivar

---

[1]Report of the Tomato Genetics Cooperative, Department of Horticulture, Purdue University, West Lafayette, IN 47907.

FIGURE 4.3. Machine harvest of processing tomatoes in California.
Courtesy of Johnson Farm Machinery Co., Woodland, CA.

characteristics. The new crop type had a small vine, concentrated fruit set, and adequate firmness to withstand machine handling in addition to having the required yield, disease resistance, and quality characters. Accompanying changes in crop culture and management and the design of an efficient machine to harvest the crop were equally important elements in the system. Success in modern-day plant breeding requires imagination, cooperation with other disciplines, and an intimate knowledge of the environmental and cultural aspects of production of the crop to be improved.

Several "highlights" mark the short history of tomato improvement and its dramatic increase in popularity to the status of a food staple. Foremost among these is the fact that fruit quality has remained a major focus of most tomato breeding programs (despite contrary assertions in the popular press) while cost to the consumer has declined in real terms as a result of increased efficiencies of production.

Several simply inherited genetic characters have had important impacts on the improvement of new cultivars and the changes in culture made possible by their development (73). One such gene is *sp* (self-pruning), which appeared as a spontaneous mutation in Florida in 1914 and has been used in the vast majority of cultivars developed during the past two decades. This recessive allele conditions determinate habit of growth, which results in two or fewer nodes between inflorescences (as opposed to three or more in indeterminate types). This results in more compact growth and early, more concentrated flowering, a feature essential for machine-harvestable cultivars. This single feature made possible the use of higher plant populations to increase yield per unit area and the reduction of the number of times the crop had to be harvested. The same innovative process, popularly referred to in the early 1970s as "the green revolution" for wheat and rice, had been applied to tomato improvement and production several years earlier.

Numerous simply inherited characteristics are part of the architecture of modern-day tomato cultivars (Table 4.5). Single genes also control resistance to many of the common diseases, and dominance of resistance has facilitated the development of F_1 hybrids with resistance to as many as eight different pathogens (35). Much of these disease resistances originated in the wild relatives of the tomato and were transferred by recurrent backcrossing to adapted cultivars of *L. esculentum*. Bohn and Tucker (15) pioneered work that identified the dominant allele *I*, controlling resistance to fusarium wilt in *L. pimpinellifolium*. The first resistant cultivar developed from this interspecies hybrid was appropriately named Pan American to reflect the North and South American parentage in its pedigree (69).

TABLE 4.5.   Examples of Single Genes That Have Been Useful for Tomato Improvement

Gene designation	Gene symbol	Variety
	Growth habit	
Self-pruning	*sp*	Many
Brachytic	*br*	Redbush
Dwarf	*d*	Epoch, Tiny Tim
Potato leaf	*c*	Geneva 11
Jointless pedicel	*j-1*	Penn Red
	j-2	Many
	Pest resistance	
Leaf mold resistance	*Cf-1*	Sterling Castle
	Cf-2	Vetamold
	Cf-3	V_{121}
	Cf-4	Purdue 135
Fusarium immunity		
race 1	*I-1*	Pan American
race 2	*I-2*	Walter
Verticillium resistance	*Ve*	VR Moscow
Septoria resistance	*Se*	Targinnie Red
Late blight resistance	*Ph-1*	New Yorker
Alternaria resistance	*Ad*	Southland
Stemphylium resistance	*Sm*	Tecumseh, Chico III
Tobacco mosaic resistance	*Tm,Tm-2,Tm-2²*	
Curly top virus	?	C5, Columbia
Spotted wilt virus	Several genes, race specific	Pearl Harbor, Rey de los Tempranos
Nematode resistance	*Mi*	Rossoll, VFN Bush
	Fruit characters	
Uniform ripening	*u*	Heinz 1350
High pigment	*hp*	Redbush
Green stripe	*gs*	Tigerella (novelty)
High beta	*B*	Caro-Rich
Old gold crimson	*og^c*	Vermillion
Low total carotene	*r*	Snowball
Tangerine	*t*	Sunray, Jubilee
Colorless peel	*y*	Traveller
Nonripening	*nor^A*	Long Keeper
Male sterility	many genes	Some F_1 hybrids
Parthenocarpic fruit	*pat-2*	Severianin

Pan American and its offspring served as a primary source of resistance to this disease until 1960, when a new race of the pathogen appeared in Florida. Resistance to this new race (designated race 2) was quickly identified in a plant introduction (PI 126915), and the cv. Walter with resistance to both races was released in 1969 (*103*). Recently, a third race of the *Fusarium* organism has been reported in Australia for which tolerance has been identified in *L. pimpinellifolium* (PI 124034) as well as two *L. esculentum* breeding lines, US 629 and US 638 (*57,116*).

Disease resistance has been a major contribution of past breeding efforts, and current varieties generally possess resistance to one or more pathogens. Host resistance has been particularly important for control of soil-borne pathogens such as verticillium and fusarium wilts, for which chemical control has been relatively ineffective or costly. Virtually all important present-day cultivars possess resistance to one or both of these diseases.

The use of F_1 hybrid varieties has increased dramatically in recent years, particularly for fresh-market and home garden production. Hybrid cultivars generally do not show large yield advantages when compared with inbred varieties; their advantage appears to derive from improved earliness and better consistency of performance, particularly under less than optimal growing conditions (*123*). Virtually all hybrid seed production is done manually, thus requiring skilled yet inexpensive labor. Taiwan has become a leading producer of hybrid tomato seed (*7*).

BREEDING GOALS

Effective crop improvement programs require clearly defined objectives and a well-conceived breeding strategy to accomplish established goals. Improved yield and quality are universal goals of most breeding programs; however, selecting for yield per se is seldom very effective. Instead, the plant breeder must often define the production system and the individual components that contribute to yield or quality and emphasize selection for those individual attributes. This may mean that primary emphasis is placed on selection for such characteristics as disease resistance, earliness, habit of growth, or some novel feature rather than yield per se. Frequently, plant improvement involves adapting the crop to changes in culture and management, to the vagaries of weather and pests, or to anticipated future needs of the producer, processor, or consumer. The effective plant breeder must therefore be intimately acquainted with industry and consumer needs (as well as the genetic diversity of the crop to be improved) to establish relevant and realistic goals.

Four distinct uses and/or methods of culture characterize the tomato industry, and breeding objectives will depend upon the intended use of the new cultivar. Whereas a decade or so ago, most tomato cultivars served multiple purposes, modern cultivars are developed specifically for processing, fresh-market, greenhouse, or home garden use. This has occurred largely because the quality and/or cultural requirements may be quite distinct for each of these four uses.

Processing Tomatoes

Processing tomatoes are grown on large acreages with highly mechanized production systems (Fig. 4.3). Direct field seeding and harvest mechanization, which require high plant populations to achieve the concentrated fruit set needed for mechanical harvest, have fostered the development of compact, highly determinate processing cultivars to fit the systems of culture and harvest used. These features must be combined with other essential horticultural characteristics—disease resistance, firm fruit, earliness, ability to set fruit at adverse temperatures, resistance to rain-induced cracking of fruit, tolerance to major ripe-fruit rots, ease of fruit separation from the vine, and adequate vine cover—which are needed for adaptation to the environment in which the cultivar will be grown.

Fruit quality has also been a very important consideration in processing-tomato breeding programs. Several individual parameters of quality—color, pH, total acidity, soluble solids, total solids, and viscosity—are recognized and their relative importance depends upon the processed product for which the cultivar is to be used. Improved fruit quality has been a major objective of breeding programs supported by the food processing industries since it influences both quality and "case yield" (number of cases of processed product per unit of raw fruit) of the finished product.

Case yield, in turn, depends upon specific quality attributes that influence the amount of fruit required to produce a unit of processed product. Tomato paste standards, for example, are based upon final soluble-solids content of the finished product. As a consequence, high-soluble-solids cultivars yield more cases per ton and require less energy in concentration than do low-solids cultivars. For a product such as catsup, in contrast, viscosity (or consistency) may be the primary quality attribute influencing the number of cases of finished product produced per ton of fruit. In the highly competitive food industry, varietal or location differences in case yield may be the difference between success and failure in producing a particular processed product.

Fresh-Market Tomatoes

Quality requirements and methods of crop management for fresh use have become sufficiently distinct from processing use that cultivars seldom serve both purposes. The marketplace demands large, round fruit (to fit a hamburger bun conveniently?) with adequate firmness and shelf life for shipping to distant markets; uniform fruit size, shape, and color; and freedom from external blemishes or abnormalities. These features must be combined with the required horticultural characteristics—earliness, growth habit, disease resistance, and adaptation to environment—to make a successful cultivar.

In recent years, fresh-tomato quality has come under criticism by the consuming public, and the tomato breeder has been held accountable for many of the deficiencies of this favored vegetable crop. The issue illustrates the complex, yet subjective, nature of our perception of quality. Taste panel studies carried out at the University of California at Davis have clearly shown that free sugars, organic acids, and the sugar : acid ratio are the major identifiable determinants of flavor preference. Color, appearance, and texture, however, also contribute to perception of quality. In a study at Purdue University, taste panelists judged orange-fruited cultivars as inferior in flavor to red cultivars unless the color differences were masked by colored lights. We apparently perceive flavor with our eyes as well as our taste buds and have been conditioned by past experience to decide what represents superior quality.

Appearance has probably received more emphasis in breeding programs than flavor or other sensory aspects of quality. The recent view that modern tomato cultivars are inferior in quality to their predecessors is, however, difficult to document. The expectation that this fresh commodity be harvested (usually "green mature") and shipped thousands of miles during the winter months and still have a taste equivalent to a fully mature fruit freshly picked from the home garden may be more than modern technology can provide.

Greenhouse Production

Cultivar requirements for controlled-environment or glasshouse tomato production are quite different from those for outdoor culture. Whereas the majority of cultivars for outdoor commercial production are determinate in growth habit and produce for a relatively short period of time, greenhouse cultivars are generally indeterminate and will produce for several successive months (Fig. 4.4). In addition, certain disease problems that do not frequently cause serious losses in outdoor culture—tobacco mosaic virus (TMV), leaf mold (*Cladosporium fulvum*), gray mold (*Botrytis cinerea* Pers.), and whitefly (*Trialeurodes vaporariorum* Westwood)—may be serious problems under intensive greenhouse production.

Escalating energy costs for greenhouse heating have prompted recent efforts to develop cultivars that would perform at lower temperatures and light intensities. European and Canadian workers have been particularly active in glasshouse tomato breeding, largely because protected-environment production in these areas represents a major source of supply of this commodity during times of the year when outdoor production is not possible.

Home-Gardening Cultivars

Home-gardening cultivars constitute a particularly diverse assortment of types. This diversity provides a measure of the environmental as well as personal preference differences that characterize home gardens and their keepers throughout North America. The tomato is unquestionably the most popular and pampered garden vegetable, and many cultivars have been developed to meet the unique needs and desires of the home gardener. Several fresh-market and processing cultivars have achieved popularity among the home-gardening population. Earliness, appropriate disease resistance, large fruit size, high fruit quali-

FIGURE 4.4. Hydroponic production of greenhouse tomatoes.

ty, and continuous production throughout the gardening season are important attributes of cultivars destined for home garden use. In addition, novelty is frequently a desirable characteristic, and the use of yellow- and orange-fruited cultivars (some of which are nutritionally superior to red-fruited cultivars) has been largely restricted to home gardens. The American consumer has apparently been conditioned to believe that tomatoes should be red and the marketplace offers, in the vast majority of cases, only red-fruited types. The home gardener, on the other hand, is often willing to experiment with novel variants, and several unique cultivars are available strictly for such home garden use. Extremely dwarf cultivars have been developed for use by "high-rise" gardeners whose only area may be a window box or hanging basket. For those who wish to savor the fruits of their gardening efforts well after the last frost, long-storing cultivars are being developed to extend fresh-fruit shelf life.

SPECIFIC BREEDING OBJECTIVES

Earliness

Earliness is of particular importance in short-season areas to extend the production or gardening season. The recent trend to establish commercial plantings by direct field seeding rather than by transplants has further increased the importance of earliness. Three main components contribute to earliness: time from planting to flowering, time from flowering to the initiation of ripening, and the concentration of flowering (or number of flowers produced per unit of time). Wide variation exists for each of the above components of earliness (*50*). The earliest cultivars will, under optimal conditions, produce mature fruit less than 90 days after seeding.

Growth Habit

Many genes affecting growth habit have been identified and described. Certainly the most important has been the self-pruning (*sp*) gene, which conditions determinate habit of growth. This recessive gene has been used in the vast majority of cultivars released in the past two decades. It contributes to smaller vine size, since inflorescences are borne closer than at every third node (or more), as in indeterminate types. Genes controlling dwarf (*d*) and brachytic (*br*) habits of growth have also been used; however, their acceptance has been limited to unique production situations (e.g., pot culture). These very compact vine types may offer greater potential in cultivars of the future, particularly where use of high plant populations and/or machine harvest is anticipated.

Machine Harvestability

The development of tomato cultivars for machine harvest has involved a major redesign of plant structure to fit both the machine and the cultural systems required for harvest mechanization. The most important design changes have involved development of cultivars with compact growth habit and concentrated fruit set. Vine storability, or the ability of fruit to remain sound and usable on the vine following ripening yet retain adequate firmness to withstand the rigors of mechanical handling, has also been essential for successful machine harvest cultivars. These changes to accommodate machine harvest compromised certain quality attributes (particularly fruit soluble solids), and a major focus of recent improvement efforts has been directed toward enhancement of this component of processing quality.

In humid areas, the challenge of developing tomato cultivars for machine harvest has been considerably more difficult than in the arid west. Resistance to rain-induced fruit cracking (72), tolerance to major fruit rots (11,12), and improved concentration of fruit set have been necessary in these areas to minimize field losses for once-over destructive harvest. Very compact vine types used at high populations seem to offer the greatest potential for success in such areas, where rainfall may interfere with harvest scheduling.

Disease Resistance

Without question, the greatest contribution of modern plant breeding to tomato improvement has been through the development of cultivars resistant to common pathogens. For certain organisms (notably soilborne fungi responsible for fusarium and verticillium wilts), production would be considerably more difficult and costly in the major production areas without host resistance. Breeding for resistance remains a major goal as new diseases achieve significance or new races of existing pathogens become established. The dynamic interplay of host and pathogen provides job security and a guaranteed challenge to tomato breeders serving major production areas.

Host resistance has been identified and described for most of the major pathogens of tomato, and in many cases the inheritance of resistance is clearly known (Table 4.6). Early progress was fostered by cooperative efforts in the early 1950s to screen germplasm collections for reaction to major tomato diseases. Voluntary cooperative screening by cooperators in both the public and private sector resulted in a major national program to exploit variability for disease resistance within the wild relatives of tomato (3). Single dominant genes specifically confer resistance to several of the major tomato diseases, and inbred cultivars and F_1 hybrids with multiple resistance are now widely used where host resistance is necessary. A noble (but presently unachieved) goal would be the elimination of the extensive need for pesticides in tomato production by use of host resistance. Such an objective would require considerably more effort and expenditures in tomato breeding than is currently available.

Host resistance has often been derived from the wild relatives of tomato and incorporated into adapted cultivars by backcross breeding. For certain pathogens, several distinct races of the organism are known and new types may appear following the introduction of resistant cultivars. Surprisingly rapid evolution of new pathotypes of the organism responsible for the common greenhouse disease "leaf mold" (*Fulvia fulvum*) has provided greenhouse tomato breeders a "high degree of job security" (51). Fortunately, this disease is not widespread under field conditions and when it occurs, chemical control is relatively effective and is the chosen method for control. Such has not been the case for new races of the organism causing fusarium wilt for which the only effective means of control involves host resistance. In such cases, the identification of a new biotype is followed immediately by redirection of program objectives to locate and incorporate sources of resistance.

The wealth of genetic diversity in the genus *Lycopersicon* is clearly evident in the host resistance that has been reported for this extensively studied crop. Three examples will serve to illustrate how this resistance is identified and utilized.

Fusarium Wilt

The recent appearance in Australia of a third race of the organism causing fusarium wilt has fostered international cooperation to locate sources of resistance. It is expected that this new race (or a similar pathotype) will inevitably appear in other major tomato

TABLE 4.6. Sources of Resistance and Inoculation Techniques for Screening Tomato Diseases[a]

Disease	Causal organism	Source of resistance	Inoculation techniques	References
Fusarium wilt	*Fusarium oxysporum* f. sp. *lycopersici* (Sacc.)	Pan American (race 1), Walter (race 2)	Dip roots in inoculum and transplant to flats	69,103
Verticillium wilt	*Verticillium albo-atrum* Reinke & Berth.	VR Moscow	Dip roots in inoculum and transplant to flats	117
Late blight	*Phytophthora infestans* (Mont.) DBy.	West Virginia 63	Inoculate leaves with swarmspore suspension by atomization	30
Early blight	*Alternaria solani* (Ell. & G. Martin) Sor.	(on foliage) 68B134	Inoculate with atomized spore suspension	10
		(collar rot) Southland	Dunk tops and stems of seedlings into inoculum and transplant to a depth of 3 in.	71
Septoria leaf spot	*Septoria lycopersici* Speg.	Targinnie Red	Atomize conidial suspension onto plants	5,14
Gray leaf spot	*Stemphylium solani* Weber	Manalucie	Atomize spore suspension onto plants	42
Leaf mold	*Fulvia fulvum* Cke.	Sterling Castle	Inoculate by spraying lower-leaf surfaces with spore suspension	51
Foot and stem rot	*Didymella lycopersici* Kleb.	*L. hirsutum* 66087 (IVT 61292)	Apply 3 ml of inoculum around the stem base of 4-week-old seedlings	16
Brown root rot (corky root)	*Pyrenochaeta lycopersici*	PI 260397	Plant to fields infested with the causal organism	45,115
Anthracnose	[*Colletotrichum phomoides* (Sacc.) Chester] *C. coccodes* (Wallr.) Hughes	PI 272636	Inoculate fruit by atomization or puncture	12
Rhizoctonia soil rot	*Rhizoctonia solani*	PI 193407	Place mature green fruit on infested soil	11
Bacterial wilt	*Pseudomonas solanacearum* (E. F. Sm.) E. F. Sm.	PI 127805A, Saturn, Venus	Cut roots on one side; pour bacterial suspension into soil trench	1,41
Bacterial canker	*Corynebacterium michiganense* (E. F. Sm.) H. L. Jens.	Bulgarian 12, Utah 737	Inoculate cut made by excising first true leaf at point of attachment with bacterial suspension	24,108
Bacterial speck	*Pseudomonas tomato*	Ontario 7710	Atomize suspension of *P. tomato* on both sides of leaf	4,68
Tomato mosaic	Tobacco mosaic virus	Ohio M-R9	Apply expressed inoculum with an air brush; inoculate again 10 days later	4,65
Spotted wilt	Spotted wilt virus	Pearl Harbor	Place seedlings in a disease nursery and encourage thrips	29
Curly top	Curly top virus	CVF4	Release viruliferous leafhoppers into screened greenhouse 2 or 3 times at 1-week intervals	55
Yellow leaf curl	Yellow leaf curl virus	*L. pimpinellifolium* (LA 121)	Release viruliferous whitefly females onto caged tomato seedlings; allow feeding for 72 hr	67

[a]Adapted from Webb *et al.* (*120*).

producing areas of the world, thus justifying cooperative efforts to stay ahead of the organism's ability to evolve new races capable of infecting resistant cultivars. Several sources of resistance have been identified, and genetic studies are in progress to establish the breeding strategy most appropriate to utilize this resistance and identify the most effective resistance sources (*57,116*).

Anthracnose Fruit Rot

The adoption of mechanical harvest has had a major influence on the importance of resistance and/or tolerance to fruit rots, since ripe fruit must be held on the vine for up to 3 weeks for once-over destructive harvest. The most significant ripe fruit rot in the mid-western United States is anthracnose, caused by several species of *Colletotrichum*. An ambitious program to locate sources of resistance, develop screening procedures, and transfer resistance from small-fruited wild species to commercially acceptable germplasm was initiated by the USDA (*12*). This program has resulted in promising germplasm sources, which are currently being used by both public and private tomato breeders to develop cultivars with high levels of field tolerance to this disease.

Tobacco Mosaic Virus

The tomato is susceptible to many virus diseases. Greenhouse tomatoes are particularly prone to losses from tomato mosaic, caused by the tobacco mosaic virus (TMV), which may be mechanically transmitted during growing and harvesting operations. Host resistance has proven to be a most effective and reliable means of control (*46*). Several distinct strains of the virus are recognized, and the resistances identified for these strains are race specific as well as environment dependent (*21*). Dominant genes at two loci (*Tm-1* and *Tm-2*) have been used in tomato improvement. A third allele (*Tm-2*[a]) at the *Tm-2* locus is completely dominant at 20°C, but mild necrosis of hybrids occurs with certain virus strains at 30°C (*39*). Tolerance to TMV is widespread among the wild relatives of the tomato; these species will undoubtedly serve as a valuable reservoir of germplasm to meet future needs in breeding for mosaic resistance.

Insect Resistance

Insect resistance has received considerably less attention than disease resistance breeding, and very few commercial cultivars have been developed with specific resistances to problem insect pests. This situation is not a result of inadequate genetic variability but rather the low priority given to insect resistance in applied breeding programs and the difficulty of developing breeding and selection procedures to use this variability effectively. Use of pesticides has efficiently controlled most major insect pests and the tomato crop value justifies extensive use of this method. As an unfortunate consequence, little effort has been expended to utilize genetic resistance to facilitate insect control. We hope this situation will change as integrated pest management research attempts to develop viable alternatives to the use of pesticides for insect control.

Resistance or tolerance has been reported to most of the major insect pests of tomato (Table 4.7). McKinney, in 1938, was the first to report insect resistance, which he attributed to entanglement of the insects (aphids and thrips) in a gumlike exudate from the tomato foliage (*59*). Gilbert *et al.* (*36*) reported that certain Hawaiian cultivars showed resistance to spider mites (*Tetranychus telarius*). Reduced oviposition was later shown to be related to the frequency of glandular hairs on the foliage (*100*). A similar mechanism appears to influence resistance to whiteflies (*33,119*) and flea beetles (*Epitrix hirtipennis*) (*32*). In the latter case, glandular hair secretions appear also to influence the observed resistance, since washing leaves with 75% ethanol reduced resistance. In subsequent work, resistance to both mites and tobacco hornworm (*Manduca sexta*) has been associated with leaf trichome frequencies, type, and levels of 2-tridecanone (a methyl ketone) secreted from glandular leaf trichomes. Knowledge of the nature of resistance to specific insect pests will certainly facilitate screening and selection to utilize this variability (*89*).

The tomato fruitworm (*Heliothis zea*), also known as corn earworm, cotton bollworm, and soybean podworm, may be a devastating pest in commercial tomato plantings

TABLE 4.7.   Sources of Insect Resistance Reported in the Genus _Lycopersicon_

Insect pest	Resistance source	Species	Reference
Flea beetle	PI 126449	_L. hirsutum_	32,79
(_Epitrix hirtipennis_)		f. glabratum	
Potato aphid	PI 129145	_L. peruvianum,_	_31,70_
(_Macrosiphum euphorbiae_)		_L. pennellii_	
Spider mite	Anahu	_L. esculentum_	_36,101_
(_Tetranychus telarius_ L.)			
Carmine spider mite	Several cultivars	_L. hirsutum_	_100_
(_Tetranychus cinnabarinus_)			
Colorado potato beetle	PI 134417	_L. hirsutum_	_87_
(_Leptinotarsa decimlineata_)			
Pinworm	PI 127826	_L. hirsutum_	_89_
(_Keiferia lycopersicella_)			
Leaf miner	PI 126445	_L. hirsutum_	_90,121_
(_Liriomyza munda_)	PI 126449	_L. hirsutum_	
		f. _glabratum_	
Fruitworm	PI 126449	_L. hirsutum_	_27,28_
(_Heliothis zea_)		f. _glabratum_	
Tobacco hornworm	PI 134417	_L. hirsutum_	_49_
(_Manduca sexta_ L.)		f. _glabratum_	
Whitefly	IVT 74453	_L. hirsutum_	_25_
(_Trialeurodes vaporariorum_)	IVT 72100	_L. pennellii_	

throughout North and South America. Fery and Cuthbert found that leaves of _Lycosperi-con hirsutum_ contain a factor highly detrimental to development of fruitworm larva, resulting in high larval mortality (_28_). Since early larval stages feed on leaf tissues as their primary food source, it was concluded that this would be a valuable source of resistance.

The species _hirsutum_ and _pennellii_ appear to be particularly valuable sources of insect resistance, and efforts to exploit this variability deserve greater future support to reduce losses to insect pests. Such efforts offer particular promise in tropical and developing areas, where insect control via chemical means may be difficult and costly and insect transmission of virus diseases represents an added threat from inadequate control (_48_).

Nematode Resistance

Nematodes may cause devastating losses where these pests are endemic. Further, they may be widely transported on transplants and cause serious losses from introduction on infected planting stock. Seven species of the root-knot nematode (_Meloidogyne_ spp.) are known to attack tomato. Yield losses from severe infestations may be almost complete and predisposition to attack by other pathogens may be increased by the presence of nematodes.

Resistance to the root-knot nematode _Meloidogyne incognita_ was first reported in _L. peruvianum_ (PI 128657) by Bailey (_8_). Through the use of embryo culture, this resistance was successfully transferred by backcrossing to adapted _L. esculentum_ cultivars (_92_). Resistance derived from this source is conditioned by a single dominant gene (_Mi_) located on chromosome 6 (_9,34_). This gene fortunately also provides resistance to three other prevalent nematode species (_Meloidogyne javanica, Meloidogyne arenaria_ and _Meloidogyne acrita_). Many fresh-market cultivars and hybrids now possess this resistance, and new efforts are under way to incorporate nematode resistance into processing and green-house cultivars where it is necessary or desirable (99).

Fruit Quality

Fruit quality must be an important consideration in any tomato improvement program (94). In certain cases, efforts are made simply to maintain quality while emphasis is given to higher priority objectives. Recent criticism of fresh-tomato quality by the popular press and long-term interest by processing industries in improving quality and increasing ''case yield'' have encouraged recent efforts to improve genetically both fresh- and processing-tomato quality.

Breeding for improved fruit quality initially requires a definition of the major param-eters that contribute to it. Since perception of quality is highly subjective and a result of both visual and sensory stimuli, taste panel evaluations to quantify the importance of individual quality parameters must precede improvement efforts to define the relative importance of each.

Tomato fruit are 94–95% water. The remaining 5–6% is predominantly organic con-stituents, which give the fruit its characteristic flavor, aroma, and texture (Fig. 4.5). During fruit maturation, dramatic changes in chemical composition of the fruit occur. As a consequence, careful sampling to ensure that fruit to be compared are at approximately similar physiological maturity is required for meaningful comparisons of different genotypes.

Appearance

Size, shape, external color, smoothness, uniformity, and freedom from defects are of major concern for fresh use but are of less significance in tomatoes to be crushed and concentrated for pizza sauce or other similar products.

The tomato is highly susceptible to rain-induced fruit cracking which may render fruit unmarketable for fresh or processing use. Significant progress has been made to develop firm cultivars highly resistant to fruit cracking (72,91).

Fruit Color

Many genes affecting tomato fruit color have been identified and described (23). For practical breeding purposes, the crimson (og^c) and the high-pigment (hp) genes have been of particular interest to enhance fruit color (107). The og^c gene increases lycopene at the expense of β-carotene, resulting in fruit with lower vitamin A levels. The hp gene, in contrast, increases total fruit carotenoids, resulting in excellent color and improved vi-tamin A levels. Unfortunately, several undesirable apparent pleiotropic effects associated with hp (slow germination and growth, premature defoliation) have limited its use for tomato improvement (85). Both of these genes can be identified visually (with practice); however, colorimetric measurements may be preferred when quantitative measures of color are desired. The Hunterlab Color Difference Meter provides a measure of redness (a), yellowness (b), and lightness (L) of raw juice (47). These values may be used to calculate standard estimates of juice color.

Texture and Firmness

Fruit texture, notably firmness and the ratio of fruit wall to locular contents, plays an important role in quality as perceived by the consumer of fresh tomatoes (91). This particular facet of tomato quality has been soundly criticized by the popular press during

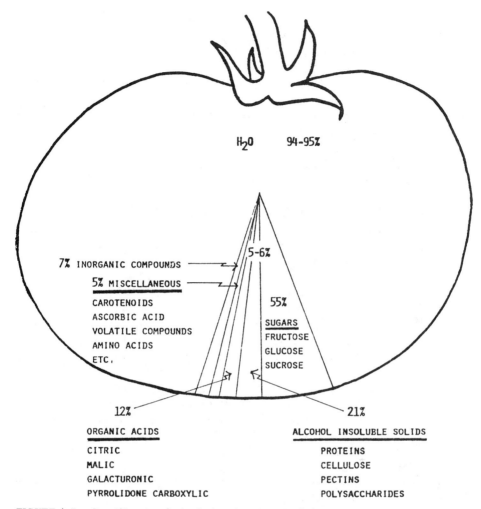

FIGURE 4.5. Constituents of a typical mature tomato fruit.

the past decade and plant breeders have borne the brunt of the blame. A 1974 article reflects opinions of fresh-tomato quality that have been restated in numerous subsequent articles appearing in the popular press. For example, "Not so long ago, tomatoes were soft and juicy and tasted of tomato. Several varieties available in today's supermarket are rubbery gobs of cellulose that taste of nothing. They are bred that way for mechanical picking" (*60*, p. 240). Such articles have reinforced the view that firm fruit precludes flavor and quality.

This view represents an oversimplification of a problem commonly encountered in the handling and marketing of perishable commodities. Consumers would like fresh tomatoes during midwinter with the same fresh quality they find in fruit harvested from their home gardens during the long days of summer, but this is still not possible. Harvesting immature fruit for long-distance shipping and producing the crop under the short photoperiods of winter are equally important factors that contribute to winter tomato quality. Criticism by consumers has fostered cooperative efforts by postharvest physiologists and plant breeders to find solutions to this problem.

Flavor

Studies by Stevens *et al.* (*96,98*) have clearly established that sugars and organic acids are important determinants of tomato flavor. Apparently the proper balance of these fruit constituents (sugar:acid ratio) is required to give optimal flavor, whereas intensity of flavor (sweetness or sourness) is a result of the relative levels of each. Volatile constituents also contribute to "tomato-like" flavor. The importance of the individual fruit volatiles to flavor is still poorly understood, due in large measure to the very complex array of these constituents in tomato fruit. Since a single constituent is not responsible for the predominant flavor identified with fresh or processed tomatoes, it is difficult to assign minimum acceptable levels of any single constituent for "good flavor."

Nutritional Value

The tomato is a significant source of vitamins A and C in human nutrition. Wide genetic variation exists in tomato for the levels of these nutrients, and conscious efforts to exploit this variation have resulted in several nutritionally superior cultivars.

Plant carotenoids, which represent the major pigments in tomato fruit, are the primary dietary source of vitamin A. For example, oxidation of β-carotene (an orange pigment) yields two molecules of vitamin A. Compounds that are converted in vivo to vitamin A are termed provitamin A. Certain carotenoids in tomato fruit also may be converted to vitamin A, but, lycopene, the major pigment of red-fruited cultivars, has no provitamin A activity. Some orange-fruited cultivars, on the other hand, have much higher vitamin A activity than red-fruited cultivars. A single dominant gene *B* favors β-carotene synthesis (at the expense of lycopene) and results in orange fruit with provitamin A levels 8–10 times higher than in red-fruited cultivars (*109,114*). Genes that enhance total carotenoids may also increase provitamin A activity [for example, high pigment (*hp*)] and genes that enhance lycopene [for example, the crimson (*og^c*) gene] decrease provitamin A content. The widely held consumer view that "redder is better" presents a dilemma to the tomato breeder concerned with nutritional value.

There is also a wide range of fruit ascorbic acid (vitamin C) levels in the genus *Lycopersicon* (10–120 mg/100 g fresh wt) (*43,52*). Linkage or pleiotropy between high ascorbic acid and small fruit size has limited use of this wide variability largely to maintenance of acceptable levels. The high-pigment (*hp*) gene offers possibilities to enhance both vitamins A and C; however, undesirable linkage or pleotropic effects of this gene on growth rate, yield, and fruit size have severely limited its use for simultaneous improvement of color and nutritional value (*85*).

Processing Quality

Fruit characters that contribute to processing quality and case yield (cases of final product per ton of fruit) have been well studied and defined. Five distinct parameters are commonly used to evaluate processing quality. The purpose of each is to quantify raw fruit quality to meet standards established for specific processed products. Careful fruit sampling is important in obtaining reliable measures of fruit quality. Undermature and/or overripe fruit may give erroneous values for certain quality parameters since fruit is continuously changing during ripening and senescence. A moderate-sized sample (5–7 lb) of uniform fruit is desirable to minimize environmental variation in estimating quality.

Color

Fruit color is often a key quality parameter used in grading raw fruit to reimburse producers. In addition to providing a measure of fruit maturity, color also influences the grades and standards of the processed commodity. Colorimetric measurement of raw color is now a standard practice in most tomato processing establishments.

Fruit pH

Fruit pH affects the heating time required to achieve sterilization of the processed commodity. Longer times are required as product pH increases. Values above pH 4.5 are considered unacceptable for fruit destined for unconcentrated products in which sterilization is achieved by preprogrammed heating times. For cultivar and breeding line comparisons, pH is measured directly with fresh juice prepared from a uniform sample of fully ripe fruits. Overmature fruit will give erroneously high pH values.

Titratable Acidity

Titratable acidity provides a measure of organic acids (total acidity) present in a fruit sample, which in turn estimates tartness. Total acidity and pH are not always closely correlated due to differences in the degree of buffering of pH by other fruit constituents. To determine titratable acidity, 10 ml of fresh raw juice is diluted to 50 ml with distilled water. The volume of 0.1 N NaOH required for titration to pH 8.1 is multiplied by a correction factor (0.064) to estimate titratable acids as percentage of citric acid (*62*).

Soluble Solids

Quality standards for processed tomato pulp and paste are defined in terms of soluble-solids content. This parameter of quality directly influences flavor and the degree of concentration required to manufacture products in which standards of quality are determined by solids content. High-soluble-solids cultivars give more cases of finished product per ton of raw fruit and thus require less energy in concentration. As a consequence, this parameter of quality has been of major interest to the processing industries that manufacture concentrated tomato products (*53*). Soluble solids are measured by placing two or three drops of filtered juice on the prism of a refractometer and directly reading the percentage of soluble solids.

Tomato fruit solids content is influenced by both environmental and genetic factors. High light intensity, long photoperiods, and dry weather at harvest favor high fruit solids. Small fruit size and indeterminate habit of growth also favor high solids content (*26*). As a consequence, selection for high yield or compact growth habit frequently results in sacrifices in solids content.

Tomato solids are comprised of a soluble and insoluble fraction. The soluble fraction is made up largely of free sugars and organic acids (see Fig. 4.5). The insoluble fraction (made up of proteins, pectins, cellulose, and polysaccharides) contributes to the viscosity (consistency) of processed tomato products. Stevens and Paulson have shown that the polygalacturonides are the most important component of the insoluble fraction contributing to viscosity (*95*).

Viscosity (Consistency)

For many processed tomato products, viscosity is an important parameter of established grades and standards. Perceived quality of items such as juice, catsup, tomato sauces, soup, and tomato paste is influenced by consistency. Advertising emphasis for products such as catsup reflects the importance of this quality attribute. Viscosity potential of raw fruit will influence processed product consistency and the amount of raw product required to achieve a desired consistency (case yield).

Several methods are used to determine viscosity potential (*37*). The acid efflux method is commonly used to evaluate raw fruit samples originating from variety trials or breeding plots. Approximately 2 kg of fully mature fruit is blended in 30 ml of concentrated HCl to inactivate pectic enzymes. This preparation is then passed through an extractor fitted with a 400-mesh screen to remove skin and seeds. The resulting extract (juice) is then deaerated under vacuum for 3 min and used immediately to estimate viscosity. The flow rate through a standard viscometer is timed and viscosity expressed as the time required for 100 ml to flow through the viscometer column. Alcohol-insoluble solids (AIS) also provide a measure of fruit viscosity potential (97). A major challenge has been to combine high soluble solids with improved alcohol-insoluble solids and high yield. Apparently, increasing solids occurs at the expense of yield.

Several other quality attributes are recognized for specific processed products. Crack resistance and fruit rot tolerance are required in humid production areas. Firm fruit indirectly improves processing quality by reducing mechanical damage. For canned whole tomatoes, uniform color, size and shape, small core size, and jointless pedicel *j-2* are essential attributes where a high percentage of raw fruit is desired for peeling purposes.

Physiological Traits

Many physiological characters contribute to the wide adaptation of the tomato. Early tomato improvement efforts emphasized disease resistance; however, more recently, attempts have been made to understand and utilize the more subtle variation contributing to crop adaptation and development. Considering the diversity of ecological niches occupied by the wild relatives of tomato, it is not surprising that substantial variation exists for a variety of adaptive features.

Low-Temperature Germination and Growth

Low-temperature germination and growth have been examined to improve emergence at the low soil temperatures frequently encountered with direct field seeding (63). The glasshouse tomato industry has had a major interest in developing cultivars adapted to lower light and temperature environments to reduce energy inputs into winter crop production. Patterson *et al.* (64) have examined *L. hirsutum* introductions originating from different elevations and have shown marked quantitative differences in germination, growth, and susceptibility to chilling injury as a function of the elevation of origin. Obvious variation thus exists, and techniques are being developed to exploit this variation in applied breeding programs.

Fruit Set

Fruit set under temperature extremes has been an improvement goal in production areas where high and/or low temperatures may interfere with pollination and fruit development. Schaible (86) was among the first to show that genotypic variation for ability to set fruit at low temperatures also favors high temperature setting. Not surprisingly, much of the known variation in cultivars of *L. esculentum* has originated from breeding programs in regions with very short seasons or with extreme summer temperatures. Such regions have apparently permitted rigorous field screening, which is not always possible in areas with temperatures more favorable to tomato fruit set.

Recently, the use of genetic parthenocarpy has been examined as a method to alleviate environmental limitations on fruit set (66). Several sources of parthenocarpy have been identified in *L. esculentum;* the most promising originated in the Russian cv. Severianin and is controlled by a single recessive gene, *pat-2*.

Chilling Injury

Chilling injury may occur under field conditions but is most commonly regarded as a fruit storage disorder that severely limits tomato shelf life. In addition, seedling injury may result from prolonged exposure to low temperatures. Patterson *et al.* (*64*) have examined various low-temperature responses in *L. hirsutum* originating from different altitudes to identify potentially useful variability for improving crop performance at low temperatures.

Salt Tolerance

Salt tolerance is commonly observed among certain wild relatives of the tomato in their native habitat (*77*). Rush and Epstein (*82,83*) have compared the tolerance of susceptible and tolerant species and efforts have been initiated to introgress the tolerance from wild species into adapted cultivars (*84*).

Drought Tolerance

Drought tolerance is found in *Lycopersicon chilense* and *L. pennellii*, both of which occur in native habits of low annual rainfall. Rick (*80*) suggests that the physiological basis for drought tolerance in *L. chilense* may be related to its deep vigorous root system. *L. pennellii,* in contrast, has a limited root system and the basis for its tolerance to drought is presumably related to its ability to conserve moisture during periods of limited rainfall.

Fruit Ripening

Fruit ripening is under genetic control and recent studies have identified and described several genes with striking effects on the ripening process (*111,113*). Two of these [ripening inhibitor (*rin*) and non-ripening (*nor*)] virtually inhibit the changes that accompany ripening, including color development, fruit softening, ethylene production, and respiratory changes. Their effects on enhancing fruit shelf life in homozygotes or heterozygotes have prompted interest in their use in applied breeding programs, particularly for fresh tomatoes destined for long-distance shipping (*102,112,113*).

BREEDING PROGRAM DESIGN

Hybridization followed by pedigree selection has been the most commonly used breeding method for tomato improvement. Backcross breeding has been the method of choice in wide crosses or for interspecies gene transfer. In certain situations, a combination of pedigree selection and backcross breeding has proven useful to exploit the advantages of each method.

In recent years, time has been recognized as an important element in plant improvement efficiency and off-season breeding nurseries have become an integral part of many tomato improvement programs. In this way, tomato breeding becomes a year-round activity with two (or in some cases three) generations each year. Winter programs are now virtually essential in the competitive race to develop improved cultivars; it is no longer possible to store seed for 7 or 8 months and keep current with changing needs.

The use of single-seed descent (SSD) has been examined as an alternative to pedigree selection for use where facilities or funds do not permit maintenance of winter breeding nurseries. We have compared pedigree selection and SSD and found combinations of early-generation pedigree selection followed by SSD to be most efficient in both time and progress under selection (*18*). Obviously, the merit of each method will depend upon breeding objectives and heritability of the trait(s) under selection (*110*). Computer simulation studies indicate pedigree selection to be most efficient with high heritabilities, whereas SSD is favored for characters of low heritability since a broader genetic base is maintained to advanced generations (*19*). Where several characters are under selection simultaneously, which is frequently the case, a combination of the two methods appears desirable. Pedigree selection is practiced in early generations (F_2 ad F_3) for highly heritable traits, followed by selection among SSD-derived inbred F_6 or F_7 lines for characters of lower heritability. We have used this procedure for several years and feel it combines the desirable features of the two methods. Obviously, the plant breeder must choose the appropriate breeding method to fit the established improvement objectives.

Currently, there is considerable interest in the use of genetic engineering and related biotechnologies for tomato improvement. The popular press has optimistically proposed some rather spectacular possibilities from this technology, including improvement of tolerance to drought and salinity, low temperature adaptation, improved disease resistance, and increased soluble solids. These suggestions of short-term benefits and unusual accomplishments minimize the problem of physiological limitations and techniques that remain to be developed before this technology can be applied to crop improvement. Techniques evolving from research in biotechnology may provide useful adjuncts to, rather than replacements for, conventional plant breeding (*93*). Sex is not only the spice but also the essence of life, for without sexual reproduction, evolutionary progress in higher organisms is severely restricted. Since plant breeding is simply a directed form of plant evolution, use of the sexual cycle appears essential and will remain the future technique of choice to exploit induced or natural variability for crop improvement purposes.

The organization and methods for handling breeding populations and maintaining records vary considerably from program to program. A typical scheme used for tomato improvement is illustrated in Fig. 4.6. Several key elements in program organization and implementation deserve comment.

Choice of Parents

In virtually all cases, one of the parents used to develop hybrid populations for selection is an adapted cultivar that requires improvement for one or more characteristics. The second parent should complement the weaknesses of the adapted cultivar to ensure adequate segregation for the character(s) under selection. Wise choice of parents is a key decision, which requires an intimate knowledge of germplasm sources available for tomato improvement.

F_1s may be evaluated for horticultural and quality attributes in their area of intended use, and only promising combinations retained for selection in F_2 and subsequent generations.

Selection in Segregating Generations

Single plant selection is initiated in F_2 and is continued through successive generations until stable lines are obtained (generally F_7–F_{10}). Since within-line segregation decreases with each generation of inbreeding, with pedigree selection, population sizes for selection are reduced by 50% each generation. During generation advance, selection emphasis is shifted from individual plant performance in early generations to line performance in more advanced generations. In practice, selection in early generations generally emphasizes

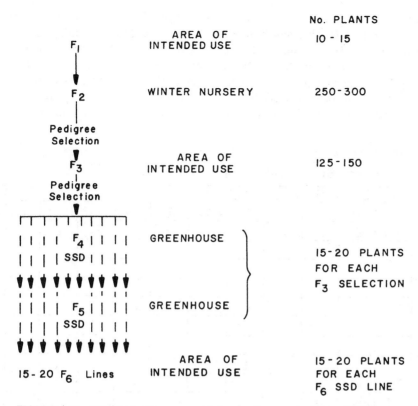

FIGURE 4.6. Breeding scheme combining pedigree selection and SSD.

highly heritable characters (disease resistance, growth habit, fruit characteristics, etc.), whereas selection emphasis may be shifted to less highly heritable characters in more advanced generations. For this reason, it is important to maintain a broad genetic base through early generations to maintain adequate variability among selected lines for selection in later generations. Combining pedigree selection in early generations with SSD following F_3 or F_4 seems to accomplish the desired balance between stringent selection and maintaining a broad genetic base. This balance is important since it is relatively easy to accumulate more breeding lines than can be properly evaluated in a short period of time or with limited space.

Maintaining Pedigrees and Records

Efficient, simple, and rapid methods of record keeping are essential in an active breeding program. Various systems have been devised and used by tomato breeders. As an example, the system used at Purdue University illustrates the essentials of an efficient system.

All field records for a single line or selection are maintained on a single record page (Fig. 4.7). These sheets are printed on light blue heavy paper to facilitate field work. (Light blue is easier on the eyes in bright sunlight; heavy paper does not get mutilated on windy days!) Preprinted columns for all characteristics to be evaluated are included on each record page to simplify record keeping. These are kept in a hard-backed, two-ring notebook until season's end, when they are cut and filed (Fig. 4.8).

Pedigrees are identified by an arbitrary field number assigned to each line prior to planting (See Fig. 4.7). This number (e.g., 83-2-406) identifies the planting year (1983), the specific project (2 = vitamin C breeding), and a sequential field number (406). Its pedigree identifies it as cross 102 made in 1980 and the seed source as SSD plant 17-2 grown in the greenhouse (prefix 6) during 1982–1983. This system provides a simple method of tracing records of selections for their entire period of development.

Field records must be taken in the relatively short time period during the growing

Pedigree F_6 of cross 80102 (V 6724 x PU 74-32)										83-2-406	
Seed Source 82-83-6-17-2 SSD											
	Earli-ness	Cracking	Firm-ness	Color	Vine Cover	Stylar Scar	Stem Scar	Core	Texture	(Set)	Genes
①	3.5	2	2	3	2	3	3	2	2	4	j$^+$
	①		①		Vit. C	-272					
pH	4.27	v	R								
SS	5.1	F$_1$	R								
Col	71.2	F$_2$	–								Discard
Visc	112.0	SI	S								
Pedigree F_6 of cross 80102 (V 6724 x PU 74-32)										83-2-407	
Seed Source 82-6-17-3 SSD											
	Earli-ness	Cracking	Firm ness	Color	Vine Cover	Stylar Scar	Stem Scar	Core	Texture	(Set)	Genes
pH		v									
SS		F$_1$									
Col		F$_2$								Discard	
Visc		SI									

FIGURE 4.7. Field record sheets used to evaluate breeding lines and selections.

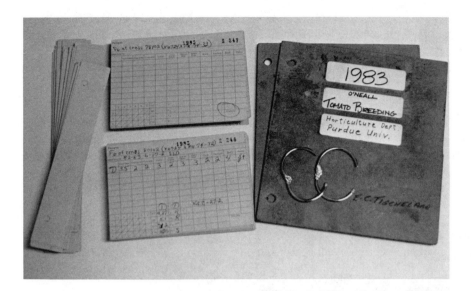

FIGURE 4.8. Filing system used to maintain field records.

season when expression of the characters to be selected is optimal. Breeding populations are usually observed several times before final selections are taken. Field records generally involve subjective evaluations against a standard cultivar using a simple number system [e.g. 1 (poor) to 5 (excellent)]. In this way, selected lines may be compared with a standard as well as with each other and progress measured over successive years.

Fruit Quality Evaluation

Environment and stage of maturity influence the major parameters of quality, thus necessitating careful sampling for reliable measures of fruit quality. Ideally, all fruit should be at an approximately similar stage of maturity for valid comparisons of cultivar quality differences. This may be difficult to achieve with material of widely differing maturities; the simplest procedure involves harvesting all ripe fruit and then waiting several days and harvesting again for quality determinations. This ensures that fruit is relatively uniform and largely eliminates the confounding effects of cultivar earliness on quality determinations.

Disease Resistance

Breeding for disease resistance generally requires artificial screening procedures that ensure uniform exposure to the pathogen. This seldom occurs under field conditions, and so artificial inoculation is commonly used (either in the field or laboratory) to identify resistant genotypes. The common procedures used for screening the major tomato diseases are given in Table 4.6.

Nematode Resistance

Breeding for resistance to the root-knot nematode (*Meloidogyne* spp.) has been simplified by the recent observation of isozyme differences in resistant and susceptible genotypes (*78*). Electrophoretic separation of seedling proteins provides a rapid, nondestructive test to screen large populations for the *Mi* gene, which controls resistance.

Early- versus Advanced-Generation Selection

In view of the large number of individual attributes that must be considered, early-generation selection frequently emphasizes characters with relatively high heritability and for which single-plant selection is relatively effective. If a broad genetic base can be maintained to advanced generations, selection for traits of lower heritability may be most effective in advanced generations when selection emphasis can be shifted to family performance. We consider this to be a major advantage of combining early-generation pedigree selection with SSD. Two or three generations of pedigree selection minimize the number of undesirable lines retained, whereas SSD effectively maintains a broad genetic base to advanced generations. Complex characters such as yield and quality are evaluated at this time. Where several hundred advanced SSD lines are developed each year, defect elimination (i.e., discarding lines that do not meet standards) throughout the season may be an effective method of selecting for specific characteristics and of maintaining manageable numbers of lines from one season to the next.

Monitoring Progress

Records are maintained during generation advance to monitor progress and establish the merits or faults of a particular parent, population, or individual breeding line. Simple regression of offspring on parent provides a measure of the progress with selection for a particular selected trait and yields useful information on the specific merits of a line before final testing and evaluation are initiated.

Trials of Advanced Lines

Testing of breeding lines begins when a line appears stable and shows sufficient merit to warrant yield and quality trials (generally F_7-F_{10}). Seed from a single-plant selection is generally adequate to perform transplanted trials at several locations and also to initiate a preliminary small seed increase. Testing should be performed at as many locations as is feasible within the intended area of use before release is considered. Cooperative testing programs involving exchange of advanced breeding lines for evaluation over a broad geographical area are conducted annually for cultivars intended for fresh-market or processing use. The Southern Tomato Exchange Program (STEP), Northern Tomato Exchange Program (NTEP), and All-American Vegetable Trials are designed for this purpose. The STEP trials involve largely fresh-market cultivars, whereas the NTEP trials emphasize processing cultivars. The All-America Vegetable Trials serve both fresh and home garden cultivars. Such broad-based trials expose potential new cultivars or hybrids to many different environments and evaluators. This broad exposure prior to release is essential to ensure wide adaptation and consistent performance over different seasons.

RELEASE PROCEDURES

Both public and private institutions are actively involved in tomato improvement. The seed industry has historically provided the delivery system for improved cultivars via multiplication and sale of new cultivars. Public institutions, on the other hand, have been involved in basic genetic and breeding methods research as well as cultivar development. The major tomato processing industries have also maintained active and productive tomato cultivar development programs for several decades.

Few, if any, standard guidelines exist regarding release procedures for new tomato cultivars. Virtually anyone can develop and/or release a new variety; however, to obtain plant variety protection (PVP) to patent a new cultivar, uniqueness must be clearly demonstrated. Many public institutions have established variety release procedures; however, in the final analysis, the marketplace determines the ultimate usage of a new cultivar.

The true merit of a new cultivar is ultimately determined by the grower and consumer. As a consequence, small-scale grower trials should be an integral part of prerelease testing. A truly superior cultivar "sells itself" and rapidly establishes its position in the marketplace. Consistency of performance under diverse environments is the mark of a worthy cultivar.

FUTURE PROSPECTS

The vast reservoir of genetic variability within the genus *Lycopersicon* and the "favored status" of the tomato as a crop for genetic and physiological studies have been factors that have fostered the impressive improvements that have occurred in the past. Despite this progress, plant breeders have merely scratched the surface in redesigning this crop to meet the needs of the grower, processor, and consumer. An almost inexhaustible supply of unexplored diversity exists within the wild taxa of *Lycopersicon* (2,22,77), which to date have served primarily as sources of resistance to major disease pests. This germplasm source has been virtually unexploited in tomato improvement for insect resistance, tolerance to environmental stresses, fruit quality, or other valuable traits. The building blocks for future improvements are clearly available with the genus *Lycopersicon* and expanded efforts to conserve and evaluate this variability are crucial to its future utilization.

The recent view expressed by Griesbach *et al.* (*38*) that new techniques are required in genetic manipulation "due to lack of sufficient gene reserves" grossly underestimates the vast diversity found in the wild species of most of our cultivated crops. The array of variation in cultivated *L. esculentum* is limited when compared with the wild taxa of *Lycopersicon* (*56,81*) and greater future efforts will be required to exploit the useful diversity from the wild species. The long-term nature of such programs will require greater public support for conventional genetic and physiological studies of this diversity to facilitate its use. Continued progress is certain if improvement goals are clearly defined and established and appropriate breeding strategies are employed. Mother Nature has been exceedingly generous in providing the raw materials for improvement of this favored crop.

ACKNOWLEDGMENTS

Special thanks to Deb Altman for typing the manuscript, to Chris Thoe for her artistic skills in the drawing of figures, and to Dr. R. W. Robinson for his constructive review.

REFERENCES

1. Acosta, J. C., Gilbert, J. C., and Quinon, V. L. 1964. Heritability of bacterial wilt resistance in tomato. Proc. Am. Soc. Hortic. Sci. *84,* 455.
2. Alcazar-Esquinas, J. T. 1981. Genetic Resources of Tomatoes and Wild Relatives. International Board for Plant Genetic Resources, Rome.
3. Alexander, L. J., and Hoover, M. M. 1955. Disease resistance in wild species of tomato. Ohio, Agric. Exp. Stn., Res. Bull. *752.*
4. Alexander, L. J. 1963. Transfer of a dominant type of resistance to the four known Ohio pathogenic strains of TMV from *Lycopersicon peruvianum* to *L. esculentum.* Phytopathology *53,* 869.
5. Andrus, G. F., and Reynard, G. B. 1945. Resistance to Septoria leafspot and its inheritance in tomatoes. Phytopathology *35,* 16–24.
6. Anon. 1979. The Gallup Organization. National Gardening Survey. Publ. from Gardens for All, The National Association of Gardening, Burlington, VT.
7. Anon. 1979. Production and marketing of F_1 hybrid seed in Taiwan. *In* Tropical Tomato. R. Cowell (Editor), pp. 254–256. Asian Veg. Res. Dev. Cent., Tainan, Taiwan.
8. Bailey, D. M. 1941. The seedling test method for root-knot nematode resistance. Proc. Am. Soc. Hortic. Sci. *38,* 573–575.
9. Barham, W. S., and Winstead, N. N. 1957. Inheritance of resistance to root-knot nematodes in tomato. Proc. Am. Soc. Hortic. Sci. *69,* 372–377.
10. Barksdale, T. H., and Stoner, A. K. 1973. Segregation for horizontal resistance to tomato early blight. Plant Dis. Rep. *57,* 964–965.
11. Barksdale, T. H. 1974. Evaluation of tomato fruit for resistance to Rhizoctonia soil rot. Plant Dis. *58,* 406–408.
12. Barksdale, T. H., and Stoner, A. K. 1975. Breeding for tomato anthracnose resistance. Plant Dis. Rep. *59,* 648–652.
13. Barksdale, T. H., and Stoner, A. K. 1977. A study of the inheritance of tomato early blight resistance. Plant Dis. Rep. *61,* 63–65.
14. Barksdale, T. H., and Stoner, A. K. 1978. Resistance in tomato to *Septoria lycopersici.* Plant Dis. Rep. *62,* 844–847.
15. Bohn, G. W., and Tucker, C. M. 1940. Studies on *Fusarium* wilt of the tomato. Immunity in *Lycopersicon esculentum* Mill. and its inheritance in hybrids. Res. Bull.—Mo., Agric. Exp. Stn. *311.*
16. Boukema, I. W. 1982. Inheritance of resistance to *Didymella lycopersici* Klebb. in tomato (*Lycopersicon esculentum* Mill.) Euphytica *31,* 981–984.

17. Brandt, J. R., French, B. C., and Jesse, E. V. 1978. Economic performance of the processing tomato industry. Univ. Calif. Bull. *1888.*

18. Casali, V. W. D., and Tigchelaar, E. C. 1975. Breeding progress in tomato with pedigree selection and single seed descent. J. Am. Soc. Hortic. Sci. *100,* 362–364.

19. Casali, V. W. D., and Tigchelaar, E. C. 1975. Computer simulation studies comparing pedigree, bulk and single seed descent selection in self pollinated populations. J. Am. Soc. Hortic. Sci. *100,* 364–367.

20. Chinn, J. T., Gilbert, J. C., and Tanaka, J. S. 1967. Spider mite tolerance in multiple disease resistant tomatoes. J. Am. Soc. Hortic. Sci. *89,* 559–562.

21. Cirulli, M., and Alexander, L. J. 1969. Influence of temperature and strain of tobacco mosaic virus on resistance in a tomato breeding line derived from *Lycopersicon peruvianum.* Phytopathology *59,* 1287–1297.

22. Clayberg, C. D. 1971. Screening tomatoes for ozone resistance. HortScience *6,* 396–397.

23. Darby, L. A. 1978. Isogenic lines of tomato fruit color mutants. Hortic. Res. *78,* 73–84.

24. DeJong, J., and Honma, S. 1976. Inheritance of resistance to *Corynebacterium michiganense* in the tomato. J. Hered. *67,* 79–84.

25. DePonti, O. M. B., Pet, G., and Hogenboom, N. G. 1975. Resistance to the glasshouse whitefly (*Trialeurodes vaporariorum* Westw.) in tomato (*Lycopersicon esculentum* Mill.) and related species. Euphytica *24,* 645–647.

26. Emery, J. C., and Munger, H. M. 1970. Effects of inherited differences in growth habit on fruit size and soluble solids in tomato. J. Am. Soc. Hortic. Sci. *95,* 410–412.

27. Fery, R. L., and Cuthbert, F. P., Jr. 1974. Resistance of tomato cultivars to the fruitworm, *Heliothis zea* (Boddie). HortScience *9,* 469–470.

28. Fery, R. L., and Cuthbert, F. P., Jr. 1975. Antibiosis in *Lycopersicon* to the tomato fruitworm (*Heliothis zea*). J. Am. Soc. Hortic. Sci. *100,* 276–278.

29. Frazier, W. A., Dennett, R. K., Hendrix, J. W., Poole, C. F., and Gilbert, J. C. 1950. Seven new tomatoes: Varieties resistant to spotted wilt, Fusarium wilt and gray leaf spot. Hawaii, Agric. Exp. Stn., Bull. *103.*

30. Gallegly, M. E. 1960. Resistance to the late blight fungus in tomato. Proc., Plant Sci. Symp., pp. 113–135.

31. Gentile, A. G., and Stoner, A. K. 1968a. Resistance in *Lycopersicon* and *Solanum* species to the potato aphid. J. Econ. Entomol. *611,* 1152–1154.

32. Gentile, A. G., and Stoner, A. K. 1968b. Resistance in *Lycopersicon* spp. to the tobacco flea beetle. J. Econ. Entomol. *611,* 1347–1349.

33. Gentile, A. G., Webb, R. E., and Stoner, A. K. 1968c. Resistance in *Lycopersicon* and *Solanum* to greenhouse whiteflies. J. Econ. Entomol. *611,* 1355–1357.

34. Gilbert, J. C., and McGuire, D. C. 1956. Inheritance of resistance to severe root knot from *Meloidogyne incognita* in commercial type tomatoes. Proc. Am. Soc. Hortic. Sci. *68,* 437–442.

35. Gilbert, J. C., McGuire, D. C., and Tanaka, J. 1961. Indeterminate tomato hybrids with resistance to eight diseases. Hawaii Farm Sci. *9,* 1–3.

36. Gilbert, J. C., Chinn, J. T., and Tanaka, J. S. 1966. Spider mite tolerance in multiple disease resistant tomatoes. J. Am. Soc. Hortic. Sci. *59,* 559–562.

37. Gould, W. A. 1983. Tomato Production Processing and Quality Evaluation, 2nd Edition. AVI Publishing Co., Westport, CT.

38. Griesbach, R. J., Koivuniemi, P. J., and Carlson, P. S. 1981. Extending the range of plant genetic manipulation. BioScience *31,* 754–756.

39. Hall, T. J. 1970. Resistance at the Tm_2 locus in the tomato to tomato mosaic virus. Euphytica *29,* 189–197.

40. Heiser, C. J. 1969. Love apples. *In* Nightshades: The Paradoxical Plants. pp. 53–105. Freeman, San Francisco, CA.

41. Henderson, W. R., and Jenkins, S. F. 1972. Venus and Saturn. N.C., Agric. Exp. Stn., Bull. *444*.

42. Hendrix, J. W., and Frazier, W. A. 1949. Studies on the inheritance of Stemphylium resistance in tomatoes. Tech. Bull.—Hawaii, Agric. Expt. Stn. *8*, 24.

43. Hobson, G. E., and Davies, J. N. 1971. The tomato. *In* The Biochemistry of Fruits and Their Products. A. C. Holme (Editor), Vol. 2, pp. 437–482. Academic Press, New York.

44. Hogenboom, N. G. 1971. Inheritance of resistance to corky root in tomato. Euphytica *19*, 413.

45. Hogenboom, N. G. 1972. Breaking breeding barriers in *Lycopersicon*. 1. The genus *Lycopersicon*, its breeding barriers and the importance of breaking these barriers. Euphytica *21*, 221–227.

46. Holmes, F. O. 1960. Control of important viral diseases of tomatoes by the development of resistant varieties. Proc., Plant Sci. Symp. pp. 1–17.

47. Hunterlab 1971. Instructions for Hunterlab Model D25 Color Difference Meter. Hunter Associates Laboratory, Inc., Fairfax, VA.

48. Kennedy, G. C. 1976. Host plant resistance and the spread of plant viruses. Environ. Entomol. *5*, 827–831.

49. Kennedy, G. C., and Henderson, W. R. 1978. A laboratory assay for resistance to the tobacco hornworm in *Lycopersicon* and *Solanum* spp. J. Am. Soc. Hortic. Sci. *103*, 334–336.

50. Kerr, E. A. 1955. Some factors affecting earliness in the tomato. Can. J. Agric. Sci. *35*, 300–308.

51. Kerr, E. A., Kerr, E., Patrick, Z. A., and Potter, J. W. 1980. Linkage relation of resistance to Cladosporium leaf mold (Cf) and root-knot nematodes (Mi) in tomato and a new gene for leaf mold resistance (Cfii). Can. J. Genet. Cytol. *22*, 183–186.

52. Lambeth, V. N., Straten, E. F., and Fields, M. L. 1966. Fruit quality attributes of 250 foreign and domestic tomato accessions. Res. Bull.—Mo., Agric. Exp. Stn. *908*.

53. Lower, R. L., and Thompson, A. E. 1967. Inheritance of acidity and solids content of small-fruited tomatoes. Proc. Am. Soc. Hortic. Sci. *91*, 486–494.

54. Magoon, C. E. 1969. Fruit and Vegetable Facts and Pointers. Tomatoes. United Fresh Fruit and Vegetable Association, Washington, DC.

55. Martin, N. W. 1969. Inheritance of resistance to curly top in the tomato breeding line C/F_4. Phytopathology *59*, 1040.

56. Maxon-Smith, J. W. 1977. *Lycopersicon hirsutum* and a source of genetic variation for the cultivated tomato. Proc. Eucarpia Congr., 8th, 1977, pp. 119–128.

57. McGrath, D. J., and Toleman, M. A. 1983. Breeding and Selection for Resistance to Fusarium Wilt Race 3, Proc. 4th Tomato Qual. Workshop, Res. Rep. VEC 83-1, p. 104. Univ. of Florida, Gainesville.

58. McGuire, D. C., and Rick, C. M. 1954. Self incompatibility in species of *Lycopersicon* Sect. Eriopersicon and hybrids with *L. esculentum*. Hilgardia *23*, 101–124.

59. McKinney, K. B. 1938. Physical characteristics on the foliage of beans and tomatoes that tend to control some small insect pests. J. Econ. Entomol. *30*, 630–631.

60. Miller, J. 1974. Agriculture: FDA seeks to regulate genetic manipulation of food crops. Science *185*, 240–242.

61. Morrison, G. 1938. Tomato varieties. Mich., Agric. Exp. Stn., Spec. Bull. *290*.

62. National Canners Association 1968. Laboratory Manual for Food Canners and Processors, Vol. 2. AVI Publishing Co., Westport, CT.

63. Ng, T. G., and Tigchelaar, E. C. 1973. Inheritance of low temperature seed sprouting in tomato. J. Am. Soc. Hortic. Sci. *98*, 314.

64. Patterson, B. D., Paull, R., and Smillie, R. M. 1978. Chilling resistance in *Lycopersicon hirsutum*, a wild tomato with a wide altitudinal distribution. Aust. J. Plant Physiol. *5*, 609–617.

65. Pelham, J. 1966. Resistance in tomato to tobacco mosaic virus. Euphytica *15*, 258.

66. Philouze, J., and Maisonneuve, B. 1978. Heredity of the natural ability to set parthenocarpic fruits in the Soviet variety Severianin. Rep. Tomato Genet. Coop. *28*, 12–13.

67. Pilowski, M., and Cohen, S. 1974. Inheritance of resistance to tomato yellow leaf curl virus in tomatoes. Phytopathology *64*, 632.

68. Pitblado, R. E., and Kerr, E. A. 1979. A source of resistance to bacterial speck—*Pseudomonas tomato*. Rep. Tomato Genet. Coop. *29*, 30.

69. Porte, W. S., and Walker, H. B. 1941. The Pan American tomato, a new red variety highly resistant to fusarium wilt. U.S., Dep. Agric., Circ. *611*.

70. Quiros, C. F., Stevens, M. A., Rick, C. M., and Kok-Yokomi, M. L. 1977. Resistance in tomato to the pink form of the potato aphid (*Macrosiphum euphorbiae* Thomas): The role of anatomy, epidermal hairs, and foliage composition. J. Am. Soc. Hortic. Sci. *102*, 166–171.

71. Reynard, G. B., and Andrus, C. F. 1945. Inheritance of resistance to the collar rot phase of Alternaria solani on tomato. Phytopathology *35*, 25–36.

72. Reynard, G. B. 1960. Breeding tomatoes for resistance to fruit cracking. Proc., Plant Sci. Symp. pp. 93–112.

73. Reynard, G. B. 1961. A new source of the j_2 gene governing jointless pedicel in tomato. Science *134*, 2102.

74. Rick, C. M., and Butler, L. 1956. Cytogenetics of the tomato. Adv. Genet. *8*, 267–382.

75. Rick, C. M. 1960. Hybridization between *Lycopersicon esculentum* and *solanum pennellii:* Phylogenetic and cytogenetic significance. Proc. Natl. Acad. Sci. U.S.A. *46*, 78–82.

76. Rick, C. M. 1967. Exploiting species hybrids for vegetable improvement. Proc. Int. Hortic. Congr., 17th, 1967 Vol III, pp. 217–229.

77. Rick, C. M. 1973. Potential genetic resources in tomato species: Clues from observations in native habitats. *In* Genes, Enzymes and Populations. A. SrB (Editor), pp. 255–269. Plenum, New York.

78. Rick, C. M., and Fobes, J. 1974. Association of an allozyme with nematode resistance. Rep. Tom. Genet. Coop. *24*, 25.

79. Rick, C. M., Quiros, C. F., Lange, W. H., and Stevens, M. A. 1976. Monogenic control of resistance in the tomato to the tobacco flea beetle. Probable repellence by foliage volatiles. Euphytica *25*, 521–530.

80. Rick, C. M. 1978. The tomato. Sci. Am. *239*, 76–87.

81. Rick, C. M. 1979. Tomato germplasm resources. *In* Tropical Tomato. R. Cowell (Editor), pp. 214–224. Asian Veg. Res. Dev. Cent. Tainan, Taiwan.

82. Rush, D. W., and Epstein, E. 1976. Genotypic responses to salinity: Differences between salt-sensitive and salt-tolerant genotypes of tomato. Plant Physiol. *57*, 162–166.

83. Rush, D. W., and Epstein, E. 1981. Breeding and selection for salt tolerance by the incorporation of wild germplasm into a domestic tomato. J. Am. Soc. Hortic. Sci. *106*, 699–704.

84. Sacher, R. F., Staples, R. C., and Robinson, R. W. 1982. Saline tolerance in hybrids of *Lycopersicon esculentum* × *Solanum pennellii* and selected breeding lines. Biosaline Res.: A Look to the Future, pp. 325–336. Plenum Press, New York.

85. Sayama, H. 1979. Morphological and physiological effects associated with the crimson (*og^c*), high pigment (*hp*), and other chlorophyll intensifier genes in tomato (*Lycopersicon esculentum* Mill.). Ph.D. Thesis, Purdue University, Lafayette, IN.

86. Schaible, L. W. 1962. Fruit setting responses of tomatoes to high night temperatures. Proc., Plant Sci. Symp. pp. 89–98.

87. Schalk, J. M., and Stoner, A. K. 1976. A bioassay differentiates resistance to the Colorado potato beetle on tomatoes. J. Am. Soc. Hortic. Sci. *101*, 74–76.

88. Schalk, J. M., Sinden, S. L., and Stoner, A. K. 1978. Effects of daylength and maturity of tomato plants on tomatine content and resistance to the Colorado potato beetle. J. Am. Soc. Hortic. Sci. *103*, 596–600.

89. Schuster, D. J. 1977. Resistance in tomato accessions to the tomato pinworm. J. Econ. Entomol. *70*, 434–436.

90. Schuster, D. J., Waddill, V. H., Augustine, J. J., and Volin, R. B. 1979. Field comparisons of *Lycopersicon* accessions for resistance to the tomato pinworm and vegetable leafworm. J. Am. Soc. Hortic. Sci. *104*, 170–172.

91. Scott, J. W. 1983. Genetic sources of tomato firmness. Proc. 4th Tomato Qual. Workshop, Univ. Fla. Res. Rep. *VEC 83-1*, 60–67.

92. Smith, P. G. 1944. Embryo culture of a tomato species hybrids. Phytopathology *34*, 413.

93. Sprague, G. F., Alexander, D. E., and Dudley, J. W. 1980. Plant breeding and genetic engineering: a perspective. BioScience *30*, 17–21.

94. Stevens, M. A. 1973. The influence of multiple quality requirements on the plant breeder. Raw product evaluation of horticulture products. HortScience *8*, 110–112.

95. Stevens, M. A., and Paulson, K. N. 1976. Contribution of components of tomato fruit alcohol insoluble solids to genotypic variation in viscosity. J. Am. Soc. Hortic. Sci. *101*, 91–96.

96. Stevens, M. A., Kader, A. A., Albright-Holton, M. and Algazi, M. 1977. Genotypic variation for flavor and composition in fresh market tomatoes. J. Am. Soc. Hortic. Sci. *102*, 680–689.

97. Stevens, M. A. 1979. Breeding tomatoes for processing. *In* Tropical Tomato, pp. 201–213. Asian Veg. Res. Dev. Cent., Tainan, Taiwan.

98. Stevens, M. A. 1979. Tomato quality: Potential for developing cultivars with improved flavor. Acta Hortic. *93*, 317–329.

99. Stevens, M. A. and Medina Filho, H. P. 1980. Tomato breeding for nematode resistance: Survey of resistant varieties for horticultural characteristics and genotype of acid phosphates. Acta Hortic. *100*, 383–393.

100. Stoner, A. K., and Gentile, A. G. 1968. Resistance of *Lycopersicon* species to the Carmine spider mite. U.S., Dep. Agric., Prod. Res. Rep. *101*.

101. Stoner, A. K. 1970. Selecting tomatoes resistant to spider mites. J. Am. Soc. Hortic. Sci. *93*, 532–538.

102. Strand, L. L., Morris, L. L., and Heintz, C. M. 1983. Taste life of *rin* and *nor* hybrids. Proc. 4th Tomato Qual. Workshop, Univ. Fla. Res. Rep. *VEC 83-1*, 68–77.

103. Strobel, J. W., Hayslip, N. C., Burgis, D. S., and Everett, P. H. 1969. Walter, a determinate tomato resistant to races 1 and 2 of the Fusarium wilt pathogen. Circ.—Fla., Agric. Exp. Stn. *S-202*.

104. Sullivan, G. H. 1977. Fresh market tomatoes; production trends and industry organization. Indiana, Agric. Exp. Stn., Res. Bull. *945*.

105. Sullivan, G. H. 1977. Tomatoes for processing; production and economic trends. Indiana, Agric. Exp. Stn., Res. Bull. *947*.

106. Thomas, B. R., and Pratt, D. 1981. Efficient hybridization between *Lycopersicon esculentum* and *L. peruvianum* via embryo callus. Theor. Appl. Genet. *59*, 215–219.

107. Thompson, A. E. 1961. A comparison of fruit quality constituents of normal and high pigment tomatoes. Proc. Am. Soc. Hortic. Sci. *78*, 464–473.

108. Thyr, B. D. 1968. Resistance to bacterial canker in tomato and its evaluation. Phytopathology *58*, 279–281.

109. Tigchelaar, E. C., and Tomes, M. L. 1974. Caro-Rich tomato. HortScience *9*, 82.

110. Tigchelaar, E. C., and Casali, V. W. D. 1976. Single seed descent: Applications and merits in breeding self pollinated crops. Acta Hortic. *63*, 85–90.

111. Tigchelaar, E. C., McGlasson, W. B., and Buescher, R. W. 1978. Genetic regulation of tomato fruit ripening. HortScience *13*, 508–513.

112. Tigchelaar, E. C., and Grazzini, R. A. 1981. Use of Tomato Fruit Ripening Mutants to Extend Tomato Shelf Life, Proc. 3rd Tomato Qual. Workshop. Dept. Hortic., Univ. of Maryland, College Park.

113. Tigchelaar, E. C., Jarret, R., and Delaney, D. 1983. The Genetics of Ripening, Proc. 4th Tomato Qual. Workshop, Res. Rep. VEC 83-1, ppl 78–100. Univ. of Florida, Gainesville.

114. Tomes, M. L., Quackenbush, F. W., and Kargl, T. E. 1956. Action of the gene B in biosynthesis of carotenes in the tomato. Bot. Gaz. (Chicago) *117*, 248–253.

115. Volin, R. B., and McMillan, R. T. 1977. Inheritance of resistance to *Pyrenochaeta lycopersici* in tomato. Euphytica *24*, 75–79.

116. Volin, R. B., and Jones, J. P. 1983. Progress in Developing Resistance to Fusarium Race 3 in Florida, Proc. 4th Tomato Qual. Workshop, Res. Rep. VEC 83-1, p. 105. Univ. of Florida, Gainesville.

117. Walter, J. M. 1967. Hereditary resistance to disease in tomato. Annu. Rev. Phytopathol. *5*, 131–162.

118. Wann, E. V., and Johnson, K. W. 1963. Intergeneric hybridization involving species of *Solanum* and *Lycopersicon*. Bot. Gaz. (Chicago) *124*, 451–455.

119. Webb, R. E., Stoner, A. K., and Gentile, A. G. 1971. Resistance to leaf miner in *Lycopersicon* accessions. J. Am. Soc. Hortic. Sci. *86*, 65–67.

120. Webb, R. E., Barksdale, T. H., and Stoner, A. K. 1973. Tomatoes. *In* Breeding Plants for Disease Resistance. R. R. Nelson (Editor), pp. 344–360. Pennsylvania State Univ. Press, University Park.

121. Wolfenbarger, D. A. 1966. Variations in leaf miner and flea beetle injury in tomato varieties. J. Econ. Entomol. *59*, 65–68.

122. Yamaguchi, M. 1983. World vegetables: Principles, Production and Nutritive Values. AVI Publishing Co., Westport, CT.

123. Yordanov, M. 1983. Heterosis in the tomato. Monogr. Theor. Appl. Genet. *6*, 189–219.

5
Cucumber Breeding

R. L. LOWER
M. D. EDWARDS

THE CUCUMBER INDUSTRY

Cucumbers are one of the most popular members of the Cucurbitaceae (vine crop) family. Like most cucurbits, the cucumber is a warm-season crop and has little or no frost tolerance. Growth and development are favored by temperatures above 20°C. Cucumbers were once a very seasonal crop and a single crop per year was common in most locations, but now several crops are grown in many areas within a single year. This change in cropping systems has occurred through changes in cultivar maturity and adaptability, in processing technology, and in consumer preference for the final product.

Pickling and Fresh-Processed Cucumbers

Tapley *et al.* (*85*) prepared an excellent treatise on cucumber classification and history prior to 1935. Currently, cucumber cultivars are generally classified as either pickling or slicing types. Inherent in this classification are many differences directly attributable to final product usage.

173

Quality attributes may vary for pickling and slicing cucumbers. Pickling types have smaller length/diameter ratios (L/D) than slicers, and usually have lighter colored skin with more pronounced warts (tubercles) at the immature stage when they are most frequently harvested (Figure 5.1a). The L/D for pickles is expected to be in the range 2.8–3.2. Normally it is greater for smaller than for larger fruit. Also, fruit at the base or crown of the plant have smaller L/D than those borne on laterals or on the upper portions of the plant. L/D is also influenced by plant density and plant size. Increases in planting density usually result in reduction of size of the plant frame and this is frequently accompanied by shorter fruit (lower L/D).

Pickles are available with pointed, rounded, or very blocky ends and with either tapered or cylindrical shape. Cylindrical shapes are preferred for chipping or stripping of the processed fruit, while slightly tapered fruit are favored for whole-pack products, particularly in the smaller sizes, as they are easier to load into glass jars or containers.

Until the mid-1960s the majority of pickling cucumber cultivars grown in the northern United States were black spined and those grown in the southern United States were white spined. Since that time, important processing characteristics once associated with black spine color have become available in white-spined cultivars and resulted in less regional difference in spine color. In the last two decades, major changes in cultural practices coupled with the availability of new germplasm have led to the development of white-spined cultivars with multiple disease resistance. These cultivars have been accepted throughout the pickling industry, and the use of black-spined types once associated with northern production areas has declined dramatically.

The processing industry is very small (limited) outside the United States, although

FIGURE 5.1. Samples of cucumbers used for (a) pickling, (b) slicing, (c) European parthenocarpic, and (d) Middle East fresh consumption.

interest is intensifying in Europe and other areas. The pickling cucumber is usually fresh-processed (quick-pasteurization), or brined and then processed. Processed cucumbers are available in several sizes and different forms (whole, half, strips, chips, cubed, and diced) and flavors (dill, garlic, sweet, and sour). Preservation via brining was an art until the food industry developed regimented schedules and precise techniques for salting and fermentation procedures. Numerous refinements of this method of preservation are now proven. These include new concepts such as nitrogen purging of the brine solution to eliminate carbon dioxide, which causes deterioration of quality; and seeding or lacing of the brining medium with a fermenting organism (*Lactobacillus* spp. or *Pediacoccus* spp.) that ensures a high-quality brined product at completion of the fermentation process. Brined products must be desalted, then flavored, and processed before they are sold. Those interested in learning more about these processes should study the reviews of Etchells *et al.* (24), Fleming *et al.* (27), Jones *et al.* (38), and others (3).

Fresh-pack processing of cucumbers has received rapid consumer acceptance due to the high quality of the final product and its attractive color, flavor, texture, and crispness. Less than 20% of the pickling cucumbers were fresh processed in the 1950s, while today slightly more than half of the processed product is prepared by the fresh-pack method. These changes are attributable to advances in processing technology and to a nearly continuous supply of fresh product.

A very recent trend in consumer preference is for the refrigerated or cold-pack product. Like the other fresh-processed products, shelf life is limited but quality factors are favored over those of brined fruit. The industry would fresh-process even a higher percentage of its pickling cucumbers if they were available on a year-round basis; however, the perishable nature of the fresh cucumber prevents economical storage of fruit for any length of time. To the contrary, the preservation of large quantities of fruit via brining allows for tremendous intake of produce over a relatively limited time frame. Thus, the long storage life and the product flexibility inherent to the brined cucumber make the brining process important to the industry. Nonetheless, the outlook for brining has been bleak, because the salt and waste from the process were major problems and the reclamation of salt is extremely expensive. A recent report of the development of a closed tank brining system offers hope that these problems may be resolved through advances in technology (25a).

Slicing Cucumbers

All slicing cucumbers, commonly called fresh-market varieties, are white spined and most possess dark green exterior color (Fig. 5.1b). Numerous cultivars are now available that possess genes for uniform color, which essentially eliminate lighter colored striping as well as stippled or mottled color patterns over the lower one-half to two-thirds of the fruit. The darker, uniform-colored fruit are preferred by buyers in nearly all markets because of their even, attractive color. Most slicers have slightly rounded ends and taper slightly from the stem to blossom end, although cylindrical-shaped fruit with blocky or even rounded ends are also available. Slicing cucumbers are frequently sold in lengths from 6 to 10 in., and diameter varies from 1½ to nearly 3 in. Generally slicers have $L/D \geq 4.0$ and are thicker skinned than pickles. This difference is accentuated because slicers are harvested at later stages of maturity.

The fresh-market cucumber is also a highly perishable product and, prior to shipping, is frequently washed, waxed, and cooled to prolong its marketable value and quality. Its primary use is as a salad ingredient or green table vegetable. It is rarely consumed in other than a fresh or raw form. Although slicing cultivars may be processed, they generally are not acceptable substitutes for the pickling cucumber. On the other hand, fresh pickling cucumbers do make good salad additions, as they are generally thinner-skinned and more succulent than slicing cucumbers at comparable maturities.

A third and lesser known class of cucumber is the parthenocarpic slicing type (Fig. 5.1c). These cultivars have been bred primarily for the glasshouse trade. Unlike most field-grown cultivars, they set and develop fruit without pollination. Although this kind of cucumber has received only limited attention in the United States, it is the predominant fresh-market cucumber in Western Europe. The fruit are seedless and non-bitter, are usually in excess of 12 in. in length, and have tender, smooth, uniform, green-colored skin. The European consumer treasures these high-quality fruit, which are frequently marketed at a price 10 times greater than seeded cucumbers of similar size and age. The American consumer does not yet appreciate the quality attributes of the parthenocarpic slicer and is unwilling to pay the high cost for the seedless fruit. The rising costs of energy and other factors associated with greenhouse vegetable production will likely keep parthenocarpic cultivars from playing a major role in the U.S. industry in the near future.

The European and Middle Eastern populace consume large quantities of immature seeded cucumbers as well as the greenhouse parthenocarpic types (Fig. 5.1d). Nearly all of these are field-grown and serve as a regular component of the diet during the outdoor growing season. The fruit of these cultivars are usually fine spined rather than warted, have large seed cavities, and may be either mottled or of uniform color. Reliable parthenocarpic cultivars for field production are unavailable at this time.

Commercial Production and Yield

Yield in pickling cucumbers is based on weight and fruit diameter. Although minor differences in grade or fruit size might be found in various growing areas, basically the breakdown is as follows:

> *Size 1:* diameter up to 1$\frac{1}{16}$ in., \$12–16/cwt
> *Size 2:* diameter 1$\frac{1}{16}$ to 1$\frac{1}{2}$ in., \$6–10/cwt
> *Size 3:* diameter 1$\frac{1}{2}$ to 2 in., \$3–5/cwt
> *Size 4:* diameter in excess of 2 in., no value

Values per hundredweight also vary with growing area. Because the value is a function of fruit size, yield reports for pickling cucumbers frequently present both weight and dollar value per acre. Five tons of size 4 fruit are of no value, while 1 ton of sizes 1 and 2 might be worth \$300 at current prices.

Slicing cucumbers are subject to USDA marketing standards and grades. A premium is usually paid for fancy grades. The basis of sale is frequently by volume (bushel) and/or weight. In contrast to pickling-cucumber marketing systems, where price is relatively stable for an entire season, the value of slicing cucumbers fluctuates greatly and is primarily subject to supply and demand. Proximity to market is extremely important, as transportation costs are frequently as high as growing costs.

Yield of pickling cucumbers is usually greater per acre with multiple harvest than once-over harvest. This is understandable as plants normally set additional fruit after the removal of developing fruit. Thus, with multiple harvests the total number of fruit harvested per plant should increase as long as the plant remains healthy and free of seeded

TABLE 5.1.   Area, Yield, and Value (per cwt) of Fresh and Processed Commercial Cucumbers in the United States (1949–1979)[a]

Measure of production	Type	1949	1959	1969	1979
Area (acres)	Fresh	55,500	48,900	50,570	50,680
	Processing	134,530	100,500	131,020	131,780
Yield (×1000 cwt)	Fresh	3,224	3,658	4,678	5,818
	Processing	5,611	6,705	10,159	13,458
Value ($/cwt)	Fresh	4.46	5.71	7.25	14.70
	Processing	2.94	2.56	4.60	7.25

[a]Agricultural Statistics, USDA, 1950–1980.

fruit. Slicing cucumbers are nearly all hand-harvested, although mechanical harvesting aids (nondestructive) such as tractor-drawn sleds for workers to ride on are frequently used. Slicers are commonly harvested as long as it is economical. Frequently the harvest period is relatively short. As fields closer to the market area come into production, the produce from more distant fields, although of similar grade and quality, is more expensive due to transportation, time, and energy considerations.

The acreage of fresh and processed cucumbers has remained relatively stable over the past 30 years; however, both yield and marketable value have increased substantially (Table 5.1). Pickling and slicing cucumbers are commonly grown in most home gardens, and yield and acreage data from this segment of the population would greatly inflate the data presented in Table 5.1, which is for general farming operations.

The major U.S. production of pickling cucumbers is centered in the Midwest and the southeastern states owing to the excellent growing season, location of food processing facilities, and population distribution. During 1980–1982, Michigan, North Carolina, Wisconsin, and California were the leading production states, although as many as 10–15 states make significant contributions in some years (Table 5.2). Marketable yields per

TABLE 5.2.   Top Five States in Production of Fresh and Processed Cucumbers in the United States[a]

1949			1959			1969			1979		
						Processing					
MI	1530	(46.3)	MI	2100	(22.8)	MI	1760	(21.6)	MI	2360	(25.5)
WI	980	(22.7)	WI	930	(16.1)	NC	1440	(28.0)	OH	1600	(6.4)
CA	510	(3.2)	CA	600	(3.0)	CA	980	(5.2)	NC	1600	(28.5)
NC	400	(9.2)	NC	590	(14.8)	WI	860	(15.0)	CA	1380	(4.4)
CO	290	(2.7)	TX	280	(5.0)	OH	680	(4.8)	WI	1280	(10.3)
						Fresh					
FL	823	(11.7)	FL	1300	(15.3)	FL	1761	(17.3)	FL	2214	(15.0)
NY	360	(6.0)	CA	644	(3.2)	CA	741	(2.8)	CA	891	(3.3)
CA	303	(2.5)	SC	234	(4.2)	SC	379	(6.2)	TX	588	(7.2)
SC	290	(6.4)	VA	228	(4.3)	NC	336	(6.1)	NC	487	(7.9)
NJ	276	(3.6)	NJ	216	(1.8)	NY	273	(2.6)	SC	425	(4.8)

[a]Figures for yield are in thousands of hundredweight; figures for acreage (in parentheses) are in thousands of acres. (Agricultural Statistics, USDA, 1950–1980.)

acre are extremely variable, and fruit grades based on diameter frequently change from state to state and year to year, as does the market value of each grade.

On a commercial scale, fresh-market slicers are grown primarily in the southern United States when it is too cool for satisfactory plant growth in the north (Table 5.2). The fertile coastal plains of the east coast and the valleys in California are also major areas of concentration of fresh cucumbers. They are generally sold by grade (see USDA Handbook) at auctions or on consignment. Supply and demand control the pricing structures, which may rise or fall severalfold in less than 48 hr. Florida and California have been the leading states in production.

Historic Preference for Spine Color

As cultural practices have changed and mechanization of the harvesting process has become prominent, the use of black-spined types has decreased since the exterior of the fruit turns yellow–orange or bronze as they increase in size and approach maturation. The bronzing process essentially renders the fruit unusable for processing and is associated only with black-spined cultivars. The rate and intensity of the yellow–orange coloration are related to temperature, light exposure, and genetic factors associated with the black-spined character. With the advent of gynoecious hybrids and once-over mechanical harvesting, the period of time that fruit were "stored" on the vine was sometimes increased, which contributed to the development of unacceptable color. This physiological aging process occurs in white-spined cultivars also, but they remain light green or cream color.

Although many processors feel that black-spined fruit have a thinner and, subsequently, more tender skin than white-spined fruit, the data to support this concept are incomplete. Processors now use primarily white-spined cultivars. The relationship between spine color and skin toughness is not fully understood. Flesh or carpel wall thickness and the size and rate of maturation of the seed cavity are important in palatability assays related to toughness. Processors preferred the external color of brined or processed black-spined fruit, but this preference has been minimized by the availability of high-quality white-spined hybrids.

Another common association was between spine color and disease resistance in cucumbers. The prominent diseases in northern production areas were cucumber scab and cucumber mosaic virus (CMV), which both affected all parts of the plant. The first resistance was developed in black-spined cultivars. These two diseases occurred less frequently and less severely in the South; and thus, resistance in white-spine cultivars was not as common or necessary. The major diseases in the warmer and more humid climes are foliage diseases and fruit rots, and initial sources of resistance were developed on white-spined backgrounds. As the development of cultivars became more intensive and the processing industry became more sophisticated, the regional differences associated with spine color and disease resistances were countered with the availability of white-spined cultivars possessing resistance to both northern and southern disease complexes.

ORIGIN AND BOTANY

The cucumber is believed to be native to India or southern Asia, and has apparently been cultivated there for 3000 years (*19*). The plant was carried westward to Asia Minor, North Africa, and southern Europe long before written history (*92*).

The cucumber *Cucumis sativus* L. is commonly a monoecious, annual, trailing or climbing vine (5). Unlike *Cucumis melo,* the muskmelon or cantaloupe, cucumbers have hirsute or scabrous stems. The cucumber is distinct from *Cucumis anguria* in having triangular–ovate leaves with shallow and acute sinuses, unlike the deep, rounded lobes of the West Indian gherkin. Cucumber flowers are usually from 1 to 2 in. in diameter. Staminate flowers are borne either singly or in clusters, but pistillate flowers are often solitary in the leaf axes and have shorter and stouter pedicels (peduncles) than staminate flowers. Cucumber fruit are spiny when young and develop into a nearly globular to oblong or cylindrical shape as they mature, turning to a creamy light yellow or deep orange at seed maturity.

The cucumber is distinct from other *Cucumis* species (*18,79,92*) in that it has seven pairs of chromosomes ($2n = 2x = 14$), whereas most other *Cucumis* species have 12 pairs of chromosomes, or multiples of 12 (i.e., $2n = 2x = 24$, $2n = 4x = 48$, etc.) (*18*). This suggests that most other uncultivated *Cucumis* species are more genetically similar to the muskmelon, which also has 12 pairs of chromosomes, than to the cucumber.

Another related cucumberlike *Cucumis* species is the West Indian gherkin, or burr (bur) cucumber, *Cucumis anguria* var. *anguria*. It grows in the southern United States and tropical America, south to Brazil, and the gherkin fruit are edible and can be used for pickles (*92*). The var. *anguria* has 12 pairs of chromosomes ($2n = 2x = 24$) and is not cross compatible with *C. sativus* (*18*). The var. *anguria* was once thought to be native to North America; however, Meeuse (*54*) found that a wild *Cucumis* species from southern Africa, *Cucumis anguria* var. *longipes,* had many morphological features in common with var. *anguria* and was cross compatible. He suggested that the var. *anguria* was a non-bitter variant of var. *longipes* that had reached the New World in the seventeenth century through the slave trade.

Although the cucumber is known only as a cultivated plant, a *Cucumis* form, *Cucumis sativus* var. *hardwickii* R. (Alex.) ($2n = 2x = 14$), will cross readily with the cucumber and sets fruit with full complements of plump seeds from reciprocal crosses (*18,35*). DeCandolle (*19*) states that Sir Joseph Hooker collected specimens of the var. *hardwickii* along the southern foothills of the Himalayas and found that its range of variability fell within that of *C. sativus.* This has led to the conclusion (*19,92*) that the var. *hardwickii* is either a feral or progenitor form of the cultivated cucumber *C. sativus* L.

The var. *hardwickii* line LJ 90430 is currently being used in many breeding programs. It differs from *C. sativus* cultivars in several morphological and flowering characteristics (*35*). Typically, LJ 90430 plants are larger than *C. sativus* cultivars, seem to lack apical dominance, and have more and larger lateral branches. Seed are one-sixth the size of seed of *C. sativus* cultivars. The fruit are bitter, ellipsoidal in shape, and weigh 25–35 g when mature. LJ 90430 is a short-day plant that produces flowers only when the photoperiod is less than 12 hr at 30°C day/20°C night temperatures. However, it did not flower even in an inductive photoperiod after 72 days when the night temperature was reduced to 15°C.

One of the most potentially useful characteristics of var. *hardwickii* is its ability to set large numbers of seeded fruit per plant in a sequential fashion. Fruit with developing seed do not inhibit development of later fertilized flowers, as is common with *C. sativus* cultivars (*52,87*). Horst and Lower (*36*) reported that LJ 90430 frequently averaged 80 mature fruit per plant under North Carolina fall conditions.

Interspecific hybridization between the cucumber and the muskmelon has been attempted numerous times. The pollen germinates and pollen tubes traverse the length of the style, sometimes entering the ovule, but the embryos do not develop (*40*). Embryo culture or other techniques may eventually allow exchanges of genes between cucumber and muskmelon, as well as all the other cultivated or feral species of *Cucumis*.

Many of the changes that accompanied domestication of the cucumber relate to fruit morphology. Fruit sizes and shapes have been diversified for specialized uses. A more recent change accompanying domestication of the cucumber is the transition from monoecious to gynoecious or predominantly female sex expressions for commercial cultivars. Female sex tendencies are used to obtain greater uniformity of fruit maturity at harvest and greater early-harvest yields.

Continued domestication of the cucumber may lead to major alterations in plant morphotype. Several vegetative mutants are being developed for specialized commercial applications. Among these are determinate vine types and compact or bush plant types. Both adaptations are of interest for application to high-density, mechanically harvested production schemes (20,39). Another potentially useful morphotype is "little-leaf," a highly branched plant type with small leaves, potentially possessing some drought adaptation (33). Finally, glabrous plant types that lack foliar trichomes are under consideration primarily for greenhouse slicing-cucumber production. The lack of trichomes is thought to facilitate control of the whitefly *Trialeurodes vaporariorum* Westwood in the greenhouse (68) and the pickleworm *Diaphania nitidalis* Stoll in field plantings (73).

FLORAL BIOLOGY AND CONTROLLED POLLINATION

Cucumbers exhibit a fascinating range of floral morphologies. Within *C. sativus*, staminate, pistillate, and hermaphroditic flowers occur in various arrangements, yielding several types of sex expression. Furthermore, these types are influenced greatly by environmental conditions, producing a virtually continuous spectrum of sex expressions.

Although most cucumber cultivars bear diclinous flowers, embryonic flower buds possess both staminate and pistillate initials. The development of these initials may be selective, thus producing staminate or pistillate flowers; or unselective, giving rise to hermaphroditic flowers. Multiple-staminate flowers often develop in a single leaf axil. Pistillate flowers frequently occur singularly, although occasionally staminate or additional pistillate flowers subsequently develop at the same leaf axis. A gene conditioning multiple-pistillate flowers per node has been reported (57).

In monoecious forms, the succession of flower types may vary with location on the plants. The main stem of a monoecious cucumber cultivar is typically characterized by three phases of sex expression, each of variable duration. Only staminate flowers occur in the first phase; followed by a phase of irregularly alternating female, male, or mixed nodes; and finally a phase of only pistillate nodes. Lateral shoots of monoecious cultivars usually have stronger female tendencies, possibly even to the degree of bearing only pistillate flowers, while the main stem is producing only staminate flowers.

Mature flowers of different sex types also differ in the relationship of the receptacle to the other floral parts (Fig. 5.2). Pistillate flowers are epigynous, and hermaphroditic flowers are basically perigynous but may range from nearly hypogynous to nearly epigynous. Epigynous and perigynous flowers develop into differing fruit shapes. Fruit from perigynous flowers are more rounded, as opposed to elongated, and have a pronounced "scar" after shedding of the receptacle. These fruit are usually of poorer horticultural quality due to much larger seed cavities than those of longer fruit associated with epigynous flowers.

FIGURE 5.2. Hermaphroditic fruit representing varying degrees of epi-
gyny/perigyny. Scale in centimeters.
Courtesy of T. P. M. den Nijs.

Genetic Control of Sex Expression

Sex expression in cucumbers has been reported to be under the control of at least three
major loci.

1. m^+, m governs the specificity of the stimulus to develop primordial staminate
   and pistillate flower parts. Homozygotes for the m allele exhibit nonspecific development of staminate and pistillate flower parts, resulting in
   hermaphroditic flowers. $m^+/-$ genotypes are strictly diclinous.
2. F^+, F controls the degree of female tendency. The F allele is partially
   dominant and intensifies female expression. The locus is apparently
   subject to strong environmental influence and background genetic
   modification.

3. a^+,a Homozygotes for the a allele intensify male tendency. The effects of this locus are subordinate to the F locus (hypostatic to F), and so male intensification is contingent upon F^+/F^+. The genotypes $m^+/-$, F^+/F^+, a/a and m/m, F^+/F^+, a/a are typically androecious.

Combinations of genotypes at these loci yield the basic sex types (Table 5.3).

A number of other major loci have been implicated in regulation of sex expression; however these have not generally been included in the basic model (37,42,88). The loci conditioning determinate (de/de) and compact (cp/cp) plant types have been reported to have pleiotropic effects on sex expression (30,49,71). It is apparent that genetic effects, other than the three major loci, do influence sex expression. The marked variation observed in sex expression of genotypes identical for major loci implicates the presence of considerable background genetic influence.

Part of the difficulty in specifying the source(s) and degree(s) of genetic influence(s) in control of sex expression lies in the extreme instability of the sex type of cultivars grown in different environments. Climatic alteration of sex expression in cucumbers was reported as early as 1819. Photoperiod and temperature are two environmental factors specifically known to influence sex expression. With the exception of *C. sativus* var. *hardwickii,* nearly all cucumbers are day neutral, requiring no specific photoperiod for flowering. However, cucumbers respond to different photoperiodic regimes with alterations in flower number and type (63). Generally, shorter days promote increased female tendencies, both from the standpoint of reduced time to the appearance of the first female node and with regard to the frequency of female nodes. Lower temperatures, whether constant or simply lower in the dark period, also generally increase female tendencies (10,47). Conversely, high temperatures and/or long days tend to promote maleness. With sufficient stress, many genetically "gynoecious" cultivars will produce staminate flowers, sometimes abundantly. Sex expression of gynoecious hybrids that are heterozygous at the F locus is particularly subject to environment influence (29,46). In conclusion, though a simple three-locus Mendelian model is descriptive of the basic regulation of sex types in cucumber, interactions with background genotype and environmental effects exert significant influences,. which confound the basic model.

Male sterility has been reported in cucumbers. It is of little practical significance because of genetic control of dicliny vs. monocliny and staminate vs. pistillate flowering. Two genic male steriles *ms-1* and *ms-2* condition the abortion of staminate flowers (79). Cytoplasmic male sterility has not been reported in *C. sativus.*

TABLE 5.3. The Phenotypes and Genotypes of Basic Sex Types in Cucumber

	Genotype, locus		
Phenotype	m	F	a
Androecious	$-/-$	F^+/F^+	a/a
Monoecious	m^+/m^+	F^+/F^+	$-/-$
Hermaphroditic	m/m	F/F	$-/-$
Gynoecious	m^+/m^+	F/F	$-/-$

Unstable Sex Expression in Gynoecious Lines

Although the associations between some environmental parameters and sex expression have received considerable attention, little is known about the specific effects of environments in the broad sense on sex expression of cucumbers. Many commercial cultivars are grown in various regions across North America and at different times of the year. These diverse environments represent differences in both temperature and day length. They also differ for many other environmental factors and certainly differ drastically when considered as groups of potentially complex and poorly understood interactions between numerous climatic and edaphic factors. Yet, it is these complex sets of often unpredictable environmental parameters that constitute the "real" environments of concern regarding sex expression. In a study involving several commercial sources of several hybrids replicated across the United States, it was determined that most gynoecious × monoecious F_1 hybrids vary widely for the ratio of female : male flowers (46). Performance at a given location did not accurately predict performance in other locations. However, year-to-year variations within locations in female:male ratio over the 3-year study was of a relatively small magnitude when compared with location-to-location variability. A gynoecious inbred (m^+/m^+, F/F, a^+/a^+) and a gynoecious × hermaphroditic hybrid (m^+/m, F/F, a^+/a^+) were also evaluated and found to be far more consistent in sex expression across locations. Some male flowers were present in all environments, but at a very low frequency ($<1\%$).

Another pronounced result of the previous study was the dramatic difference in sex expression between different commercial sources of the same hybrid. Presumably, differences in management of inbred stocks by seed producers have resulted in inadvertent modification of sex tendencies, probably due to changes in background genotype rather than to alterations of major genes controlling sex expression. As pointed out earlier, plants with monoecious tendency occur at a low frequency among progeny of gynoecious × hermaphrodite crosses. Elimination of these "off-types" or rogues that "throw" occasional male flowers is important in increases of gynoecious inbred lines since a single individual can become a pollen parent of a large number of offspring in a bee-pollinated field situation. Proper cultural care as regards inducement and timing of staminate flowers can greatly reduce the impact of off-types (see later discussion). Thus, the development of inbred lines with satisfactorily stable sex expression often requires more than the transfer of three Mendelian alleles. Lines must be tested while inbreeding to select for favorable genetic background and should be subjected to various environmental conditions and stresses to select for stability of sex expression across environments.

At present, development of stable gynoecious inbreds and hybrids is a topic of considerable commercial interest. Roguing of monoecious segregates during increases of gynoecious lines is important both to development of stable inbreds and to maintenance of sex expression purity in established inbreds. Artificially stressing gynoecious lines has also been suggested as a means of more stringent selection for stability. Cultivation at high density under high temperatures and long days (such as broadcasting seed in a flat in a summer greenhouse environment) strongly promotes male tendencies and may allow considerable selection pressure for gynoecious stability.

Another approach to achieving stability of gynoecious hybrids is to constitute hybrids from gynoecious and hermaphroditic inbreds as female and male parents, respectively. In this fashion, the genotypic constitution at major loci becomes m^+/m, F/F, $a^+/-$ rather than the typical gynoecious × monoecious hybrid genotype m^+/m^+, F/F^+, $a^+/-$. Since the *F* locus is only partially dominant in some backgrounds, this hybrid (m^+/m, F/F) is thought to result in a more stable gynoecious performance. Unfortunately, hermaphroditic cucumbers are characterized by perigynous flowers, which overrides other fruit characteristics and gives rise to rounded rather than elongated fruit. In the heterozygous hybrid this factor is recessive, allowing elongated fruit shape and expression of underlying quality factors. Since the expression of fruit characteristics is obscured in a hermaphroditic background, evaluation of hermaphroditic inbreds for the fruit characteristics they will confer in hybrid combinations is difficult. Of course, potential hermaphroditic parents can be screened based upon testcross performance, although this requires additional effort and expense.

Testcrossing may also be an effective means of selecting for stable gynoecious inbreds. Monoecious inbreds with strong male tendencies may be crossed onto gynoecious lines and the subsequent hybrid progeny evaluated for the ability of gynoecious parents to confer strong female tendencies to their progeny. In this fashion, specific hybrid combinations with exceptional stability or gynoecious inbreds with good general combining ability can be identified.

Chemical Regulation of Sex Expression

Over the last 30 years, various growth regulators and synthetic compounds have been shown to be effective in regulation of cucumber sex expression. Some of these compounds are analogs of endogenous substances, which apparently mediate genetic control of sex expression. The first report in any plant species of alteration of sex expression via exogenous chemicals was in 1949: auxin application shifted sex expression in cucumbers toward femaleness (*43*). Twenty years later, the unsaturated hydrocarbons ethylene and acetylene were shown to be very potent promoters of female sex expression. The synthetic, ethylene-releasing compound 2-chloroethylphosphonic acid (ethephon) was found to provide an effective means of female flower promotion in both monoecious and andromonoecious cultivars (*53,77*).

Male flower promotion is also possible in cucumber. In 1958 Wittwer and Bukovac (*94*) reported that foliar applications of the gibberellin GA_3 promoted male flowers in a monoecious cultivar at nodes where female flowers would have normally occurred. Subsequently, gibberellins were found to be effective in promotion of male flowers in otherwise strictly female-flowering lines, and the various gibberellins were determined to differ in their effectiveness for male flower promotion. GA_4 and GA_7 were several times more effective than GA_3. Concomitant with staminate induction was the lengthening of internodes and increasing brittleness of stems. Other substances also promote male flowering in cucumbers. Silver nitrate ($AgNO_3$), silver thiosulfate [$Ag(S_2O_3)_2$], and aminoethoxyvinylglycine (AVG) all induce male flowering (*7,50,61,62*). The mode of action is by interfering with endogenous ethylene production. Some of these compounds, particularly $AgNO_3$ and $Ag(S_2O_3)_2$, have advantages over gibberellins, in that they are less expensive to apply, more stable in solution (less likely to break down), and more effective in converting strongly gynoecious lines (*50,62*). If applied at excessively high concentrations, these compounds may induce phytotoxic reactions (*62*). Foliar applications of $AgNO_3$ and $Ag(S_2O_3)_2$ are usually recommended over other treatments for reliable induction of staminate flowering, with minimum risks of phytotoxicity. Also, silver compounds do not cause the internode elongation or brittleness associated with gibberellins (*50*).

The ability to chemically alter the sex expression of cucumber lines, whether gynoecious, monoecious, or androecious, by exogenous application of growth regulators is of tremendous advantage in cucumber breeding programs. Production of gynoecious × monoecious hybrid cultivars is contingent upon the ability to maintain and increase gynoecious parental stocks. Only with the advent of growth regulators that promote male flowering has it become possible to self-pollinate genetically gynoecious plants and subsequently develop gynoecious inbred lines.

Alteration of sex expression is also tenable in field-scale hybrid or inbred production because of the ease and inexpensiveness of silver nitrate or ethephon treatments. Spraying rows of gynoecious inbreds with $AgNO_3$ will induce sufficient pollen production for gynoecious lines to serve as male parents in gynoecious × gynoecious hybrids. Conversely, monoecious × monoecious hybrids may be produced without contamination due to undesired self- or sib pollination, by spraying the designated ''female'' inbred regularly with ethephon beginning at an early developmental stage (one or two leaves).

Population improvement breeding strategies may be equally facilitated by exogenous regulation of sex expression. An initial planting of gynoecious lines may be treated with a silver compound to serve as male parents. Sex expression of these lines should be monitored through the first four or five nodes to ensure pistillate expression prior to treatment. A second planting of a heterogeneous gynoecious population may be interplanted some 2–3 weeks later and rogued of staminate-flowered segregates. The initial planting of desirable gynoecious lines then serves as the pollen source for the heterozygous population. Likewise, random intermating between families may be optimized in a recurrent selection scheme by randomizing ''male'' and ''female'' plots of families that have been treated with $AgNO_3$ and ethephon, respectively (*48*).

Controlled Pollination

Pollination techniques for controlled crosses in cucumbers are simple and easy to conduct. If bees and natural pollen vectors can be excluded, the breeder need not be concerned about measures to prevent selfing or other pollen contamination because of the diclinous nature of cucumbers and the stickiness or adherence of pollen to its source flower. There is no wind dissemination of pollen. Pistillate flowers are receptive in the morning or up to midday on the day they open. Hot, dry conditions reduce pistil receptivity and pollen viability.

The success of controlled pollinations is enhanced by removal of any previously pollinated female flowers. The development of fruit generally proceeds according to the sequence of pollination and the first fertilized flower inhibits the development of subsequent fruit. For this reason, controlled pollinations in the field are most easily conducted as soon as possible after flowering begins. In outdoor plantings or others unprotected from natural pollination vectors, male and female flowers to be used in pollination must be identified the day before they open and covered with half of a size 000 gelatin capsule or a section of wire tie to prevent them from opening, as depicted in Fig. 5.3a. The following day these flowers are relocated via field markers, such as prominent flags, and pollination is conducted.

A rapid-pollination technique is diagrammed in Fig. 5.3. A newly opened male flower should be selected and removed, keeping a portion of the pedicel attached for easy handling (Fig. 5.3b). The corolla may be gently torn from the point of attachment with the sepals by grasping it between the thumb and forefinger of each hand. By using one hand to grasp the calyx, the corolla may then be removed by tearing it from its attachment all the

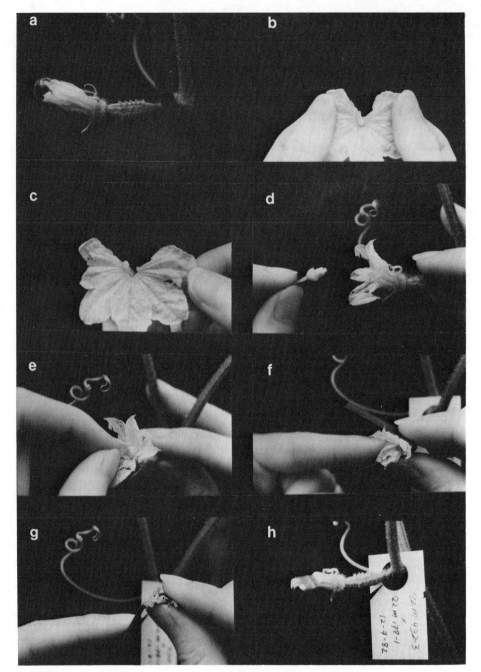

FIGURE 5.3. A rapid-pollination technique for cucumber. See text for explanation.

way around the calyx (Fig. 5.3c). This leaves only the pedicel and the calyx with the stamens protruding. The pedicel may then be used as a handle to transfer pollen from the staminate flower, stamens first, onto the pistil of the open female flower (Fig. 5.3d). When in this position, gently rotating the male flower helps to dislodge pollen onto the pistil to assure thorough pollination (Fig. 5.3e). The male flower may be left in contact with the pistil. The female flower must be covered or closed after pollination to prevent contamination during subsequent bee visits. This can be accomplished with a wire tie (Fig. 5.3f,g). Gelatin capsules are sometimes used, but frequently they cause abortion of the flower due to excessive heat and/or humidity. Tagging of the female flower completes the procedure (Fig. 5.3h). Tags detailing the pollination should be durable enough to withstand the elements because seed is seldom mature in less than 45 days, and 60 are preferred for optimum quality.

MAJOR BREEDING ACHIEVEMENTS

Disease and Insect Resistance

Perhaps the major achievements of cucumber plant breeders have been in the area of disease resistance. Cultivars are available today that possess resistance to as many as eight diseases, nearly any of which would have eliminated a crop in a given environment prior to incorporation of resistance. Plant introductions (PIs) and foreign accessions have been exploited in the search for sources of resistance. Peterson (64) offers an excellent review of this topic. Procedures for screening for resistance to numerous diseases have been thoroughly documented (1,93). Sources of resistance and the genes controlling resistance are discussed by Walker (89), Barnes (6), Sitterly (81), and Robinson et al. (79).

Principal insect pests of cucumber include cucumber beetles (*Diabrotica undecimpunctata howardi* Barber and *Acalymma vittatum* Fabricius), pickleworm (*D. nitidalis* Stoll), aphid (*Aphis gossypii* Glover), and leaf miner (*Lyriomyza sativae* Blanchard) on field-grown plantings and two-spotted spider mite (*Tetranychus urticae* Koch), whitefly (*T. vaporariorum* Westwood), and leaf miner in greenhouse production. Insect resistance in cucumber had received relatively little attention from plant breeders until the past two decades.

The role of cucurbitacins in the feeding habits of cucumber beetles and spider mites has been investigated by several workers (13–16,34,70,75). The presence of the non-bitter gene *bi* suppresses the development of these bitter terpenoid compounds. Lack of these terpenoids conditions a level of resistance to cucumber beetles, but may increase susceptibility to spider mites (15), although this situation is under continuing investigation and at present is unclear (70). Quisumbing has thoroughly documented research in this area (74).

The pickleworm has been the major insect pest in many of the Southern production areas and frequently is the limiting factor in late summer or fall plantings because of damage to fruit. A discussion of this insect and its effect on production is provided by Pulliam (72). Recent investigations by Pulliam et al. (73) and Day et al. (17) indicate that the glabrous (*gl*) character may be beneficial in limiting or overcoming fruit infestation. Glabrous plants lack trichomes, thus eliminating the favored oviposition site of the adult moth. Unfortunately, the glabrous gene affects numerous plant characters (76) and, to date, an acceptable glabrous cultivar is unavailable.

The leaf miner is the major insect common to both field and greenhouse crops. This pest has great adaptability and persists in spite of numerous pesticides that are available for its control. It is frequently controlled in field plantings by predators until the predator population is either reduced or eliminated by pesticides that are ineffective against the leaf miner. Without this check-and-balance system, the leaf miner multiplies rapidly and becomes a major pest. The actual damage is difficult to measure in terms of dollar value and may, in fact, be purely cosmetic in most cases. Eason (21) has reported on screening techniques, effects of infestation, and levels of resistance available in both greenhouse and field cultivars.

De Ponti (68,69) has developed methods of screening for resistance to both whitefly and spider mite. These insects have tremendous reproductive capacities, and populations build up rapidly in greenhouse plantings without the use of pesticide or predator control.

Several other insects attack cucumber and may cause damage to nearly all portions of the plants (in some areas); however, most can be adequately controlled with prudent use of insecticides.

Cucumber crops are frequently in competition with weeds throughout their growing season. Thus, herbicides are used to control weeds and are a necessary part of most modern-day cultural practices. Unfortunately, most herbicides are extremely phytotoxic to cucumbers; thus the choices of herbicides have been limited and, in many cases, their effectiveness has been less than desired. However, a few herbicides have contributed positively to stand establishment and have substantially reduced weed competition. The use of these herbicides has become routine and they are of particular benefit in once-over harvest operations.

Adaptation for Mechanical Harvest

The pickling cucumber industry underwent tremendous changes in the 1960s with the decline of available field labor and the concomitant increase in mechanization of the harvesting process. This change in harvesting procedure was complemented by a series of developments:

1. New high-density cultural systems (in excess of 50,000 plants/acre) that made once-over destructive harvesting more plausible and profitable than the single-row spacings more commonly used for multiple-harvest regimes (15,000–20,000 plants/acre).
2. Release of gynoecious inbreds and the development of gynoecious hybrid cultivars that were earlier, more productive, and more uniform than their monoecious progenitors, and thus more adaptable to machine harvest.
3. Development of once-over destructive harvest machinery that would replace field labor, which was both scarce and expensive.

Once-over harvesting of pickling cucumbers placed demands on plant breeders and horticulturists as well as on agricultural engineers; and these groups of scientists worked hand in hand to develop a system that allowed for economical once-over harvesting. The major concept to evolve was to develop extremely uniform plant populations through plant breeding and cultural manipulations that would allow for a maximum number of marketable fruit to be harvested by a machine in one destructive pass.

Prior to mechanical harvesting, the lack of uniformity was of less importance since each plant would be hand-harvested many times (as many as 15) during the season. With once-over harvesting the goal was for all plants to have at least one marketable fruit at harvest time; otherwise they were weeds or competition to fruited plants.

The limiting factor in cucumber yield is related to the inability of the plant, despite its sometimes large vine size, to support more than one or two marketable fruit at a time. The development of seed in pollinated fruit severely limits the ability of the plant to set additional fruit (52,87). Only after initial fruit(s) have either been harvested or approach maturity (too large to be of market value) will the plant set additional fruit. This phenomenon greatly limits the fruit production within a given area and makes uniformity extremely critical to success of the once-over harvest procedure. Increasing plant densities usually results in decreasing numbers of fruit per plant. The balance between plant number, plant density, and number of marketable fruit per acre is delicate and subject to cultivar, as well as to environmental variables such as water, nutrients, and light. Under warm, humid conditions the growth of cucumber fruit is continuous and rapid. Since oversized fruit (2 in. or more in diameter) are of essentially no value (and actually are a detriment), the timing of the harvest operation is extremely critical. A shift in the predominant fruit size may occur in 12 hr. This results in minimum vine storage time and consequently a minimum time frame for maximum dollar return via a once-over destructive harvest.

Increased plant densities result in reduction of fruit quality (decreased firmness and color), increased incidence of fruit rots (belly rot and cottony leak), reduction in number of fruit/plant, decreased stability of the gynoecious character, and greatly increased cost of cucumber culture on a per acre basis. Decreases in fruit quality were accentuated by the rigors of mechanization. Nonetheless, mechanization was an absolute necessity to the industry and once-over harvesting equipment is utilized today in approximately 50% of the acreage in the United States and Canada, but only in very limited instances in other countries.

Yield has been more intensively researched with pickling than with slicing cucumbers. Numerous reports indicate that yield of pickling cucumbers is a complex character and narrow-sense heritability is probably <20% (23,31,52,84). Attempts to increase yield by increasing the numbers of pistillate flowers per plant, e.g., either by changing sex expression or by increasing the number of flowers per node, have met with little success because of the effect of the seeded fruit on subsequent fruit set.

Improved Fruit Quality

Plant breeders in cooperation with food scientists and the processing industry have been successful in identifying causal relationships between various components of fruit quality and fruit morphology or structure. Breeders have made positive contributions to processing quality by reducing seed cavity size, seed maturation rate, and placental hollowness. Selecting against brining defects, such as soft centers and various types of fruit bloating, has also been successful. Food scientists have made nearly quantum gains in brining technology to eliminate processing anomalies further.

The greenhouse industry in Western Europe is unique in its development and has relied on the availability of high-quality parthenocarpic cucumbers for its livelihood. The non-bitter cultivars possess excellent fruit quality and are sometimes cropped in excess of 150 days. Yields may reach 100 fruit/m^2 over a long season. The fruit are long (>12 in.), slender, thin-skinned, smooth, and of a uniform dark green color (Fig. 5.1c). They are shipped throughout Western Europe from major production centers like Holland. Both yield and quality are enhanced by lack of seed set. The cultivars are exclusively gynoecious hybrids. Sex expression is extremely stable and parthenocarpic fruit set under greenhouse conditions is no problem. Dutch plant breeders have excelled in the development of cultivars for this industry.

CURRENT GOALS OF BREEDING PROGRAMS

Although each breeding or improvement program is unique, nearly all have several common goals, namely, increases in yield, pest resistance, cold tolerance and stand establishment, and improved fruit quality.

Tapley et al. (85) reported in 1937 that improved cultivars were more productive than their progenitors. Yield has continued to increase because of improved pest resistances, pesticides, and other cultural practices. Direct selection for yield has probably not played a major role in this yield increase. Rather, improved plant health and reduced competition by pests (weeds, insects, pathogens) were important in lengthening the harvest period and thereby increasing yield. The major limitation to increasing yield is related to lack of genetic variability.

Smith et al. (84) suggested that sequential-fruiting cultivars might contribute positively to yield. Horst (35) reported that C. sativus var. hardwickii might be a potential source of the sequential-fruiting trait. Several breeders are exploring the utilization of this characteristic. Other solutions to the inhibition of fruit set by developing seed are the use of genetically parthenocarpic cultivars (66,67) or the induction of parthenocarpy via morphactin-like compounds such as chlorflurenol (11,78). Neither of these alternatives has met with much success to date.

Multiple-Pest Resistances

Pest resistance is extremely important in both pickling and slicing cucumbers. Resistances and procedures for screening and incorporation are the same in both cucumber types.

Resistance to some troublesome diseases is still being pursued because present levels of field tolerance are inadequate. This is particularly true for the following diseases: watermelon mosaic virus (WMV-1 and -2), cucumber green mottle virus, gummy stem blight (Didymella melonis), target spot (Corynespora cassiicola), and several fruit rot organisms, namely, Rhizoctonia, Pythium, and Phomopsis. Damping-off, associated with Pythium aphanidermatum, is a major problem under cool or unfavorable germination conditions. Resistance to nematodes is totally lacking (25) and would be extremely beneficial, particularly in southern growing areas, where soil nematicide treatments are a routine and expensive cultural practice. Recent information (4) indicates that some cultivars are susceptible to Fusarium sp., and perhaps resistance to this pathogen will be necessary in future years, although there appears to be a positive and complete relationship with scab resistance.

The resistance to downy mildew in U.S. cultivars does not appear to hold up in the Middle East (32), and this disease is responsible for severe losses in that area.

Cold tolerance and stand establishment are beneficial to uniformity, which is one of the cultural keys to enhancing yield. Considerable genetic variation exists for germinability in less than optimum soil temperatures, and this trait offers insurance against cold soils or unfavorable germination conditions following seeding (45,60). Good germinability and seedling growth are both favored by temperatures of 20°C or more, and planting in a warm soil is conducive to rapid uniform stands. Weak, slow-growing plants are more subject to pests and environmental stresses.

Miller *et al.* (*55*) and Werner and Putnam (*91*) reported that some vine crops do possess limited resistance to the triazine herbicide compounds. The potential apparently exists for the development of herbicide-tolerant cultivars. Even more exciting is the potential for development of allelopathic cultivars that essentially contain built-in mechanisms for weed control (*44*).

Improving Fruit Quality

Fruit quality encompasses a multitude of variables. Many are under simple genetic control, e.g., skin color, spine color, presence or absence of warts and spines, and fruit shape, and are related to cosmetic needs or desires of various segments of the industry. These characters are frequently changed to accommodate new marketing trends. Unfortunately, from a plant breeder's standpoint, cultivar development often takes longer than current marketing trends allow, i.e., it is easier to change the size of a jar from 3½ to 4 in. deep than to develop another cultivar.

Flesh firmness is a positive attribute, as is skin (exocarp) tenderness; however, the latter character represents a dilemma to the breeder. Skin that is very tough is unpalatable in either pickling or slicing products but offers considerable protection against the rigors of harvesting, sorting, grading, and shipping. Tender skin is desirable in the edible product but offers little protection against rough handling and also is conducive to considerable weight loss (H_2O). Genetic variation exists for fruit shape, flesh firmness, skin toughness, and numerous color genes.

Fruit firmness is related to both flesh firmness and seed cavity (endocarp) or locule size. As seed cavity diameter increases in relation to overall fruit diameter, the fruit becomes less firm. This should not necessarily be construed as a decrease in flesh firmness, but more usually is related to an increase in seed matrix area, which is less firm than the carpel walls (mesocarp). Slender (large L/D) fruit generally have a greater fraction of diameter in mesocarp and exocarp than in endocarp. Fruit with lesser L/D have proportionally larger endocarp areas.

Another factor of major importance in pickling cucumber is the rate of fruit enlargement and the role that the seed cavity size increase plays in growth in diameter (Fig. 5.4). Cultivars also vary greatly in their rates of maturation and seed development. There is a positive linear relationship between rapid fruit maturation and seed cavity growth rate. These two characters are negatively correlated with fruit firmness and vine storage time for marketable fruit (diam. > 2 in.). Extreme variation in overall fruit diameter and seed cavity diameter relationships is available to the plant breeder in the PI collection.

Increases in fruit firmness and decreases in seed cavity size are frequently accompanied by placental hollowness (Fig. 5.5, top). This condition is an unattractive and undesirable void or hole in the placental area and may lead to processing problems. Carpel separation (Fig. 5.5, bottom) is associated with rapid fruit enlargement and less-firm cultivars in large-sized fruit. This anomaly leads to "bloaters" in brined stock and is unattractive in both fresh and brine products. Increases in flesh firmness tend to reduce occurrence of carpel separation.

Stable Gynoecious Lines

Cucumber seedsmen and growers have been concerned for some time over the apparent lack of stability of the gynoecious character in inbred lines and more principally in hybrids. Interest in stabilizing the gynoecious character has intensified and become a

FIGURE 5.4. Differences in seed cavity size and
maturation rate between cultivars Carolina (left) and
Calico (right).

common goal of numerous breeding programs. Many combinations of sex expression
types have been recommended at one time or another as superior (in femaleness) to the
stability of the straight gynoecious character (*80,86*).

Recently Lower and den Nijs (*49*) have shown that plant type genes *de* and *cp* (which
were known to reduce the size of the plant frame) also result in reduced male or staminate
expression in gynoecious hybrids. This enhanced femaleness may translate into increased
gynoecious stability and higher fruit numbers.

SELECTION PROCEDURES FOR SPECIFIC CHARACTERS

Disease Resistance

Resistance to many economically significant diseases of cucumbers is conferred by single
genes. For many of these diseases, incorporation of major resistance genes into cucumber

cultivars has resulted in stable resistance rather than proliferation of new, virulent races of the pathogen. For these reasons, considerable effort has been directed toward development of rapid and efficient seedling-screening methods for major disease resistance genes in cucumber populations.

The advantages of testing at an early developmental stage and the ability to conduct multiple-disease evaluations on single seedlings make this screening procedure a powerful tool for the plant breeder. The effectiveness of such procedures depends upon achieving repeatability by standardization of the inoculum concentration and the infection and incubation conditions. Disease reactions under controlled conditions must also correspond well with reactions under highly variable field environments. The seedling-screening method appears to meet this criterion for a number of diseases.

Procedures for seedling screening have currently been developed for eight diseases (*1,93*). Five of the pathogens are fungi, two are bacteria, and one is a virus, as listed in Table 5.4. With the exception of fusarium wilt, powdery mildew, and CMV, all inocula-

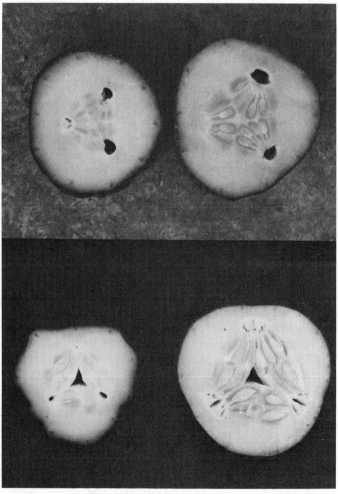

FIGURE 5.5. Placental hollowness (top) and carpel separation (bottom) in pickling cucumbers.

TABLE 5.4.   Diseases with Established Procedures for Seedling Screening

Type of causal organism	Disease	Organism
Fungus	Anthracnose	*Colletotrichum lagenarium*
	Downy mildew	*Pseudoperonospora cubensis*
	Fusarium wilt	*Fusarium oxysporum*
	Scab	*Cladosporium cucumerinum*
	Powdery mildew	*Sphaerotheca fuliginea*
Bacterium	Bacterial wilt	*Erwinia tracheiphila*
	Angular leaf spot	*Pseudomonas lachrymans*
Virus	Cucumber mosaic virus	CMV

tions may be made at the cotyledonary stage and evaluated within 1 week. Inoculum preparation and inoculation procedures vary, but none require complicated or expensive equipment or materials. Fusarium wilt resistance is evaluated by planting seed in flats with the pathogen incorporated in the sand. CMV and powdery mildew inoculations are conducted at the first or second true leaf stage.

One of the more expedient features of seedling screening is the ability to evaluate resistance to several diseases on the same seedlings. For example, when evaluating for anthracnose resistance, it is possible simultaneously to inoculate the scab, downy mildew, angular leaf spot, and bacterial wilt and subsequently inoculate survivors with CMV and powdery mildew. Because of the interaction of disease symptoms, there are limitations to the combinations of resistance that can be evaluated. Continued experimentation may allow for increased localization of disease tests and further expand the number of diseases that may be simultaneously evaluated on single individuals (Figure 5.6).

A screening procedure such as the one described may be integrated into a general breeding effort in several fashions. It may be used to identify disease-resistant individuals in either a backcrossing or pedigree breeding scheme.

Alternatively, it may be incorporated into population improvement approaches involving recombination of selected lines. Here, initial disease screening may enrich the frequency of genes for disease resistance in a heterogeneous population. Or, disease screening may be combined with simultaneous selection for other traits and provide an input into index selection. Seedling screening may also be conducted after population improvement of other traits is completed. Disease-resistant individuals would provide a superior subpopulation from which to begin the extraction of inbred lines.

In any of these uses, it must be remembered that the resistance identified in the seedling screen may not represent all the useful resistance to the organisms considered. Powdery mildew is an example of a disease of cucumbers to which several major genes for resistance may be involved. Discrimination between lines under natural infestations in either field or greenhouse trials should be practiced whenever possible. Such natural host–parasite interactions are not dependent upon the many assumptions necessary in a seedling-screening procedure.

Yield and Other Quantitative Traits

Many important features of successful cucumber cultivars are not associated with discrete, Mendelian traits, but are of a continuous or quantitative nature. Yielding ability is a prime example of such a trait and is of obvious importance. Ultimately many factors play a role in yield. Some of these may be environmental adaptation, vegetative vigor, and resource partitioning within the plant. Yield and the many factors that affect it are subject to considerable environmental influence (*12,22,31,82*). Other quantitative characters include seed cavity characteristics, fruit firmness, stability of sex expression, and environmental adaptability. Thus, differences between breeding lines may be difficult to detect.

Selection for quantitative traits requires replication of breeding lines within and/or between environments. This need for replication reduces the number of breeding lines that may be evaluated. When line performance varies considerably due to environment, replication should be conducted across a sample of production environments. Specific details such as plot size, number of replications, the replication unit, and the number of environments will depend upon the heritability of the trait, the degree of variability present in the population, and the available resources.

In our program, selection for yield has generally been conducted by replicating within one testing location. Sampling of environments has been achieved by planting at different times during the growing season. Two to four replications of each breeding line have been evaluated at each planting. The replication units have been S_1 lines, full-sib families, or testcross progenies, depending upon the population and breeding scheme. Plots are generally 30 plants each, with the middle 25 plants harvested. This plot size has been established to be near optimum in previous research (*83*). Yield is usually measured as total fruit number based upon one to several harvests. This criterion correlates well with dollar value of the crop, is more stable over time, and is easier to measure since it does not require sizing, weighing, and grading of the fruit.

Recurrent selection is frequently used for the improvement of quantitative traits. This type of selection is time consuming and must proceed for some duration to make maximum gains. Thus, establishment of a suitable base population warrants considerable attention.

Cucumbers have been identified as a crop with limited genetic variability (*58*). Populations derived from adapted breeding lines have been found to contain little genetic vari-

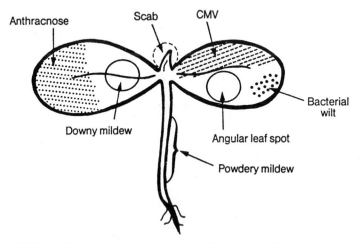

FIGURE 5.6. Hypothetical diagram of "target sites" for multiple-disease inoculations on single seedlings. Further investigations into disease interactions and symptom expression will be required for implementation.
Courtesy of P. H. Williams.

ance for yield (*84*). Selection within adapted material would be expected to progress slowly. Because of this narrow genetic base, a source population with a broad genetic base should be more responsive to selection for quantitative traits. Such a population has been established and designated "gynoecious synthetic" (*59*). It encompasses cucumber germplasm from cultivars and breeding lines worldwide and has served as a base population for several breeding programs.

We are utilizing two selection procedures for increasing fruit number of pickling cucumbers. Both procedures have been applied to populations derived from gynoecious synthetic (Figs. 5.7 and 5.8).

Fruit Appearance and Quality Traits

Among horticultural crops, the cucumber epitomizes the requirement for rigid conformity to numerous appearance and quality criteria that are exactly defined. Whether waxed, individually wrapped, and marketed as an attractive slicer for fresh consumption, or brined and neatly displayed in jars for pickled consumption, cucumbers owe much of their appeal to numerous esthetic traits. Many other traits influence cucumber product palatability and processability.

Selection for all the necessary quality attributes is a sizable task, particularly when utilizing unadapted or exotic materials. Selection for traits that are fairly simply inherited is not difficult. Such traits include fruit and spine color, spine type, and fruit conformation (L/D). Heterogeneous populations may be easily "enriched" with these traits by recurrent mass selection. Extraction of inbred lines with all the fruit quality attributes is considerably more difficult.

These attributes include features of the skin, outer wall, and seed cavity (exo-, meso-,

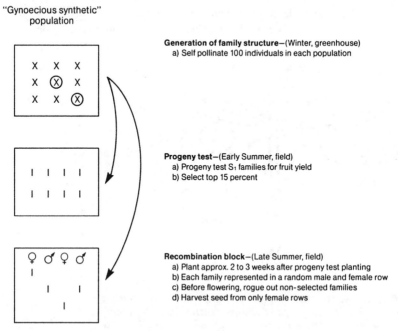

FIGURE 5.7. Outline of S_1 progeny selection for fruit yield in cucumber. After Nienhuis (*59*).

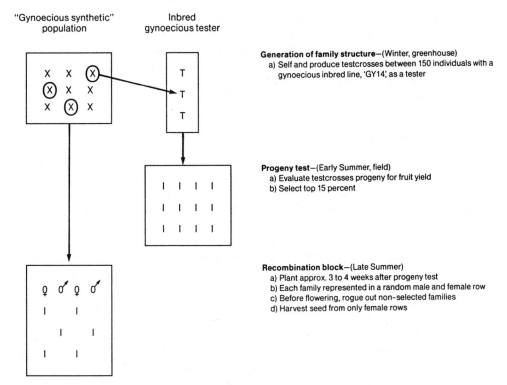

"Gynoecious synthetic"
population

Inbred
gynoecious tester

Generation of family structure—(Winter, greenhouse)
a) Self and produce testcrosses between 150 individuals with a
gynoecious inbred line, 'GY14', as a tester

Progeny test—(Early Summer, field)
a) Evaluate testcrosses progeny for fruit yield
b) Select top 15 percent

Recombination block—(Late Summer)
a) Plant approx. 3 to 4 weeks after progeny test
b) Each family represented in a random male and female row
c) Before flowering, rogue out non-selected families
d) Harvest seed from only female rows

FIGURE 5.8. Outline of recurrent selection for specific combining ability for fruit yield in a cucumber population.

and endocarp, respectively). Skin characteristics include thickness, stomate number, and other factors that may influence palatability and fruit processing (*9,26,41*). Outer-wall flesh thickness (*2,51*) and color are important aspects of fruit firmness and quality. Acceptability of the seed cavity depends on many characteristics, some of which are (1) the ratio of cavity diameter to fruit diameter, (2) the incidence of carpel separation, (3) the incidence of placental hollowness, (4) the occurrence of two- or four-carpel "off-types," and (5) the rate of seed maturation. Pressure test readings are often taken for either fresh fruit or brine stock as a measure of firmness. These readings are influenced by a combination of exo-, meso-, and endocarp. The Magness Taylor Fruit Pressure Tester is the instrument most frequently used for measuring firmness. The combined attributes of skin, outer wall, and seed cavity are also important in the brining process.

Many fruit quality characteristics are influenced by the growing environment. Cultivars with acceptable fruit quality for a northern growing region may be entirely unacceptable for a southern one. Warmer climates result in more rapid fruit growth and demand a higher level of fruit quality characteristics such as fruit firmness, seed cavity size and maturation rate, and external color. These measures of quality are difficult to evaluate adequately and may be quite variable from fruit to fruit, even within highly inbred lines. Consequently, large samples of fruit must be taken to detect differences between lines. For advanced lines, 100 plants are usually sufficient to produce the numbers of fruit necessary to evaluate quality characters. At earlier breeding stages, lines will often be culled if inferior for numerous quality criteria.

New lines must compare favorably with present cultivars for quality criteria. If breeding lines are outstanding for some characteristics, but moderately deficient for others, they may be evaluated in hybrid combinations with inbreds possessing excellent fruit quality to see if any combinations adequately express the best features of both parents. Failure to meet all quality criteria has undermined many breeding lines with excellent yields and disease resistances.

Sex Expression in Gynoecious Hybrids

Numerous researchers have investigated the genetic control of sex expression in cucumbers. Frankel and Galun (28) present an excellent review of this area. Inheritance of female tendency is basically under the control of a few Mendelian loci. It is relatively simple and straightforward to transfer and select for these genes and the desired qualitative types.

Most commercial hybrids of cucumbers use a gynoecious line as the female parent. The male parent in these hybrids has most commonly been monoecious. The resultant hybrid is heterozygous at the F locus and is not totally gynoecious, but more appropriately termed "predominantly female" since the locus exhibits incomplete dominance for gynoecy. Evaluation of a number of hybrids all possessing the F/F^+ genotype will reveal marked differences in ratios of male to female flowers and in total flower number. These differences are due to quantitative modification of the basic sex types. This genetic variation is sufficient to yield a nearly continuous distribution of sex types in cucumbers. Heterozygotes at the F locus are subject to greater environmental variation of sex expression than are homozygous F/F types (28,46).

Uniform sex expression within a cultivar is desirable for conventional cucumber production. A major advantage of gynoecious cucumbers is the potential for earlier harvest, which is important to assure a premium value for the crop. Even as the harvesting season progresses, uniformity in sex expression is important to obtain uniform maturity. For once-over mechanically harvested plantings of pickling cucumber, excess monoecious plants, whether they result from unstable sex expression or were added for pollination purposes, fail to contribute to yield and are essentially weeds. Attempts to compensate for this later-fruiting type by allowing the field to mature a few more days results in a preponderance of oversized, valueless fruit.

Because of concern for stability of gynoecious sex expression, some cultivars have been developed from (gynoecious × hermaphrodite) × monoecious three-way crosses $(m^+/m^+, F/F \times m/m, F/F) \times m^+/m^+, F^+/F^+$. The gynoecious × hermaphroditic hybrid confers stability to the gynoecious sex expression of the seed parent and simplifies increase of the seed parent stock. Three-way crosses may be less uniform for flowering and fruit production because of segregation of gametes from the hybrid seed parent. Some seed production problems have resulted in occasional hermaphroditic plants in cucumber production fields. Such problems are avoidable, however, and some successful three-way hybrids have been developed. These hybrids require maintenance of more inbreds.

Another scheme used to ensure all-female hybrids involves releasing gynoecious × hermaphroditic hybrids as cultivars. Such cultivars have expressed extremely stable sex expression and require blending seed of a monoecious pollinator with the hybrid seed to assure adequate fruit set in commercial production. This procedure imposes a few limitations. First, the 10–15% pollinator plants in the grower's field fail to contribute to yield in the early harvests and may exert a substantial competitive influence. Second, development of satisfactory hermaphroditic lines is more difficult than development of mono-ecious inbreds. Hermaphroditic fruit exhibit very different fruit characteristics, having rounded fruit and large blossom scars. Only by testcrossing can such lines be evaluated to determine what quality characteristics they will impose upon their offspring in the absence of the *m* locus effects.

Previous to the advent of silver compounds, $F/F \times F/F$ hybrids were not feasible due to unsatisfactory seed yields. This hybrid scheme is now gaining favor for the seed production of all female cultivars.

A satisfactory level of stability in sex expression may also be achieved with conven-tional gynoecious $\times$ monoecious hybrids. The success of such a scheme largely depends upon maintaining the purity of the seedstock for the gynoecious inbred. The female line should be selected during inbreeding for stability of sex expression in adverse environ-ments (such as hot, dry environments and under high interplant competition—all factors that promote male flower formation). Once a stable gynoecious line has been developed, it should be rogued for off-types during seed increases. Individuals or lines that easily convert to ''predominantly female'' or monoecious sex expression when treated with male-promoting substances are likely ''weak'' female types and should be rogued from the inbred increase prior to pollen shedding. In this manner, good gynoecious inbreds may be developed and maintained. Selected lines will combine with monoecious lines to produce predominantly female hybrids exhibiting uniform sex expression.

Although uniformity of sex expression has been established as an important criterion of pickling cucumber hybrids for production purposes, hybrids need not be all-female types. A sizable study was recently conducted in 10 production regions of the United States for three consecutive years to measure the variability in sex expression due to cultivars, seed lots, and locations (*46*). Hybrids homozygous at the *F* locus tended to be much more uniform for percentage of female flowers, with almost 100% observed in all locations and years. Conversely, locations and years exerted a large influence on percentage of female flowers for hybrids heterozygous at the *F* locus. Importantly, however, there was no apparent relationship between percentage of female flowers and multiple-harvest yield of the cultivars. This kind of data fails to substantiate industry concerns about having ''truly gynoecious'' hybrids. Instead, it suggests that the relationship between number (or per-centage) of female flowers and yield is weak or nonexistent. High-yielding cultivars possess a wide range of male : female flower ratios and owe their productivity to features other than sex expression.

BREEDING METHODS

Cucumber improvement employs a virtually unlimited array of breeding methods. Several examples are mentioned in the following discussions, although these are by no means all inclusive.

Routine backcross programs have been extremely successful for transferring single gene characters with promising disease resistances or quality aspects from donor lines into more acceptable recurrent parents (*8,56,64,90*).

Selection of single-plant segregates is commonly practiced in segregating F_2, F_3, or backcross populations that are derived from crosses between parents. Desirable recombi-nants are frequently subjected to inbreeding to stabilize and true-up prospective new lines. Selection and screening procedures dictate population sizes and numbers of selections saved.

One of the most successful breeding programs relies on selections of disease-resistant segregates from large segregating populations (65). Selections that are resistant to several diseases are frequently recombined in early generations and then subjected to additional screening for higher levels of resistance or multiple resistances. Inbreeding of resistant lines is practiced through F_6 to F_8 generations, at which time selection for other horticultural qualities such as gynoecious sex expression is undertaken. Fruit number and type are then assayed in both inbreds and hybrids.

Population Improvement Followed by Inbred Extraction

The breeding programs mentioned above deal with relatively narrow germplasm bases, involving primarily qualitative characters, and thus short-term objectives are usually accomplished. An equally successful, but more long-range approach is one of population improvement designed to capitalize on quantitative variation. Population improvement approaches based on recurrent selection theoretically allow the greatest long-term gains and may be required to improve traits of low to moderate heritability substantially. The greatest gains may be realized in populations with large genetic variances. Populations having a narrow germplasm base may exhibit little response to such breeding procedures. Since most adapted cucumber breeding lines share a very narrow germplasm base (58), population improvement within adapted material will likely not bear returns that justify the sizable undertaking.

Incorporation of unadapted and exotic germplasm into breeding populations should allow far more progress to be made for traits under quantitative genetic regulation. Yield is an example of such a trait, and promising exotic germplasm has been identified (48). Figure 5.9 provides an example of a program designed to utilize a broad-based population to achieve maximum long-term gains for yield while also selecting for other essential horticultural traits.

This program allows considerable flexibility for specific objectives and priorities, which may be tailored to either the attributes or deficiencies of the population. In our breeding program, several cycles of recurrent selection for yield were conducted before other traits were considered in the selection process. Yielding ability was emphasized initially because it was the trait for which increased genetic variance was deemed most necessary to realize maximum long-term gains.

This approach required that variance for other traits be maintained in the population. Several cycles of recombination without selection were conducted in the base population to minimize linkage disequilibrium. This allowed subsequent selection pressure without loss of favorable genes that were initially in undesirable backgrounds. Preliminary evidence would indicate that the necessary variance for quality traits and major disease resistances remains after three cycles of recurrent selection for yield.

Selection may now progress toward enhancement of fruit quality, disease resistance, and other traits, independently or in conjunction with each other and/or yield. At the same time recurrent selection for yield in the improved population may result in further gains. In this way, work may proceed toward long-term goals while intermediate-range goals are also pursued.

Before extracting inbreds, it is important to ensure that the frequency of genes for other desirable traits is great enough to recover them at an early inbreeding stage. It would be pointless to advance lines to an inbred status and then be forced to abandon them due to insufficiency of necessary traits. This gene enrichment may take the form of population improvement with random matings or of strategic crosses (pedigree breeding).

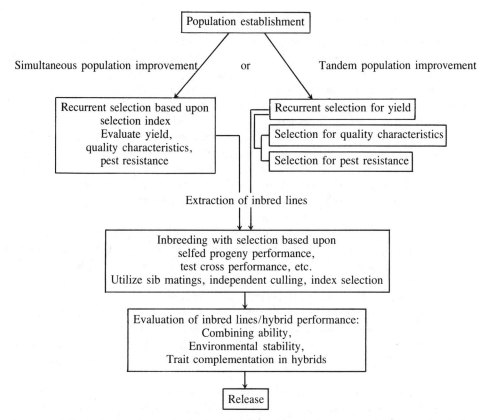

FIGURE 5.9. Schematic of typical population improvement programs in cucumber. Choice of simultaneous vs. tandem depends on genetic variation available and industry needs.

Select lines, or desired segregates from select lines (e.g., disease-resistant individuals), may then be advanced to the inbred extraction phase of the breeding program. Evaluation and selection during inbreeding is important to achieve maximum gains. Yield evaluation may be facilitated by testcrossing with gynoecious inbreds. Testcross progeny are of fairly uniform maturity and more easily harvested without biasing selections due to differences in sex expression. As inbreeding progresses, plot sizes must increase to allow discrimination of subtle fruit quality differences. Replication across environments should also be expanded to evaluate adaptability and stability of the lines. Selection during inbreeding must be rigorous. Deficiency or mediocrity for important traits cannot be allowed if the inbred lines are to compete with the many good existing lines.

Finally, a small number of exceptional inbred lines may be evaluated in a systematic crossing scheme to identify the best-combining lines (general combining ability, GCA) or most favorable specific crosses (specific combining ability, SCA). Evaluations should be conducted at locations and times to confirm the stable performance necessary for general acceptance.

Obviously, a breeding program such as the one just outlined is a long-term investment of effort and resources. If the base population contained elite germplasm, this approach is fairly certain to isolate improved types. If short-term gains are also sought, other breeding schemes with a narrow germplasm base may be conducted simultaneously.

Development of Inbred Lines from F_2 Populations

The development of advanced breeding lines (and subsequent hybrids) usually entails screening for numerous characters that are under rather simple and straightforward genetic control, such as some pest resistances, plant type, some fruit quality characteristics, and other desirable horticultural characters. Screening for many of these attributes is simple and can be done routinely by trained personnel. Identification and isolation of simply inherited qualitative characters is commonly achieved in F_2 and F_3 generations. The development of lines that contribute positively to less discrete quantities such as adaptability, stability, resistance to environmental stress, overall fitness, and yield is less routine and requires more stringent testing over a diversity of environmental conditions. Traditionally, contributions of successful breeding programs have been tested in numerous locations selected for their role as major cropping areas.

Parental materials of hybrid cucumber cultivars are maintained as inbred lines. These lines may be developed in several different ways as discussed previously. Testing of these lines is truly a paradox. If the original parents of the segregating generation were both of acceptable but different types, a breeder may choose to develop as many as 10–25 F_3 lines from the same F_2 population and compare them in the field with a commercial cultivar that has gained acceptance. Generally in our pickling-cucumber program, we use 40–50 plants/plot grown at common commercial spacing, and we harvest these at the optimum once-over stage. In our slicing-cucumber program, the harvest is made at the optimum fruit size. The entries are nonreplicated, except that we might have 15 F_3 selections from the same F_2 population. Thus, one might screen as many as 10 F_2 populations and develop 10–25 F_3s from each.

Prior to harvesting the fruit, we attempt to pollinate (self or sib) three or four plants in each plot, taking care to mark these carefully so the hand-pollinated fruit are not harvested. Pollinations are usually made indiscriminately at the time of initial flowering. We eliminate at least 50% of the plots on the basis of the first-harvest fruit characteristics; and if time permits a second harvest, we remove another 75%. By the end of the fruit harvest period we can begin to zero in on additional horticultural characters of major importance, and we attempt to identify the most desirable pollinated plants in the selected rows. Many times this can be done at a glance, but most often all controlled pollinations are saved and progeny are further screened for pest resistance, cold tolerance, or other attributes. Survivors may be carried to the next generation through selfing or sibbing, testcrossed with desirable inbreds to assess hybrid performance if the line appears uniform, or backcrossed to pick up a desirable qualitative trait. Sometimes lines are simply assessed for presence or absence of desirable characters (e.g., scab resistance, spine color, non-bitterness); in these cases selections may be advanced and further field testing begins.

The above procedure continues through F_4 and F_5 generations with increasing pressure on the inbred material for quality characters (fresh and brine stock evaluation), stable sex expression, seedling vigor, etc.

F_5 and F_6 lines are then used to develop hybrid combinations for testing. Initial hybrid developments are nonreplicated, except where several desirable sister inbreds may be evaluated on a common tester. We plant the same series of crosses on two or three different dates in the same season in an attempt to compare the hybrids across more than one environment. In effect this is replication in time and protects the breeder in the event of an environmental calamity that could eliminate a single replicated planting.

The use of parental material that is quite diverse results in a slight alteration of the above program. Isolation of lines that are true-breeding for many of the simply inherited characters generally takes either much larger F_2 populations or more recombination in the F_3 and F_4 lines. Thus the program with diverse parents may be two or more generations behind the program that involves parents of more similar genetic background. Crossing of desirable lines to a recurrent parent (backcross) or a new parent (outcross) for a positive genetic contribution simply delays the testing stage but is an important aspect of inbred or parental development.

Trials of Experimental Hybrids

Hybrid combinations are frequently developed using proven inbred testers, which may be either gynoecious or monoecious, or they may be developed exclusively from new lines (tests for GCA and SCA can be developed). Regardless, hybrids should be tested across several environments. To identify a desirable or worthy hybrid we rely only on data from replicated tests in different seasons and different locations.

Plots of hybrid combinations are usually of 100 uniformly spaced plants grown at a commercially feasible population density and in the usual season. Initial harvest is at the optimum once-over harvest stage. We find that rows of less than 100 plants frequently do not yield enough fruit to assess all of the different quality attributes at each of the four fruit grades of pickling cucumbers. In early testing of hybrids, the major yield characteristic we are interested in is fruit number. Hybrids that compare favorably with commercial cultivars but have nothing more to contribute are usually eliminated after the first season of testing. Only those hybrids that have greater potential than the checks are tested again. Inbred parents of desirable combinations are further evaluated to eliminate minor within-line variations, i.e., to true up the inbred. Advanced testing is an extremely important prerequisite to cultivar release. In the initial stages of hybrid testing, a small number of hand-pollinations usually serves the need for seed. On the average, about 100 seeds can be expected per pollination, but a range of 100 to more than 200 seeds is common. A replicated test at the site of development might take a minimum of 200 seeds per plot to ensure a thinned uniform stand of 100 plants. Deviations in the number of plants per plot are critical, and attempts should be made to keep plant number and spacing per plot consistent. Stand adjustments are difficult and not precise enough to warrant nonuniform plantings. Commonly we have two or three replications in each of three plantings, and this requires 1800 (3 × 200 × 3) seeds. The amount of seed necessary for regional or national trials increases substantially, and hand-pollination becomes a poor substitute for an isolation cage increase or a field increase (isolated) of desirable new combinations. In field increases, frequently the pollen parent may be used on several different seed parents, and care is exercised to avoid pollen production in seed parent lines.

Superior hybrids should be further tested in regional and national trials to assess fitness across numerous environments. Regional testing is common in the United States, e.g., the Southern Cooperative Cucumber Trials, both pickling and slicing, and the Northern Cooperative Pickling Cucumber Trials. Trials are commonly conducted at several agricultural experiment stations and seed company trial grounds. Favorable responses should lead to commercial trials with either pickling or slicing cucumber growers. Also, at this stage, seed companies (producers) should be alerted to the development phase of the hybrid and care should be taken to make certain that commercial quantities of seed can be produced in the traditional manner. Excellent hybrids are of little value if seed cannot be grown because of poor parental performance in a seed production field. Although this situation is seldom encountered in cucumbers, it is still worthy of consideration, as unstable sex expression, poor seed quality, and other detrimental characters may be highly heritable and may result in an inferior seed crop.

REFERENCES

1. Abul-Hayja, Z. M. 1975. Multiple disease screening and genetics of resistance in cucumber. Ph.D. Thesis. University of Wisconsin, Madison.
2. Achahboun, M., Davis, D. W., and Breene, W. M. 1980. Influence of some growing and raw product storage variables on fresh-pack quality of pickling cucumber cultivars. HortScience *15*, 377.
3. Anon. 1967. U.S. Food Fermentation Laboratory—Publications. 1938–1967. Raleigh, NC.
4. Armstrong, G. M., and Armstrong, J. K. 1978. Fusarium wilt resistance of cucumbers. Cucurbit Genetics Coop. Rep. *1*, 3.
5. Bailey, L. H. 1969. Manual of Cultivated Plants. Macmillan Company, New York.
6. Barnes, W. C. 1966. Development of multiple disease resistant hybrid cucumbers. Proc. Am. Soc. Hortic. Sci. *89*, 390–393.
7. Beyer, E., JR. 1976. Silver: A potent antiethylene agent in cucumber and tomato. HortScience *11*, 195–196.
8. Bliss, F. A. 1981. Utilization of vegetable germplasm. HortScience *16*, 129–132.
9. Breene, W. M., Davis, D. W., and Chou, H. 1972. Texture profile analysis of cucumbers. J. Food Sci. *37*, 113–117.
10. Cantliffe, D. J. 1981. Alteration of sex-expression in cucumber due to changes in temperature, light intensity and photoperiod. J. Am. Soc. Hortic. Sci. *106*, 133–136.
11. Cantliffe, D. J., and Phatak, S. C. 1975. Use of ethephon and chlorflurenol in a once-over pickling cucumber production system. J. Am. Soc. Hortic. Sci. *100*, 264–267.
12. Cantliffe, D. J., and Phatak, S. C. 1975. Plant population studies with pickling cucumbers grown for once-over harvest. J. Am. Soc. Hortic. Sci. *100*, 464–466.
13. Chambliss, O. L., and Jones, C. M. 1966. Chemical and genetic basis for insect resistance in cucurbits. Proc. Am. Soc. Hortic. Sci. *89*, 394–405.
14. Chambliss, O. L., and Jones, C. M. 1966. Cucurbitacins. Specific insect (*Diabrotica undecim-punctata, Apis mellifera, Vespula*) attractants in Cucurbitaceae. Science *153*, 1392–1393.
15. Da Costa, C. P., and Jones, C. M. 1971. Cucumber beetle resistance and mite susceptibility controlled by the bitter gene in *Cucumis sativus* L. Science *172*, 1145–1146.
16. Da Costa, C. P., and Jones, C. M. 1971. Resistance in cucumber to three species of cucumber beetles. HortScience *6*, 340–342.
17. Day, A., Nugent, P. E., and Robinson, J. F. 1978. Variation of pickleworm feeding and oviposition on muskmelon and cucumber. J. Am. Soc. Hortic. Sci. *13*, 286–287.
18. Deakin, J. R., Bohn, G. W., and Whitaker, T. W. 1971. Interspecific hybridization in *Cucumis*. Econ. Bot. *25*, 195–211.
19. DeCandolle, A. 1886. Origin of Cultivated Plants. Kegan, Paul, Trench and Company, London.
20. Denna, D. W. 1971. Expression of the determinate habit in cucumber. J. Am. Soc. Hortic. Sci. *96*, 277–279.
21. Eason, G. 1982. Procedures in screening *Cucumis sativus* L. for resistance to the vegetable leaf miner *Liriomyza sativae* Blanchard. M.S. Thesis. North Carolina State University, Raleigh.
22. Edwards, M. D., and Lower, R. L. 1982. Comparative yields of compact and vining plant type isolines in cucumber at four densities. Cucurbit Genet. Coop. Rep. *5*, 6–7.
23. El-Shawaf, I., and Baker, L. R. 1981. Combining ability and genetic variances of G × H F_1 hybrids for parthenocarpic yield in gynoecious pickling cucumber for once-over mechanical harvest. J. Am. Soc. Hortic. Sci. *106*, 365–370.

24. Etchells, J. L., Bell, T. A., Fleming, H. P., Kelling, R. E., and Thompson, R. L. 1973. Suggested procedure for the controlled fermentation of commercially brined pickling cucumbers—The use of starter cultures and reduction of carbon dioxide accumulation. Pickle Pak Sci. *3*, 4–14.

25. Fassuliotis, G. 1970. Resistance of *Cucumis* spp. to the root-knot nematode, *Meloidogyne incognita acrita*. J. Nematol. *2*, 174–178.

25a. Fleming, H. P., Humphries, E. G., and Macon, J. A. 1983. Progress on development of an anaerobic tank for brining of cucumbers. Pickle Pak Sci. *7*, 3–15.

26. Fleming, H. P., and Pharr, D. M. 1980. Mechanism for bloater formation in brined cucumbers. J. Food Sci. *45*, 1595–1600.

27. Fleming, H. P., Thompson, R. L., Etchells, J. L., Kelling, R. E., and Bell, T. A. 1973. Bloater formation in brined cucumbers fermented by *Lacto-bacillus plantarum*. J. Food Sci. *38*, 499–503.

28. Frankel, R., and Galun, E. 1977. Pollination Mechanisms, Reproduction and Plant Breeding. Springer-Verlag, New York.

29. Galun, E. 1959. The role of auxins in sex-expression of the cucumber. Physiol. Plant. *12*, 48–61.

30. George, W. L., Jr. 1970. Dioecism in cucumbers *Cucumis sativus* L. Genetics *64*, 23–28.

31. Ghaderi, A., and Lower, R. L. 1981. Estimates of genetic variance for yield in pickling cucumbers. J. Am. Soc. Hortic. Sci. *106*, 237–239.

32. Globerson, D. 1983. Agricultural Research Organization, The Volcani Center, P.O.B. 6, Bet Dagan, 50250 Israel. Personal communication.

33. Goode, M. J., Bowers, J. L., and Bassi, A., Jr. 1980. Little-leaf, a new kind of pickling cucumber plant. Arkansas Farm Res. *29*, 4.

34. Gould, F. 1978. Resistance of cucumber varieties to *Tetranychus urticae:* Genetic and environmental determinants. J. Econ. Entomol. *71*, 680–683.

35. Horst, E. K. 1977. Vegetative and reproductive behavior of *Cucumis hardwickii* R. and *Cucumis sativus* L. as influenced by photoperiod, temperature and planting density. M.S. Thesis. North Carolina State University, Raleigh.

36. Horst, E. K., and Lower, R. L. 1978. *Cucumis hardwickii*, a source of germplasm for the cucumber breeder. Cucurbit Genet. Coop. Rep. *1*, 5.

37. Iezzoni, A. F., and Peterson, C. E. 1980. Linkage of bacterial wilt resistance and sex expression in cucumber. HortScience *15*, 257–258.

38. Jones, I. D., Etchells, J. L., and Monroe, R. J. 1954. Varietal differences in cucumbers for pickling. Food Technol. *8*, 415–418.

39. Kauffman, C. S., and Lower, R. L. 1976. Inheritance of an extreme dwarf plant type in the cucumber. J. Am. Soc. Hortic. Sci. *101*, 150–151.

40. Kho, Y. O., Nijs, A. P. M. den, and Franken, J. 1980. *In vivo* pollen tube growth as a measure of interspecific incongruity in *Cucumis* L. Cucurbit Genet. Coop. Rep. *3*, 52–54.

41. Kingston, B. D., and Pike, L. M. 1975. Internal fruit structure of warty and non-warty cucumbers and their progeny. HortScience *10*, 319.

42. Kubicki, B. 1980. Investigations on sex determination in cucumbers *Cucumis sativus* L. IX. Induced mutant with the recessive character of gynoecism Genet. Pol. *21*, 409–424.

43. Laibach, F., and Kribben, F. A. 1949. The influence of growth regulators on the development of male and female blossoms in a monoecious plant (*Cucumis sativus*). Bot. Ges. *62*, 53–55 (in German).

44. Lockerman, R. H., and Putnam, A. R. 1981. Mechanisms for differential interference among cucumber (*Cucumis sativus* L.) accessions. Bot. Gaz. (Chicago) *142*, 427–430.

45. Lower, R. L. 1974. Measurement and selection for cold tolerance in cucumber. Pickle Pak Sci. *4*, 8–11.

46. Lower, R. L. 1979–1982. Unpublished data. Cooperative study involving seedsmen, A.E.S. workers and Pickle Packers International, Inc., St. Charles, IL.

47. Lower, R. L., McCreight, J. D., and Smith, O. S. 1975. Effects of temperature and pho-
    toperiod on plant growth and sex-expression of cucumber. HortScience *10*, 318 (abstr.).
48. Lower, R. L., and Nienhuis, J. 1985. Prospect for increasing yield in cucumber via *Cucumis
    sativus* var. *hardwickii* germplasm. *In* Biology and Chemistry of the Cucurbitaceae. R. W.
    Robinson (Editor). Cornell Univ. Press, Ithaca, NY. (In press)
49. Lower, R. L., and Nijs, T. P. M. den 1979. Effect of plant type genes on growth and sex
    expression of pickling cucumber. HortScience *14*, 435 (abstr.).
50. Lower, R. L., Pharr, D. M., and Horst, E. K. 1978. Effects of silver nitrate and gibberellic
    acid on gynoecious cucumber. Cucurbit Genet. Coop. Rep. *1*, 8–9.
51. Mather, J., and Lower, R. L. 1980. The effect of fruit size on various fruit quality charac-
    teristics. Cucurbit Genet. Coop. Rep. *3*, 15–16.
52. McCollum, J. P. 1934. Vegetative and reproductive responses associated with fruit develop-
    ment in the cucumber. Mem.—N.Y., Agric. Exp. Stn. (Ithaca) *163*.
53. McMurray, A. L., and Miller, C. H. 1968. Cucumber sex expression modified by 2-chlo-
    roethane-phosphonic acid. Science *162*, 1396–1397.
54. Meeuse, A. D. J. 1958. The possible origin of *Cucumis auguria* L. Blumea, Suppl. 4 (H. J.
    Lam Jud. Vol.), 196–204.
55. Miller, J. C., Jr., Baker, L. R., and Penner, D. 1973. Inheritance of tolerance to chloramben
    methyl ester in cucumber. J. Am. Soc. Hortic. Sci. *98*, 386–389.
56. Munger, H. M., and Newhall, A. G. 1953. Breeding for disease resistance in celery and
    cucurbits. Phytopathology *43*, 254–259.
57. Nandgaonkar, A. K., and Baker, L. R. 1981. Inheritance of multi-pistillate flowering habit in
    gynoecious pickling cucumber. J. Am. Soc. Hortic. Sci. *106*, 755–757.
58. National Academy of Science 1972. Genetic Vulnerability of Major Crops. NAS, Washington,
    DC.
59. Nienhuis, J. 1982. Response to different selection procedures for increased fruit yield in two
    pickling cucumber populations. Ph.D. Thesis. University of Wisconsin, Madison.
60. Nienhuis, J., Lower, R. L., and Staub, J. 1983. Selection for improved low temperature
    germination in cucumber (*Cucumis sativus* L.). J. Am. Soc. Hortic. Sci. *108*, 1040–1043.
61. Nijs, T. P. M. den 1980. Effectiveness of AVG for inducing staminate flowers on gynoecious
    cucumbers. Cucurbit Genet. Coop. Rep. *3*, 22–23.
62. Nijs, T. P. M. den, and Visser, D. L. 1979. Silver compounds inducing male flowers in
    gynoecious cucumbers. Cucurbit Genet. Coop. Rep. *2*, 14–15.
63. Nitsch, J. B., Kurtz, E., Jr., Liverman, J., and Went, F. W. 1952. The development of sex
    expression in cucurbit flowers. Am. J. Bot. *39*, 32–43.
64. Peterson, C. E. 1975. Plant introduction in the improvement of vegetable cultivars. Hort
    Science *10*, 575–579.
65. Peterson, C. E. 1983. Carrot, Onion and Cucumber Investigations. USDA-ARS. University of
    Wisconsin, Madison.
66. Pike, L. M., and Peterson, C. E. 1969. Inheritance of parthenocarpy in the cuke (*Cucumis
    sativus* L.). Euphytica *18*, 101–105.
67. Ponti, O. M. B. de 1975. Breeding parthenocarpic pickling cucumbers (*Cucumis sativus* L.):
    Necessity, genetical possibilities, environmental influences and selection criteria. Euphy-
    tica *25*, 29–40.
68. Ponti, O. M. B. de 1979. Breeding glabrous cucumber varieties to improve the biological
    control of the glasshouse whitefly. Cucurbit Genet. Coop. Rep. *2*, 5.
69. Ponti, O. M. B. de 1979. Development and release of breeding lines of cucumber with
    resistance to the two-spotted spider mite, *Tetranychus urticae* Koch. Cucurbit Genet. Coop.
    Rep. *2*, 6.
70. Ponti, O. M. B. de, Kennedy, G. G., and Gould, F. 1983. Different resistance of non-bitter
    cucumbers to *Tetranychus urticae* in the Netherlands and the USA. Cucurbit Genet. Coop.
    Rep. *6*, 27–28.

71. Prend, J., and John, C. A. 1976. Improvement of pickling cucumber with the determinate (*de*) gene. HortScience *11*, 427–428.

72. Pulliam, T. L. 1969. The effects of plant type and morphology of the cucumber *Cucumis sativus* L., on infestation by the pickleworm, *Diaphania nitidalis* Stoll. M.S. Thesis. North Carolina State University, Raleigh.

73. Pulliam, T. L., Lower, R. L., and Wann, E. V. 1979. Investigations of the effects of plant type and morphology of the cucumber on infestation by the pickleworm, *Diaphania nitidalis* Stoll. Cucurbit Genet. Coop. Rep. *2*, 16.

74. Quisumbing, A. R. 1975. Host-plant resistance in cucumbers; influence of plot size and seeding rate in screening for field resistance to cucumber beetles. Ph.D. Thesis. North Carolina State University, Raleigh.

75. Quisumbing, A. R., and Lower, R. L. 1978. Influence of plot size and seeding rate in field screening studies for cucumber resistance to cucumber beetles. J. Am. Soc. Hortic. Sci. *103*, 523–527.

76. Robinson, R. W. 1978. Pleiotropic effects of the glabrous gene of the cucumber. Cucurbit Genet. Coop. Rep. *1*, 14.

77. Robinson, R. W., Shannon, S., and Guardia, M. D. De la 1969. Regulation of sex expression in the cucumber. BioScience *19*, 141–142.

78. Robinson, R. W., Cantliffe, D. J., and Shannon, J. 1971. Morphactin-induced parthenocarpy in the cucumber. Science *171*, 1251–1252.

79. Robinson, R. W., Munger, H. M., Whitaker, T. W., and Bohn, G. W. 1976. Genes of the Cucurbitaceae. HortScience *11*, 554–568.

80. Scott, J. W., and Baker, L. R. 1975. Inheritance of sex expression from crosses of dioecious cucumber (*Cucumis sativus* L.). J. Am. Soc. Hortic. Sci. *100*, 457–461.

81. Sitterly, W. R. 1972. Breeding for disease resistance in cucurbits. Annu. Rev. Phytopathol. *10*, 471–490.

82. Smith, O. S. 1977. Estimation of heritabilities and variance components for several traits in a random-mating population of pickling cucumbers. Ph.D. Thesis. North Carolina State University, Raleigh.

83. Smith, O. S., and Lower, R. L. 1978. Field plot techniques for selecting increased once-over harvest yields in pickling cucumbers. J. Am. Soc. Hortic. Sci. *703*, 92–94.

84. Smith, O. S., Lower, R. L., and Moll, R. H. 1978. Estimation of heritabilities and variance components in pickling cucumbers. J. Am. Soc. Hortic. Sci. *103*, 222–225.

85. Tapley, W. T., Enzie, W. D., and Van Eseltine, G. P. 1937. The Vegetables of New York. L. B. Lyon Company, Albany, NY.

86. Tasdighi, M., and Baker, L. R. 1980. Sex expression and yield of single and 3-way cross hybrids of pickling cucumber. HortScience *15*, 419.

87. Tiedjens, A. A. 1928. Sex ratios in cucumber flowers as affected by different conditions of soil and light. J. Agric. Res. *36*, 720–746.

88. Uzeategui, N. A., and Baker, L. R. 1979. Effects of multiple-pistillate flowering on yields of gynoecious pickling cucumbers. J. Am. Soc. Hortic. Sci. *104*, 148–151.

89. Walker, J. C. 1952. Diseases of Vegetable Crops. McGraw-Hill Book Co., New York.

90. Wasuwat, S. L., and Walker, J. C. 1961. Inheritance of resistance in cucumber to cucumber mosaic virus. Phytopathology *51*, 423–428.

91. Werner, G. M., and Putnam, A. R. 1978. Mechanism for differential atrazine tolerance in cucumbers. Proc., North Cent. Weed Control Conf. *33*, 49 (Abstr.).

92. Whitaker, T. W., and Davis, G. N. 1962. Cucurbits: Botany, Cultivation and Utilization. Interscience Publ., NY.

93. Williams, P. H. 1983. Department of Plant Pathology, University of Wisconsin, Madison, Wisconsin 53706. Personal communication.

94. Wittwer, S. H., and Bukovac, M. J. 1958. The effects of gibberellin on economic crops. Econ. Bot. *12*, 213–255.

6
Squash Breeding

THOMAS W. WHITAKER
R. W. ROBINSON

Squash and pumpkin are lay terms having no precise botanical meaning and are quite often used interchangeably. The term ''squash'' was evidently derived from a northeastern American Indian word indicating a fruit, apparently *Cucurbita pepo* L., eaten raw as an immature fruit or consumed for the mature seed. It is now employed to designate the forms of *C. pepo* L. that are used immature, all baking cultivars of *Cucurbita maxima* Duch., and the cushaw-type cultivars of *Cucurbita mixta* Pang., used mature. Squash is also applied to certain baking cultivars of *C. pepo* (e.g., Acorn) and *Cucurbita moschata* (Duch.) Duch. ex Poir. (e.g., Butternut) that are used in the mature state. Squash usually has a fine-grained flesh with mild flavor. The term ''pumpkin'' is normally applied to the edible fruit of any species of *Cucurbita* utilized when ripe as a table vegetable or in pies. The flesh is somewhat coarse and strongly flavored, and for this reason is not usually served as a baked vegetable. It is clear that the terms squash and pumpkin are not associated with any particular species of *Cucurbita,* and there is no reason for thinking of them in this sense.

The genus *Cucurbita* is native to the Americas. There is good evidence from archaeological sites in southwestern United States, Mexico, and northern South America that *C. pepo, C. moschata, C. mixta,* and *C. maxima* were widely cultivated in pre-Colombian times (prior to 1492 A.D.).

It is extremely difficult to arrive at accurate estimates of the importance of squashes and pumpkins in the economy of the United States and Canada since statistics for produc-

209

tion are not generally available. Squashes and pumpkins are usually grown on small acreages or in backyard gardens, and occasionally they are interplanted with corn. Thus, unless a specified area is contracted to processors, data on acreage and yields are not likely to be reported. Furthermore, these crops do not often enter into interstate commerce except for winter shipments from Florida, but are consumed by growers and their neighbors, or are sold through roadside or local markets. Jaycox *et al.* (*42*) noted that Illinois produces more canning pumpkin annually than the remaining states combined. Nearly 10,000 acres are devoted to the crop.

Despite lack of documentation, squashes and pumpkins make a significant contribution to American horticulture. The winter squashes especially have become a frequently used vegetable, particularly during the late summer, fall, and winter months. It is probable that acreages will gradually increase as more Americans discover the culinary and nutritional values of these crops.

Squash is widely cultivated in several European countries, for example, Germany, where it is often grown as an oilseed crop. In India, cultivars of *C. moschata* are a basic crop ingredient of the so-called riverbank culture practiced on the flood plain, a characteristic of the meandering rivers of India.

The U.S. canned pack of squashes and pumpkins in 1979 was 4,641,000 cases. Since 1964, the canned pack has never been lower than 2,641,000 cases. In the bumper crop year of 1975, the harvest reached a peak of 5,805,000 cases (Agricultural Statistics, 1979, USDA). The annual U.S. per capita consumption of squashes and pumpkins was almost static from 1963 to 1978 (about 0.22 kg). These figures are for processed products only and do not include the vast amount of summer squashes, pumpkins, and winter squashes used as a fresh vegetable or for home processing.

ORIGIN AND GENERAL BOTANY

The Cucurbitaceae is a tropical or semitropical family, although some species are found as far north as Indiana in the midwest and Oregon in the northwest. They do not tolerate below-freezing temperatures.

The cultivated species of *Cucurbita,* to which the squashes and pumpkins belong, are nearly all grown in the subtropical or temperate latitudes. Cultivars of *C. pepo* and *C. maxima* are adapted to the long days of summer and the cooler, shorter days of fall for best fruiting and development. Cultivars of *C. moschata* do best in warm tropical lowlands, but some cultivars have been developed by selection for adaptation to the temperate, short-growing season of the northern tier of states bordering Canada. The squash cultivars of *C. mixta* are mostly adapted to a tropical or subtropical environment.

A majority of the wild species of *Cucurbita* are found in the area south of Mexico City extending to the Mexico–Guatemala border. On the basis of this evidence it is suggested that this area is the center of distribution of the genus. Not surprisingly, the wild species, *Cucurbita lundelliana* Bailey and *Cucurbita martinezii* Bailey, which are evidently closely related to the cultivated species, occur in this area (see Fig. 6.1).

There are about 25–27 species of *Cucurbita,* all with 20 pairs of small, dotlike chromosomes. The cucurbits have traditionally been difficult material to study using conventional cytological methods. Varghese (*91*), however, has improved and expanded these methods to obtain excellent cytological preparations. Morphological differentiation of species appears to be based upon gene mutation, rather than differences in chromosome

FIGURE 6.1. Some species of *Cucurbita*. The five species on the bottom row are cultivated.

numbers or polyploidy. As far as is known, chromosome translocations, deletions, or inversions do not have a role in species differentiation in *Cucurbita*.

The four species of *Cucurbita* that produce the squashes and pumpkins are annual herbaceous vines with numerous runners, except for a number of cultivars of *C. pepo* (summer squash) and *C. maxima* (winter squash) that have short internodes (bush type). Runners may reach 10 m in length. They are prickly or spiny, either rounded or angled, and often root at the nodes. They have mostly long, branched tendrils, and the leaves are large, alternate, shallow to deeply lobed, and palmate. Hunziker and Subils (*36*) have detected foliar glands on leaf surfaces of wild and cultivated species of *Cucurbita*, which they claim have taxonomic value. If confirmed, these peculiar glands could be of much taxonomic importance.

The flowers are large, showy, with yellow or creamy corollas, and are unisexual. They occur singly in the axils of the leaves. The staminate flowers are located near the center of the plant and are borne on long, slender pedicels. The pistillate flowers are borne on short, ridged pedicels, distal to the staminate ones.

Root growth is characterized by the development of a taproot that may penetrate the soil to a depth of 1.83 m or more and by a network of lateral roots that are positioned slightly below the surface (2.34–6.70 cm)(*98*).

Squash develops an efficient root system at an early age; young plants with tops of only 30 cm or more have a taproot 0.76 m deep with horizontal laterals 6.3–11.8 cm in length. Mature plants have an intricate and extensive root system that may occupy as much as 28.32 m^3 of soil and grow at the rate of 0.984 cm/day. The tremendous absorbing power

TABLE 6.1. Key to the Annual Cultivated Species of *Cucurbita*

Key	Characteristics	Species
A	Stems soft, round, moderately bristly; peduncle soft, round, enlarged by soft cork	*C. maxima*
AA	Stems hard, angular; peduncle basically hard, angular, grooved	
B	Stems and leaves with spiculate bristles; peduncle hard, sharply angular, grooved, not flared at fruit attachment	*C. pepo*
BB	Stems and leaves lacking bristles; peduncle hard, smoothly grooved, flared at fruit attachment	*C. moschata*
BBB	Stems and leaves lacking bristles; peduncle hard, angular, greatly enlarged in diameter by hard cork; round at maturity; not flared at fruit attachment	*C. mixta*

of such extensive root systems probably accounts for the rapid, vigorous growth of squash, provided an abundant supply of water and nutrients is available.

The most conspicuous vegetative characters that separate the four cultivated species of *Cucurbita* are given in Table 6.1. The cultivated species can be distinguished on the basis of trichome, stem, and peduncle characteristics. The wild species that appear to be closely related to the cultivated ones are annual or perennial as contrasted to the cultivated species, which are all annuals, except *Cucurbita ficifolia*. For example, *Cucurbita okechobeensis* (syn. *martinezii*) is an annual, whereas *C. lundelliana* is perennial, as are all xerophytic species of *Cucurbita*. In many areas, perennial species are facultative annuals, being killed by frost in their first year of growth.

Wild species differ markedly from cultivated ones in the character of the fruit, as shown in Table 6.2. Diversity in fruit characters is probably the result of selection under cultivation, and cultivars differ considerably in fruit type (Fig. 6.2).

FLORAL BIOLOGY AND CONTROLLED POLLINATION

Floral development was described by Kirkwood (*43*) and has not been substantially modified by recent investigations. The essential features of his developmental scheme for pistillate and staminate flowers follow. The sepals arise first around the margin of the

TABLE 6.2. Variation of Fruit Character in *Cucurbita*: Cultivated vs. Wild Species

Fruit	Cultivated species	Wild species
Size	Large, 0.453–146.5 kg (1–320 lb)	Small, 0.170–0.255 kg (6–9 oz)
Shape	Globular, oval, bell, pyriform	Round, oval, pyriform
Exterior	Grooved, usually	Smooth
Shell	Moderately hard at maturity, non-durable	Hard at maturity
Skin color	Various, mostly solid colors	Usually green with stripes or mottling, yellowish at maturity
Flesh	Flesh non-stringy (few exceptions); orange, white, or creamy white	Flesh stringy; white, yellowish, or greenish
Tast	Non-bitter	Bitter
Seeds	Large (comparatively few)	Small, many

FIGURE 6.2. Diverse fruits of the cultivated *Cucurbita* species. Cultivars and PIs are given from left to right. Top left, *C. pepo:* Gourmet Globe, Vegetable Spaghetti, Pankow's Field pumpkin, Small Sugar (upper row); Gold Rush, Scallopini, Scallop, Table Queen, Vegetable Gourd, Sun Gold (lower row). Top right, *C. maxima:* Boston Marrow, Mammoth King, NK580 (upper row); PI 458741, Turk's Turban, Buttercup, Queensland Blue (lower row). Bottom left, *C. moschata:* White Cushaw, Libby Select, Melonsquash (upper row); PI 173894, Ponca, Butternut, PI 442279 (lower row). Bottom right, *C. mixta:* Green Striped Cushaw, PI 438546, Gold Striped Cushaw.

receptacle. The petals follow on an inner circle, each petal alternating with a paired sepal. The carpels arise from the receptacle, extending upward to form the pistil. There are usually three carpels, but occasionally four or five. The reflexed margins of the carpels form the placentae. They appear as longitudinal, parietal ridges to which the ovules are attached. The ovary is inferior and is divided into three locules. The relatively short, stout stigma placed above the ovary usually has three lobes, equal to the number of carpels. The calyx tube and corolla tube are five-lobed. The staminate and pistillate flowers are of about equal size. In the staminate flowers the filaments are free, but the anthers are united into a more or less columnlike mass (see Fig. 6.3). Both pollen and nectar are produced by the staminate flowers, nectar by the pistillate ones.

Sex expression in *Cucurbita* is relatively stable. All species are monoecious. This means that the heavy, sticky pollen grains, characteristic of *Cucurbita*, must be transported by some agent other than wind from the staminate flower to the receptive stigma of the pistillate flower for fertilization to be accomplished. The domestic honey bee acts as the pollinator. Originally, *Cucurbita* was pollinated by species of the gourd or squash bees

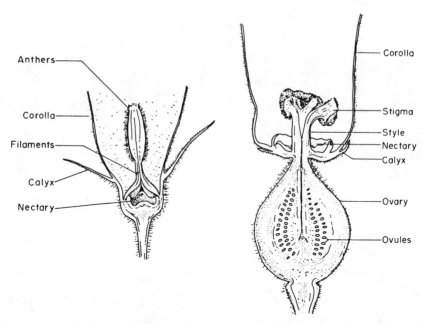

FIGURE 6.3. Monoecious flowers of *Cucurbita;* left, staminate; right, pistillate.

(*Peponapis* spp. and *Xenoglossa* spp.). These wild bees are better adapted to and much more efficient pollinators of *Cucurbita* than the honey bee (*37*). They are not numerous enough, however, to be an adequate pollen vector for pollination of most commercial *Cucurbita* plantings.

Kubicki (*44*) found an androecious mutant in *C. pepo* that differs from the normal monoecious condition by a single recessive gene, but thus far it has had no practical application in squash breeding. Dossey *et al.* (*22*) have isolated gynoecious lines of the wild, xerophytic *Cucurbita foetidissima*, which have been used to produce hybrid seed of the Buffalo gourd. A gynoecious gene for the cultivated species of *Cucurbita* would be very useful for hybrid seed production, but such a gene has not yet been found for *C. maxima, C. moschata,* or *C. pepo,* and incompatibility has prevented transfer of the *G* gene for gynoecious sex expression from *C. foetidissima.*

There are no well-documented examples of self-incompatibility in *Cucurbita*. Cummings and Jenkins (*17*) inbred Hubbard squash (*C. maxima*) for 10 generations without ill effects. Vigor and reproductive capacity were maintained without resort to outcrossing. Other investigators have found support for the idea that inbreeding in *Cucurbita* does not decrease vigor (*10,30,74*).

Schuster (*71*), however, indicates that he and his colleagues (*72*) have found clear evidence of inbreeding depression in *C. pepo,* particularly in seed yield. These results appear contradictory to those of Cummings and Jenkins (*17*). Since cultivars of two different species were used in these experiments, direct comparisons are not possible. Cultivars of different species may respond differently to inbreeding.

Absence of inbreeding depression does not signify that hybrid vigor in *Cucurbita* is lacking. A number of investigators have found significant evidence for hybrid vigor (*18,19,40*). Advantages that F_1 hybrids may have over open-pollinated squash cultivars include earlier and more uniform maturity and increased vigor and uniformity. Curtis (*19*) reported that an F_1 hybrid summer squash produced 87% more yield early in the season than the higher yielding parent. A number of genes of economic importance are dominant, at least to some degree, and a hybrid will combine the desirable traits of each of its parents.

Commercial F_1 Hybrid Seed Production

Robinson *et al.* (*63*) performed experimental work that has provided a basis for the economical production of hybrid seed. They applied the chemical 2-chloroethylphosphonic acid (ethephon) to young *Cucurbita* (several species) seedlings. Applications of 250 ppm prevented the development of staminate flowers for extended periods, but had no effect on the production of pistillate flowers (see Fig. 6.4). Thus, by arranging the seed field into alternate rows of treated and untreated plants and harvesting only fruits from treated rows, hybrid seed can be produced in abundance with little hand labor (see Fig. 6.5).

Rudich *et al.* (*65*) obtained similar results with *C. pepo*. Shannon and Robinson (*75*) in a further extension of these results, recommend the use of two applications of 400–600 ppm of ethephon. This rate of application produced the greatest reduction in staminate flowers without reducing seed yield or quality. Inbred lines differ in their response to ethephon, and the ethephon treatment for optimum results may need to be adjusted by seedsmen when producing seed of different hybrids. Ethephon is widely used in the production of hybrid seed today, particularly for summer squash (*C. pepo*). Under field conditions, where there may be variation in rate of seedling emergence and other factors influencing response to ethephon, there may be occasional male flowers produced by the female parent treated with ethephon; when this happens, to avoid contamination of the

FIGURE 6.4. Summer squash (*C. pepo*) untreated (left) and treated with 250 ppm ethephon to promote pistillate flower formation.

FIGURE 6.5. Squash field for seed production. One row of the staminate inbred line is planted earlier than the intervening two rows of the pistillate inbred parent, which is treated with ethephon to inhibit staminate flower formation. Note hives of honey bees in background for pollination.

hybrid seed with selfed or sibbed seed of the female parent, seedsmen often have the male flowers manually removed before they shed pollen. Rows of the pollen parent are often disced before seed harvest, to avoid seed mixture.

The problem of preventing self-pollination in the seed parent during hybrid seed production can also be solved by using male sterility. Unfortunately, a source of cytoplasmic male sterility has not been found in *Cucurbita*. Two different single-recessive genes for male sterility have been reported for *C. pepo* (*24*), and male sterility has also been found in *C. maxima* (*73*). Male sterility eliminates the need for emasculation or defloration in the production of hybrid seed, but it is necessary to remove male-fertile plants from among plants of the pistillate parent since it segregates for male sterility. The highest proportion of male-sterile plants that can be expected is 50%, produced by crossing plants heterozygous (*ms*/+) with those homozygous (*ms*/*ms*) for the male-sterile gene. Despite the need for roguing, male sterility is now being used to a limited extent for producing F_1 hybrid seed of *C. maxima*.

F_1 hybrids represent a growing proportion of squash seed planted for the commercial crop. This trend will most likely continue, particularly in the summer squash cultivars of *C. pepo*. F_1 hybrids allow the seedsman to invest research dollars in the development of inbred lines for providing F_1 hybrids and yet to retain control of the product.

Curtis (*19*) suggested growing cultivars that are the F_2 generation, since they would have much of the heterosis without the high seed cost of F_1 hybrids, but this practice has not been adopted. In order to reduce costly hand-pollination in the production of F_1 hybrid seed, Curtis (*18*) proposed manual defloration (the removal of male flower buds by hand from the female parent of a hybrid) to permit the use of open-pollination. This method is in common use today. Seedsmen grow several rows of the female parent for each row of the male parent and harvest seed from only the female parent. Since the two parents are grown in an isolated field and male flowers are not allowed to open on the female parent, hybrid seed is produced by bees bringing pollen from the male inbred grown in adjoining rows. Hives of honey bees are often placed in seed production fields to ensure pollination. In order to reduce the labor of removing male flower buds from the female parent of a hybrid, inbred lines with a high degree of female sex expression have been developed, producing a high ratio of female to male blossoms.

Manual Pollination Technique

Techniques for selfing or crossing *Cucurbita* are simple. Surveyors' flags (small plastic strips attached to wire stakes) are a convenient way to indicate which plants or lines are to be pollinated. Flags of different colors can be used to indicate the pollen parent. Pistillate and staminate flowers that are to be used in pollination the following morning are located the afternoon prior to anthesis by the appearance of a slight touch of yellow at the apex of the corolla tube (Fig. 6.6). They are prevented from opening by tying the tips of the corolla tube. Both pistillate and staminate flowers are secured to protect the flowers from insect pollination. The following morning, as soon as the pollen sacs dehisce, transfer of the pollen from anther to stigma can commence (see Fig. 6.7). Pollinations can be made from anthesis until about noon. There is some evidence to show that the percentage of successful pollinations is greater shortly after anthesis and then decreases gradually until about midday (*32*). The parentage of the pollination can be indicated on a tag attached to

FIGURE 6.6. Male (left) and female (right) flowers one day before anthesis. At this stage the petals are tied shut to prevent pollinating insects from visiting hand-pollinated flowers.

FIGURE 6.7. Artificial pollination of *Cucurbita* flowers. Breeder is about to apply pollen from the staminate flower to the receptive stigma of the pistillate flower.

the pedicel of the pollinated flower; when selfing or making many cross-pollinations with the same male parent, it is convenient to indicate the pollen parent by a plastic-covered wire ("twist tie") coded with a different color for each source of pollen.

It is desirable to pollinate the first pistillate flowers to develop on a plant, since fruit set is usually better if the plant has not previously produced fruit. If open-pollinated fruit are already formed, it is best to remove them to improve the fruit set of controlled pollinations.

After pollination, the pistillate flowers are labeled and tied shut or enclosed in a small bag to prevent insect visitations. A slender stake 3–4 ft long is placed adjacent to the pollinated fruit to mark its location. The fruits are usually harvested shortly before frost and then stored in a cool (but above 10°C), dry place until the seed is extracted and cleaned.

It is easy to lose the identity of pollinated fruit after harvest, if the peduncle with the attached tag detaches from the fruit or if rot obliterates the number written on the fruit. To avoid this problem, the pedigree number can be written on masking tape attached to the fruit.

Interspecific Hybridization

Species crossing *Cucurbita* has been investigated by a number of researchers [see Whitaker and Bemis (*101*)] and is for the most part well understood. With respect to the four annual species (*C. pepo, C. mixta, C. moschata,* and *C. maxima*) from which the squashes and pumpkins originate, at least three conclusions are obvious: (1) F_1 hybrids can be obtained from most interspecific matings, usually with difficulty. Such hybrids, however, are normally highly sterile because of the impaired ability of the staminate flowers to produce functional pollen (see Fig. 6.8). (2) The breeding data suggest that the four annual species are arranged like a wheel with *C. moschata* as the hub, the other three species as the spokes. (3) There is no evidence for spontaneous hybridization among these four species, despite the fact that they have been grown side by side in fields and gardens for many generations.

Interest in species crossing has centered around transferring several desirable traits between species, for example, good-quality flesh from *C. maxima* to *C. moschata.* The latter species is resistant to the squash bug and squash vine borer, but *C. maxima* is susceptible to these insects and would benefit from genes of *C. moschata.* The work of Pearson *et al.* (*57*) is typical of such efforts. These authors found they could attain good-quality fruits and insect resistance in the diploid F_1 hybrids of the cross *C. maxima* × *C. moschata.* However, sterility prevented the stabilization of the desirable character combinations in diploid lines. The F_1 species hybrids were polyploidized by the use of colchicine. Some of the amphidiploid lines resulting from this treatment were reasonably fertile and produced fruits comparable in quality to several commercial cultivars of baking squash.

Seed of the interspecific hybrids of *C. moschata* × *C. maxima* is offered for sale by the Sakata Seed Company, Japan. Fruits of the hybrids (Fig. 6.9) are reported to combine favorable characters from both species, and there is considerable heterosis for degree of female sex expression and for yield. Usually, this cross is difficult to make, but crossability depends on the particular accessions of the two species used as parents.

Bemis (*3*) gives a schematic diagram for producing interspecific aneuploids, as shown below. The original interspecific crosses were made at the tetraploid level. The amphidiploid hybrid was backcrossed to the diploid cultivated parent, and the resulting triploid

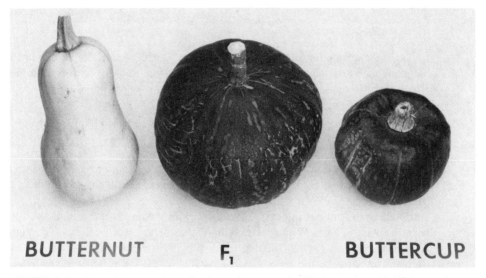

BUTTERNUT F_1 BUTTERCUP

FIGURE 6.8. *Cucurbita moschata* (left) can be crossed with *C. maxima* (right) to produce a very productive but highly sterile hybrid (center).

FIGURE 6.9. Fruit of interspecific hybrid *C. maxima* × *C. moschata* cultivars Aiguri (left), Kikusui (center), and Tetsakabuto (right).

was backcrossed again to the same parent to produce trisomics, having 40 chromosomes of *C. moschata* and one of *Cucurbita palmata*. This procedure has promise as a method for transferring genes for desirable traits on one chromosome of wild species to cultivated ones without simultaneously transferring undesirable genes on other chromosomes. In the following diagram, M is the *C. moschata* genome (cultivated) and P the *C. palmata* genome (wild):

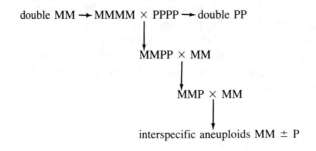

One of the difficulties of using wild *Cucurbita* species for interspecific hybridization is that many are late to flower, and it may not be possible to make crosses with them in the field in time for the seed to mature before frost. They may require a short photoperiod for flower initiation and may not flower in the field until fall, when the days become short. In some cases, it may be possible to find an accession that flowers earlier. *Cucurbita ficifolia* is normally very late to flower, and most accessions were still vegetative when killed by frost in New York. But one *C. ficifolia* accession, grown commercially as a root stock for cucumbers, is day neutral and produced flowers early enough in New York to make field pollinations possible. Grafting wild species to summer squash may induce them to flower earlier (*55*).

It is difficult, but not impossible, to grow wild species in the greenhouse to permit a longer growing season. Wild species require considerable space, high light intensity, and warm temperatures. A convenient alternative is to grow the wild species in the field and use them to pollinate the more compact plants of the cultivated species grown in the greenhouse.

The fruit should be harvested before frost. Fruits from interspecific crosses made in the field are stored in a cool, dry place for several weeks before extracting the seed. The seed will continue to mature in the fruit even after harvest.

Bridging species can be used to circumvent sterility in interspecific crosses (see Fig. 6.10). Whitaker (*99*) found that *C. lundelliana* could be crossed with each of the cultivated species of *Cucurbita*. Rhodes (*60*) developed an interbreeding population from crosses involving *C. lundelliana, C. pepo, C. mixta, C. moschata,* and *C. maxima.* Thus, *C. lundelliana* served as a bridge to transfer genes between species that are difficult to cross.

In recent research at Cornell University, Butternut (*C. moschata*) was used as a bridge to transfer disease resistance genes from *C. martinezii* to *C. pepo*. It is difficult to cross *C. pepo* and *C. martinezii,* although this can be accomplished with the aid of embryo culture (*92*). *Cucurbita martinezii* crosses more easily with *C. moschata,* and *C. pepo* will cross more readily with this hybrid than directly with *C. martinezii.* This three-way cross is being used to transfer resistance to powdery mildew and cucumber mosaic virus (CMV)

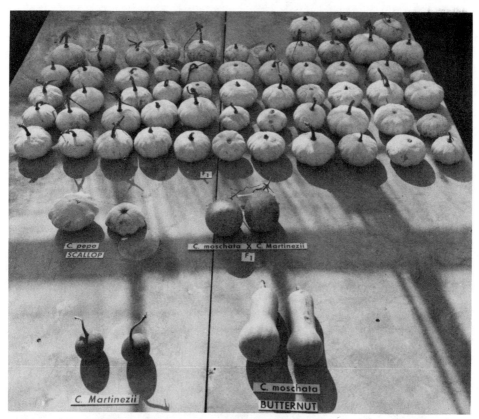

FIGURE 6.10. Bridging species can be used to transfer disease resistance and other desirable genes from wild to cultivated *Cucurbita* species. It is difficult to cross *C. pepo* with *C. martinezii,* although this can be accomplished with the aid of embryo culture, but the cross can be facilitated by using *C. moschata* as a bridge. *C. martinezii* and *C. moschata* (below) were crossed to produce the F₁ hybrid (center), which was then crossed with *C. pepo* cv. Scallop to produce the three-way cross. All 57 fruit illustrated for the three-way cross (above) were from a single plant, exhibiting an extraordinary degree of heterosis.

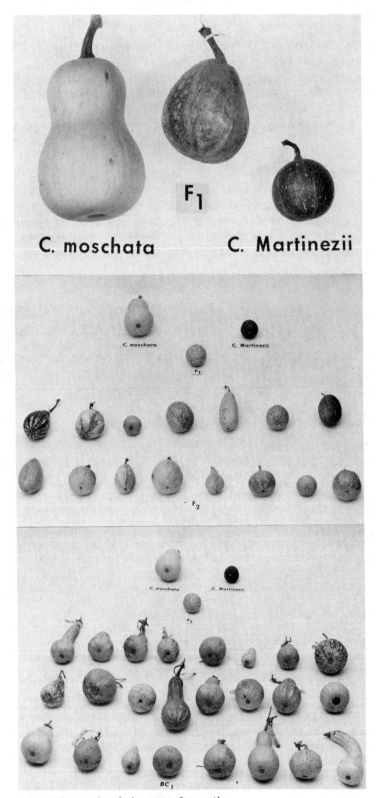

FIGURE 6.11. See facing page for caption.

from *C. martinezii* to *C. pepo* (*51*). Genes for good fruit quality, insect resistance, and other desirable traits of *C. moschata* can also be transferred to *C. pepo* by this cross.

Another problem that commonly occurs in the F_1 and early succeeding generations of distant crosses is sterility and poor seed development. Often the embryo does not abort, but the nutritive tissue of the seed fails to develop normally. Embryo culture (*93*) may be required in such cases. Wall (*93*) used embryo culture to facilitate the cross *C. moschata* × *C. pepo*. Recently, Washek (*96*) achieved a cross of *C. pepo* with the multiple-disease-resistant *Cucurbita ecuadorensis* through the use of embryo culture. The poorly developed seed of interspecific hybrids and their progeny often germinates better if the seed coats are removed. Pearson *et al.* (*57*) used colchicine-induced amphidiploidy to combat the problem of sterility in the cross *C. maxima* × *C. moschata*.

Genetic variability exists for fertility in interspecific matings. Wall and York (*95*) reported that gametic diversity is helpful in making interspecific crosses. For example, an F_1 hybrid of *C. pepo* that produced gametes of many different genotypes because of its heterozygosity was easier to cross with *C. moschata* than were more homozygous *C. pepo* cultivars. Cultivars of the same species may vary greatly in crossability with another species. Crosses with *C. moschata* cv. Butternut were more easily made with 'Scallop' than with other *C. pepo* cultivars in tests at Geneva, New York.

Unacceptable horticultural types usually predominate in the F_2 generation of interspecific crosses. The undesirable traits of the wild parent are often dominant, and a large number of segregating genes have an influence on economic characters. Consequently, the proportion of desirable plants in the F_2 may be small, and the desired type of plant may not be found in the size of population the squash breeder can manage. Backcrossing selected plants to the cultivated parent is desirable in such cases. In the squash-breeding program at Cornell University, only a single backcross to Butternut with the F_1 of Butternut × *C. martinezii* was required to obtain good fruit type. Selfing and selection after the first backcross resulted in the development of breeding lines with the disease resistance of the *C. martinezii* parent and the horticultural characteristics of 'Butternut' (see Fig. 6.11). Several backcrosses to *C. pepo*, however, were required to develop disease-resistant summer squash with good fruit and plant type from the three-way cross *C. pepo* × (*C. moschata* × *C. martinezii*) F_1. The number of backcrosses made to the recurrent parent depends on the breeding objective and the nature of the germplasm. If the aim is to develop an isogenic line, similar to the recurrent parent in every way except for the character transferred from the wild parent, a minimum of six backcrosses should be made. If the objective is to breed a cultivar with commercially acceptable type, but not necessarily exactly like the recurrent parent, then fewer backcrosses will be sufficient. Generally, backcrossing is continued until satisfactory types are found; then they are selfed until they become uniform.

MAJOR BREEDING ACHIEVEMENTS OF THE RECENT PAST

The introduction of the squash cv. Butternut (*C. moschata*) by Breck's Seed Company of Boston in 1936 initiated a chain of events that contributed much to plant breeding methods

FIGURE 6.11. *Cucurbita martinezii,* a source of resistance to powdery mildew and CMV, can be crossed with *C. moschata* (top). The F_2 generation (center) contains very few plants with desirable horticultural type. Better horticultural type, combined with disease resistance, is more easily recovered in the BC_1 generation, produced by crossing the F_1 with *C. moschata* (bottom).

in this group (*56*). Butternut is a high-quality cultivar grown over the entire country. It was evidently a selection within a population of the well known cv. Canadian Crookneck. The original seedstock was unstable, producing both 'Crookneck' and 'Butternut' fruit types, but an even more serious defect was a tendency to produce dimorphic plants, i.e., those having both Crookneck and Butternut types on the same plant. Strenuous efforts were made through selection to eliminate the ''Crookneck' rogue and obtain lines free of this troublesome defect. These efforts were mostly unsuccessful. Another approach has been to use material largely unrelated to the original Butternut to synthesize a cultivar similar to Butternut but without the undesirable Crookneck rogue. These efforts were to some extent successful and resulted in the release of cvs. Waltham Butternut (*56*) and New Hampshire Butternut (*104*). Coyne (*14*) has suggested technical procedures that will eliminate the Crookneck rogue, and he and his colleagues have released two cultivars of the Butternut type: Butternut Ponca (*15*) and Butternut Patriot (*16*). Each of these cultivars has specific characters that recommend it over the original cultivar, in addition to being free of the Crookneck rogue.

F_1 hybrids superior to open-pollinated cultivars have been developed. Hybrid cultivars have become especially important for summer squash. There is significant heterosis for early yield.

The bush habit found in some cultivars of *C. pepo* and *C. maxima* is assuming much importance as we attempt to produce more food in a limited space, particularly in small gardens. Denna and Munger (*21*) have studied the bush habit in these species. They found that the major alleles for bush or vine habit are located at the same genetic locus in both species, and Munger (*51*) has transferred the *Bu* allele from *C. pepo* to Butternut, a cultivar of *C. moschata*. Bush pumpkin cultivars have been developed from crosses with summer squash. Backcrossing has been used to develop bush counterparts of vining cultivars of winter squash; Emerald is a bush version of Buttercup (*C. maxima*), and Table Ace and Table King are bush cultivars similar to Acorn (*C. pepo*). Action of the *Bu* allele is unusual in that it undergoes a developmental reversal of dominance. In *C. pepo,* the *Bu* allele is almost completely dominant to the vine allele during early development, but *Bu* is incompletely dominant during latter stages of development (*79*). In *C. maxima, Bu* is completely dominant during early growth, but is completely recessive during later stages of growth.

The *B* gene has been transferred to summer squash from the bicolored gourd (*81*). The resulting cultivars have fruit with very deep yellow color and high vitamin A content. Progress has been made in selecting against the occurrence of green areas in the fruit that were originally associated with the *B* gene.

CURRENT GOALS OF BREEDING

Disease Resistance

Disease resistance is a major goal of many squash breeding programs. This is not unusual, considering that diseases are often a limiting factor in the production of squash and pumpkin. It is surprising, however, that disease-resistant cultivars were not developed earlier. Muskmelons resistant to powdery mildew, watermelons resistant to anthracnose, and cucumbers resistant to scab have been available for more than 50 years, but squash cultivars resistant to these and other fungal diseases have not yet been introduced. CMV-resistant cultivars have been available for cucumbers for decades but not for squash. All squash cultivars are susceptible to CMV and to several other important viral diseases.

Demski and Sowell (20) tested all introductions of *C. pepo* then available for resistance to watermelon mosaic virus-2 (WMV-2), but they found only susceptible plants in each collection. The lack of resistant material was attributed to the fact that only two of the 300 plant introductions of *C. pepo* tested were from Mexico, the center of origin for the species. Salama and Sill (66) found only moderate levels of resistance to squash mosaic virus in *C. pepo, C. maxima,* and *C. moschata.* Sowell and Corley (89) tested 292 introductions of *C. pepo;* all were susceptible to the powdery mildew fungus.

The lack of disease-resistant squash cannot be attributed to lack of effort. Rather, it is probably due to the absence of good sources of resistance in available germplasm of cultivated *Cucurbita.* There are isolated exceptions, but in most cases the level of resistance to diseases found in the cultivated species is less than desirable. Disease resistance in the cultivated species of *Cucurbita* has not matched that of *Cucumis,* where good powdery mildew resistance for muskmelon and CMV resistance for cucumber were found within the cultivated species. Perhaps the reason for the lack of satisfactory levels of disease resistance in *Cucurbita* in the past is the limited amount of germplasm sampled from the center of origin for the species. Resistance is most likely to be found in collections from this area. Fortunately, a recent expedition to Mexico, where most *Cucurbita* species originated, has provided 183 new introductions of *Cucurbita* for breeders searching for disease resistance and other traits (*102*).

Breeding squash for disease resistance has obviously lagged behind that for other cucurbits, but this situation is changing rapidly. Interspecific hybridization is providing squash breeders with the germplasm they need to develop disease-resistant cultivars (see Fig. 6.12). *Cucurbita lundelliana* is resistant to powdery mildew (99). Resistance is governed by a single dominant gene that can be transferred to *C. moschata* (61). Sitterly (88) transferred powdery mildew resistance to *C. pepo* from a gene pool developed by Rhodes (60) from interspecific hybridizations involving *C. lundelliana.* Contin (12) concluded that *C. martinezii* has the same single dominant gene for powdery mildew resistance as *C. lundelliana,* and *C. martinezii* also possesses additional modifier genes influencing the level of resistance. Munger (52) found that *C. martinezii* is resistant to CMV as well as powdery mildew. This species has a potentially important role in the development of disease-resistant squash (see Fig. 6.13).

Virus resistance, although rare among the cultivated species of *Cucurbita,* appears to be the rule rather than the expectation in the wild species of *Cucurbita* (Table 6.3). Provvidenti *et al.* (59) tested 14 wild *Cucurbita* species and found that all but three were resistant to CMV. They developed only localized reaction on the inoculated leaf and did not become systemically infected. Two species—*C. ecuadorensis* Cutler and Whitaker and *C. foetidissima* HBK—were resistant to WMV-1 and WMV-2 as well as to CMV. Eleven species had a localized reaction without systemic infection when inoculated with tobacco ring spot virus (TRSV), and they were considered resistant. Several species were also resistant to tomato ringspot virus (TMRSV). A high level of resistance to the severe strain of bean yellow mosaic virus (BYMV) was found among several wild species of *Cucurbita.* The only virus tested for which a high level of resistance was not found was squash mosaic virus, but *C. ecuadorensis* and *C. martinezii* were tolerant; they became infected, but later recovered.

Breeders have used other approaches to protect squash from disease, namely, breeding

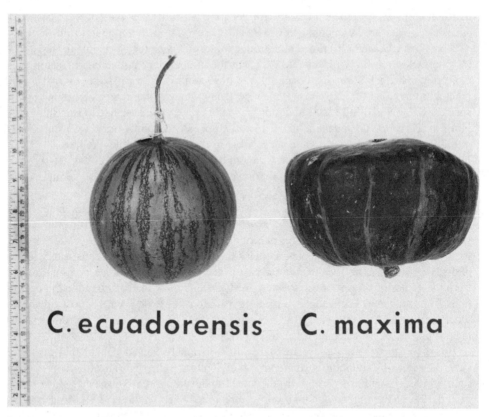

FIGURE 6.12. *Cucurbita ecuadorensis* (left) can be crossed with *C. maxima* (right) to breed squash that is resistant to multiple viruses.

for escape mechanisms or for different symptoms of infection. Shifriss (*82*) has proposed the use of ''silvering'' of the leaves, due to the dominant gene *M* and modifier. He suggested that the reflected light from silvered leaves would deter aphids from feeding and transmitting viruses, but this theory has not yet been adequately tested. Shifriss (*80,81*) also popularized the use of the *B* gene in breeding *C. pepo*. This gene, which produces deep-golden fruit color, has been transferred from the bicolor gourd (*C. pepo* var. *ovifera*) to several cultivars of summer squash. These cultivars often retain the yellow fruit color even when infected with WMV-2, although other yellow-fruited cultivars develop green blotches on the fruit when infected with this virus (see Fig. 6.14). The yield of *B* cultivars can be reduced as much as other susceptible cultivars by virus infection, but they may produce more marketable fruit because of their better fruit color.

Although less common than in the wild species, disease resistance has been found in some cultivated species of *Cucurbita*. Strider and Konsler (*90*) tested 661 *Cucurbita* introductions and found that most (633) were susceptible and none were immune to the fungal disease scab (*Cladosporium cucumerinum*). Several accessions of *C. maxima* and one of *C. moschata*, however, were rated resistant. Other introductions of these species and of *C. pepo* were reported to have a lower level of resistance to scab. Resistance to the bacterial wilt *Erwina tracheiphila* was reported by Watterson *et al.* (*97*) for *C. maxima* and *C. pepo* as well as for *Cucurbita andreana* Naud., *C. ficifolia* Bouché, and *C.*

lundelliana. Resistance to bacterial wilt was more common in *Cucurbita* than in other genera of the Cucurbitaceae surveyed. Resistance to the powdery mildew fungus, *Sphaerotheca fuliginea,* was found in introductions of *C. moschata* by Sowell and Corley (*89*).

Insect Resistance

The prospects are good that squash can be bred for resistance to several important insects. Although no cultivar has been introduced that was specifically developed for insect resistance, researchers have discovered differences in cultivar susceptibility and have found other sources of resistance to several insects. Brett and Sullivan (*7*) reported that squash cultivars differed in susceptibility to the pickleworm, striped cucumber beetle, spotted cucumber beetle, and serpentine leaf miner. Hall and Painter (*31*) screened plant introductions for resistance to the squash bug and cucumber beetle, and found combined resistance in several accessions of *C. pepo, C. maxima,* and *C. moschata.* Lal (*46*) tested

FIGURE 6.13. Breeding lines with powdery mildew resistance derived from *C. martinezii* (background); susceptible *C. pepo* defoliated by powdery mildew (foreground).

TABLE 6.3.   Reaction of *Cucurbita* Species to Six Viruses[a]

Species	CMV	TRSV	BYMV	TmRSV	WMV-1	WMV-2
C. andreana	S	R	O	S	S	S
C. cordata	R	R	O	S	S	S
C. cylindrata	R	R	O	S	S	S
C. digitata	R	S	S	R	S	S
C. ecuadorensis	R	R	S	S	O	O
C. foetidissima	R	R	O	S	O	O
C. gracillior	R	R	O	R	S	S
C. lundelliana	R	S	S	S	S	S
C. martinezii	R	R	O	S	S	S
C. palmata	R	R	R	R	S	S
C. palmeri	R	R	R	R	S	S
C. sororia	S	S	R	R	S	S
C. texana	S	R	S	S	S	S
C. maxima	S	R	R	S	S	S
C. moschata	S	R	O	S	S	S
C. pepo	S	R	S	S	S	S

[a]O, No infection; S, systemic symptoms; R, resistant, developing local reaction but not systemic symptoms; CMV, cucumber mosaic virus; TRSV, tobacco ringspot virus; BYMV, bean yellow mosaic virus; TmRSV, tomato ringspot virus; WMV-1, watermelon masaic virus-1; WMV-2, watermelon mosaic virus-2. [After Provvidenti and Robinson (58)].

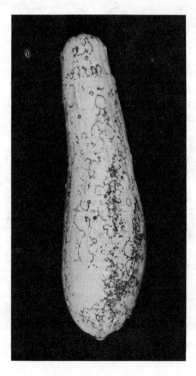

FIGURE 6.14.   Fruit of Yellow Straight-neck summer squash, with green rings produced by infection with WMV-2.

37 squash cultivars and hybrids for aphid resistance and found that three had a relatively high level of resistance. Benepal and Hall (6) reported that *C. pepo* cultivars differed in susceptibility to the squash bug *Anasa tristis*.

Domestication of Wild Species

The wild species of *Cucurbita* have hitherto been used in breeding programs only as a source of genes for specific traits, notably disease resistance, to transfer to cultivated species. Bemis *et al.* (4,5) have proposed the domestication and utilization of the feral "Buffalo gourd" *C. foetidissima* as a crop plant. The abundant seeds produced by this xerophytic species are rich in protein and oils, and its large storage root is a good source of carbohydrates.

Fruit Color

Many different fruit colors occur in *Cucurbita,* and several genes that determine fruit color have been investigated (62). Selection is usually made for orange skin color when breeding *C. maxima* cultivars for processing. Portions of the skin may inadvertently be included in the product during commercial canning or freezing of winter squash, and orange skin color is preferred because it is less conspicuous when mixed with the orange flesh. Deep-orange-colored flesh is desirable for winter squash. It not only is attractive in appearance, but also is nutritious since the orange pigment is partly due to β-carotene, the precursor of vitamin A.

"Naked Seed"

A mutant of *C. pepo* with tender seeds, due to inhibition of thickening and lignification of the cell walls of the testa, is being used to breed cultivars with edible seeds. Actually, all squash and pumpkin seeds are edible; they have long been a popular food in Mexico and were used by American Indians centuries ago. Palatibility, however, is enhanced when the seeds can be eaten without removing the tough seed coat. The seeds are the most nutritious part of the squash fruit because they are rich in protein and oil. Grebenščikov (29) concluded that a single recessive allele (*n*) determines the "naked seed" character, but modifying genes may also influence it. Lady Godiva, Eat All, and other naked-seeded cultivars of *C. pepo* have been developed.

SELECTION TECHNIQUES FOR SPECIFIC CHARACTERS

The genetics of *Cucurbita* and other cucurbits was reviewed by Robinson *et al.* (62). The genus *Cucurbita* is very diverse, but little is known about the genetic basis for this variation. Only 37 traits have been determined to be simply inherited (Table 6.4). The large space requirement for growing mature *Cucurbita* plants is a barrier to genetic investigations. There has not been an organized effort to produce mutants in squash, perhaps because it is not a high-value economic crop, and also because there appears to be much natural variation available for study.

Most of the economic characters in *Cucurbita* are not simply inherited. Singh (83) concluded that two or more genes in *C. maxima* determine each of several important traits, including growth habits, days from sowing to flowering, fruit color, fruit weight, and total solids in the fruit. Nath and Hall (54) concluded that resistance to *C. pepo* to the

TABLE 6.4. Simply Inherited Characters in Cucurbita and Their Gene Symbols[a]

Preferred symbol	Character description	Cucurbita species	Reference
Corolla color			
cr	*Cream* corolla	*okeechobeensis*	64
Gb	*Green band* at base of petal	*pepo*	23
i	*Intensifier* of the *cr* gene for cream-colored flowers	*okeechobeensis*	64
ly	*Light-yellow* corolla	*pepo*	68
Seed coat			
n	*Naked* seeds, lacking a lignified seed coat	*pepo*	29,70
Fruit color			
B	*Bicolor* fruit, deep-yellow fruit pigmentation that develops before anthesis	*pepo*	80,81
bl	*Blue* external color of mature fruit	*maxima*	38
C	*Colored* fruit; green fruit color	*pepo*	28
l	*Light* fruit color; uniformly light-colored fruit	*pepo*	87
r	*Recessive* white; white fruit color	*pepo*	28
Rd	*Red* skin; red external color of mature fruit	*maxima*	48
St	*Striped* fruit; longitudinal stripes on fruit surface	*pepo*	67
W	*White* external fruit color	*pepo*	87
wf	*White flesh*; white internal fruit color	*pepo*	87
Y	*Yellow* external fruit color	*pepo*	87
ygp	*Yellow–green placenta*	*pepo*	23
Fruit shape, surface			
Di	*Disc* fruit shape	*pepo*	84–86
Hr	*Hard* rind	*pepo*	49
Wt	*Warty* fruit; bumpy fruit surface	*pepo*	87
Fruit bitterness			
Bi	*Bitter* fruit; high cucurbitacin content	*pepo*	29,99
cu	*Cucurbitacin* content; reduced content of cucurbitacin B	*pepo*	77
Isozymes			
Est	*Esterase*; electrophoretic form of α-naphthyl acetate esterase	*maxima, ecuadorensis*	94
Lap	*Leucine aminopeptidase*; electrophoretic form of leucine aminopeptidase	*maxima, ecuadorensis*	94
Plant habit			
Bu	*Bush* habit; short internodes; influenced by modifier genes	*pepo, maxima*	21,79,87
Stem color			
D	*Dark* green stem	*pepo*	28

TABLE 6.4. (*cont.*)

Preferred symbol	Character description	*Cucurbita* species	Reference
Tendril shape			
lt	*Leafy tendril*; tendrils with laminae	*pepo*	67
Leaf color, shape			
M	*Mottled* leaves; silver-gray areas in leaf axils	*pepo, maxima, moschata*	13,67
ro	*Rosette* leaf, with spiraled lobes	*pepo*	49
ys	*Yellow seedling*; lacking chlorophyll; lethal	*pepo*	49
Fertility			
ms-1	*Male sterile*-1; male flowers abort before anthesis	*pepo*	24
ms-2	*Male sterile*-2; male flowers abort	*pepo*	24
s	*Sterile*; male flowers small and without pollen; female flowers sterile	*maxima*	39
Sex expression			
a	*Androecious*; produces only male flowers	*pepo*	44
G	*Gynoecious*; only female flowers	*foetidissima*	22
Disease resistance			
Pm	*Powdery mildew* resistance	*lundelliana*	61
Herbicide tolerance			
I-T	Interaction of the *T* gene for trifluralin tolerance	*moschata*	1
T	*Trifluralin* (a herbicide) tolerance	*moschata*	1

[a]Adapted from Robinson *et al.* (*62*).

spotted cucumber beetle is not simply inherited. Chambliss and Jones (*11*) reported that cucumber beetle resistance is dominant but not simply inherited, and they suggested that a relatively small number of genes may be involved. Sharma and Hall (*76,77*) determined that spotted cucumber beetle resistance and total sugar content of squash cotyledons were controlled by polygenic mechanisms. At least three genes were considered by Benepal and Hall (*6*) to be involved in resistance to squash bug.

Regional differences in preference have an influence on the type of squash developed by breeders for different parts of the country. In the northern states, for example, Yellow Straightneck summer squash is preferred, but squash with a curved neck such as Yellow Crookneck is in demand in southern states. *Cucurbita maxima* cultivars with elongated fruits, such as Pink Banana, are grown in the western states but are seldom seen in eastern markets. Hubbard squash was probably brought to New England from Chile. It was introduced in 1842 and is still more popular in the northeast than in other parts of the country.

One of the difficulties in breeding *Cucurbita* is that pollinations are made before it is possible to make selections for fruit type. Summer squash fruit, however, can be evaluated when immature. For this reason, the open-pollinated fruit from selected plants can be removed and they can be self-pollinated. Winter squash and pumpkins cannot be judged for fruit type until mature, when it is too late to make additional pollinations. Breeders, therefore, often make more pollinations than required for the desired number of selections, and save seed from only the best fruit. If no controlled pollinations were successful on the most desirable plants, open-pollinated seed may be saved. Although there is a high rate of natural cross-pollination, some of the open-pollinated progeny may have resulted from selfing, and it may be possible to distinguish selfs from crosses by disease resistance or a similar marker.

Insect Resistance

Selective evaluations of feeding injury are often used in comparative tests of insect resistance (*76*). Also, counts may be made of the number of insects per plant. Hall and Painter (*31*) placed 4 × 8 in. wooden blocks beside plants being tested for resistance to squash bugs. The overwintering adults generally hid under the board adjacent to the plant to which they migrated, making it possible to make counts of adults at intervals throughout the season. Counts were also made for the number of nymphs on the foliage of each plant and on the survival rate of plants. Ivanoff (*41*) described a method for testing squash and other cucurbits for aphid resistance. It is based on the slight curling of primary leaves of susceptible plants after feeding by aphids.

Breeding for insect resistance is difficult because of the intricate difficulties of three phases of the work: (1) rearing the insects or obtaining natural infestation, (2) exposing test plants at the appropriate stage to a uniform number of insects, and (3) accurately assaying the response. It would be very helpful for breeders if objective techniques were developed for measuring directly the biochemical or physical factor essential for resistance. Indirect evaluation for resistance on the basis of observed effects of insect feeding is inherently less reliable due to greater experimental error. The best known such case is the association between cucurbitacins and resistance to cucumber beetles. Cucurbitacins are bitter, tetracyclic triterpene compounds that are often present in species of the Cucurbitaceae. Chambliss and Jones (*11*) demonstrated that susceptibility of squash to the spotted cucumber beetle was correlated with concentration of cucurbitacins; the higher the cucurbitacin content, the greater was the damage done by beetles (see Fig. 6.15). Howe *et al.* (*35*) reported a similar association between cucurbitacins and susceptibility to the western corn root worm.

Sharma and Hall (*76*) reported that cucurbitacins played a major role in seedling susceptibility of various cucurbits to the spotted cucumber beetle, but they also found correlations between susceptibility and total sugars, palmitic acid, and linolenic acid. Brett *et al.* (*8,9*) found no correlation between amino acid or carbohydrate composition and susceptibility of squash to pickleworm. They found, however, that galacturonic acid occurred only in Butternut 23, the most resistant of the cultivars investigated. They concluded that further studies of the possible role of this carbohydrate in susceptibility to the pickleworm were warranted. Brett *et al.* (*9*) found higher concentrations of D-glucose in fruit of cultivars resistant to the pickleworm.

Howe (*34*) concluded that mechanical and nutritive factors were associated with resistance to the squash vine borer. The relatively resistant *C. moschata* and *C. mixta* plants

FIGURE 6.15. Cucurbitacin content and insect resistance. Cucurbitacins are an attractant for cucumber beetles. Cucumber beetle resistance is associated with low cucurbitacin content. *Cucurbita* fruit of interspecific origin, such as above, may be very high in cucurbitacins and preferred by cucumber beetles.

were characterized by having hard, woody stems with comparatively little parenchyma tissue, in contrast to the susceptible *C. maxima,* which had softer stems and more widely scattered vascular bundles.

In selecting for insect resistance, squash breeders should be aware that the basis for resistance may be nonpreference. If so, a cultivar may appear resistant when grown in mixed populations, but susceptible when grown alone. Wisemann *et al.* (*103*) reported that *C. pepo* cultivars differed in susceptibility to cucumber beetles when different cultivars were grown in the same field, but the resistant cultivars were susceptible when the beetles had no choice among cultivars. Resistance based on antibiosis, reported for the squash vine borer (*34*), will be retained even when the resistant cultivar is grown alone.

Disease Resistance

In screening tests for virus resistance, Provvidenti *et al.* (*59*) inoculated *Cucurbita* seedlings by dusting carborundum abrasive on young leaves and rubbing them with inoculum prepared by triturating infected squash or cucumber leaves with 0.05 *M* phosphate buffer at pH 7.0. Resistance was evaluated on the basis of symptoms and by the ability of the plant to avoid systemic infection, as indicated by recovery tests in which extracts from the

FIGURE 6.16. CMV resistance in *Cucurbita pepo*. Uninoculated (top left) and inoculated (top right) plants of the susceptible cultivar Straightneck. Low degree (lower left) and high degree of resistance (lower right), derived from *C. martinezii*.

plants were used to inoculate susceptible squash or cucumber plants (see Figs. 6.16 and 6.17).

Strider and Konsler (*90*) tested for scab resistance by spraying squash seedlings with *C. cucumerinum* spores from potato–dextrose–agar cultures. The inoculated plants were kept at 18°–23°C at near 100% relative humidity (RH) for 18–24 hr and then at 50–70% RH and the same temperature for 2–5 additional days. Watterson *et al.* (*97*) used a hypodermic needle to inject a suspension of *E. tracheiphila* spores into the midrib of the cotyledon or first leaf of seedlings 2–5 days old to test squash and other cucurbits for resistance to bacterial wilt. Susceptibility was evaluated 5 days after inoculation and evaluated again 1 week later on the basis of wilting of the inoculated leaf or cotyledon. Contin (*12*) and others have used aerial transmission of conidiophores from infected plants to inoculate squash leaves with powdery mildew. He used a subjective rating from 1 to 9 to classify plants for powdery mildew resistance.

DESIGN OF THE COMPLETE BREEDING PROGRAM

Although Bailey (*2*) believed that self-incompatibility occurs in *Cucurbita,* this report has been contradicted by later investigators. Self-pollination is possible, and inbreeding does not usually depress vigor. Self-pollination can be utilized to develop uniform strains from heterogenous, open-pollinated cultivars. Cummings and Jenkins (*17*) were able to isolate uniform lines of different types by inbreeding Hubbard squash, and one of the inbred selections was released as the cv. Kitchenette Hubbard. Pedigree and mass selection procedures have been used to breed improved, more uniform strains of open-pollinated cultivars.

Kubicki and Walczak (45) increased the β-carotene content as much as 70% by selfing cultivars of *C. maxima* and making selections for high β-carotene. Cultivars differ, however, in their ability to be improved by selfing and selection. C. A. John (unpublished) was unable to improve the β-carotene content of cv. Golden Delicious by inbreeding.

Parents chosen to combine the characteristics desired in a new cultivar are often crossed; then for several generations individual plants are selected for those characteristics and self-pollinated. When one of the parents is of much better horticultural type than the

FIGURE 6.17. Squash seedlings in cold frame (above), after virus inoculation. Susceptible seedlings are discarded and resistant plants are transplanted to the field. Note leaf distortion of susceptible plants (close-up, below).

other, it may be desirable to backcross to the better parent for one or more generations, followed by several generations of selfing. When the breeding line is reasonably uniform, seed increases may be obtained by bulking selfed or sibbed seed from a number of plants. The bulk increase may be produced by open pollination in a field isolated from other plantings of the same species, or it may be obtained by more laborious hand-pollinations.

Germplasm that has been used as parents in controlled squash pollinations includes cultivars, breeding lines, and plant introductions. Breeders may obtain seed of introductions from the USDA Plant Introduction Stations at Geneva, New York, for *C. maxima,* and at Experiment, Georgia, for *C. pepo, C. moschata,* and other species.

Inbred lines developed by self-pollination have been utilized as parents of F_1 hybrids. F_1 hybrid cultivars are generally produced by the direct cross of two inbreds. Three-way crosses, double-cross hybrids, and other types of crosses have not been used commercially.

Squash breeding is best accomplished in the field. The large space requirement is a hindrance to growing an additional generation in the winter greenhouse, although this is sometimes done with high-priority breeding materials. Tests of disease resistance may be done in the greenhouse with young plants, and quality tests with frozen samples from field-grown plants can be analyzed in the winter; but squash breeders in temperate climates usually grow a single generation per year in the field during summer. Some squash breeders grow an extra generation each year in the field in southern Florida or Central America during the winter.

Squash breeders may plant seed directly in the field in spring after soil temperatures are warm and frost unlikely, or they may transplant greenhouse-grown plants. Transplanting is often preferred if seed supplies are limited or germination precarious, as often happens in early generations of interspecific crosses. The plants should be young, about 3 weeks old, and hardened off in a cold frame before transplanting. They should not be transplanted bare root, but benefit from having intact roots enclosed in growing media. The young plants should be given water but not starter solution when transplanted. The use of transplants may enable the breeder to inoculate seedlings of segregating populations, and transplant only disease-resistant plants to the field; this procedure is being used at Geneva, New York, to breed for resistance to several viruses. It is usually necessary to use older transplants when testing for disease resistance, to allow sufficient time for incubation and symptom development.

Spacing for segregating material should be considerably wider between plants than is customary for a commercial planting. Wider spacing permits selection of individual plants and makes it easier to find a staminate and pistillate flower on the same plant for self-pollination. In one squash-breeding program, bush breeding lines are spaced 3–5 ft and vine types 5–8 ft apart, but other squash breeders use different plant spacing. The use of black plastic mulch, to increase soil temperature, is particularly beneficial for wild species of tropical origin. Standard practices for fertilization, irrigation, and application of herbicides and pesticides for squash are followed. It is particularly important to apply an insecticide soon after transplanting or emergence of direct-seeded squash to control cucumber beetles.

If it has not been possible to self-pollinate a particularly desirable plant, cuttings can be taken for greenhouse propagation. Vine tip cuttings should be taken before the plant is weakened by disease, fruit load, or the onset of cold weather. Foster (*25*) successfully rooted cuttings by immersing them in nutrient solution with 4 ppm indolebutyric acid.

TRIALS OF ADVANCED LINES

In early generations, selection of flesh color in winter squash is usually made visually, with preference given to fruit with bright-orange flesh color. More objective measurements of color, carotenoid content, and other quality factors are desirable in advanced generations when a decision must be made regarding release of breeding lines as cultivars. Color determinations of baked squash were more closely correlated with taste panel scores than were other objective measurements of quality in experiments by Schayles and Isenberg (69). Flesh color was not well correlated with total carotene and chlorophyll content in tests made by Francis (27).

Marx et al. (50) blanched and froze samples and later used them in chemical and physical determinations of quality. The amount of fiber was evaluated by washing pulped samples through a 30-mesh screen with hot water, followed by drying and weighing the residue. A Gardener Color and Color Difference Meter was used to measure color. Total carotenoids were determined by chromatography. The difference between fresh weight and dry weight was used to determine total solids content, and soluble solids were estimated with a hand refractometer. Lana and Tischer (47) recommended using the Adams consistometer to predict the quality of breeding lines for canned pumpkin.

Cultivars differ in quality changes in storage (33) and in storage life. Storage tests are recommended when evaluating breeding lines of winter squash.

The ultimate measure of quality is taste, and taste tests are recommended before releasing a winter squash cultivar. In sensory evaluations of C. maxima breeding lines, Murphy et al. (53) used 1-lb composite samples from four or five fruit of different plants for each breeding line. They steamed the samples over boiling water for 30 min and served 8–16 samples at a time, using balanced lattice or incomplete block designs and 25–28 judges.

Bitterness due to cucurbitacin compounds can occasionally be a problem. Environmental factors influence the development of objectionable bitterness, but genetics is also important. Sharma and Hall (77) reported that cv. Black Zucchini summer squash had a higher cucurbitacin content than cv. Early Golden Bush Scallop. Quantitative analysis of cucurbitacin B was accomplished by spectrophotometry of ethanol and chloroform extracts.

When advanced breeding lines are compared with standard cultivars, to determine if some are sufficiently superior to merit release as a new cultivar, commercial cultural practices should be used. Thus, they should be direct seeded rather than transplanted. Row width in advanced trials varies, but some squash breeders plant bush lines in rows 4–5 ft apart and vining lines in 6–8 ft wide rows, and thin seedlings so they are 12–24 in. apart.

Research on efficiency of different yield trial designs is lacking for squash. Often, much of the evaluation is observational, based on fruit color and shape, uniformity, disease resistance, and subjective estimates of yield. When yield data are taken, a randomized complete block design is often used. The number of replications and plants per plot varies, depending on the resources available to the breeder. More labor is needed for summer squash trials, which require multiple harvests, than for winter squash trials, which are harvested only once. Hutchins and Croston (40) used four replications, each with five plants, in yield tests with 21 lines and cultivars of C. maxima. A breeding line should be tested in more than one season and in different locations; again, squash breeders differ, and some test breeding lines in more trials than others, but often cultivars are introduced if a line is consistently superior in trials for two or three seasons. There is no organized system for testing the same squash breeding line in coordinated trials in different states, similar to the Southern Tomato Exchange Program (STEP) and Northern Tomato Exchange Program (NTEP), but squash breeders informally exchange seed of breeding material and provide information on the performance of a breeding line in another area. Information on breeding lines available for trial is provided in the *Vegetable Improvement Newsletter* (published by H. M. Munger at Cornell University, Ithaca, New York).

Squash breeders usually produce seed themselves, extracting seed by hand from mature fruits, until a line is nearly ready for release. If a large quantity of seed is needed for advanced trials, and a public breeder does not have a vine thresher or other facilities to handle large quantities of seed, a seedsman may help. Excellent cooperation exists between public and private squash breeders, with the public breeders often providing germplasm of early generation breeding lines to seed company breeders, who may aid the public breeder by obtaining a winter generation of that germplasm and by making bulk seed increases of advanced lines. When a squash cultivar is named and released by a public breeder, it is usually announced in *HortScience* and seed is provided at no charge or nominal cost to seed producers. Seed of a new cultivar should also be sent to the National Seed Storage Laboratory, Fort Collins, Colorado, for preservation.

REFERENCES

1. Adeoye, A. A., and Coyne, D. P. 1981. Inheritance of resistance to trifluralin toxicity in *Cucurbita moschata* Poir. HortScience *16*, 774–775.
2. Bailey, L. H. 1890. Experiences in crossing cucurbits. Bull.—N.Y. Agric. Exp. Stn. (Ithaca) *25*.
3. Bemis, W. P. 1973. Interspecific aneuploidy in *Cucurbita*. Genet. Res. *21*, 221–228.
4. Bemis, W. P., Curtis, L. D., Weber, C. W., and Berry, J. 1978. The feral Buffalo gourd, *Cucurbita foetidissima*. Econ. Bot. *32*, 87–95.
5. Bemis, W. P., Berry, J. W., Weber, C. W., and Whitaker, T. W. 1978. The Buffalo gourd: A new potential horticultural crop. HortScience *13*, 295–340.
6. Benepal, P. A., and Hall, C. W. 1967. The genetic basis of varietal resistance of *Cucurbita pepo* to squash bug, *Anasa tristis* DeGeer. Proc. Am. Soc. Hortic. Sci. *90*, 301–303.
7. Brett, C. H., and Sullivan, M. J. 1970. The use of resistant varieties and other cultural practices for control of insects of *Cucurbita* in North Carolina. N.C., Agric. Exp. Stn., Bull. *440*.
8. Brett, C. H., McCombs, C. L., and Daugherty, D. M. 1961. Resistance of squash varieties to the pickleworm and the value of resistance to insecticidal control. J. Econ. Entomol. *54*, 1191–1197.
9. Brett, C. H., McCombs, C. L., Henderson, W. R., and Rudder, J. O. 1965. Carbohydrate concentration as a factor in the resistance of squash varieties to the pickleworm. J. Econ. Entomol. *58*, 893–896.
10. Bushnell, J. W. 1922. Isolation of uniform types of Hubbard squash by inbreeding. Proc. Am. Soc. Hortic. Sci. *19*, 139–144.
11. Chambliss, O. L., and Jones, C. M. 1966. Chemical and genetic basis for insect resistance in cucurbits. Proc. Am. Soc. Hortic. Sci. *89*, 384–405.
12. Contin, M. E. 1978. Interspecific transfer of powdery mildew resistance in the genus *Cucurbita*. Ph.D. Thesis. Cornell University, Ithaca, NY.

13. Coyne, D. P. 1970. Inheritance of mottle-leaf in *Cucurbita moschata* Poir. HortScience *5*, 226–227.

14. Coyne, D. P. 1971. A new procedure to develop hybrid 'Butternut' squash relatively stable for fruit shape. Hortic. Res. *11*, 183–187.

15. Coyne, D. P. 1976. 'Butternut Ponca' squash. HortScience *11*, 617.

16. Coyne, D. P., and Hill, R. M. 1976. 'Butternut Patriot' squash. HortScience *11*, 618.

17. Cummings, M. B., and Jenkins, E. W. 1928. Pure line studies with ten generations of 'Hubbard' squash. Vt., Agric. Exp. Stn., Bull. *280*.

18. Curtis, L. C. 1939. Heterosis in summer squash (*Cucurbita pepo*) and the possibility of producing F_1 hybrid seed for commercial planting. Proc. Am. Soc. Hortic. Sci. *37*, 827–828.

19. Curtis, L. C. 1941. Comparative earliness of first and second generation squash (*Cucurbita pepo*) and the possibility of using second generation seed for commercial planting. Proc. Am. Soc. Hortic. Sci. *38*, 596–598.

20. Demski, J. W., and Sowell, G., Jr. 1970. Susceptibility of *Cucurbita pepo* and *Citrullus lanatus* introductions to watermelon mosaic virus-2. Plant Dis. Rep. *54*, 880–881.

21. Denna, D. W., and Munger, H. M. 1963. Morphology of the bush and vine habits and the allelism of the bush genes in *Cucurbita maxima* and *C. pepo* squash. Proc. Am. Soc. Hortic. Sci. *82*, 370–377.

22. Dossey, B. F., Bemis, W. P., and Scheerens, J. C. 1981. Genetic control of gynoecy in the Buffalo gourd. J. Hered. *72*, 355–356.

23. Dutta, L. P., and Nath, P. 1976. Inheritance of flower and fruit characters in squash, *Cucurbita pepo* L. Proc. Int. Symp. Sub-Trop. Trop. Hortic., 3rd, 1972 Vol. 1, pp. 69–74.

24. Eisa, H. M., and Munger, H. M. 1968. Male sterility in *Cucurbita pepo*. Proc. Am. Soc. Hortic. Sci. *92*, 473–479.

25. Foster, R. E. 1964. Vegetative propagation of cucurbits. J. Ariz. Acad. Sci. *3*, 90–93.

26. Frances, R. R., and Bemis, W. P. 1970. A cytomorphological study of male sterility in a mutant of *Cucurbita maxima* Duch. Econ. Bot. *24*, 325–332.

27. Francis, F. J. 1952. Relationship between flesh color and pigment content in squash. Proc. Am. Soc. Hortic. Sci. *81*, 408–414.

28. Globerson, D. 1969. The inheritance of white fruit and stem color in summer squash, *Cucurbita pepo* L. Euphytica *18*, 249–255.

29. Grebenščikov, I. 1954. Zur Vererbung der Dünnschaligheit bei *Cucurbit pepo* L. Zuechter *24*, 162–166.

30. Haber, E. S. 1928. Inbreeding in the Table Queen (Des Moines) squash. Proc. Am. Soc. Hortic. Sci. *25*, 111–114.

31. Hall, C. V., and Painter, R. H. 1968. Insect resistance in *Cucurbita*. Kans., Agric. Exp. Stn., Tech. Bull. *256*.

32. Hayase, H., and Ueda, T. 1965. *Cucurbita* crosses. IX. Hybrid vigor of reciprocal F_1 crosses in *Cucurbita maxima*. Hokkaido Natl. Agric. Stn., Res. Bull. *71*.

33. Hopp, R. J., Merrow, S. B., and Elbert, E. M. 1960. Varietal differences and storage changes in β-carotene content of six varieties of winter squash. Proc. Am. Soc. Hortic. Sci. *76*, 568–576.

34. Howe, W. L. 1949. Factors affecting the resistance of certain cucurbits to the squash borer. Econ. Entomol. *42*, 321–326.

35. Howe, W. L., Sanborn, J. R., and Rhodes, A. M. 1976. Western corn root worm adult and spotted cucumber beetle association with *Cucurbita* and cucurbitacins. Environ. Entomol. *5*, 1043–1048.

36. Hunziker, A. T., and Subils, R. 1975. Sobil la importancia taxonomica de los nectarios foliares en especies silvestres y cultarado de *Cucurbita*. Kurtziana *8*, 43–47.

37. Hurd, P. D., Lindsley, E. G., and Whitaker, T. W. 1971. Squash and gourd bees (*Peponapis, Xenoglossa*) and the origin of the cultivated *Cucurbita*. Evolution *25*, 218–234.

38. Hutchins, A. E. 1935. The interaction of blue and green color factors in Hubbard squash. Proc. Am. Soc. Hortic. Sci. *33*, 514.

39. Hutchins, A. E. 1944. A male and female sterile variant in squash, *Cucurbita maxima* Duch. Proc. Am. Soc. Hortic. Sci. *44*, 494–496.

40. Hutchins, A. E., and Croston, F. E. 1941. Productibility of F_1 hybrids in squash (*Cucurbita maxima*). Proc. Am. Soc. Hortic. Sci. *39*, 332–336.

41. Ivanoff, S. S. 1945. A seedling method for testing aphid resistance. J. Hered. *36*, 357–361.

42. Jaycox, E. R., Gwynn, G., Rhodes, A. M., and Vandemark, J. S. 1975. Observations on pumpkin pollination in Illinois. Am. Bee J. *115*, 139–140.

43. Kirkwood, J. C. 1905. The comparative embryology of the Cucurbitaceae. Bull. N.Y. Bot. Gard. *3*, 313–402.

44. Kubicki, B. 1970. Androecious strains of *Cucurbita pepo* L. Genet. Pol. *11*, 45–51.

45. Kubicki, B., and Walczak, B. 1976. Variation and heritability of β-carotene content in some cultivars of the *Cucurbita* species. Genet. Pol. *17*, 531–544.

46. Lal, O. P. Relative susceptibility of some cucumber and squash varieties to melon aphid, *Aphis gossypi*. Indian J. Plant Prot. *5*, 208–210.

47. Lana, E. P., and Tischer, R. G. 1951. Evaluation of methods of determining quality of pumpkins for canning. Proc. Am. Soc. Hortic. Sci. *58*, 274–278.

48. Lotsy, J. P. 1920. *Cucurbita* strijduragen. II. Eigen onderzolkingen. Genetica 2, 1–21.

49. Mains, E. B. 1950. Inheritance in *Cucurbita pepo*. Pap. Mich. Acad. Sci., Arts Lett. *36*, 27–30.

50. Marx, G. A., Robinson, W. B., Massey, L. M., and Vittum, M. T. 1961. Breeding squash for improved processing quality. Farm Res. *27*, 13.

51. Munger, H. M. 1981. Personal communication. Cornell University, Ithaca, NY.

52. Munger, H. M. 1976. *Cucurbita martinezii* as a source of disease resistance. Veg. Improv. Newsl. *18*, 4.

53. Murphy, E. F., Hepler, P. R., and True, R. H. 1966. An evaluation of the sensory quality of inbred lines of squash (*Cucurbita maxima*). Proc. Am. Soc. Hortic. Sci. *89*, 483–490.

54. Nath, P., and Hall, C. V. 1963. Inheritance of resistance to the spotted cucumber beetle in *Cucurbita pepo*. Indian J. Genet. Plant Breed. *23*, 337–341.

55. Neinhuis, J., and Rhodes, A. M. 1977. Interspecific grafting to enhance flowering in wild species of *Cucurbita*. HortScience *12*, 458–459.

56. Pearson, O. H. 1968. Unstable gene systems in vegetable crops and implications for selection. HortScience *3*, 271–274.

57. Pearson, O. H., Hopp, R., and Bohn, G. W. 1951. Notes on species crosses in *Cucurbita*. Proc. Am. Soc. Hortic. Sci. *57*, 310–322.

58. Provvidenti, R., and Robinson, R. W. 1978. Multiple virus resistance in *Cucurbita*. Cucurbit Genet. Coop. Rep. *1*, 26–27.

59. Provvidenti, R., Robinson, R. W., and Munger, H. M. 1978. Resistance in feral species to six viruses infecting *Cucurbita*. Plant Dis. Rep. *62*, 326–329.

60. Rhodes, A. M. 1959. Species hybridization and intrerspecific gene transfer in the genus *Cucurbita*. Proc. Am. Soc. Hortic. Sci. *74*, 546–552.

61. Rhodes, A. M. 1964. Inheritance of powdery mildew resistance in the genus *Cucurbita*. Plant Dis. Rep. *48*, 54–55.

62. Robinson, R. W., Munger, H. M., Whitaker, T. W., and Bohn, G. W. 1976. Genes of the Cucurbitaceae. HortScience *11*, 554–568.

63. Robinson, R. W., Whitaker, T. W., and Bohn, G. W. 1970. Promotion of pistillate flowering in *Cucurbita* by 2-chloroethylphosphonic acid. Euphytica *19*, 180–182.

64. Roe, N. E., and Bemis, W. P. 1977. Corolla color in *Cucurbita*. J. Hered. *68*, 193–194.

65. Rudich, J., Kedar, N., and Halevy, A. H. 1970. Changed sex expression and possibilities for F_1 hybrid seed production in some cucurbits by application of Ethrel and Alar (B-995). Euphytica *19*, 47–53.

66. Salama, E. A., and Sill, W. H., Jr. 1968. Resistance to Kansas squash mosaic virus strains among *Cucurbita* species. Trans. Kans. Acad. Sci. *71*, 62–68.
67. Scarchuk, J. 1954. Fruit and leaf characters in summer squash. J. Hered. *45*, 295–297.
68. Scarchuk, J. 1974. Inheritance of light yellow corolla and leafy tendrils in gourd (*Cucurbita pepo* var. *ovifera* Alef.). HortScience *9*, 464.
69. Schayles, F. D., and Isenberg, F. M. 1963. The effect of curing and storage on the chemical composition and taste acceptability of winter squash. Proc. Am. Soc. Hortic. Sci. *83*, 667–674.
70. Schöniger, G. 1952. Vorläufige Mitteilung uber das Verholten der Testa and Farbgene bei vescheidenen Kreuzungen innerhalb der Kürbisart, *Cucurbita pepo* L. Zuechter *22*, 316–337.
71. Schuster, W. 1977. Der Ölkürbis (*Cucurbita pepo* L.). Adv. Agron. Crop Sci., Suppl. J. Agron. Crop Sci. *4*, 1–53.
72. Schuster, W., Haghdadi, M. R., and Michael, J. 1974. Imzüchtwirking und Heteriosiseffekt beim Ölkürbis (*Cucurbita pepo* L.). Z. Pflanzenzucht. *73*, 112–124.
73. Scott, D. H., and Riner, M. E. 1946. Inheritance of male sterility in winter squash. Proc. Am. Soc. Hortic. Sci. *47*, 375–377.
74. Scott, G. W. 1934. Observations on some inbred lines of bush types of *Cucurbita pepo*. Proc. Am. Soc. Hortic. Sci. *32*, 480.
75. Shannon, S., and Robinson, R. W. 1979. The use of ethephon to regulate sex expression of summer squash for hybrid seed production. J. Am. Soc. Hortic. Sci. *104*, 674–677.
76. Sharma, G. C., and Hall, C. V. 1971. Cucurbitacin B and total sugar inheritance in *Cucurbita pepo* L. related to spotted cucumber beetle feeding. J. Am. Soc. Hortic. Sci. *96*, 750–754.
77. Sharma, G. C., and Hall, C. V. 1971. Influence of cucurbitacins, sugars and fatty acids on cucurbit susceptibility to spotted cucumber beetle. J. Am. Soc. Hortic. Sci. *96*, 675–680.
78. Shifriss, O. 1945. Male sterility and albino seedlings in cucurbits. J. Hered. *36*, 47–52.
79. Shifriss, O. 1947. Developmental reversal of dominance in *Cucurbita pepo* L. Proc. Am. Soc. Hortic. Sci. *50*, 330–346.
80. Shifriss, O. 1955. Genetics and origin of the bicolor gourd. J. Hered. *46*, 213–222.
81. Shifriss, O. 1981. Origin, expression and significance of gene *B* in *Cucurbita pepo* L. J. Am. Soc. Hortic. Sci. *106*, 220–232.
82. Shifriss, O. 1981. Do *Cucurbita* plants with silvery leaves escape virus infection? Cucurbit Genet. Coop. Rep. *4*, 42–43.
83. Singh, D. 1949. Inheritance of certain economic characters in squash, *Cucurbita maxima* Duch. Minn., Agric. Exp. Stn., Tech. Bull. *180*.
84. Singh, S. P., and Rhodes, A. M. 1961. A morphological and cytological study of male sterility in *Cucurbita maxima*. Proc. Am. Soc. Hortic. Sci. *78*, 375–378.
85. Sinnott, E. W. 1922. Inheritance of fruit shape in *Cucurbita pepo*. I. Bot. Gaz. (Chicago) *74*, 95–103.
86. Sinnott, E. W. 1927. A factorial analysis of certain shape characters in squash fruits. Am. Nat. *1*, 333–343.
87. Sinnott, E. W., and Durham, G. B. 1922. Inheritance in summer squash. J. Hered. *13*, 177–186.
88. Sitterly, W. R. 1972. Breeding for disease resistance in cucurbits. Annu. Rev. Phytopathol. *10*, 471–490.
89. Sowell, G., Jr., and Corley, W. L. 1973. Resistance of *Cucurbita* plant introductions to powdery mildew. HortScience *8*, 492–493.
90. Strider, D. L., and Konsler, T. R. 1965. An evaluation of the *Cucurbita* for scab resistance. Plant Dis. Rep. *49*, 388–394.
91. Varghese, B. M. 1973. Studies in the cytology and evolution of South Indian Cucurbitaceae. Ph.D. Thesis. Kerala University, India.

92. Vaulx, R. D. De, and Pitrat, M. 1979. Interspecific cross between *Cucurbita pepo* and *C. martinezii*. Cucurbit Genet. Coop. Rept. *2*, 35.

93. Wall, J. R. 1954. Interspecific hybrids of *Cucurbita* obtained by embryo culture. Proc. Am. Soc. Hortic. Sci. *63*, 427–430.

94. Wall, J. R., and Whitaker, T. W. 1971. Genetic control of leucine aminopeptidase and esterase isozymes in the interspecific cross *Cucurbita ecuadorensis* × *C. maxima*. Biochem. Genet. *5*, 223–229.

95. Wall, J. R., and York, T. L. 1960. Genetic diversity as an aid to interspecific hybridization in *Phaseolus* and *Cucurbita*. Proc. Am. Soc. Hortic. Sci. *75*, 419–420.

96. Washek, R. 1982. Personal communication. Cornell University, Ithaca, NY.

97. Watterson, J. C., Williams, P. H., and Durbin, R. D. 1971. Response of cucurbits to *Erwinia tracheiphila*. Plant Dis. Rep. *55*, 816–819.

98. Weaver, J. E., and Bruner, W. E. 1927. Root Development of Vegetable Crops. McGraw-Hill Book Co., New York.

99. Whitaker, T. W. 1959. An interspecific cross in *Cucurbita* (*C. lundelliana* Bailey × *C. moschata* Duch.). Madrono *15*, 4–13.

100. Whitaker, T. W. 1960. Breeding squash and pumpkins. Handb. Pflanzenzuecht. *6*, 331–350.

101. Whitaker, T. W., and Bemis, W. P. 1965. Evolution in the genus *Cucurbita*. Evolution *18*, 553–559.

102. Whitaker, T. W., and Knight, R. J., Jr. 1980. Collecting cultivated and wild cucurbits in Mexico. Econ. Bot. *34*, 312–319.

103. Wisemann, B. R., Hall, C. V., and Painter, R. H. 1961. Interactions among cucurbit varieties and feeding responses of striped and spotted cucumber beetles. Proc. Am. Soc. Hortic. Sci. *78*, 379–384.

104. Yeager, A. F., and Meader, E. M. 1957. Breeding new vegetable varieties. N.H., Agric. Exp. Stn., Bull. *140*.

7
Snap Bean Breeding

M. J. SILBERNAGEL

The terms snap beans (*Phaseolus vulgaris* L.), string beans, garden beans, and fresh beans are more or less synonymous, referring primarily to beans produced for consumption as a fresh or processed vegetable as opposed to a dry bean seed (pulse). Snap bean seed can also be used in the dry state like the dry bean types (pinto, kidney, pink, small red, etc.).

Snap beans are an important and stable component of the vegetable diet consumed by Americans (about 7 lb/capita), exceeded only by sweet corn, tomatoes, cabbage, and green peas(*114*). Over the past 20 years fresh per capita consumption has declined from 3 to 1.5 lb, whereas processed usage has increased from 3.8 to 5.5 lb. Canned consumption increased from 3 to 4 lb, whereas frozen usage went from 1 to 1.5 lb per person.

While the indeterminate tall climbing vine is genetically dominant and adaptatively superior in the wild, most snap beans grown in the United States today are determinate bush types. Home gardeners and some fresh-market growers still use a few vine types; however, vine types probably constitute less than 5% of the total acreage. Tall vine types are sometimes called pole beans because poles are often used as trellises, whereas short vine types are also referred to as half-runner types. Scarcity of labor and the high cost of hand picking led to the development of mechanical harvesters in the mid 1950s. Virtually all commercial operations are now mechanically harvested.

In 1980 about 370,000 acres of snap beans were harvested commercially in the United States with an approximate farm value of $192 million(*113*). Of that, canned beans were

243

harvested from 214,970 acres (av. 5460 lb/acre) for a value of $82,580,000. Beans for freezing were grown on 59,580 acres (5700 lb/acre) at a worth of $27,095,000. Fresh-market beans (3200 lb/acre) were produced on 95,700 acres with a farm value of $82,541,000.

Beans as a vegetable are produced and used in a number of different ways. Most of the beans for processing (canned, frozen, freeze-dried) are round podded, while fresh-market cultivars are often flat or oval podded. Yellow-podded cultivars (wax beans) comprise about 15% of the total pack. Home gardeners, particularly in the northeast, also use certain cultivars in the green shell (shell bean) stage, i.e., large but still soft immature seeds.

The major U.S. processing areas (113) are Wisconsin (83,900 acres), New York (49,000 acres), Oregon (32,000 acres), Michigan (14,600 acres), and Tennessee (13,700 acres). The rest of the acreage is scattered throughout the country. Most of the fresh-market beans are produced in Florida (48,000 acres) and seed production (40,000 acres) is concentrated in south-central Idaho (112). The total annual production of snap bean seed, about 80 million lb, has an approximate farm value of $32 million. Most of the U.S. seed crop is used domestically, but an increasing proportion has been going to Europe in recent years.

Europeans generally consume more beans (especially fresh) than Americans, and European seedsmen traditionally have produced seed for Europe in East Africa. During the past ten years, several large European seedhouses have also established production operations in south-central Idaho. Most U.S. cultivars are not suited to European conditions because of susceptibility to anthracnose and halo blight, and the European preference for smaller sieved pods than the American processors use.

ORIGIN AND GENERAL BOTANY

The common bean is of New World origin, principally Central and South American (60). In the wild state, beans or near relatives are found from the lowland, warm, humid tropics, to the cold, high-altitude, short-season mountains, and the hot, arid deserts. Generally, however, the common beans with which we are most familiar are those cultivars that fit into a relatively narrow ecological zone.

Smartt (107, p.19) in describing the domestication of *Phaseolus* species, states that "there is no doubt that *Phaseolus vulgaris* is the most successful American bean followed by *P. lunatus* L. (lima), *P. coccineus* L. (scarlet runner), and *P. acutifolius* A. Gray (tepary) in that order. It is perhaps no coincidence that there is a rough correlation between the extent to which their habitat preferences and those of man coincide and their advance under domestication." The beans grown in North America are usually day-neutral, determinate bush types (91) that fit the temperate-zone requirements of warm soil (13°–21°C), moderate air temperatures (24°–29°C), especially during bloom, adequate moisture (10–18 in.) distributed more or less evenly throughout the growing season, a relatively neutral, fertile, well-drained soil, and adequate sunlight. Outside this optimal zone in North America for the common bean, other species of *Phaseolus,* some of which are near relatives but adapted to wider environmental extremes, are often substituted. In the Southeast during the hot humid part of the summer, lima beans and cowpeas [*Vigna unguiculata* (L.) Walp.] are grown more often than the common bean. In the hot semiarid zones of the Southwest, the tepary bean produces more reliably than the common bean, and at the colder moist extremes (the Northeast), selections within *P. coccineus* like the scarlet runner bean are used as garden varieties.

Several closely related species of *Phaseolus* (all $2n = 22$) can be hybridized to common bean (*107*). Honma (*55*) succeeded in crossing tepary with common bean and the resulting cv. Great Northern 27 sel. 1 carries resistance to common blight (CB) incited by *Xanthomonas campestris* pv. *phaseoli* (Smith 1897) *comb. nov.* from the cross. This germplasm was used by Schuster and Coyne (*92*) as a source of CB resistance. Giles Waines, at the University of California, Riverside (personal communication), has crossed tepary to common bean to transfer drought tolerance into common bean. Many breeders have utilized *P. coccineus* to obtain bacterial disease resistance, tolerance to colder climates, root rot resistance, and bean yellow mosaic virus (BYMV) resistance. Lorz (*65*) working in Florida showed common bean could be crossed with a number of closely related species of *Phaseolus*. Mok *et al.* (*76*) studied the barriers to interspecific hybridization using tissue culture and biochemical techniques.

In view of the largely unutilized genetic diversity within common bean for commercially desirable characteristics, bean improvement by interspecific hybridization should be left to scientists who specialize in this area of complex basic research. The same must be said of induced mutations, tissue culture screening, and protoplasmic fusion as means of crop improvement.

Domestication of Natural Mutations

Bush-type beans are almost never found in the wild. The determinate bush habit, so widely adapted to mechanical-harvester requirements, was most certainly derived from mutants as was the apical dominance found in commercial types (*107*). Similarly the convenience and commercial value of the stringless character (*1*) and the round pod shape (*2*) were quickly recognized and incorporated into breeding programs within the past 100 years. As little as 20 years ago, garden beans were usually referred to as string beans, a term still used today even though virtually all cultivars are now stringless. The term "snap" originated from the way fresh garden beans were broken or snapped into short segments by hand in preparation for cooking.

The high pod wall fiber content of most wild legumes is necessary for the way in which legume pods dehisce easily (even forcibly) when dry. This is a natural dispersal mechanism that favors survival of the species in the wild (*107*). However, a high pod wall fiber content is undesirable in a table vegetable, and premature shattering of a seed crop is economically unacceptable. Plant breeders have systematically selected against these wild traits. Again, most likely low fiber content originated in mutations that ancient civilizations were able to recognize, perpetuate, and utilize (*107*).

Some wild forms of *Phaseolus* like *Phaseolus polystachyus* can be either perennial or annual. A few perennial forms (none of which are found in the United States) form large fleshy tubers. Many of the related species, as well as many forms of *P. vulgaris* in the centers of genetic diversity, have a short-day photoperiod requirement, which precludes their culture in North America since they would not flower until fall and thus be killed by frosts before maturity. Steve Temple and Jeremy Davis, Plant Breeders at CIAT (Centro Internacional de Agricultura Tropical, Cali, Colombia), are transferring useful genetic characteristics like cold tolerance and flood tolerance into day-neutral germplasm lines so they can be more easily utilized by temperate-zone breeders (personal communication).

Gigantism (i.e., larger stems, leaves, pods, and seeds) is another distinction usually differentiating the wild from cultivated forms of the *Phaseolus* species. "Plants in the wild will by natural selection tend to adopt the strategy of producing the largest number of seeds possible" (*107*, p. 15). The gigantism found in domesticated forms is usually only compensatory. The increased pod and seed size in the domestic cultivars is offset by a reduction in the number of pods and seeds. Final total yield limits are about the same as the wild form.

The survival value of hard-seededness, a type of dormancy induced by the water impermeability of the testa (*107*), is another wild trait that the domestic cultivars do not require. In fact, the variability in emergence and the difficulties hard seed present in cooking make this characteristic a disadvantage in the domesticated forms.

Plant Patent Requirements

Excellent botanical descriptions of *P. vulgaris* L. (chromosome number $2n = 22$) and its related species are presented by Smartt (*107*), and so they are not repeated here. Of more immediate relevance to the North American bean breeder is the description required by the USDA Agricultural Marketing Service (*111*) which administers the National Plant Variety Protection Act of 1970. This legal description (for garden beans and/or dry beans) becomes part of the registration procedure and eventually the basis for granting a patent.

This four-page description [Form LPGS-470-12(2-79)] is basically comparative rather than absolute. This is in recognition of the effect local environmental factors can have in the expression of genetic potential. It is also important that the descriptions be comprehensible, not only by nonscientific people involved in the production and commercialization of beans, but also by the legal profession and juries in the event of litigation over patent infringements.

The essence of patentability is a recognizable novelty, the variability of which can be described. Production performance or end-product quality characteristics are not part of the currently required description. Included are relative (compared to a known standard cultivar) descriptions of maturity, physical measurements, and comparative descriptions of the plant habit, leaves, flowers, pods, seeds, and pigmentation. Disease, insect, and physiologic resistance factors are also included. It was felt that requirements for grow-out trials or performance characteristics would be expensive, could be hard to administer, and would unduly delay release of new cultivars, thus cutting several commercially productive years from the 17-year patent.

FLORAL BIOLOGY AND CONTROLLED POLLINATION

The American Society of Agronomy and the Crop Science Society of America have recently published *Hybridization of Crop Plants* (*9*). "Each crop chapter specifically discusses parental material; plant culture; floral characteristics; artificial hybridization or self-pollination; natural hybridization; seed development, harvest and storage; and techniques for special situations."

The introductory chapters, plus the detailed chapter by Bliss on the common bean, make any attempt to repeat such information here completely unnecessary.

MAJOR BREEDING ACHIEVEMENTS OF THE RECENT PAST

Since the release of the first round-podded cultivar in 1865 and the first stringless cultivar in 1870 (*131*), the most significant breakthrough in snap bean improvements came with Bill Zaumeyer's (USDA, retired) release of the cv. Tendercrop (*134*) in 1958. Tendercrop set new industry standards for several plant and pod characteristics, and is still used today by the frozen bean processing industry. In 1970, 46% of the green-podded bush types had

Tendercrop germplasm in their ancestry (*132*). Besides having large, fleshy, bright-green, smooth, straight, round pods, Tendercrop has a relatively slow rate of seed and/or fiber development once it reaches maturity, thus giving some "holding ability" to harvest operations if delayed by weather or equipment problems. Tendercrop has a strong upright, relatively narrow plant habit, which holds pods primarily in the upper two-thirds of the canopy. This feature reduces spoilage of pods. Tendercrop was the first high-quality processing cultivar with a concentrated pod maturity, i.e., a majority of the pods were ready to pick at the same time. This is a prime requirement for the success of mechanical harvesting of green beans, which was commercially perfected about the same time.

Another factor that influenced cultivar development in the past three decades was the rapid increase in popularity of frozen foods. Frozen vegetables in the United States have become popular primarily since World War II. Most canning or fresh market cultivars used in the 1940s and 1950s were light to medium green, which were quite unattractive as a frozen product. With the introduction of Tendercrop, the frozen food processors had a product with a uniform and very appealing bright-green color.

In the early 1950s, bean canners became aware of a darkening of the liquid in cans of dark-seeded cultivars that had been harvested on the late side of optimal maturity. This represented a potential buyer's barrier, and so when Mel Parker of Gallatin Valley Seed Company discovered and released an off-white seeded mutant (GV-50) from Tendercrop, it became an immediate success (*131*). Earlier white-seeded cultivars had not been accepted because of poor seed quality. The cv. Gallatin 50 was the leading canning cultivar for about 10 years, until replaced by the white-seeded cv. Early Gallatin.

Prior to the late 1950s when beans were picked by hand two to six times during the season, yields for bush types ranged from 3000 to 8000 lb/acre (*41*), while pole beans went as high as 18,000 lb/acre (*66*). At the time of the introduction of the first mechanical harvesters, a yield of 5000 lb/acre was considered necessary from a once-over destructive harvest in order to be economically feasible. While national average processing yields are only slightly above that today, beans for processing are not as mature as they were 25 years ago and this makes yield comparisons difficult. Progressive grower returns are usually 10,000–12,000 lb/acre at the "optimal" harvest stage of 50% one- to four-sieve pods, and 50% five-sieve and larger. The integration of recent improvements in cultivars, equipment, disease control, and cultural practices indicate yields as high as 40,000 lb/acre are possible. Therefore, as these advances are commercialized, average yields of 10,000–15,000 lb/acre should be realized by the turn of the century, at least for those areas with supplemental irrigation.

The recently developed Multi-D Harvester (Chisholm Ryder Company) made possible high-density bean culture as proposed by Andy Duncan, Oregon State University (OSU), in the mid-1960s, and described by Mack and Stang (*67*). High-density (174,200 plants/acre) culture (narrow rows with widely spaced plants within the row) is contingent upon reliable chemical weed and disease control. Besides high-density planting arrangements, which make optimal use of available sunlight, water, and nutrients, other cultural practices like amount and frequency of irrigation (*73*), and deep soil chiseling to reduce soil compaction (*75*) also seem to be important, especially under conditions favorable to root disease.

Two additional areas of cultivar improvement had significant impact on a national scale. The first was the transfer of pod quality characteristics from pole-type Blue Lake cultivars like FM-1, to bush types adapted to mechanical harvesting. The second was the revolution in market garden (fresh-market) cultivars led by Harvester.

W. A. Frazier, OSU, was one of the prime movers behind the transfer of the Blue Lake pod to a machine-harvestable bush. His OSU-1604 is still widely used in the Willamette Valley 10 years after its release. Yields of 8–10 tons/acre are not unusual for OSU-1604, and it has most of the Blue Lake pod quality factors for which the Willamette Valley is famous. The pole-type Blue Lake cultivars commanded a market premium for many years, not only because of the distinctive Blue Lake flavor, but they were also well suited for the institutional trade. Cut-style Blue Lake canned beans remain firm and with relatively little carpel separation or skin sloughing after long hours on a restaurant or institutional steam table. By contrast, canned Tendercrop types have a bland flavor and do not retain an attractive appearance for long under steam table conditions.

Art Sprague, Del Monte Corp., and Walt Pierce, Asgrow Seed Co., were also among the early proponents of the Blue Lake movement. While the Del Monte cultivars are available only to their own growers, emulation of their unquestionable high quality and yield ability became the objective of many other bean breeders. Two other distinctive features of the Del Monte cultivars included a tendency for a large number of fruiting lateral branches (instead of a few strong central stalks as in Tendercrop), and a profusion of flowers that bloomed over an extended period. The heavy branching provided many flowering nodes, thus the high yield potential and the extended flowering period provided some protection against "split sets" (expanded and uneven distribution of pod maturities) caused by blossom drop during periods of high temperatures during bloom.

Asgrow cvs. BBL 47, 240, 274, and 290 generally introduced Blue Lake types to the rest of the U.S. industry (Midwest and East). As mentioned earlier, the Del Monte cultivars were only available to Del Monte growers, and the OSU cultivars were generally not well adapted outside the Willamette Valley, primarily because of their extreme sensitivity to heat during bloom.

Pierce's cultivars were also successful outside the Willamette Valley because of their distinctive improvements in machine harvestability and in plant efficiency over the earlier OSU and Del Monte cultivars, although they were still not as easy to harvest as Tendercrop. Processors and growers, being used to the high field recovery rate (90–95%) of machine-harvested Tendercrop types, were appalled at leaving 2–3 tons/acre in the field (70%) recovery) when using the early OSU or Del Monte cultivars. Even though machine-harvested yields of the early Blue Lake cultivars were higher than those of the Tendercrop types, the increase in broken-off branches, clusters of unseparated pods, and dirt (from trying to harvest too close to the ground) for a long time kept the midwestern and eastern processors from adopting the Blue Lake types. The increased trashiness often slowed through-the-plant flow by several tons per hour thus increasing unit production costs considerably.

Pierce's cv. Harvester, released in the mid-1960s, was well adapted to mechanical harvesting and replaced many of the previously used hand-picked types, thus becoming the fresh-market and shipping bean industry standard for many years. Fresh-market pods are generally light to medium green, round (a few are oval or flat) in cross section, and heavily pubescent. The pubescence presumably reduces pod blemishes from wind scarring and damage during shipping. A convenient field test of this criterion is the ability of a pod to cling to a cotton shirt.

The requirement for an attractive appearance on a Chicago grocer's shelf, after shipment from Florida and being handled several times, imposes different quality standards on fresh-market cultivars, which generally make them unsuitable for processing. A fresh-market or shipping pod must have enough pod wall fiber to retain its shape and fresh appearance 7–10 days after harvest, even with some desiccation incurred during shipment, storage, and display.

A review of major breeding achievements would not be complete without mention of Mel Anderson (Rogers Brothers Seed Co.). Anderson's cultivars dominated the industry for many years and were often produced in larger volume than all others combined. In 35 years of bean breeding, he developed more than 40 new cultivars, including Earligreen, Earliwax, Slendergreen, Slimgreen, and Improved Tendergreen. In 1970, the Bean Improvement Cooperative presented Anderson, Frazier, Parker, Pierce, and Zaumeyer with the Meritorious Service Award, in recognition of their outstanding contributions to the U.S. snap bean industry for about 40 years.

CURRENT GOALS OF BREEDING PROGRAMS

To be successful, a new snap bean cultivar must please the grower, the seedsman, the processor, and finally the consumer. Excellence in one or more of these categories can create a temporary demand for a new cultivar, but a serious deficiency invariably invites replacement. The fact that few new cultivars remain popular for more than 5–7 years attests that we have not yet found the "perfect bean for all seasons."

There can never be a perfect bean, because (1) different end uses have different requirements, and (2) as we approach the current objectives for any particular end use, we broaden our horizons and set higher goals. With the above in mind and before a cross is ever made, bean breeders need to determine not only what is needed by the particular segment of the market they are aiming for now, but what else will be needed in 10–15 years when the new cultivar is introduced. Few breeders have the time or opportunity to acquaint themselves thoroughly with all the important aspects of each segment of a particular market, and all of its regional idiosyncrasies dictated by such things as climate, available labor or equipment, cultural practices, and the time demands of other crops, to name just a few. Thus breeders must rely heavily on a variety of information sources from which to define 10-year objectives. In the commercial arena, they are assisted by regional seeds salespeople who are most familiar with the problems encountered by local growers, shippers, and/or processors. They are also assisted by managerial economic considerations of such factors as present- and future-market analysis, anticipated population shifts, and anticipated transport and labor costs. Current literature and attendance at meetings of scientific societies and the Bean Improvement Cooperative biennial meetings, keep breeders abreast of technological advances in all the related scientific fronts. Finally, commercial breeders must integrate all this information into an ongoing program that focuses maximum effort on those goals most likely to succeed financially in the shortest possible time. This may leave many research voids, especially in areas of long-range needs. It is the role of public breeders to address these problems.

Hence, the following analysis of current goals will touch on the diversity of types needed to satisfy the many and varied end uses, consider some of the common objectives sought by all breeders, and speculate on current commerical goals of private breeders and some of the long-term futuristic goals of public breeders.

Seed Characteristics

Improved seed quality is one of the greatest needs of the bean industry. Good seed quality, uniform emergence, and early seedling vigor are prerequisites to consistent and maximum production at harvest. Variable seedling emergence and vigor can result from inherent genetic characteristics or from improper seed harvest, storage, and/or handling conditions (*17,104*).

Dickson and Boettger (*25,26*) reported that seed coat thickness, tightness of adherence of seed coat to cotyledons, and a solid contact between cotyledon faces contribute to resistance to mechanical damage. Hoki (*54*) showed that the size and shape of the seed are important factors in resistance to mechanical injury.

Beside breeding for seed characteristics that contribute directly to resistance to mechanical injury, breeders might consider selection for plant and pod characteristics that indirectly lead to improved seed quality. Westermann and Crothers (*122*) showed that snap beans grown for seed also respond to high-density culture. Silbernagel (*102*) suggested that direct harvesting of snap bean seed grown under high-density culture would eliminate many of the problems contributing to decreased seed quality that are associated with the present windrow system. Windrowed beans are cut below the soil line and laid on the soil surface to dry, where they may be exposed to moisture, causing molds following rains, or subsequent overdrying. The rubber-belt thresher proposed by Silbernagel strips dry pods from standing mature plants. The rubber belts extract seed with a minimum of mechanical damage; and since the plants are not windrowed, there is less seed spoilage from stains and molds during rainy weather. To facilitate optimization of the system for high-density culture followed by direct seed harvest, breeders should select for a very concentrated pod maturity, numerous small vertically oriented leaves, and a strong, upright, narrow plant habit.

Plant Characteristics

The term "Tendercrop plant type" is synonymous with a strong upright, relatively narrow habit, stiff enough to remain upright with the weight of a mature crop, but with very few pods touching the ground.

With the current trend to high-density culture, the "ideal" plant type would seem to be a miniaturized Tendercrop but with Blue Lake pod quality, that is, about 16–18 in. instead of 20–24 in. tall, and correspondingly upright and narrow with numerous small leaves (5 × 8 cm) instead of the larger leaves on Tendercrop. The smaller plant should give a slightly higher harvest index, even with "luxury levels" of fertilizer to obtain maximum pod yields. The small, vertically oriented leaves should present a greater total photosynthetic surface because light can penetrate deep into the canopy.

The force of the mechanical harvester, which strips pods and leaves off the main stems and branches, requires a strong plant with a well-anchored root system (*96*). However, some breeding lines, even with a very thick main stem, show a tendency to break easily at the primary leaf node area. This is often noticed when handling plants in making single-plant selections. The heritability of this character has not been studied, but these "weak-kneed" types should be ruthlessly eliminated.

Small leaf size can be found in the USDA Plant Introduction Service Accession (PI 165426. This PI line was reported by McClean *et al.* (*70*) to be resistant to rhizoctonia root rot and the root-knot nematode, and is the genetic basis for several improved bush-bean-type breeding lines from Charleston by Deakin and Dukes (*22*) and Wyatt *et al.*, (*126, 128*), with resistance to these problems. A small black-seeded selection from this PI line (which has a mixture of seed types) was later found also to be resistant to the root rot organisms, *Fusarium, Pythium, Thielaviopsis,* some races of rust, and cold–wet emergence conditions (*98*). However, in numerous crosses with this line, small leaves have not been recovered in commercial-type breeding lines with otherwise acceptable horticultural characteristics. There seems to be a strong association between small leaf size and small pod size. Dickson (*33*) released a small-leafed bush type (L-1). However, it is not known whether the general association between small leaves and small pods has ever been broken.

Disease Resistance

Recent reviews (*94, 106, 133*) on breeding beans for disease resistance, which list sources of resistant germplasm and disease-screening techniques, are thorough and so only a brief summary of current highlights will be covered here.

Viruses

Most snap bean cultivars in the United States carry dominant *I* gene resistance to bean common mosaic virus (BCMV) (*106*). In view of the presence of new strains of BCMV capable of causing a lethal systemic necrotic reaction in these cultivars (*35*), breeders should combine the *I* gene resistance with either *bc-1²*, *bc-2²*, or *bc-3* resistance (*34*) in order to have broad resistance to both the necrotic and mosaic mottle reactions.

Curly top virus (CTV) resistance is needed for some of the western seed production and processing areas. Resistance is probably due to two epistatic dominant factors (M. J. Silbernagel, unpublished). Sources of resistance include Apollo, Blue Mountain, Goldcrop, and Wondergreen.

Peanut stunt virus (*72*), bean strains of the cucumber mosaic virus (CMV), and BYMV strains are occasionally epiphytotic on the east coast. Resistance to a few strains of BYMV is known (*86*), but more studies are needed on the genetics of resistance to a wide range of specific strains that have not even been identified.

Foliar Fungi

Rust *Uromyces phaseoli* (Reben) Wint. is a serious problem in East Coast fall crops of snap beans. Several sources of resistance to different strains are known (*133*). A recent USDA germplasm release PR-190 by Freytag in Puerto Rico, and BARC-1 by Meiners and Silbernagel (USDA, Prosser, Washington) are resistant to most of the prevalent strains. Another USDA breeding line by Silbernagel, 8BP-3 (a small-sieve whole-pack type), is also resistant to some races of rust, as well as anthracnose (*ARE* gene), BCMV (*I* gene), CTV, and some strains of BYMV. The Rogers Brothers Seed Co. cv Resisto is tolerant to several races of rust. The Wisconsin breeding line BBSR-130 is resistant to rust as well as to four other diseases (*42*).

Root Rots

Fusarium root rot, *Fusarium solani* (Mart.) Appel & Wr. f. sp. *phaseoli* (Burk.) Snyd. & Hans., is widespread and can reduce yields in the Northwest by as much as 25–50% (*100*). It is also serious in New York (*83*). No resistant commercial snap bean cultivars are yet available, but Dickson and Boettger (*28*) released a number of breeding lines with a moderate degree of tolerance to fusarium and/or pythium root rot. Resistance to fusarium root rot seems to be due to several quantitative genes (*10, 50*), and it is apparently independent of resistance to thielaviopsis (*52*) and pythium root rots (*129*). Cultural practices, biological control practices, and seed treatment can reduce the severity of root rot injury and increase production levels (*97*). Breeder–pathologists are tending toward the opinion that a high degree of physiologic resistance may not be necessary for effective field tolerance.

Several breeding lines with resistance to rhizoctonia root rot caused by *Rhizoctonia solani* Kuhn [*Thanatephorus cucumeris* (Frank) Donk] have been released over the past several years from the Charleston, South Carolina, USDA Vegetable Breeding Laboratory by McLean *et al.,* (*70*), Deakin and Dukes, (*22*), and Wyatt *et al.* (*128*). Most of these are colored-seeded types (*21*). Several also carry resistance to the root-knot nematodes (*126*). *Rhizoctonia* is prevalent in many southeastern production areas (*77*) with warm soils. Prasad and Weigle (*85*) reported on a number of breeding lines and cultivars with tolerance to *R. solani.*

Resistance to several *Pythium* spp. has been found in white-seeded sources by York, Dickson, and Abawi (*24, 129*) and by Hagedorn and Rand (*43, 45*). The Wisconsin breeding lines RRR-46 and -36 are resistant to *Pythium* and to the recently described aphanomyces root rot of beans caused by *Aphanomyces euteiches* f. sp. *phaseoli* Phend. & Hag. (*81*). Resistance to *Pythium* is needed almost anywhere snap beans are grown.

Resistance to *Thielaviopsis basicola* (Berk. & Br.) Ferr. seems to be available in several sources of root rot resistance (*51, 52*). However, since it does not usually cause serious damage by itself, little effort has been devoted to incorporation of this resistance into snap beans, even though it can be found in most production areas. Someone needs to determine if early season injury by thielaviopsis root rot predisposes plants to more serious injury by other root rots.

Bacteria

Repeated halo blight epidemics incited by race 2 of *Pseudomonas syringae* pv. *phaseolicola* (Burkholder 1926) *comb. nov.* in the major U.S. seed production areas (Idaho and California) in the past 15 years have finally persuaded U.S. breeders to emulate the European breeders in the development of resistant cultivars. Most resistant snap bean cultivars, like Noorinbee (*87*) and RH-13 (*40*), are foreign introductions, except for the halo-blight-resistant breeding line Nebraska HB-76-1 (*19*). Little or no effort has been made to develop snap bean types with resistance to bacterial wilt [*Corynebacterium flaccumfaciens* pv. *flaccumfaciens* (Hedges 1922) Dowson 1942], or common blight *Xanthomonas campestris* pv. *phaseoli* (E. F. Smith) Dowson, because they are rarely serious problems in the United States. Brown spot (*P. syringae* pv. *syringae* van Hall 1902), however, is severe and widespread in Wisconsin and Minnesota. Recent germplasm releases by Hagedorn and Rand of Wis BBSR-130 (*42*), and WB-BSR-17 and -28 (*44*) are highly and moderately resistant, respectively. This resistance needs to be combined with aphanomyces and pythium root rot resistance for reliable production in the Midwest. Epoch, a Wilbur Ellis Blue Lake type cultivar, is tolerant to brown spot.

Insect Resistance

Very little is known about resistance to insects in beans, and I know of no cultivars with identified resistances. Insects are usually controlled through the use of insecticides; however, with governmental regulatory procedures for the production and use of pesticides becoming progressively more restrictive, we may increasingly have to turn to genetic control. Seedcorn maggot resistance has been reported in New York (*116*) and Washington (*47*). Cultivar differences in insect preference have been reported for mites, thrips, and aphids (*48*). Likewise, sources of resistance to leaf-hopper burn (*Empoasca*) are also known (*31*). Resistance to Mexican bean beetle is being studied in the Southeast (*127*) and differences in tolerance to lygus bug stings on pods have been observed in Washington (*46*). These factors should be bred into production cultivars to reduce the need for pesticide applications.

Environmental Stress Tolerance

Relatively little effort has gone into the introduction of factors for environmental stress resistance in snap beans. There is, however, a growing awareness among breeders and research administrators that environmental stress extremes directly cause a large part of the annual fluctuations in legume production. When production levels fluctuate, supplies, demand, and prices often vary out of proportion to actual crop losses. Thus, environmental stress tolerance is needed as much for market stability as for ultimately raising area and national yield levels.

Environmental stresses for which tolerance is needed include cold and/or wet emergence conditions, temporary drought and/or flooding, acid or alkaline soils, high or low temperatures during bloom, photoperiod sensitivity, N_2 deficiency, and air pollution. A cultivar with such tolerances likely would have a wide range of adaptation.

Breeders can tentatively identify drought-tolerant materials by selective water management (*135*) in trial plots. Usually the better materials are coming from root rot breeding programs, where a low nitrogen stress is combined with water stress to identify those lines also able to fix more of their own nitrogen (*11*). Studies in Idaho indicate some cultivars may lose the ability to fix atmospheric N_2 when all selection takes place in nurseries with high levels of supplemental nitrogen (*123*). The snap bean cv. Canyon was found to have the lowest rate of N_2 fixation, while Viva Pink had the highest (*124*). Viva, whose parentage includes a wild bean from Mexico (PI 203958) with a high rate of N_2 fixation ability, was developed by selection in root rot nurseries stressed for supplemental nitrogen and water (*12*). Viva was also reported by D. H. Wallace (Cornell University, personal communication) to have one of the highest harvest indices (*118*) of several dry bean cultivars tested, even though it was not purposely selected for harvest index, only production under multiple-stress conditions.

In recent years, many wild sources of disease resistance have been crossed with modern snap bean cultivars. Thus the opportunity exists for recovery of not only root rot resistance, but root vigor, drought and flood tolerance, and effective N_2 fixation by rhizobium nodulation as well. This field stress system also lends itself well to the identification of germplasm that does not require seed treatment fungicides and/or insecticides. This ability will become more important as environmental concerns limit the availability and/or use of agricultural pesticides.

Soil Compaction Tolerance

A vigorous root system may be less restricted by the so-called plow pan often found at the 10- to 12-in. depth in field soils compacted by heavy equipment. When roots of snap bean cultivars are confined to this zone where most root rot organisms are found (*13*), the severity of root rot damage is increased, and unless frequent irrigations are maintained during hot periods, yields are reduced. The frequent irrigations are deemed necessary by growers to promote secondary root formation to replace those rotted off by root rots. However, too frequent irrigation also aggravates the severity of root rot damage (*100*), and keeping the soil surface moist for long periods of time promotes white and gray molds

(75). Thus deep-penetrating, rot-resistant roots could contribute toward water conservation and at the same time reduce the potential losses due to white and gray molds.

Air Pollution Tolerance

Tolerance to air pollutants is almost mandatory under some eastern seaboard conditions. Fortunately, tolerance is found in a wide array of commercial snap bean cultivars (89). Screening of segregating populations under metropolitan eastern or California conditions would be highly desirable. Alternatively, choosing parents that are well adapted to polluted conditions would greatly increase the probability of finding tolerant materials in unscreened late-generation selections.

Tolerance to Low and High Temperatures

Since photosynthetic surface area and fruiting nodes are prerequisites for pod production, early seedling vigor is essential to realization of maximum yield potential. In the cool environment of a spring planting, the ability to imbibe, emerge, transport water, and photosynthesize at slightly lower than normal temperatures would support the desired early branching and rapid leaf surface area development.

The need for cold tolerance in beans has been recognized for some time, especially in places like England (3) and Canada (61). It now appears that research on cold tolerance in bean should be divided into at least three apparently independent phases: (a) the imbibition–emergence phase, (b) the vegetative development phase, and (c) the reproductive phase.

The effects of temperature on reproductive development in beans have been known for some time. However, only recently have any serious efforts been initiated toward the incorporation of genetic tolerance to both high- and low-temperature stresses during bloom into improved cultivars. Farlow (37) found that the difference in ability to tolerate cold during flower development between two Australian cultivars was primarily due to a difference in incidence of ovule abortion. The rate of failure of the female reproductive organs was progressively higher as temperatures were reduced from 21° to 10°C. This resulted in fewer pods per plant, fewer seeds per pod, and more crooked pods. Dickson and Boettger (29) also observed that low night temperatures (8.5°C) cause "split set" in snap beans. They found that fewer pods and/or seeds per plant were produced at a night temperature of 8.5°C than at 18°C.

Sensitivity to high temperatures during bloom is one of the principal reasons why beans are not grown in much of the Southeast during June, July, and August. According to Farlow et al. (38) high daytime temperatures (>35°C) reduce pollen production and/or viability. The resulting split sets can be a serious problem to both the bean processor and the bean seed producer. Much additional information is needed to clarify the effects of high night temperature and the possible interactions with relative humidity and to identify cultivar × time × temperature threshold differences. Marsh et al. (68) and Weaver et al. (120, 121) are developing cost-and space-effective screening procedures to identify resistant individuals in segregating populations. Sources of tolerance to high temperatures during bloom have been reported (6, 78, 103, 125), but little is known about the mode of inheritance. It is possible to recover heat-tolerant single-plant selections from advanced-generation hybrid populations derived from a heat-tolerant breeding line. Silbernagel (99) released the heat-tolerant breeding line 5BP-7 in 1979. In 1982 (103) he released a pair of isogenic lines (derived from a cross with 5BP-7), one of which is sensitive and the other resistant to high temperature during bloom. These are being studied by Marsh *et al.* (*68*) to determine the nature and inheritance of heat tolerance. Tolerance appears to be simply inherited.

Numerous mechanisms and/or modifying factors for heat tolerance may be involved. Work by Farlow (*37, 38*) on the nature of temperature stress tolerance during reproduction in beans should have a great impact on future bean breeding. The widespread application of the information provided by these two papers in the development of new commerical cultivars may be a major contribution toward breaking the so-called yield barrier in bean production.

Flood Tolerance

The sensitivity of bean roots to oxygen starvation has been known for some time. Miller and Burke (*74*) and Noor *et al.* (*79*) have shown that the stress of temporary near-anaerobic conditions induced by flooding can alter root metabolism and greatly increase sensitivity to fusarium root rot. Resistance to temporary flood conditions is available (*63,79*), but little is known about its genetic inheritance or economic importance.

Some confusion exists as to the effects of sensitivity to cold emergence conditions vs. sensitivity to flooding and/or oxygen starvation. Ladror *et al.* (*63*) recently studied the interacting effects of cultivar, initial seed moisture content, temperature, oxygen content, and flooding. Much of the reduced emergence and seedling vigor, presumed to be due to O_2 starvation or low temperature during seed imbibition, appears to be due to flooding.

As the importance of temporary flooding injury to emergence and/or root rot resistance is better understood, tolerance to flooding should receive increasing attention from breeders.

Response to Cultural Practices

Response to Optimum Fertility

While yield stability due to disease and insect resistance and tolerance to environmental stresses is important, it is only meaningful when combined with high yield and processed quality. All too often, pest- and/or stress-resistant cultivars are not impressive in the absence of those pests and/or stresses. The cv. Red Mexican UI-36 is very sensitive to fusarium root rot. However, in soils free of root rot it responds extremely well to high fertility levels (*14*). Therefore, growers keep using it because without severe root rot it often outyields similar cultivars with root rot tolerance. Snap bean breeders need to maintain genetic factors for maximum productivity and quality under good growing conditions, as well as stress and disease conditions.

Response to High-Density Culture

In the early 1960s workers at Oregon State University and Cornell University (Geneva) began asking why beans were grown in 30- to 36-in. rows, thus wasting all that space between rows. They concluded that originally the wide rows were for the passage of horses, and later tractor tires and equipment used in cultivation. Now that we have effective herbicides and fungicides that can be applied by air or through overhead sprinkler systems, tractors are often not needed in the field between planting and green bean harvest. Spacing studies by Mack and Stang (*67*) showed that maximum production was obtained when each plant had an average of 36 in.^{2} of space in a nearly equidistant arrangement. Subsequent commercial experience has shown that populations of 160,000–170,000 plants/acre in various spacing arrangements can yield 8–12 tons/acre. This is a dramatic increase from the 4–6 tons/acre that the better Willamette Valley growers currently obtain with 32-in. rows. Current cultivars are not ideally suited to this production practice, and the characteristics that presumably would contribute to even more efficient and higher production levels are being identified.

Rapid uniform emergence and seedling development are the first requirements. The ideal plant type for high-density culture is a strong central upright stalk, with three or four narrow-angled branches, a high yield of pods all close to the same maturity, borne high on the outer periphery of the bush for ease of mechanical harvest. Pods should separate easily (unbroken), with no pod clustering or broken-off branches being carried into the picker. Leaves should be small, and oriented toward the sun to allow maximum light interception and penetration through the canopy. Plants should be 16–18 in. tall and strong enough not to lodge when heavy with crop. Vigorous roots resistant to diseases should anchor the plant well enough not to be pulled up by the harvester. The roots should nodulate profusely and be capable of high rates of early-season nitrogen fixation. At the early pin bean stage, when natural nodulation declines, the plant should respond to supplemental nitrogen fertilizer by maximizing pod development and yields, instead of renewing vegetative growth. Finally, the ideal bean should be compatible to minimum tillage practices. This would probably require *Pythium* resistance, since *Pythium* populations seem to increase under minimum tillage practices.

SELECTION TECHNIQUES FOR SPECIFIC CHARACTERS

General Considerations

Since there are many different markets for which beans are produced, there is no one set of selection criteria for specific characters that is applicable to all needs. Also, the environment under which the breeding, screening, and selection work is done may differ from the commercial production area environment; and there may be genotype–environment interactions. Therefore, it is essential to maintain trial nurseries for disease screening and plant/pod-type evaluation in both the seed production and the vegetable crop production areas.

After intensive early-generation screening for resistance to locally prevailing diseases in each area, there should be enough genetic diversity left for plant and pod characteristics in segregating populations to recover most of the desired agronomic–horticultural recombinants. Another reason for alternating early-generation disease screening between seed production and processing (or shipping) areas is that each area may have a completely different set of diseases related to the prevailing respective environments. In the western desert area where most seed production is concentrated, resistance to BCMV and CTV is highly desirable, although CTV is never a problem in the major processing or market garden (shipping) areas. Likewise, fusarium root rot causes the most serious root rot in the major northwest seed production areas, but in Wisconsin *Aphanomyces* and *Pythium* are the major root rot incitants. Rhizoctonia root rot is quite widespread in all production areas, but is apparently only economically serious in California, Arkansas, and the eastern seaboard from Maryland to Florida. Tolerance to cold–wet late-April planting conditions in Oregon may not be needed in an Arkansas mid-May planting. Blossom drop (split set), due to high temperatures during the flowering period, may prevent bean production through much of the Southeast during July and August, but is only an occasional problem in Wisconsin or New York. The ability to grow and set well under cool growing-season conditions may be desirable in cultivars destined for Canada or England, but is almost never needed in Arkansas. However, the Florida shippers who grow beans in December and January could use cold-tolerant cultivars. Thus, since no one test location represents all growing areas, most commercial breeders maintain trial grounds in four or five locations around the country. And since a breeder can only be in one place at a time, considerable regional technical support and collaboration is needed to plant and evaluate the early, midseason, and late crop responses within any given major production area.

Detailed multiple-harvest and location information is needed to evaluate advanced-generation materials properly (this will be covered in the section on Trials of Advanced Lines). The local evaluator assists the breeder in screening segregating populations for disease resistance and in selecting for desired plant and/or pod characteristics. However, the green-pod production area is usually not a good seed production area because of wet fall weather.

If seed from wet areas is returned to the breeders' trial grounds in the seed production area, there is a possibility of contaminating the rest of the breeding lines with seedborne pathogens, some of which may represent new strains and/or higher degrees of virulence, because they may have been screened on resistant or tolerant germplasm populations. Moreover, most evaluators in the green-pod production areas are not equipped with seed harvesting, drying, cleaning, and storage facilities. There are two ways of handling this problem: (1) seed being returned from any trial ground outside the breeder's nursery should first go through an isolation nursery (always a good basic procedure), or (2) only the trial information and not the seed should be returned. Option 2 (progeny testing) involves making numerous single-plant selections (SPS) in segregating populations in the breeder's (seed production area) trial grounds. The following generation each SPS is subdivided. Part of it remains in the breeder's trial grounds, and part goes to a green-pod production area trial. The evaluator in the green-pod production area trial records disease reactions and relevant plant and/or pod characteristics. If a particular SPS is well adapted and homozygous resistant in the green-pod production area trial, then further SPS can be made in the breeder's own trial grounds (in the seed-producing area), since the breeder knows SPS are resistant to a particular disease in the green-pod production area trial, and that the seed harvested will be free of seedborne problems. Furthermore, the green-pod production area trial evaluator does not need to worry about harvesting poor-quality seed in wet weather.

Cooperative and Regional Testing

Breeding programs differ in both the technically trained manpower capability and in the natural environment adequate to screen cultivars or segregating populations for more than a few factors. To offset these limitations, some public breeders, through the sponsorship of regional projects like WR-150 (*117*) (Genetic Improvement of Beans for Yield, Pest Resistance, and Nutritional Quality), screen materials for each other, usually under natural conditions. Other breeders (public and private) exchange materials for testing on a personal cooperative basis. Some provide only information, while others return resistant selections and comments to the originator. Private seed companies, large enough to afford regional test nurseries and trained personnel, do this within their own organization. Either way, regional testing of early-generation materials is important to the development of new cultivars that are not only resistant to various factors but widely adapted and highly productive.

When to Screen

Simply inherited factors can be recovered and stabilized at a relatively early (F_2 or F_3) generation. Conversely, the more complex the inheritance of the character sought and the lower its heritability, the longer (F_5 to F_8) rigorous selection should be delayed. However, even though selection efficency may be low in early generation for complex characters with low heritability, those characters should not be ignored in early generations. Postponement until F_8 with no selection pressure would build up populations to unmanageable proportions. Conversely, too much selection for other things may reduce or eliminate the genetic variability for the particular character sought. Thus selection pressure for complexly inherited characteristics, like root rot resistance, should be low in early generation to eliminate only highly sensitive individuals.

Controlled Screening Pressure

The severity or intensity of screening pressure applied to a segregating population depends both upon the level and complexity of the available resistance. If a high level of resistance is available and easy to recover, like *I* gene resistance to BCMV, then highly virulent strains like NL 2, 3, and 4 (*34*) should be used to eliminate all but the most resistant. If, however, the highest level of resistance available is an intermediate tolerance to something like halo blight, race 2, that is easily eliminated with more virulent isolates (halo blight, race 3), then the use of less virulent isolates (race 1) should be considered in order to be able to detect multiple sources of low-level tolerance, which might later be recombined in a search for accumulative or transgressive resistance. The combined higher level resistance might then be detected by screening with progressively more virulent isolates (race 2, then 3). If races or strains of a pathogen that differ in virulence are not available, then inoculum concentration or environmental conditions (temperature, moisture, inoculum density, etc.) might be manipulated to control the degree of screening pressure. Of course, under field conditions precise control of disease severity is not always possible. Greenhouse or growth chamber conditions are usually needed to obtain repeatable levels of controlled disease pressure.

Overreliance on only the highest level of resistance available to the most virulent isolate(s) might tend to shift the breeding population toward single-factor (and less stable) vertical resistance, while the use of all available factors, even resistance to low or intermediate levels of pathogenic virulence, might broaden the population resistance base toward multiple-factor or a more horizontal (and stable) form of resistance. Intercrossing of low-level tolerant lines might pyramid several sources of resistance to produce recombinant individuals with horizontal resistance levels that might then be able to tolerate more virulent pathovars under field conditions.

A screening program should be based on a thorough knowledge of the available variability in both the pathogen virulence and/or race specificity, as well as the host resistance. A lot of field-screening effort can be wasted trying to stabilize resistance to a pathogen like BYMV when the breeder is unaware there are many different strains and that they are not necessarily the same each season or testing cycle.

Field vs. Greenhouse Screening

Considerations of time, space, expense, and the requirements for large numbers often dictate that screening be done under field conditions. However, the need to know specifically to which strain(s) of BCMV, rust, halo blight, or anthracnose a breeding line has resistance eventually requires some confirmation with identified pathogens under controlled greenhouse or growth chamber conditions. Screening for certain characteristics that occur for only a short time period and/or need to be done at a specific stage of growth (like tolerance to cold–wet imbibition or heat during bloom) usually is best done under controlled conditions.

Single vs. Multiple-Disease Screening

Whether to screen for resistance to a single disease or to screen simultaneously for resistance to several diseases is a largely unresolved question. Many diseases (as well as different test conditions) are known to affect the expression of another disease; but if the purpose is to combine as many simply inherited factors (like BCMV and rust resistance) as quickly as possible, then simultaneous multiple-factor testing is sometimes the most efficient procedure.

However, with more genetically complex factors for resistance, like root rot, it may be better to rely on single-factor relay or parallel tests (*Fusarium, Pythium, Rhizoctonia, Aphanomyces*) in order to increase the probability of recovery of resistant individuals to each disease in small populations. If more than one of these resistance factors are needed in a single cultivar, eventually they have to be combined through hybridization and tested under artificial and/or natural mixed-pathogen conditions. By that time, however, the breeder knows which factors are present by previous single-disease testing, and the multiple-disease screening is merely a confirmation test, in which the emphasis can then also be turned to selection for plant and pod characteristics.

Disease Resistance

All breeding programs must at some point be concerned with diseases. A great deal of detailed information on screening and genetics is available in several current reviews [e.g., Schwartz and Galvez (*94*), Zaumeyer and Meiners (*133*), and Silbernagel and Zaumeyer (*106*)] and so only a brief summary will be presented here.

Virus Disease Screening

Most bean viruses encountered in the United States, while naturally transmitted by insect vectors, can also be mechanically transmitted. These include BCMV, BYMV, peanut stunt virus, alfalfa mosaic virus, southern bean mosaic virus, bean pod mottle virus, red node virus, and legume strains of CMV. Details of inoculum storage, preparation, and inoculation procedures are similar for all mechanically transmitted viruses. The methods described by Drijfhout *et al.* (*36*) for BCMV apply to any of the above. Basically, fresh, young, infected tissue is desiccated rapidly with silica gel (under refrigeration if possible), sealed against air and moisture, and kept frozen for 1–2 years. When needed, a small amount (¼ g) is ground in 2 ml 0.01 M neutral phosphate buffer (cold), poured through a double layer of cheesecloth, and used to inoculate a virus-susceptible buildup host. Test plants can be lightly dusted with 400- to 600-mesh carborundum prior to inoculation to facilitate cell penetration. Young tissues are best for inoculation (primary leaves when ½ to ¾ expanded). The severity of some virus diseases is intensified by holding the test plants in the dark 24–48 hr before inoculation. At greenhouse temperatures of 24°–28°C, symptoms are usually expressed in 10–14 days. When producing inoculum for screening segregating populations, susceptible virus buildup hosts usually reach the highest virus titer 2–3 days before symptoms appear (8–10 days after inoculation). When large amounts of inoculum are needed, infected tissue in buffer (1 g : 5 ml ratio) can be ground for 1 min in a Waring blender. Buffer and blender jar should be prechilled to 2°–5°C; and after screening through double-layered cheesecloth, the inoculum can be further diluted ⅟₁₀₀ with chilled buffer. For large-scale field or greenhouse inoculations of segregating populations, the carborundum can be added to the diluted inoculum at the rate of about 1 g/liter. The inoculum can be applied by a pressurized paint sprayer (50–100 psi) or with a hand-held 400- to 500-ml plastic bottle, over the neck of which a piece of Parafilm and several layers of cotton gauze are held by a rubber band [Francisco Morales, International Center for Tropical Agriculture (CIAT), personal communication]. The Parafilm is punctured with a pin until the desired amount of liquid keeps the gauze moist enough for inoculation use, but not to the point of runoff. The young leaves being inoculated are held in place by a firm plastic sponge or several layers of paper towels held in one hand by the applicator, while the damp gauze is lightly stroked over the leaf with the other hand. Under greenhouse or screenhouse conditions, susceptible individuals are easily identified in about 2–3 weeks. Symptomless plants should be inoculated a second time (on the youngest trifoliolates), especially if they are to be saved for seed production or counted in an inheritance study. Symptomless field plants should likewise be given a second inoculation (about 2 weeks after the first) to guard against escapes and late emergers being harvested as resistant. Final resistant selections (at young to full-pod stage) can be marked with wire flags, numbered tags, or simply a hand-held spray can of bright-red or fluorescent-orange paint. It is important to be sure that apparently resistant plants showing no leaf or growth depression symptoms also do not exhibit pod symptoms.

CTV is only transmitted by the sugarbeet leafhopper *Circulifer tenellus* Baker (7). It is a difficult insect to rear and work with, and so field exposure under natural conditions is the best way to screen for resistance. Susceptible sugarbeets planted in mid-March are used as a trap crop in western areas, where the insect overwinters on wild mustards in desert waste areas. As the mustards mature (mid-May), the insects migrate. Strips of two (22 in.) susceptible sugarbeet rows spaced 23 ft apart allow eight rows of beans spaced 22 in. apart to be planted between the beet strips. Beans planted during mid- to late May usually emerge in early June. By then leafhopper populations are increasing, especially if the weather has been hot and dry (*16*). By early July, a high percentage of the susceptible controls are usually infected, and resistant individuals in segregating populations are easily identified. Again, final selection is best delayed until the fresh-pod stage of maturity.

Bacterial Disease Screening

Whereas viruses are usually spread by insects, water and wind are the most common means of spreading bacterial diseases within the crop season. Between seasons, since they are all seedborne, humans transport them from one growing area to another, in spite of phytosanitary regulations.

The two *Pseudomonas*-incited diseases halo blight and brown spot both thrive under cool–wet growing conditions. However, the *Xanthomonas*-incited diseases common blight and fuscans blight [*Xanthomonas campestris* pv. *phaseoli* (var. *fuscans*)] are more prevalent under hot–humid conditions. Bacterial wilt is also favored by high temperatures, but usually under dry stress conditions.

Since the primary damage caused by the pseudomonads and xanthomonads is to leaves and pods, most screening techniques have been developed to evaluate resistance at the late-bloom or early-pod stages of growth. This requires a lot of space and time, which usually means field testing if large numbers are to be screened. In order to reduce the labor required in inoculating large segregating populations, sometimes every third row (a highly susceptible cultivar) is inoculated, which serves as a spreader source to the adjacent test rows.

Planting contaminated seed, as described by Poryazov (*84*), is an easy means of establishing infected spreader rows. If natural spread is not dependable enough, frequent use of an overhead sprinkler system or a power sprayer can be used to supplement natural spread. If the weather does not cooperate, a lot of effort can be lost in field plots.

Greenhouse or growth chamber tests can be controlled to ensure good testing conditions, but since indoor space is usually limited, it is difficult to work with large populations. Inoculation of seedlings in the crookneck stage (*101*) with a hypodermic syringe is time and space efficient for halo blight and brown spot screening (*101*). Schuster and Coyne (*92*) reviewed procedures to standardize screening of beans with the xanthomonads. Most inoculation methods use pressure sprays or vacuum to achieve leaf tissue infiltration. Needle punctures or cuts are also used in a variety of ways to achieve inoculation. Inoculum concentrations can influence disease reactions; most reported optima are in the range of 10^6–10^8 cells/ml. Symptoms develop in 10–20 days and are usually expressed on a 1 to 9 scale, with usable tolerance being in the 1 to 2 range.

Resistant and susceptible control cultivars are used as references to evaluate the degree and range of resistance found in the segregating population. In early generations, typically 5–10% of the test population might be saved for seed production. An increasing proportion of resistant survivors is saved for seed with each cycle of recurrent selection.

Inoculum is grown on YDC agar, which contains 15% agar plus 10 g yeast extract, 10 g dextrose, and 15 g $CaCO_3$ per liter. Forty-eight to 72-hr petri plate cultures are washed off the agar with sterile water or 0.01 M $MgSO_4$. Ouen Huisman (University of California–Berkeley, personal communication) found less bacterial cell disruption by osmosis in 0.01 M $MgSO_4$ than in water. For long-term storage under liquid culture, use 0.1 M $MgSO_4$ on agar slants or wash off and store without the agar. Stored inoculum should be revived and checked for virulence before being increased for screening use.

Excellent information on sources of resistance, inoculation techniques, rating systems, and heritability of resistance is reviewed by Schuster and Coyne (*92*) and Yoshii (*130*). Additional information as well as germplasm and cultures are available by contacting the authors of articles in recent annual reports of the Bean Improvement Cooperative.

Screening for Resistance to Foliar Fungi

The principal foliar infecting fungal diseases in the United States, in order of descending importance, are rust caused by *U. phaseoli* (Reben) Wint. [*Uromyces appendiculatus* (Pers.) Unger], white mold caused by *Sclerotinia sclerotiorum* (Lib.) de Bary, gray mold caused by *Botrytis cinerea* Pers. ex Fr., and anthracnose caused by *Colletotrichum lindemuthianum* (Sacc. & Magn.) Scrib. The latter three also destroy stem and pod tissue. Disease establishment and spread are favored by moderate temperatures (17°–27°C), >95% RH, and periods of wetness. Rust and anthracnose can infect plants at any stage of development when climatic conditions (cool and wet) are favorable; however, white and gray mold usually are not a problem until after bloom. For disease establishment, the mycelia produced by ascospores and conidia of white and gray mold require a food base (usually spent blossoms) before they can invade living stem, pod, or leaf tissues.

Gray Mold   No genetic gray mold resistance has been identified, but not much effort has been exerted looking for it. Some disease escape, due to plant architecture, may be operational under conditions of low disease pressure. In view of the importance of gray mold in Oregon and the lack of satisfactory chemical control, some effort should be made to find genetic resistance. The principles and methods employed to combat white mold should be a likely starting point for a campaign aimed at genetic control of gray mold. An Integrated Pest Management (IPM) bean program at Oregon State University, coordinated by Rick Weinzerrel (Corvallis, Oregon, personal communication) is attempting to develop the background information needed for practical control with minimal environmental pollution. The IMP approach, coupled with even a moderate level of genetic tolerance, should be successful for white as well as gray mold.

White Mold   It was long assumed that the only differences in genetic response to white mold in beans were due to different architectural configurations which allow a drier, warmer microclimate under the canopy (8, 53). Now that the details of environmental requirements for disease establishment and spread are more clearly understood, workers in Nebraska and New York have been able to differentiate between physiologic tolerance (58) and avoidance mechanisms related to plant architecture (20). No germplasm is entirely resistant; however, Hunter et al. (59) have been able to identify partial resistance with a limited-term inoculation method. They found that an ascospore spray procedure was less reliable than placing mycelium colonized pieces of celery or bean pod in contact with a host test plant internode for 15 hr at 21°C and >95% RH. After this disease exposure, the 3-week-old test plants are transferred to a greenhouse and rated for disease severity on a 1 to 9 scale 6–10 days later. Certain *P. coccineus* lines, and to a lesser degree some *P. vulgaris* lines, show physiologic resistance by taking longer to die. More recent work by Dickson et al. (32) shows this technique can be used to classify segregating hybrid populations for genetic tolerance to white mold. Many researchers believe a combination of genetic tolerance, architectural avoidance, judicious use of chemicals, cultural practices, and perhaps biologic control agents provide a realistic basis for long-term cost effective management of the disease.

Rust   Vargas (115) recently reviewed the status of breeding for rust resistance. The high degree of pathogenic variability in the bean rust fungus has inclined many workers toward efforts to base long-range control on a program of combining horizontal (or non-race-specific) resistance factors. The components of horizontal resistance are such factors as reduced numbers of infections, decreased pustule size and spore production, late or slow rusting, increased resistance with plant maturity, longer incubation period, and slower rate of pustule development. Research workers at CIAT in Cali, Colombia, have taken the lead in coordinating an International Bean Rust Nursery (93). Schwartz and Temple (95) suggested a CIAT breeding strategy for the development of rust resistance. In the United States, Meiners, (71) and, since his retirement, Stavely (108), in Beltsville, Maryland, have coordinated a Uniform Bean Rust Nursery for bean breeders. Efforts are underway to establish a uniform set of host differentials and a standardized disease rating system for race identification. Generally the scale developed by Ballantyne (5) is used.

Inocula collected from diseased plants can be kept frozen (sealed against moisture) for up to 2 years. The usual inoculation procedure consists of a spore suspension in water containing a few drops of Tween 20 detergent as a dispersant, sprayed onto foliage during cool–wet periods or evenings. Cordoba *et al.* (*18*) compared four inoculation methods and found that 0.12 g freshly collected uredospores diluted in 1.0 g of talc and dusted on the premoistened plants gave the most consistent results. Moderate numbers of uniform pustules are needed for effective separation of resistant segregants. In greenhouse tests, inoculated plants are held 18–24 hr in a mist chamber at 18°C before transfer to a greenhouse bench. Under field conditions, inoculation may be directly on the test plants or on highly susceptible spreader rows planted at frequent intervals (every third to tenth row). Spreader rows should be planted early and inoculated 1–2 weeks before the test plants emerge to provide ample inoculum for natural spread. A mixture of local races is usually used to screen segregating populations.

Anthracnose   Snap bean breeders or seedsmen in the United States have not been concerned with breeding for anthracnose resistance because the disease has been controlled for about 50 years by producing seed in semiarid areas of the Intermountain Region (between Rocky Mountains and Cascade Mountains) of the western states (Idaho, Washington, California). The arid summers, coupled with phytosanitary regulations restricting the introduction of infected stocks into the seed production areas, have been successful in virtually eliminating the disease from U.S. production.

In Europe, however, anthracnose is a common problem. The seed production areas of eastern Africa, where most European snap bean seed is grown, are also frequently contaminated. Consequently, the Europeans have made a concerted effort to find and breed sources of anthracnose resistance into their processing cultivars. American seedsmen need to incorporate anthracnose resistance into cultivars aimed at the European market.

Hubbeling (*56*) in 1957 found that the Cornell line 49-242 was resistant to all then known races of anthracnose. Mastenbroek (*69*) in 1960 found that resistance in C49-242 was due to a single dominant factor (*ARE* gene). For about 20 years this gene was used very extensively by the Europeans to develop a large number of resistant cultivars.

In 1973 Leakey and Simbwa-Bunnya (*64*) found a strain in Uganda that attacked the *ARE* gene. Other isolates were later discovered in Brazil (*80*) and in Germany (*90*) that also overcame the *ARE* gene. New races were also found in Malawi (*4*), but none of those overcame resistance in C49-242. There are several resistant breeding lines (Mex 222 and 227) that control the new strains however (*39, 49*); thus these new genes (*Mex-2* and *-3*) will eventually be combined with the *ARE* gene for more stable polygenic resistance.

Hubbeling (*57*) presented a useful key to anthracnose strain identification using differential bean cultivars. He suggested ways of combining the various sources of resistance to develop multigenic broad resistance. He also indicated that the old way of spray inoculation of the young seedlings just after emergence would allow better detection of more minor factors for disease resistance than seed inoculation or root dipping. Chaves (*15*) and Tu and Aylesworth (*109*) suggest various methods of inoculation and screening for anthracnose resistance.

Pure culture isolates often lose the ability to sporulate unless cultured on bean pod agar or sterilized bean pods or leaves. Use only spores suspended in sterile water (not mycelium) to transfer the culture. For inoculation, a water suspension of 200 conidiospores per milliliter is sprayed onto the emerging seedlings, which are kept at 100% RH and about 18°–21°C for 2–3 days. Symptoms develop in 8–10 days.

Field inoculation can be accomplished by sowing every third nursery row with infected seed 2 weeks before planting the test materials. Disease development is favored by daily, light overhead irrigation during dry weather periods.

Screening for Resistance to Root Rots

Root rots are probably the most ubiquitous chronic diseases of snap beans in the United States, and receive the least attention by commercial breeders for good reason—they are extremely difficult to control genetically.

There is little research data available on the actual costs of these diseases in terms of reduced yields, quality, or increased costs of production due to the use of fungicides or cultural practices like subsoiling or extra irrigations. However, in the seed production areas of the West (*100*), I would estimate yield losses alone are in the range of 500–1000 lb/acre (20–40% of potential).

Average industry snap bean seed yields are about 1500 lb/acre on old bean land. In soils free of root rot, it is not unusual to produce over 2500 lb/acre. The seed industry is so used to 1500 lb/acre that it is generally accepted as normal.

Breeding for root rot resistance is difficult because of (1) the lack of high levels of stable resistance in horticulturally acceptable plant and pod types, (2) the general association of colored seed coats and late maturity with resistance (until recently), (3) a lack of clarity as to the genetics of resistance (which generally has a low degree of heritability) or the nature of resistance, (4) and a lack of reliable screening techniques. The last item will be addressed in this section, primarily by describing the methods found most reliable over a period of years. However, much more work on screening methodology is still needed.

Environment and Medium   Most of our critical root rot screening work is done in a growth chamber where temperature, moisture, and lighting can be controlled. The chamber is kept at 60–70% RH and is unlighted until the beans emerge (about 5–6 days). High-intensity fluorescent lights provide about 1000 fc at 12 in. above the test medium surface for a 14-hr period daily.

Our testing medium is one part fine-grade quartz sand to three parts fine, horticultural-grade perlite (v/v). The medium is screened after each test, autoclaved 8 hr, and reused indefinitely. We use aluminum pans about 23 in. long × 18 in. wide × 4 in. deep. Each pan holds 20 liters of medium and has drain holes in the bottom. The predampened planting medium is 3 in. deep, and six rows of 20 seeds are planted 1.5 in. deep. The rows are 3.5 in. apart and the seeds ⅛ to ¼ in. apart in the row.

Replication and Incubation Time   Most tests with untreated seed (ten seeds per test line) are inoculated at planting and rated for disease severity 14 days later. Each pan contains test lines and two control cultivars, one resistant [NY 2114-12 (*119*)] and one susceptible (Goldcrop) to the pathogen being used. Three to six 10-seed replicates are needed for an accurate estimate of the root rot index for a test line. However, seed of a single plant selection is often in short supply and large numbers of lines need to be tested. Thus it is not always possible to do three replications per line. If the line rates well in comparison to the controls, it may be repeated a second or third time; however, highly susceptible test lines are usually discarded after the first test, unless there is another reason for retaining that particular line.

The 2-week time period from planting to evaluation is important from several points of view. Primarily the seedlings are young enough to survive transplanting for greenhouse seed production. Disease pressure must be sufficient in this time period clearly to separate susceptible from resistant lines. This is monitored by the disease index of the controls. On a 0 to 100 scale, susceptible controls are usually in the 60–80 range, while resistant controls are in the 20–30 range. Disease pressure regulation is primarily a function of inoculum density, temperature, and moisture. The requirements for each root rot disease are slightly different.

Rating Root Rot Severity At the end of the 2-week disease exposure period, emergence is recorded. The seedlings are dug, washed, and rated on a 0 to 5 scale as follows: 0, no disease; 1, trace; 2, fairly extensive hypocotyl and/or root surface lesions; 3, extensive external and slight internal decay (but still with some functional solid tissues); 4, severe external and internal decay with little or no functional hypocotyl or primary root tissue remaining (may be surviving on new secondary roots); 5, dead or dying, including those seed that rotted before emergence from *Pythium* or *Rhizoctonia*. *Fusarium* and *Aphanomyces* are not seed rotters. Be careful to distinguish between hard seeds that fail to emerge and those that are rotted due to the disease under study or secondary bacteria that invade dead seed. The disease index is calculated as follows:

$$\frac{100[0(\text{no. at }0) + 1(\text{no. at }1) + 2(\text{no. at }2) + 3(\text{no. at }3) + 4(\text{no. at }4) + 5(\text{no. at }5)]}{(\text{total plants}) \times (\text{no. of disease categories } -1)}$$

where the sum of the products [(number of plants in each disease category) × (numerical value of that category)] is multiplied by 100. That product is divided by the product of the total number of plants times one less than the number of disease categories used. This system allows for comparisons among researchers who use different disease severity ratings (e.g., 0–4, 1–5, 1–9) by putting all results on a 0 to 100 basis.

What to Save Category 0 is immune, 1 is highly resistant, 2 is moderately resistant, 3 is moderately susceptible, 4 is very susceptible, and 5 is highly susceptible. Plants with a rating of 0 or 1 are usually considered resistant enough to save for seed production. However, within a sample there may be enough variation due to various factors, especially late emergence, that the evaluator might be tempted to keep too many boderline 1s or 2s that are escapes. In order to differentiate between possible escapes and genetic resistance, it is necessary to concentrate on saving apparently resistant selections primarily from those test lines with an average disease index (DI) lower (more resistant) than the arithmetic mean of the two controls. Thus, if the resistant control DI = 25 and the susceptible control DI = 75, the control average DI =50. Those test lines with DI ≤ 50 are more likely to contain genetically resistant segregants.

The resistant survivors are transplanted to a greenhouse bench for seed production. The seed is increased the following season in a field nursery, where it is exposed to BCMV and CTV screening. Selection for plant and pod type from greenhouse-increased seed is often misleading because of the variable vigor of plants obtained from greenhouse-grown seed. Most of the seed from this field increase is grown in the root rot field the following season, where selection pressure for plant and pod type is applied. The seed harvested from the field root rot nursery is again screened for resistance to individual root rots in the growth chamber–greenhouse (winter season) to complete the testing cycle. Part of the seed of each bulk or SPS may be tested against *Fusarium* and/or *Pythium* one winter greenhouse season, then *Rhizoctonia* and/or *Aphanomyces* the following winter. We do not have the facilities or technical assistance to test for all four root rot pathogens each season.

Controls   The New York breeding line 2114-12 [Wallace (*119*)] is our most *Fusarium*-resistant control. It is also resistant to *Thielaviopsis, Rhizoctonia,* and *Pythium* (*83*), but it is sensitive to cold–wet imbibition. Line PI 165426 (black seed) is not quite as resistant to *Fusarium,* but it is also a very useful control because of its combined tolerance to cold–wet emergence conditions and root rots.

Oliver Norvell's PI 203958 (N-203) is also resistant to *Fusarium, Rhizoctonia,* and *Pythium,* but less so than NY-2114-12. The Wisconsin root-rot-resistant breeding lines 36 and 46 are resistant to *Aphanomyces* and *Pythium.* Susceptible controls might be any of the following: Early Gallatin, Tendercrop, Puregold, or Goldcrop.

Inoculum Preparation and Application   Essentially we use a modification of the Del Monte method for *Fusarium* screening (Roger Schmidt, San Leandro, California, personal communication). *Fusarium solani* f. sp. *phaseoli* sporulates readily on V-8 juice as follows: Add one part V-8 juice to four parts distilled water, autoclave, and dispense (20 ml) into sterile petri plates. Transfer a loopful of macroconidia or a small piece of sporulating culture medium aseptically to each plate. Rotate plate to mix inoculum throughout the medium. After 8–10 days at room temperature, blend the sporulating mycelial mats for 1 min (no longer) in a Waring blender, pour through double-layered cheesecloth, and adjust the macroconidia concentration in the effluent liquid to 200,000/ml (D. W. Burke, USDA, Prosser, Washington, personal communication). Usually several isolates are mixed to reduce chances of loss of virulence or strain specificity (although none has been reported for the bean *Fusarium*). This preparation is applied uniformly over the seed lying at the bottoms of the six trenches in each tray (described above) formed in the planting medium just prior to planting. A small hand-operated pressure pump is used to spray the inoculum at the rate of 1.0 ml/seed onto the trench walls and seed. After spray inoculation, the ridges are flattened causing the spore-laden trench walls to come together above the seed row. The trays are watered lightly (daily) and placed in the growth chamber at 21°C.

Aphanomyces euteiches f. sp. *phaseoli* and most *Pythium* spp. produce abundant oospores and/or sporangia on corn meal agar after 8–10 days at room temperature. For each 20-ml petri plate, add 100 ml distilled water and blend for 0.5 min in a Waring blender. The mixture is poured through double-layered cheesecloth and the oospore containing effluent is diluted to a concentration of 1000–4000 oospores per milliliter (Bill Pfender, University of Wisconsin, personal communication). This mixture is sprayed onto the seed in trenches at the rate of 1.0 ml/seed, in the same procedure described above for *Fusarium.* However, the growth chamber is kept at 26°C for *Aphanomyces* and 16°–20°C for *Pythium.* The trays are watered lightly by overhead sprinkler once daily until emergence and then to slight excess twice daily until evaluated.

Rhizoctonia solani inoculum standardization is more difficult. This pathogen can be cultured on finely granulated vermiculite particles permeated with V-8 juice, which are mixed into the perlite–sand test medium at the rate of 3% v/v. This method is also useful for *Pythium ultimum.*

The following inoculum preparation recipe is from John Kraft (USDA, Prosser, Washington, personal communication). Put 400 cc vermiculite into a 500-ml widemouthed Erlenmeyer flask, cap with aluminum foil, and autoclave 4 hr at 15 lb pressure. When flasks are cool, add 200 ml of a V-8 juice–distilled-water mix (1:4) to each flask, and autoclave another 2 hr. Inoculate the cooled flasks aseptically with a small bit of petri-plate-grown agar culture on two sides (opposite) of the vermiculite mass. Shake each flask daily to help spread the inoculum. After 10–12 days at room temperature, the flasks are well permeated by mycelial growth and the inoculum mass may be difficult to remove without cutting into smaller pieces with a long spatula. These are rubbed through a ⅛-in. mesh wire screen (hardware cloth) just prior to use. One liter (3% v/v) of this inoculum is mixed for 10 min in a small cement mixer with 30 liters of the predampened perlite–sand testing medium. The untreated seed is planted as described above, watered lightly, and placed in the growth chamber at 25°C. The *Rhizoctonia* tests are watered to slight excess daily until evaluated.

The *Pythium* tests, like the *Aphanomyces* tests, are watered once daily until emergence and then to slight excess twice daily until evaluation. The mobile zoospores produced by *Aphanomyces* and some species of *Pythium* travel in a film of water, and so slightly saturated conditions are optimal for uniform disease expression.

After the breeder has combined resistance to several of the root rots, it might be more efficient to screen for two pathogens at the same time. Dickson and Boettger (*27*) have combined *Fusarium* and *Pythium* screening. Workers in Wisconsin (*43, 81, 82*) have combined *Aphanomyces* and *Pythium* testing. Pieczarka and Abawi (*83*) tested several different disease combinations. In general, most combinations should be satisfactory except *Rhizoctonia* with *Pythium* or *Aphanomyces*. *Rhizoctonia* is a hyperparasite of *Pythium,* and so it would possibly do the same to *Aphanomyces* since it is also a member of the Pythiaceae.

The rest of this section will focus on selection methods for seed, seedling, root, plant, and pod characteristics. Included in these selection procedures will be considerations of environmental stress tolerance (heat, cold, drought), cultural practices (mechanical harvesting of green pods and/or seeds, high and/or low fertility levels, herbicide compatibility), and harvestability factors (maturity, trash, plant flow, and case recovery).

Seed Characteristics

Single plant selections and small bulks can be compared for seed yield and quality. Those lines with comparatively low yield, highly variable seed size and shape, shrunken poorly developed seed, or a high proportion (> 2%) of seed coat rupture should be categorically discarded, providing there is no other overwhelming reason to keep a particular line. The remaining lines are then given Dickson's (*23, 25*) nick test for tightness of seed coat adherence, the frequency of transverse cotyledon cracks, and thickness of the seed coat. Next, the best candidate lines are given a seed test for rate of water imbibition as recommended by Dickson and Boettger (*30*). They found a too-rapid rate of water uptake to be correlated with poor stands and weak seedlings and suggested elimination of both problems by selection of semihard seed. The procedure I have adopted is as follows: Apparently sound seed (previously dried to 6% seed moisture) is soaked for 12–24 hr at room temperature. Those lines that take up water within a few hours are discarded. Only those that imbibe after about 12 hr are saved as semihard seed with delayed imbibition, which was found to be correlated with resistance to mechanical injury (*30*). Those that are still hard after 24 hr are discarded as hard seed.

Resistance to mechanical damage is also rated with the same previously dried lot of seed (6% compared to a 14% moisture control lot, fresh-weight basis) dropped several times onto an inclined steel plate from about 2 m (*26*). The smaller the seed quality difference between dropped and not-dropped seed of the two moisture levels, the more tolerant the line is to mechanical injury. Seed damage can be estimated by comparing the percentage of broken seed and the percentage of hairline cracks found via the water test or by standard germination tests. The seed coat crack (water) test is done with several replications of 100 apparently sound seeds placed in water at room temperature. After 2–3 min, those with hairline seed coat cracks wrinkle in the vicinity of the crack. Sound seed takes much longer to begin imbibition through the micropyles or hilum (*62*).

The standard germination test (*110*) consists of several hundred seeds in wet sand (20% moisture), perlite, or vermiculite, or in rolled paper towels at about 21°C. After 7 days, those seedlings with the equivalent of at least one sound primary leaf, one cotyledon, a normal shoot and root tip, and that are at least half the normal size, are counted as germinated. If more detailed information is required, the seedlings can be classified as to the percentage of healthy, vigorous, normal (HVN) seedlings (*98*). Then the product of percentage emergence multiplied by percentage HVN seedlings is used to develop a seed quality estimate (SQE). Lines can be even more critically evaluated by the seed quality index (SQI), which is the product of the seed emergence index (percentage emergence × rate of emergence) × the percentage of perfect seedlings (*98*).

Seedling Characteristics

In the field, seedling stand counts and early-season vigor can be estimated on a 1 (excellent) to 9 (poor) scale. It is helpful to note the frequency and types of seedling abnormalities. There may also be noteworthy differences in response to herbicides, or differential reactions to an unusually cold or hot, wet or dry emergence period.

Root Characteristics

Roots are usually examined in conjunction with evaluation for disease and/or insect damage. However, this same material may be also rated for root size and vigor. Because little is known about what constitutes an ideal root system, a simple 1 (best) to 9 rating system, relative to the most successful cultivars in a given situation, may be a safe starting point. Besides the obvious absolute size (by weight, length, and/or volume), the rate of root development is perhaps its most important characteristic. This is especially true under adverse environmental extremes of heat–cold, drought–flood. The rate may be estimated in terms of gain per day relative to 1 g of initial seed weight (*3, 97*). This would differentiate between two lines with identical absolute root size and apparent vigor, but which had different initial seed weights. A large vigorous root is usually associated with a large vigorous seed, but snap bean processors do not necessarily want a large-seeded cultivar. Therefore, if a small-seeded cultivar develops as large and vigorous a root system as a large-seeded cultivar, within a limited time period, its roots had to develop at a faster rate.

Next in importance may be a cultivar's ability to rapidly regenerate secondary roots along the basal portion of the hypocotyl and to replace those lost to various biotic or abiotic factors. This characteristic may be an important supplemental or contributing factor to root rot (or insect) tolerance under field conditions.

The roots can also be given a quick evaluation, 1 (best) to 9, to estimate the relative amount and size of the rhizobial nodules.

Plant Characteristics

The single most important plant characteristic the breeder looks for is a strong upright plant habit (or architecture). The plant habit must be stiffly upright and strong enough to hold the pods well off the ground and to withstand mechanical harvesting without breaking. By growing beans in high density under sprinkler irrigation, the weak plants are identified. Care must be taken to avoid selecting those which remain standing under these conditions only by virtue of low yields.

Plant height may vary from 16 to 35 in. depending on cultivar and growing conditions. Most commercial snap beans developed for 36- to 40-in. rows are 20–22 in. tall at harvest. However, underhigh-density culture (6- to 18-in. rows and 2–6 in. between plants), many of these same cultivars are too weak and are thus unable to keep the pods off the ground just before harvest.

For high-density culture, cultivars are needed that respond to optimal fertility levels and cultural practices by increased pod fill and higher yields, rather than excessive renewed vegetative growth. The cv. Early Bird and the USDA breeding line 8BP-8 respond in this way. Under normal row spacing (32–42 in.) and fertility, these types will be 16–18 in. tall at harvest. They should be produced on good land at optimal fertility and moisture levels. On poor soils, plants may be too short and pods too close to the ground for mechanical harvesting.

Flowering should be profuse and spread over a period of about 7 days. Fruiting nodes should be high enough above the soil (mid to upper part of the plant) to keep pods free of spoilage from soilborne organisms. Flowering branches and/or peduncles should be relatively short and strong to prevent pod weight at harvest from bending them to the ground.

If the breeder is field screening for high-temperature setting ability, plantings should be timed so that most of the bloom period is likely to occur when daily maximum temperatures are over 35°C. If the breeder wants to screen for the ability to fill pods well (no seed skips), blooming plants should be exposed to low night temperatures of 10°C. Growth chamber or phytotron screening is much more reliable, but generally less available.

As a general rule, any plant that is well adapted to high-density culture will also perform satisfactorily under "normal" conditions. However, not all plant habits suitable for "normal" row spacing will respond well under high-density conditions. Therefore, the breeder should consider making selections for plant and pod characteristics under high-density, high-fertility conditions, as well as currently used plant spacings and fertility levels.

Pod Characteristics

Selection for pod characteristics in the field is complicated by differences in maturity, and so it is rarely possible directly to compare lines and/or cultivars at the same time. Moreover, even among plants within any particular line, the potential harvest period extends over 4–6 days, during which time the yield, sieve size distribution, and raw-pod quality are constantly changing. An awareness of where a line is, in terms of this harvest period, is important in order to temper the selection criteria relative to the pod characteristics. The length of this harvestable period and how rapidly the raw-pod quality changes is referred to as a cultivar's "holding" ability. From flowering to harvest, pods increase in sieve size and weight to an optimum harvest point where yields level off and then begin to decline as quality (in terms of percentage of seed, percentage of fiber, and flesh firmness) starts to deteriorate. During the early potential harvest period, quality is high (dark, small-sieved, firm pods), but yield is low. Late in the potential harvest period yield is high, but quality may be borderline or low. A breeder making selections estimates whether a line is in the early or late stages of potential harvestability, but the actual determination of yield potential, percentage sieve size distribution, and holding ability can only be accomplished by sequential replicated yield trials at several dates and plantings, in several locations, for 2–3 years.

Because the potential harvest period during which selection for pod characteristics can be made is relatively short, most breeders schedule different nursery plantings to mature at different times. In this way, they are able to look at more materials at the right time. Most breeders need extra helpers during this fleeting period to assist in making selections.

Plants in many nurseries set up for selection work are spaced 4–6 in. apart in the row to facilitate access and to allow each plant equal opportunity to express its full potential. Rows usually vary from the 22-in. spacing used in the seed production areas of the West, to the 30- to 40-in. spacings most frequently used in the major processing areas.

With the increasing trend toward high-density culture for processing, some breeders are also making selections in narrow rows (11–18 in.) with spacing of 3–6 in. within the row. Care must be exercised in any situation to avoid selection of plants that are on ends of rows or near an open space within the row. These are usually more vigorous and productive because they are able to take advantage of additional light, fertilizer, and/or water because of the reduced population competition.

Ideally each plant should be judged when the pods are at the balanced optimum between yield and quality. Since this is not always possible, a considerable amount of "guesstimation" is required by the evaluator making selections. The evaluator must imagine what the pods on this plant looked like a few days ago if it is over prime, or if the plant is slightly immature, what the pods may look like a few days later when the plant is at its prime.

The pod selector looks first for maximum yield potential, concentrated maturity, and proper placement on the bush. Pods should be well above the soil, with most borne in the mid to upper part of the plant and mostly on the outer periphery in order to be well adapted to mechanical harvesting (or even hand-picking). Once the selector determines a plant has the above qualifications, the pods are critically examined. The selector looks for pods that are straight, smooth, and round; with uniform internal and external color.

Pod cross-sectional shape is classified as round, heartshaped, slightly oval, to oval or flat. Unfortunately, a given cultivar or selection rarely retains the same cross-sectional shape in all sieve sizes. Therefore, roundness is usually relative to sieve size and type. Most Tendercrop-type cultivars are round podded in 2–4 sieve, but may be "crease-backed" (narrower suture to suture vertical diameter than in horizontal diameter) in 5–6 sieve. To avoid this, some breeders select a slightly oval to heart-shaped pod in 3–4 sieve, so that at full maturity (5–6 sieve) the pod is round in cross section instead of creasebacked. Basically, the objective is to have mostly round pods when the line "peaks-out" in terms of optimum yield, sieve size distribution, and processed quality. Cultivars differ in the rate of reversion to a flat pod shape (presumably by mutation). Seed companies spend a lot of time and effort trying to reduce the frequency of this defect, by roguing (physically removing off-type plants during the growing season), single-plant selection, and mechanical precision sizing (Calvin Lamborn, personal communication, Gallatin Valley Seed Co., Twin Falls, Idaho). Most have established tolerance limits for each stage of seed production increase, aimed at providing the processor with less than 2% flats in the final processing crop. At the breeder's seed stage, a maximum of eight plants per thousand is a good rule of thumb. Seedlots of most cultivars have to be replaced every 3–5 years, to keep the frequency of flats (and string mutants) within acceptable limits.

Processors object to flats because they are usually overmature in relation to the round-podded sieve size with which they are found. The presence of strings also seriously lowers quality grades.

DESIGN OF THE COMPLETE BREEDING PROGRAM

The typical bean breeding program includes a broad scope of specific and nonspecific objectives, each at various stages of planning, execution, revision, and completion.

The principal nonspecific objective is to identify a wide range of useful genetic diversity for plant, pod, and seed characteristics, pest and disease resistance, environmental stress tolerance, physiologic capabilities, etc.

Sources of genetic diversity include the breeders' own collections, many accessible public and/or private breeders' collections, current and old cultivars, and heirloom collections, often maintained by the National Seed Storage Laboratory (Fort Collins, Colorado), and a vast collection of exotic or wild materials available through the USDA Western Regional Plant Introduction Station (Pullman, Washington) and CIAT (Cali, Colombia).

Another nonspecific aspect of any breeding program is a working knowledge of the inheritance of the available genetic diversity. This develops through experience and study of lists of genes, and other literature citations that enable the breeder to review what is known about the inheritance of particular characters. The germplasm committee of the Bean Improvement Cooperative (BIC) has compiled such a list of bean genes and literature citations. The latest bean gene "catalog" is in the 1982 BIC Annual Report (88). This list is quite comprehensive and is in the process of being reviewed, and so it will not be repeated here.

After a specific breeding objective is defined in terms of a desired phenotype, the germplasm pool is searched for the parents needed to obtain that particular goal. Sometimes the required parents first need to be developed via a preliminary or "prebreeding" program, i.e., if one of the required characteristics is only available in another species or in a wild *P. vulgaris* line from the tropics with late maturity, black seeds, fibrous pods, and photoperiod sensitivity for flower initiation.

After the required parents are identified or developed, the breeding strategy is outlined on the basis of the available genetic information (Fig. 7.1). The segregating populations from the hybridized parents are scheduled for appropriate screening to certain diseases or environmental resistance factors at specified generations. The selection for horticultural and agronomic characteristics is also scheduled for specific generations and usually in a defined environment and sequence.

When the target combination of characteristics is finally assembled in advanced-generation single-plant selections, the most promising candidates are multiplied, evaluated under a wide range of conditions, and eventually released as a breeding line or a named cultivar.

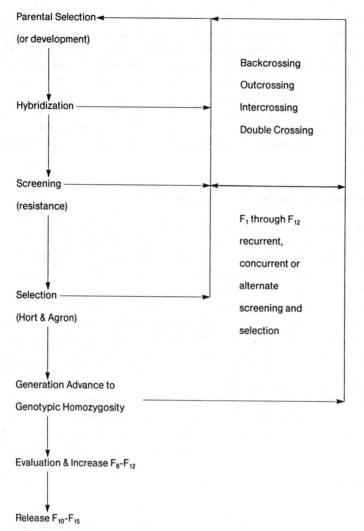

Parental Selection

(or development)

Backcrossing

Outcrossing

Hybridization

Intercrossing

Double Crossing

Screening

(resistance)

F_1 through F_{12}

recurrent,

concurrent or

alternate

Selection

(Hort & Agron)

screening and

selection

Generation Advance to

Genotypic Homozygosity

Evaluation & Increase F_6-F_{12}

Release F_{10}-F_{15}

FIGURE 7.1. Flowchart of typical breeding program.

As an example of one of many possible approaches to the development of an actual breeding program, I will outline a hypothetical program for the development of an Eagle-type canner for the Midwest with resistance to aphanomyces root rots and bacterial brown spot. Eagle (Asgrow) is a popular, high-yielding, widely adapted bush snap bean with BCMV resistance. Root rot caused by *A. euteiches* Drechs. f. sp. *phaseoli* Phend. & Hag., and bacterial brown spot are serious production-limiting diseases in Wisconsin. Resistance to aphanomyces root rot has been identified in Wisconsin Root Rot Resistant #36 and #46 (*43*). Bacterial brown spot resistance was identified in several Wisconsin germplasm releases, including BBSR-130 (*42*), -17, and -28 (*44*). Wisconsin RRR 36 and BBSR 130 both have the highest levels of resistance to aphanomyces and brown spot, respectively. However, because neither line is very close to commercial snap bean "type" in terms of plant, pod, and seed characteristics, one or two backcrosses to Eagle will be necessary to recover the characteristics required for commercial acceptance.

Objective: An Aphanomyces- and Brown-Spot-Resistant Eagle-Type Canner
for the Midwest

Cycle I.   Parental Selection and Initial Hybridization
Phase 1. Cross number
1 Eagle (BCMV resistant) × Wisconsin 36 (aphanomyces resistant)
2 Eagle × Wisconsin 130 (brown spot resistant)
Phase 2. Grow out F_1 plants to produce F_2 seed
Phase 3. Disease screening:
a. Screen cross 1 F_2 populations for aphanomyces resistance
b. Screen cross 2 F_2 populations for brown spot resistance
Phase 4. Identify best plant and pod-type single plant selections (SPS) among
resistant progeny of each population for next crossing cycle, save
resistant F_3 seed
Cycle II.   Development of Two-Factor-Resistant Eagle Populations
Phase 1. Backcross resistant F_3 SPS from each cycle I population to Eagle;
cross number
3 Eagle × cross 1 (aphanomyces-resistant F_3 SPS)
4 Eagle × cross 2 (brown spot-resistant F_3 SPS)
Phase 2. Grow out BC_1 F_1 plants to produce BC_1 F_2 seed
Phase 3. Disease screening of BC_1 F_2 populations:
a. Screen cross 3 BC_1 F_2 population for aphanomyces resistance;
save resistant F_3 seed
b. Screen cross 4 BC_1 F_2 population for brown spot resistance; save
resistant F_3 seed
Phase 4. Disease screening of BC_1 F_3 populations:
a. Screen cross 3 BC_1 F_3 aphanomyces-resistant population for
aphanomyces again, plus BCMV resistance, simultaneously
b. Screen cross 4 BC_1 F_3 brown spot-resistant population for brown
spot again, plus BCMV resistance, simultaneously
Phase 5. Identify best plant and pod-type SPS among phase 4 resistant progeny of each population to identify two-factor-resistant BC_1 F_4 seed
for next crossing cycle
NOTE: If phase 5 SPS materials do not yet resemble Eagle enough
to provide required commercial cultivar characteristics, backcross
them to Eagle again, repeating cycle II, phases 1–5 before
proceeding
Cycle III.   Development of Three-Factor-Resistant Eagle Populations
Phase 1. Intercross best BC_1 F_4 aphanomyces- and BCMV-resistant SPS
from cross 3 to best BC_1 F_4 brown-spot- and BCMV-resistant SPS
from cross 4 and reciprocal; cross number
5 Cross 3 aphanomyces- and BCMV-resistant SPS × cross 4
brown-spot- and BCMV-resistant SPS
6 Cross 4 brown-spot- and BCMV-resistant SPS × cross 3 ap-
hanomyces- and BCMV-resistant SPS
Phase 2. Grow out IC_1 F_1 plants to produce IC_1 F_2 seed
Phase 3. Sequential and dual-disease screening:
a. Screen half of each IC_1 F_2 population to aphanomyces and half
to brown spot to produce single-factor resistant IC_1 F_3 seed
b. Screen the single-factor-resistant IC_1 F_3 populations from each
test in phase 3a to the other disease to identify dual-factor resistant IC_1 F_4 seed (aphanomyces and brown spot resistance).

c. Screen the dual-factor resistant IC_1 F_4 populations from phases
3a and 3b to BCMV; save triple-factor-resistance IC_1 F_5 seed
(BCMV, aphanomyces, brown spot)
d. Screen triple-factor resistant IC_1 F_5 population to aphanomyces
and brown spot simultaneously; save resistant IC_1 F_6 seed
e. Field screen triple-factor resistant IC_1 F_6 population for BCMV;
make numerous SPS for superior plant and pod characteristics in
seed production area; save resistant IC_1 F_7 seed of superior SPS
f. Rapidly increase two generations to produce IC_1 F_9 seed
g. Field screen each SPS IC_1 F_9 population simultaneously in seed
production area for BCMV, and in Wisconsin (Hancock) for
dual resistance to aphanomyces and brown spot; after identifying
dual-resistant, well-adapted, high—yielding lines in Wisconsin,
make numerous SPS within those identical lines planted in the
seed production area nursery; save triple-factor resistant IC_1 F_{10}
SPS seed
h. Multiply best IC_1 F_{10} SPS for increasingly detailed evaluation in
Wisconsin over next 4–5 years to identify the best candidate for
eventual cultivar release.

The whole program requires a minimum turnover of about 21 generations, which at 4
months per generation would be 7 years. However, more realistically 9 or 10 years would
be considered a fast time. A lot depends on how much greenhouse space, time, and
technical assistance is available for this and the many other programs the breeder has
going simultaneously.

TRIALS OF ADVANCED LINES

Intensive disease screening from F_2 to F_6 is usually enough to stabilize factors for disease
resistance, but it also eliminates much of the phenotypic variability in the breeder's
screening trials. From F_6 onward, SPS are identified that show promise for plant and pod
characteristics in the field. Elimination of lines with low seed yield or with seed quality
defects (evaluated between field crops) helps reduce the number of lines that return to the
field each season for further seed increase and evaluation of horticultural and agronomic
requirements.

Those SPS lines that survive two seasons (F_7 and F_8) of close scrutinization in observation trials are advanced in the third year (F_9) to a small-scale preliminary processing
evaluation (one four-row × 20-ft replication), which has enough material to freeze and
can samples from two different harvest dates. If the processed products look good, the
following season (F_{10}) there should be enough seed for replicated yield trials with several
dates of planting (early–midseason–late) and several sequential harvests at 2- to 3-day
intervals of each planting. This may require 3–7 lb of seed. Single-row plots of 10–20 ft
are replicated three to six times for each planting and harvest date. The processed samples
are critically compared to standard cultivars to ensure processed quality (*41*) is as good or
better than current production cultivars (Fig. 7.2). Concurrently, enough seed has to be
increased for future testing in case the line continues to look good in the replicated
processing trials.

At least two seasons (F_{10} and F_{11}) of detailed replicated trials are required to identify
those lines that are worth sampling extensively (¼- to ½-lb observational plantings) to
cooperators in many other locations in the processing areas. If a line looks good in the

		Date Harvested _____
Variety	Sieve Size	
Location	Canned-frozen Sample	Date Evaluated _____

Style Pack: Whole-Cut

Quality Evaluation:

Appearance:	Attractive	5	4	3	2	1	Poor
Liquor:	Clear	5	4	3	2	1	Discolored
Pod Color:	Green	5	4	3	2	1	Gray
Suture Color:	Green	5	4	3	2	1	Brown
Defects:	Low	5	4	3	2	1	High
Flavor:	Good	5	4	3	2	1	Poor
Texture:	Firm	5	4	3	2	1	Mushy
String:	No String	5	4	3	2	1	Stringy
Sloughing:	None	5	4	3	2	1	Excessive
Carpels:	No Slippage	5	4	3	2	1	Excessive

Quality Rating, Total [   ]

Comments:

10 seed length	mm		
Seed Range	min-max		
Deseeded Pods Wgt.	0.00 g		
Seed Wgt.	0.00 g		
% Seed		Seed Grade	
Fiber Basket #			
After Drying Wgt.	0.0000 g		
Before Drying Wgt.	0.0000 g		
% Fiber-Blender Method (diff in g)			
Fiber Grade			

FIGURE 7.2. Processed quality evaluation form.

processing areas, a processor may next run small-scale replicated trials (10–25 lb/location) or may be ready to try a 5- to 10-acre commercial run. After 2–3 years of commercial trials (F_{12} to F_{14}) enough is usually known about the line to name and promote it officially, reselect within the line, or drop it.

Often the bulk lot that was a SPS in F_6 is increased for critical evaluations simply because there is enough seed available. If the multilocation trials (F_{10} and F_{11}) indicate there is too much phenotypic variability (often the case), then the bulk seed from the original F_6 SPS is replaced by increases from one or more SPS made in F_7 to F_{10}. These are evaluated by the same process described above, until a superior, genetically stable line is identified.

From the time an SPS increase is recognized as superior until it is adequately evaluated and increased for release takes another five to seven generations. Winter increases in the tropics or the greenhouse can speed up the process. So can double-cropping in the southwestern states, i.e., the first crop is planted in late March for harvest in July, the July-harvested seed crop is replanted immediately, and the second crop harvested in late October. Of course, there are considerable risks associated with any of these rapid-increase options.

The main objective of the small-scale processing trials is basically to be able accurately to describe the new line to a shipper or processor (potential customer). Trial data are usually expressed in terms of comparisons with a local standard. A processor needs to know the maturity, sieve size distribution, yield, and quality of a line before deciding whether it might be of use.

Yield, maturity, and sieve size data alone are absolutely worthless unless related to quality; and since different end-product uses require different quality standards, it is necessary to know the requirements of each particular customer. This is where information from area salespeople is essential to the breeders and their trial ground assistants. However, no two processors have the same requirements, and any one processor can change requirements overnight if marketing pressures warrant it. Thus an attempt has been made by Silbernagel and Drake (*105*) to enable evaluators to standardize reporting of yield, maturity, and sieve size data at the point of maximum yield and quality. This is hard to pinpoint, but basically is the point at which quality goes from fancy to extrastandard in terms of seed development (seed index). Seed development can be used because other quality factors such as suture and pod wall fiber development can be related to seed development.

Of course, quality (*41*) also includes flavor, texture, carpel separation, skin sloughing, interlocular cavitation, internal tissue breakdown, and color. However, these factors can be evaluated later in processed product trials of the better lines (Fig. 7.2) that pass the preliminary evaluations based on simpler quality-screening techniques, such as the seed index. For a 5-sieve type like Early Gallatin, this would be when the seed index for 5-sieve pods reaches 100. The seed index is the product of percentage seed by weight times average seed length in millimeters. Thus, if percentage seed in the green pods by weight is 10, and the average seed length is 10 mm, the seed index is 100. Several sequential harvests at 2- or 3-day intervals (before and after optimum harvest) are needed in order to chart the increase in yield and changes in sieve size distribution. This information, in view of changes in seed index, identifies the "optimum" harvest time in terms of any particular quality level desired by the processor. This information also suggests a line's holding ability, i.e., how long it stays in a harvestable condition. This is important if harvests are interrupted by bad weather or if for any reason the receiving plant "gets behind" during pod harvest.

Two or 3 years' data at several locations are desirable to determine if a line is worth increasing for commercial processor trials. However, most evaluators know how well their trial ground results correlate with crop development in other areas, and so local small-scale trials are not always conducted in all production areas by all companies.

Processing plants may consume 3–30 tons/hr, and so they require enough raw product (5–20 acres) to keep track of a new cultivar as it goes through the plant. Plant managers are very conscious of how many tons/hour of their standard production cultivars go through the plant, and of their case recovery, i.e., how many cases of 303 cans are recovered per ton of raw product. A new cultivar may be outstanding in all preliminary or small-scale trials, but be eliminated on the basis of tons/hour plant flow or case recovery figures. This information is generally only obtainable after a line is in fairly large volume and considerable investment has gone into its development. Nevertheless, the majority of lines that reach this level of testing are finally named and released.

In spite of all the previous seed company testing, processors must decide whether or not a new cultivar will be to their advantage. The reactions and impressions of the company field staff, plant manager, quality control manager, and sales department manager are all considered. If there is no decisive economic incentive to change, they will usually stay with what they have. The final evaluation criterion on which the decision to change cultivars usually hinges is a reliable yield of money (profit) to the processor. One outspoken breeder for a major seed company claims to know when this goal has been achieved by observing only one evaluation statistic: signed orders for seed.

REFERENCES

1. Atkin, J. D. 1972. Nature of the stringy pod rogue of snap beans, *Phaseolus vulgaris*. Search Agric. *2* (9), 1–3.
2. Atkin, J. D., and Robinson, W. B. 1972. Nature of the flat pod rogue of snap beans, *Phaseolus vulgaris*. Search Agric. *2* (9), 4–9.
3. Austin, R. B., and MacLean, M. S. 1972. A method for screening *Phaseolus* genotypes for tolerance to low temperatures. J. Hortic. Sci. *47*, 279–290.
4. Ayonoadu, U. W. U. 1974. Races of bean anthracnose in Malawi. Turrialba *24*, 311–314.
5. Ballantyne, B. 1974. Resistance to rust (*Uromyces appendiculatus*) in beans (*Phaseolus vulgaris*). Proc. Linn. Soc. N.S.W. *98*, 107–121.
6. Benepal, P. S., and Rangappa, M. 1978. Screening beans (*Phaseolus vulgaris* L.) for tolerance to temperature extremes. Annu. Rep. Bean Improv. Coop. *21*, 9–10.
7. Bennett, C. W. 1971. The Curly Top Disease of Sugarbeet and Other Plants, Monogr. No. 7. Am. Phytopathol. Soc., St. Paul, MN.
8. Blad, B. L., Steadman, J. R., and Weiss, A. 1978. Canopy structure and irrigation influence white mold disease and microclimate of dry edible beans. Phytopathology *68*, 1431–1437.
9. Bliss, F. A. 1980. Common bean. *In* Hybridization of Crop Plants. W. R. Fehr and H. H. Hadley (Editors), pp. 273–284. Am. Soc. Agron. and Crop Sci. Soc. Am., Madison, WI.
10. Bravo, A., Wallace, D. H., and Wilkinson, R. E. 1969. Inheritance of resistance to Fusarium root rot of beans. Phytopathology *59*, 1930–1933.
11. Burke, D. W. 1981. Yield response of several bean types to a complex of cold soil, wheat crop debris, drought, and Fusarium root rot. Annu. Rep. Bean Improv. Coop. *24*, 48.
12. Burke, D. W. 1982. Registration of pink beans Viva, Roza, and Gloria. Crop Sci. *22*, 684.
13. Burke, D. W., Hagedorn, D. J., and Mitchell, J. E. 1970. Soil conditions and distribution of pathogens in relation to pea root rot in Wisconsin soils. Phytopathology *60*, 403–406.
14. Burke, D. W., and Nelson, C. E. 1967. Response of field beans to nitrogen fertilization on *Fusarium*-infested and noninfested land. Bull.—Wash. Agric. Exp. Stn. *687*.
15. Chaves, G. 1980. Anthracnose. *In* Bean Production Problems. H. F. Schwartz and G. E. Galvez (Editors), pp. 37–54. Centro Internacional de Agricultura Tropical (CIAT), Cali, Colombia.
16. Clark, R. L. 1968. Epidemiology of tomato curly top in the Yakima Valley. Phytopathology *58*, 811–813.
17. Copeland, L. O. 1975. Mechanical damage in bean seed. Proc. Bean Improv. Coop. Natl. Dry Bean Counc. Meet., 1975, pp. 15–20.

18. Cordoba, J. V., Steadman, J. R., and Lindgren, D. T. 1980. Evaluation of methods for inoculating beans with rust urediospores. Annu. Rep. Bean Improv. Coop. *23*, 46–47.

19. Coyne, D. P., and Schuster, M. L. 1978. Halo blight resistant green bean line Nebr. HB-76-1. Annu. Rep. Bean Improv. Coop. *21*, 54.

20. Coyne, D. P., Steadman, J. R., and Schwartz, H. F. 1978. Effect of genetic blends of dry beans (*Phaseolus vulgaris*) of different plant architecture on apothecia production of *Sclerotinia sclerotiorum* and white mold infection. Euphytica *27*, 225–231.

21. Deakin, J. R. 1974. Association of seed color with emergence and seed yield of snap beans. J. Am. Soc. Hortic. Sci. *99*, 110–114.

22. Deakin, J. R., and Dukes, P. D. 1975. Breeding snap beans for resistance to diseases caused by *Rhizoctonia solani* Keuhn. HortScience *10*, 269–271.

23. Dickson, M. H. 1975. Inheritance of transverse cotyledon cracking resistance in snap beans (*Phaseolus vulgaris* L.). J. Am. Soc. Hortic. Sci. *100*, 231–233.

24. Dickson, M. H., and Abawi, G. S. 1974. Resistance to *Pythium ultimum* in white seeded beans (*Phaseolus vulgaris*). Plant Dis. Repr. *58*, 774–776.

25. Dickson, M. H., and Boettger, M. A. 1976. Selection for seed quality in white seeded snap bean. Annu. Rep. Bean Improv. Coop. *19*, 24–25.

26. Dickson, M. H., and Boettger, M. A. 1977. Applied selection for mechanical damage resistance in snap beans using the mechanical damage simulator. Annu. Rep. Bean Improv. Coop. *20*, 38–39.

27. Dickson, M. H., and Boettger, M. A. 1977. Breeding for multiple root rot resistance in snap beans. J. Am. Soc. Hortic. Sci. *102*, 373–377.

28. Dickson, M. H., and Boettger, M. A. 1979. Release of 12 root rot tolerant snap bean lines. Annu. Rep. Bean Improv. Coop. *22*, 102.

29. Dickson, M. H., and Boettger, M. A. 1981. Double set in beans. Annu. Rep. Beam Improv. Coop. *24*, 116–117.

30. Dickson, M. H., and Boettger, M. A. 1982. Semi-hard seed in snap beans—a tool for selection for seed quality. Annu. Rep. Bean Improv. Coop. *25*, 102–103.

31. Dickson, M. H., and Eckenrode, C. J. 1979. Resistance to leaf-hopper *Empoasca fabae* in snap beans. Annu. Rep. Bean Improv. Coop. *22*, 25–26.

32. Dickson, M. H., Hunter, J. F., Gigna, J. A., and Boettger, M. A. 1981. Resistance to white mold. Annu. Rep. Bean Improv. Coop. *24*, 126–128.

33. Dickson, M. H., and Shannon, S. 1971. Small leaved compact bush beans. Annu. Rep. Bean Improv. Coop. *14*, 27–29.

34. Drijfhout, E. 1978. Genetic Interaction between *Phaseolus vulgaris* L. and Bean Common Mosaic Virus with Implications for Strain Identification and Breeding for Resistance. Cent. Agric. Publ. Doc., Wageningen, Netherlands.

35. Drijfhout, E., and Bos, L. 1977. The identification of two new strains of bean common mosaic virus. Neth. J. Plant Pathol. *83*, 13–25.

36. Drijfhout, E., Silbernagel, M. J., and Burke, D. W. 1978. Differentiation of strains of bean common mosaic virus. Neth. J. Plant Pathol. *84*, 13–26.

37. Farlow, P. J. 1981. Effect of low temperature on number and location of developed seed in two cultivars of French beans (*Phaseolus vulgaris* L.). Aust. J. Agric. Res. *32*, 325–330.

38. Farlow, P. J., Byth, D. E., and Kruger, N. S. 1979. Effect of temperature on seed set and *in vitro* pollen germination in French beans (*Phaseolus vulgaris*). Aust. J. Exp. Agric. Anim. Husb. *19*, 725.

39. Fouilloux, G. 1976. Bean anthracnose: New genes for resistance and new physiologic races. Ann. Amelior. Plant. *26*, 443–453 (in French); Annu. Rep. Bean Improv. Coop. (Engl. Transl.) *19*, 36–37.

40. Fouilloux, G., and Bannerot, H. 1977. RH_{13}, a four disease resistant line. Annu. Rep. Bean Improv. Coop. *20*, 59.

41. Guyer, R. B., and Kramer, A. 1951. Studies of factors affecting the quality of green and wax beans. Md, Agric. Exp. Stn., Bull. *A68*.

42. Hagedorn, D. J., and Rand, R. E. 1977. The first bacterial brown spot resistant bush bean. Annu. Rep. Bean Improv. Coop. 20, 67–68.

43. Hagedorn, D. J., and Rand, R. E. 1979. Release of new *Phaseolus vulgaris* germ plasm resistant to Wisconsin's bean root rot disease complex. Annu. Rep. Bean Improv. Coop. 22, 53.

44. Hagedorn, D. J., and Rand, R. E. 1979. Release of new *Phaseolus vulgaris* germ plasm resistant to bacterial brown spot *Pseudomonas syringae*. Annu. Rep. Bean Improv. Coop. 22, 54.

45. Hagedorn, D. J., and Rand, R. E. 1979. Development of resistance to Wisconsin's bean root rot complex. Annu. Rep. Bean Improv. Coop. 22, 86.

46. Hagel, G. T., Burke, D. W., and Silbernagel, M. J. 1978. Resistance in dry beans to Lygus bug pitting of seeds. Annu. Rep. Bean Improv. Coop. 21, 62.

47. Hagel, G. T., Burke, D. W., and Silbernagel, M. J. 1981. Response of dry bean selections to field infestations of seedcorn maggot in central Washington. J. Econ. Entomol. 74, 441–443.

48. Hagel, G. T., Silbernagel, M. J., and Burke, D. W. 1972. Resistance to aphids, mites, and thrips in field beans relative to infection by aphid-borne viruses. U.S.D.A., Agric. Res. Serv., *ARS 33–139*.

49. Hallard, J., and Trebuchet, G. 1976. Bean anthracnose in Western Europe. Annu. Rep. Bean Improv. Coop. 19, 44–46.

50. Hassan, A. A., Wallace, D. H., and Wilkinson, R. E. 1971. Genetics and heritability of resistance to *Fusarium solani* f. *phaseoli* in beans. J. Am. Soc. Hortic. Sci. 96, 623–627.

51. Hassan, A. A., Wilkinson, R. E., and Wallace, D. H. 1971. Genetics and heritability of resistance to *Thielaviopsis basicola* in beans. J. Am. Soc. Hortic. Sci. 96, 628–630.

52. Hassan, A. A., Wilkinson, R. E., and Wallace, D. H. 1971. Relationship between genes controlling resistance to Fusarium and Thielaviopsis root rots in beans. J. Am. Soc. Hortic. Sci. 96, 631–632.

53. Hipps, L. E. 1977. Influence of irrigation on the microclimate and development of white mold disease in dry edible beans. Nebr., Agric. Meteorol. Prog. Rep. 77-2.

54. Hoki, M. O. 1973. Mechanical strength and damage analysis of Navy beans. Ph.D. Dissertation. Michigan State Univ., East Lansing.

55. Honma, S. 1956. A bean interspecific hybrid. J. Hered. 47, 217–220.

56. Hubbeling, N. 1957. New aspects of breeding for disease resistance in beans (*Phaseolus vulgaris* L.). Euphytica 6, 111–141.

57. Hubbeling, N. 1976. Selection for resistance to anthracnose, particularly in respect to the "Ebnet" race of *Colletotrichum lindemuthianum*. Annu. Rep. Bean Improv. Coop. 19, 49–50.

58. Hunter, J. E., Dickson, M. H., Boettger, M. A., and Cigna, J. A. 1982. Evaluation of plant introductions of *Phaseolus* spp. for resistance to white mold. Plant Dis. 66, 320–322.

59. Hunter, J. E., Dickson, M. H., and Cigna, J. A. 1981. Limited/-term inoculation: A method to screen bean plants for partial resistance to white mold. Plant Dis. 65, 414–417.

60. Kaplan, L. 1981. What is the origin of the common bean? Econ. Bot. 35, 240–254.

61. Kemp, G. A. 1973. Initiation and development of flowers in beans under suboptimal temperature conditions. Can. J. Plant Sci. 53, 623–627.

62. Kyle, J. H., and Randall, T. E. 1963. A new concept of the hard seed character in *Phaseolus vulgaris* L. and its use in breeding and inheritance studies. Proc. Am. Soc. Hortic. Sci. 83, 461–475.

63. Ladror, U., Silbernagel, M. J., and Dyck, R. L. 1982. Cold-wet imbibition injury in beans. Proc. Bean Improv. Coop. Natl. Dry Bean Counc. Meet., 1982, pp. 40–45.

64. Leakey, C. L. A., and Simbwa-Bunnya, M. 1972. Races of *Colletotrichum lindemuthianum* and implications for bean breeding in Uganda. Ann. Appl. Biol. 70, 25–34.

65. Lorz, A. P. 1952. An interspecific cross involving the lima bean *Phaseolus lunatus* L. Science 115, 702–703.

66. Mack, H. J., Boersma, L. L., Wolfe, J. W., Sistrunk, W. A., and Evans, D. D. 1966. Effects of soil moisture and nitrogen fertilizer on pole beans. Oreg., Agric. Exp. Stn., Tech. Bull. *97*.

67. Mack, H. J., and Stang, J. R. 1976. High density snap beans—what is the most desirable plant type? HortScience *11*, 322 (abstr.).

68. Marsh, L., Davis, D. W., Li, P. H., and Silbernagel, M. J. 1982. Two methods of evaluating genotypes of *Phaseolus vulgaris* under high temperature stress. Annu. Rep. Bean Improv. Coop. *25*, 55–56.

69. Mastenbroek, C. 1960. A breeding programme for resistance to Anthracnose in dry shell haricot beans, based on a new gene. Euphytica *9*, 177–184.

70. McLean, D. M., Hoffman, J. C., and Brown, G. B. 1968. Greenhouse studies on resistance of snap beans to *Rhizoctonia solani*. Plant Dis. Rep. *52*, 486–488.

71. Meiners, J. P. 1980. Results—Uniform snap bean rust nursery-1979. Annu. Rep. Bean Improv. Coop. *23*, 29–30.

72. Meiners, J. P., and Gillaspie, A. G., Jr. 1980. Screening for resistance in the field to peanut stunt virus in snap beans. Annu. Rep. Bean Improv. Coop. *23*, 26–29.

73. Middleton, J. E., and Silbernagel, M. J. 1977. Effect of irrigation frequency on snap bean production. Wash., Agric. Res. Cent., Circ. *601*.

74. Miller, D. E., and Burke, D. W. 1977. Effect of temporary excessive wetting on soil aeration and Fusarium root rot of beans. Plant Dis. Rep. *61*, 175–179.

75. Miller, D. E., and Burke, D. W. 1980. Irrigation and soil management. Controlling disease and aiding production of dry beans. Mich. Dry Bean Dig. *4*, 20–22, 33.

76. Mok, D. W. S., Mok, M. C., and Rabakoarihanta, A. 1978. Interspecific hybridization of *Phaseolus vulgaris* with *P. lunatus* and *P. acutifolius*. Theor. Appl. Genet. *52*, 209–216.

77. Moody, A. R., Benepal, P. S., and Berkeley, B. 1980. Resistance of *Phaseolus vulgaris* L. cultivars to hypocotyl inoculation with *Rhizoctonia solani* Kuehn. J. Am. Soc. Hortic. Sci. *105*, 836–838.

78. Ng, T. J., and Bouwkamp, J. C. 1978. Screening for high temperature pod setting ability in *Phaseolus vulgaris* L. Annu. Rep. Bean Improv. Coop. *21*, 39.

79. Noor, N. M., Smucker, A. J. M., and Adams, M. W. 1979. Alcohol dehydrogenase induction in *Phaseolus vulgaris* L. roots by zinc and soil flooding. Proc. Bean Improv. Coop. Natl. Dry Bean Counc. Meet., 1979, pp. 64–68.

80. Oliari, L., Vieira, C., and Wilkinson, R. E. 1973. Physiologic races of *Colletotrichum lindemuthianum* in the state of Minas Gerais, Brazil. Plant Dis. Rep. *57*, 870–872.

81. Pfender, W. F., and Hagedorn, D. J. 1982. *Aphanomyces euteiches* f. sp. *phaseoli*, a causal agent of bean root and hypocotyl rot. Phytopathology *72*, 306–310.

82. Pfender, W. F., and Hagedorn, D. J. 1982. Comparative virulence of *Aphanomyces euteiches* f. sp. *phaseoli* and *Pythium ultimum* on *Phaseolus vulgaris* at naturally occurring inoculum levels. Phytopathology *72*, 1200–1204.

83. Pieczarka, D. J., and Abawi, G. S. 1978. Effect of interaction between *Fusarium, Pythium*, and *Rhizoctonia* on severity of bean root rot. Phytopathology *68*, 403–408.

84. Poryazov, I. B. 1977. Field disease screening procedures for bacterial blights of beans. Annu. Rep. Bean Improv. Coop. *20*, 60–61.

85. Prasad, K., and Weigle, J. L. 1970. Screening for resistance to *Rhizoctonia solani* in *Phaseolus vulgaris*. Plant Dis. Rep. *54*, 40–44.

86. Provvidenti, R., and Dickson, M. H. 1981. Kelvedon Marvel: A multi-resistant cultivar of *Phaseolus coccineus* L. Annu. Rep. Bean Improv. Coop. *24*, 124–125.

87. Pryke, P. I. 1978. Release of Noorinbee snap bean. Annu. Rep. Bean Improv. Coop. *21*, 72.

88. Roberts, M. H. E. 1982. List of genes—*Phaseolus vulgaris* L. Annu. Rep. Bean Improv. Coop. *25*, 109–127.

89. Saettler, A. W. 1975. Air pollution damage to beans; presentations by several authors. (Saettler Discussion Moderator.) Proc. Bean Improv. Coop. Natl. Dry Bean Counc. Meet., 1975, pp. 1–15.

90. Schnock, M. G., Hoffmann, G. M., and Kruger, J. 1975. A new physiological strain of *Colletotrichum lindemuthianum* infecting *Phaseolus vulgaris* L. HortScience *10*, 140.

91. Schoonhoven, A. V. 1981. The CIAT Bean Program. Research Strategies for Increasing Production. Centro Internacional de Agricultura Tropical, Cali, Colombia.

92. Schuster, M. L., and Coyne, D. P. 1981. Biology, epidemiology, genetics and breeding for resistance to bacterial pathogens of *Phaseolus vulgaris* L. Hortic. Rev. *3*, 28–58.

93. Schwartz, H. F. 1980. The international bean rust nursery format. Annu. Rep. Bean Improv. Coop. *23*, 25–26.

94. Schwartz, H. F., and Galvez, G. E. (Editors) 1980. Bean Production Problems. Centro Internacional de Agricultura Tropical (CIAT), Cali, Colombia.

95. Schwartz, H. F., and Temple, S. R. 1978. Bean rust resistance strategy at CIAT. Annu. Rep. Bean Improv. Coop. *21*, 48–49.

96. Siemer, S. R., and Vaughan, E. K. 1971. A device for measuring bean plant anchorage and its relation to root rot severity. Phytopathology *61*, 590–591.

97. Silbernagel, M. J. 1977. Stabilization of genetic bean root rot resistance by combination with cold imbibition tolerance and root vigor. Proc. Bean Improv. Coop. Natl. Dry Bean Counc. Meet., 1977, pp. 27–28.

98. Silbernagel, M. J. 1977. Seed quality index as an indicator of crop production potential, and a selection tool for the genetic improvement of snap bean seed quality. Annu. Rep. Bean Improv. Coop. *20*, 40–42.

99. Silbernagel, M. J. 1979. Release of multiple disease resistant germ plasm. Annu. Rep. Bean Improv. Coop. *22*, 37–41.

100. Silbernagel, M. J. 1980. Effects of cultural practices on root rot in snap beans. Annu. Rep. Bean Improv. Coop. *23*, 84–85.

101. Silbernagel, M. J. 1980. A rapid screening technique for bacterial brown spot (*Pseudomonas syringae*) and halo blight (*Pseudomonas phaseolicola*). Annu. Rep. Bean Improv. Coop. *23*, 81–82.

102. Silbernagel, M. J. 1980. A self-propelled rubber belt harvester for fragile-seeded crops. Seed World *118*, 24–26.

103. Silbernagel, M. J. 1982. Stocks for exchange. Annu. Rep. Bean Improv. Coop. *25*, 130.

104. Silbernagel, M. J., and Burke, D. W. 1973. Harvesting high quality bean seed with a rubber-belt thresher. Bull.—Wash. Agric. Exp. Stn. *777*.

105. Silbernagel, M. J., and Drake, S. R. 1978. Seed index, an estimate of snap bean quality. J. Am. Soc. Hortic. Sci. *103*, 257–260.

106. Silbernagel, M. J., and Zaumeyer, W. J. 1973. Beans. *In* Breeding Plants for Disease Resistance. R. R. Nelson (Editor), pp. 253–269. Pennsylvania State Univ. Press, Univ. Park.

107. Smartt, J. 1976. Tropical Pulses. Trop. Agric. Ser., Longman Group Ltd., London.

108. Stavely, J. R. 1982. The 1981 bean rust nurseries. Annu. Rep. Bean Improv. Coop. *25*, 34–35.

109. Tu, J. C., and Aylesworth, J. W. 1980. An effective method of screening white (pea) bean seedlings (*Phaseolus vulgaris* L.) for resistance to *Colletotrichum lindemuthianum*. Phytopathol. Z. *99*, 131–137.

110. U.S. Department of Agriculture 1952. Manual for Testing Agricultural and Vegetable Seeds, Agric. Handb. No. 30. U.S. Government Printing Office, Washington, DC.

111. USDA-Agriculture Marketing Service 1973. U.S. Plant Variety Protection Act of Dec. 24, 1970. Plant Variety Protection Office, Washington, DC.

112. USDA-Crop Reporting Board. 1980. Vegetable Seeds. SeHy 1-1 (March 1980). U.S. Government Printing Office, Washington, DC.

113. USDA-Crop Reporting Board 1980. Vegetable-1980 Annual Summary, Acerage, Yield, Production, and Value. VG 1-2 (December 1980). U.S. Government Printing Office, WAshington, DC.

114. USDA-Economics Research Service 1973. Per Capita Consumption Table Fresh and Processed Vegetables. Vegetable Situation (July 1973). U.S. Government Printing Office, Washington, DC.

115. Vargas, E. 1980. Rust. *In* Bean Production Problems. H. F. Schwartz and G. E. Galvez (Editors), pp. 17–36. Centro Internacional de Agricultura Tropical (CIAT), Cali, Colombia.

116. Vea, E. V., and Eckenrode, C. J. 1976. Resistance to seedcorn maggot in snap bean. Environ. Entomol. *5*, 735–737.

117. Wallace, D. H. 1978. Western Regional Project-150. Annu. Rep. Bean Improv. Coop. *21*, 89–90.

118. Wallace, D. H., Sandsted, R. F., and Ozbun, J. L. 1974. Obtaining yield physiology data from standard yield trials. Annu. Rep. Bean Improv. Coop. *17*, 92–93.

119. Wallace, D. H., and Wilkinson, R. E. 1965. Breeding for Fusarium root rot resistance in beans. Phytopathology *55*, 1227–1231.

120. Weaver, M. L., Timm, H., and Gaffield, W. 1984. Possible screening procedure for temperature tolerance in common bean. Proc. Bean Improv. Coop. Natl. Dry Bean Counc. Meet., 1983, p. 67.

121. Weaver, M. L., Timm, H., Silbernagel, M. J., and Burke, D. W. 1984. Pollen viability and temperature tolerance in beans. Proc. Bean Improv. Coop. Natl. Dry Bean Counc. Meet., 1983, p. 66.

122. Westermann, D. T., and Crothers, S. E. 1977. Plant population effects on the seed yield components of beans. Crop Sci. *17*, 493–496.

123. Westermann, D. T., Kleinkopf, G. E., Porter, L. K., and Leggett, G. E. 1981. Nitrogen sources for bean seed production. Agron. J. *73*, 660–664.

124. Westermann, D. T., and Kolar, J. J. 1978. Symbiotic $N_2(C_2H_2)$ fixation by bean. Crop Sci. *18*, 986–990.

125. Wien, H. C., and Munger, H. M. 1972. Heat tolerant *Phaseolus vulgaris*. Annu. Rep. Bean Improv. Coop. *15*, 97–98.

126. Wyatt, J. E., Fassuliotis, G., Johnson, A. W., Hoffman, J. C., and Deakin, J. R. 1980. B4175 Root-knot nematode resistant snap bean breeding line. HortScience *15*, 530.

127. Wyatt, J. E., Day, A., Benepal, P. S., Sheikh, A., and Sullivan, M. J. 1980. Mexican bean beetle resistance: Field testing of breeding lines. Annu. Rep. Bean Improv. Coop. *23*, 36.

128. Wyatt, J. E., Hoffman, J. C., and Deakin, J. R. 1977. B4000-3 Snap bean breeding line. HortScience *12*, 505.

129. York, D. W., Dickson, M. H., and Abawi, G. A. 1977. Inheritance of resistance to seed decay and pre-emergence damping-off in snap beans caused by *Phythium ultimum*. Plant Dis. Rep. *61*, 285–289.

130. Yoshii, K. 1980. Common and fuscous blights. *In* Bean Production Problems. H. F. Schwartz and G. E. Galvez (Editors), pp. 155–172. Centro Internacional de Agricultura Tropical (CIAT), Cali, Colombia.

131. Zaumeyer, W. J. 1963. Some new Tendercrop mutants. Seed World March 8.

132. Zaumeyer, W. J. 1972. Snap beans. *In* Genetic Vulnerability of Major Crops, pp. 234–244. Committee on Genetic Vulnerability of Major Crops. J. G. Horsfall, Chairman. Natl. Acad. Sci., Washington, DC.

133. Zaumeyer, W. J., and Meiners, J. P. 1975. Disease resistance in beans. Annu. Rev. Phytopathol. *13*, 313–334.

134. Zaumeyer, W. J., and Thomas, H. R. 1959. Tendercrop, a new slender-podded snap bean. Seed World March 27.

135. Zimmerman, M. J. O., Waines, G., and Foster, K. 1981. Drought resistance in common beans *Phaseolus vulgaris* L. Annu. Rep. Bean Improv. Coop. *24*, 77.

8
Pea Breeding

EARL T. GRITTON

Peas are widely grown and consumed either as a fresh succulent vegetable or as dried seed that is usually used in soups. The pea (*Pisum sativum* L.) is a common garden vegetable in temperate regions, and in warmer areas peas are grown during the cooler part of the year. No figures on production or consumption from home gardens are available, though this is substantial. Human per capita consumption of the commercially produced crop on an "in pod fresh equivalent basis" in the United States in 1978 was 6.45 lb canned and 4.23 lb frozen (86). Since few peas are marketed fresh in the pod (0.3 lb/capita in 1964, 0.1 in 1967, less than 0.1 now), canned and frozen consumption figures are more meaningful. During 1964–1969, per capita consumption of canned peas averaged 4.1–4.2 lb. Consumption then began a slow decline to 3.6 lb per person in 1978. Per capita consumption of frozen peas increased slowly to a high of 2.08 lb in 1968, and then began a slow decline to 1.55 lb in 1978. Per capita consumption of dry field peas has ranged from 0.5 to 0.2 lb during the period 1964–1978, and has averaged about 0.35 lb.

Historically, there have been three major pea production areas for commercial canning and freezing in the United States. One of these is the northeastern region made up of Delaware, Maryland, and New York. The second is the north-central region comprised of Minnesota and Wisconsin. The third is the northwestern region composed of Oregon and Washington. California could be considered a fourth area, since it produces about 3–4% of the U.S. total, and other states such as Colorado, Idaho, Illinois, Iowa, Maine, Michigan, Pennsylvania, Utah, and Virginia produce smaller but important amounts. In 1981, Wisconsin produced 125,460 tons of shelled peas from 84,200 acres; Washington

BREEDING VEGETABLE CROPS

produced 98,560 tons from 61,600 acres; Minnesota produced 87,470 tons from 56,800 acres; and Oregon produced 42,310 tons from 32,300 acres 88).

Idaho and Washington produce essentially all of the dry edible field peas. In 1981 they produced 1145 tons from 108,000 acres (*87*). Peas are also one of the major processing crops in Europe as well as in most other countries with a suitably cool production period.

ORIGIN

The pea is one of the edible legumes or pulse crops (Fig. 8.1). These include many other genera of the Leguminosae and are generally considered to include, among others, peas, beans, lupines, vetches, chickpeas, and lentils, which is a broad and ambiguous classification. Such ambiguity traces back to ancient writers, who did not distinguish carefully among the pulses, which consequently clouds the picture of pea evolution. Its exact origin and progenitor are unknown, but it is one of the oldest of cultivated plants and was grown in neolithic farming villages of the Near East at least as early as 7000 to 6000 B.C. (*103*). Vavilov (*89*) listed Central Asia, the Near East, Abyssinia, and the Mediterranean as centers of origin based on genetic diversity. Blixt (*5*) states that the Mediterranean is the

FIGURE 8.1. The pea *P. sativum* L.

principal center of diversity, with secondary centers in Ethiopia and the Near East. While peas are grown in diverse environments, they are best adapted to cool, moist climates.

GENERAL BOTANY

Peas are diploid with a chromosome number of $x = 7$. Other genera included in the Vicieae tribe are *Vicia*, $x = 5$, 6, or 7 with most species diploid; *Lens*, $x = 7$ diploid; *Lathyrus*, $x = 7$, most species diploid; and *Cicer*, $x = 7$ or 8, most species diploid (*12*). To date, efforts to hybridize peas with other genera of the Vicieae sexually have been unsuccessful.

Linnaeus distinguished two species within the genus *Pisum: Pisum arvense*, the colored flower field pea and *P. sativum*, the white-flowered horticultural or garden pea. Since then, species designation has been given to *Pisum abyssinicum* (Braun), *Pisum aucheri* (Jaubert and Spach), *Pisum elatius* (Stev.), *Pisum formosum* (Alefeld), *Pisum fulvum* (Sibth and Sm.), *Pisum humile* (Boiss and Noe), *Pisum jomardi* (Schrank), and *Pisum transcaucasicum* (Gov./Stankov) (*7*). Lamprecht (*65*) separated *P. aucheri* and *P. formosum*, which are tuber-forming perennials, are synonymous, and do not cross with *Pisum*, to form the monospecific genus *Alophotropsis* (Biossier). See Lamprecht (*65*) for details on characteristics.

All forms of peas previously accorded species status (except now those placed in the genus *Alophotropsis*) have a diploid chromosome number of 14, exchange genes readily, and have few if any sterility barriers. The genus might best be considered monospecific as presented by Lamprecht (*65*), who classified the different forms as ecotypes under *P. arvense*, with the white-flowered garden pea considered ecotype *sativum*. However, the widespread use of *P. sativum* to designate the garden pea would make it difficult to change this designation.

For most purposes, peas can be divided into two classes of cultivars, garden and field peas. Garden peas generally have white flowers and their seed shapes may be round, dimpled, or wrinkled, and their seed colors green, cream, or yellow. The vines may be dwarf with internodes angled in a zigzag pattern, intermediate, or tall with internodes in a more or less straight line (Fig. 8.2). Dwarf forms are usually considered determinate, while many tall forms are indeterminate in growth habit. The seeds are usually harvested at a young, succulent stage, shelled from the pods, and eaten after boiling. The pods are inedible due to parchment on their inner surfaces and fibers along the suture lines. There are also edible-podded types called sugar peas, snow peas, or Chinese peas, which lack the membraneous coating on the inside of the pod. The pods of these are harvested soon after the seeds begin to swell. These peas usually have purple flowers although white types are also grown. Seeds may be pigmented when mature. Recently an edible podded type with thick pod walls, resembling the pod of the snap bean, has become popular.

Field peas (dry seed cultivars) often have purple, lavender, or other colored flowers. Their seeds are typically round with either yellow or green cotyledons, and seed coats may be clear or pigmented. They are harvested at the dry mature stage from pods that are not edible. Some of the production is used as human food, mostly as split peas for soup, but much is used as feed for birds or animals.

According to Hedrick (*28*), the early Greek writers confused the pulses under the names *orobos*, *erebinthos*, and *pisos*, but Theophrastus was definite in the use of *orobos* for vetch, *erebinthos* for chickpea, and *pisos* for the common pea. When the Greeks took the pea to Rome, *pisos* became *pisu*, which was passed on to the English as *peason*, then

FIGURE 8.2.   Dwarf, intermediate, and tall pea plants.

pease or peasse. The English mistook these words for plurals, dropped the "s," and thus pea became the universal name for this vegetable among English-speaking people.

English breeders were very active and successful in breeding improved cultivars, and it is probably for this reason that the term English pea came to be associated with this nutritious, tasty vegetable. Most American cultivars trace their ancestry to lines developed in England.

Pea Morphology

The pea is an annual or winter annual herbaceous plant. Germination is hypogeal with the cotyledons persisting until stored food reserves are exhausted. The taproot produces a profusion of lateral roots. Stems are slender, angular, glaucous, and in peas grown for processing, usually single in number and upright in growth. From one to many axillary stems may originate at the cotyledonary node or any superior node, especially if the apical growing point is destroyed, removing apical dominance (Fig. 8.3). Two rudimentary or primary scales or leaves are formed immediately above the cotyledonary node (Fig. 8.4), and these may be found above or below the soil surface. Stems are solid from the cotyledonary node to the node immediately above the second primary scale; they then become hollow. Leaves are alternate, distichous, and may have from one leaflet to several pairs. Leaflets of a pair are opposite or slightly alternate, with the lower leaflets larger. The rachis terminates in a simple or branched tendril. Large stipules clasp the stem and their lower parts generally overlap. Margins of leaflets and stipules may be entire or may be slightly to deeply serrated.

The inflorescence of the pea is a raceme arising from the axil of a leaf. The lowest node at which flower initiation occurs is quite constant for a given genotype in a given environment and is used in characterizing cultivars. For this purpose, the first node above the cotyledonary node, where the lower scale leaf develops, is considered node one. The upper scale leaf node is counted as node two, and so on up the stem. Under field conditions in temperate regions, the number of vegetative nodes before the first inflorescence may vary from four for the earliest lines to about 25 for the latest. Most early cultivars will produce the first flower from nodes 5 to 11, while most later cultivars will start flowering at about nodes 13 to 15.

Floral initiation begins at the lowest flowering node and proceeds sequentially up the

FIGURE 8.3. Growth of axillary branches from various nodes of the pea plant after a frost killed the growing point and removed the apical dominance that would have caused each of these plants to be single stemmed.

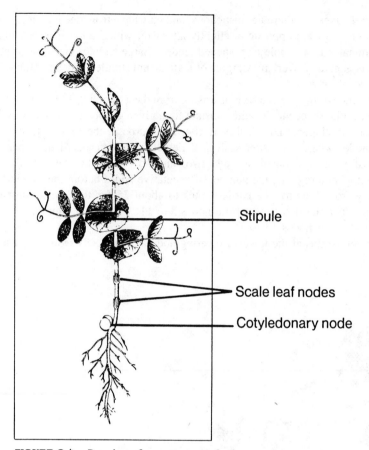

Stipule

Scale leaf nodes

Cotyledonary node

FIGURE 8.4.   Drawing of a young pea plant.

stem. One to many nodes may bear flowers, depending upon the genotype and environmental conditions. The peduncles arise from the axil of the leaf between the stem and petiole. Depending on the genotype, they vary greatly in length and bear from one to many flowers. The number of flowers is not constant on a plant, but very early cultivars are often single flowered or bear some single and some double flowers. Later cultivars are mostly double or triple flowered, and again, the number may vary from node to node. When more than one flower is borne per peduncle, the most mature flower is nearest the base of the peduncle and the least mature at the tip. It is thus possible at a single node to have a flower that has shed pollen and whose wings and standard have reflexed, a flower shedding pollen but whose wings and standard have not reflexed, and a flower whose anthers have not yet burst.

   Anthesis begins at the lowest floral node and proceeds upward. With cool, moist environmental conditions from 5 to 6 days may separate anthesis between nodes, while under hot, dry conditions the separation may be only 1 day.

FLORAL BIOLOGY AND CONTROLLED POLLINATION

The flower of the pea is characteristic of a large number of plants belonging to the family Papilionaceae or Fabaceae (Fig. 8.5). It is zygomorphic, that is, can be divided into two symmetrical parts. The green calyx is comprised of five united sepals, two behind the standard, two subtending the wings, and an anterior one subtending the keel (Fig. 8.5A). The standard has a notch at the center of the top and is keeled at the back of the base. In bud stage the standard is folded and covers the other petals. The two wings are smaller than the standard. They enclose the two lower petals that are fused along their abaxial margin to form a boat-shaped keel, which covers the stamens and pistil.

The lower portions of the filaments of nine of the 10 stamens are fused to form a staminal tube nearly the length of and surrounding the ovary (Fig. 8.5B). The tube is split by the tenth stamen, which is free throughout its length. The filaments are shorter than the style in the very young flower, but by the time the anthers dehisce they have forced the anthers tightly into the tip of the keel against the style.

The pistil is attached by a stalk to the middle of the bottom of the calyx tube. Its base is a flattened green ovary formed by a carpel whose margins are joined on the adaxial side. The ovary (Fig. 8.6) bears up to 13 ovules alternately attached to the two placentas. The slightly flattened, cylindrical style extends from the top of the ovary and is bent at nearly a right angle to the ovary. A brush of stylar hairs appears on the inner side of the style near its tip, where it recurves toward the ovary. An elliptical, sticky stigma caps the tip of the style.

The stigma of peas is receptive to pollen from several days prior to anthesis until 1 day or more after the flower wilts (*91*). Pollen is viable from the time the anthers dehisce until several days thereafter. The range in frequency of cross-pollination reported for the pea has been from 0 (*102*), for peas grown in New York up to 60% for peas grown in Peru (*27*), where several insect species were considered capable of pollinating peas. The author has observed few instances of cross-pollination among peas with normal flowers in Wisconsin, but has seen extensive outcrossing in Brazil due to an insect identified as *Paratrigona lineata* that cuts through the unopened flower petals to the stamens and pistil. In the United States, with commercial cultivars, outcrossing is generally less than 1%. The high degree of self-pollination is due to the cleistogamous nature of the pea. Pollination occurs about 24 hr before the flower opens (*11*). Pollen on the stigma germinates in about 8–12 hr and fertilization occurs about 24–28 hr after pollination (*23*).

Division of the generative nucleus into two male gametes occurs in the pollen tube, whence they are introduced into the embroyo sac. One gamete fertilizes the egg and the

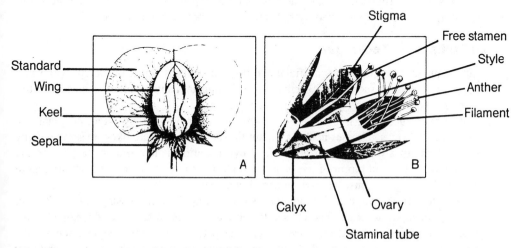

FIGURE 8.5. Pea flower. (A) Intact flower showing standard, wings, keel, and sepals. (B) Flower minus petals to show pistil and stamens.

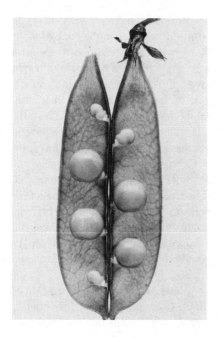

FIGURE 8.6. The pea ovary showing ovules alternately attached to the two placentas. In this ovary, only four of the seven ovules have developed.

other unites with the two polar nuclei to form a triploid endosperm that is later resorbed. The developing embryo is at first borne on a suspensor, but this later disintegrates. The mature seed contains a partly developed embryo that constitutes about 1–3% of the total seed weight, two cotyledons that constitute about 90% of the seed weight, and a testa that makes up 7–9% of the weight (8).

Because the pea is normally self-pollinated, no special arrangements are needed for selfing. Natural hybridization has not been used because of lack of pollination control. Incompatability systems are not known in *Pisum* and male sterility heretofore known has not been used in facilitating crossing. More recently, a male sterile with exposed pistils has offered promise in promoting outcrossing (76). Attempts to cross *Pisum* with other genera of the Vicieae tribe have so far been unsuccessful (72). *Vicia faba* pollen applied to *Pisum* stigmas produced hybrid embryos that developed for up to 6 days before disintegrating (23). Other purported hybridizations have not been confirmed.

Manual Pollination Technique

Little equipment is needed for artificial hybridization of peas: a pair of forceps with sharp tips, some alcohol such as 95% ethanol for killing unwanted pollen, tags to identify crosses, and a pencil to record the cross on the tag.

The most successful hybridization is likely to occur on a strong, vigorous plant just beginning to flower. Crosses made as a plant finishes flowering are likely to set few if any seed. Mornings are to be preferred for field hybridization as temperatures are lower and relative humidity and turgor of the plant are high. Crossing can be carried on throughout the day in the greenhouse or environmental chamber.

Pea flowers are large and relatively easy to manipulate in crossing (Fig. 8.7). For greatest success, the flower bud chosen to serve as female should have developed to the stage just before anther dehiscence, where the petals will extend beyond the sepals. Younger flower buds are more difficult to emasculate, are more easily damaged, and may

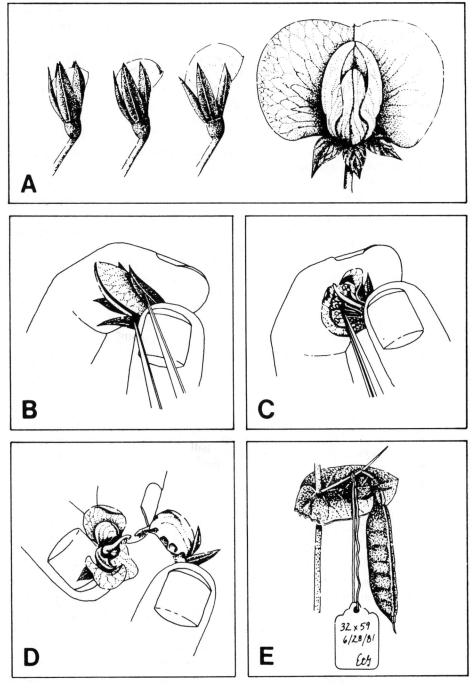

FIGURE 8.7. Pea flower stages, emasculation, pollination, and development of the pod. (A) Bud at left has not yet developed to optimum stage for emasculation, bud second from left is at about optimum stage, and bud third from left will probably have already pollinated. Freshly opened flower on right should be a good source of pollen. (B) Slitting the keel with the tip of a forceps. (C) Removing the anthers. (D) Using a pollen-laden style to transfer pollen to emasculated flowers. (E) Developing pod with tag.

set fewer seeds. Older flowers will have already self-pollinated. It may be desirable to remove other flowers from that peduncle to facilitate manipulation of the bud and to reduce competition for nutrients.

Flowers can be emasculated at any time of the day but usually are emasculated just prior to pollination. Under greenhouse conditions, or where cross-pollination is unlikely, some breeders prefer to emasculate 1 day previous to applying pollen. Usually from three to six flowers at a time are emasculated. It can be difficult to keep track of more than this number, particularly if crossing is done in the field with wind blowing the plants, and normally the pollen supplied by one flower of the pollen parent is sufficient to pollinate three to six females. Some breeders like to hang the crossing tag over the emasculated flower to make it easier to spot for pollination.

The first step in emasculation is carefully to tear away with the forceps the tip of the sepal from in front of the keel. This allows more open entry to the pistil and stamens. Then, by positioning the forefinger behind the flower and the thumb in front, a light squeezing pressure will spread the standard and wings to expose the keel. Here it may be helpful to bend the standard back and the wings out and hold them with the thumb and finger, though care must be taken not to seriously injure the flower. The keel is slit by inserting just the sharp tip of the forceps about midway down the keel and slipping it slightly up and down (Fig. 8.7B). After the keel is punctured, it can be opened by inserting both tips of the forceps and allowing them to spread. The two halves of the keel can be held down with the thumb and finger for increased exposure of the pistil and stamens, or pressure can be applied with the finger and thumb to spread the keel. The 10 anthers are removed by grasping the filaments with the forceps and carefully pulling (Fig. 8.7C). It may be wise to count them as they are being removed so as not to overlook any that may be obscured by folds in the petals.

Some breeders use an alternative emasculation procedure that they have found faster. They roll the standard and wings back and hold them with the thumb and index finger to expose the keel. Then with the thumb and index finger of the other hand, they grasp the upper part of the keel and tear off its upper half along with the anthers. If the single free anther remains it is removed with a fingernail or point of forceps or pencil.

Pollen can be obtained at any time of the day. Generally the most abundant and viable pollen is obtained from freshly dehisced anthers. The stage of flower development at which dehiscence occurs varies with environment and genotype, but dehiscence normally will occur just as the flower petals are starting to open. Flowers open throughout the day and night upon reaching the proper physiological stage. Pollen of flowers left on the plant soon loses viability, and as it does so changes from a rich moist yellow or gold color to a dry bleached yellow or white. Layne and Hagedorn (66) found that untreated pollen could be stored for up to 6 days after dehiscence and still effect fertilization. They found that vacuum drying and storage at $-25°C$ extended viability to approximately 1 year.

If the flower providing the pollen is to be preserved to set seed, a toothpick, the tip of the forceps, a cotton swab, or other instrument can be used to collect and transfer the heavy sticky pollen. In most cases it is more convenient to pick the male flower, remove the standard and wings, pull back the keel so that the style protrudes, and use the pollen-covered stylar brush as an applicator to transfer the pollen to the stigma of the emasculated flower (Fig. 8.7D). By releasing pressure on the keel, the stigma can be allowed to slip back inside to pick up another coating of pollen. The receptive surface is the elliptical stigmatic tip of the style, not the hairs on the concave side. It is necessary to tag the pollinated flower to identify the cross (Fig. 8.7E). The tag will normally include identification of the female and male parents and may include such other information as is useful. Often the date of the cross, initials of the person making the cross, and characteristics such as disease resistance are recorded. The information is written on the tag with waterproof pencil or pen. It is helpful to moisten the string of the tag by passing it through the mouth so it will slip easily while being installed on the flower. The moistened string is formed into a loop, slipped over the flower, and pulled snug around the pedicel. Possible injury to the flower can be reduced by slipping the tip of the forceps inside the loop to take the strain of pulling on the string, and then using the forceps to help move the knot up to close the loop. Care must be taken to ensure that the loop encloses only the pedicle of the pollinated flower, and not the growing point of the plant.

MAJOR BREEDING ACHIEVEMENTS

Culture and Harvesting

To understand breeding objectives and achievements, it is necessary to have some understanding of how peas are grown and harvested. Peas in the home garden are generally grown in single or double rows seeded at a rate of about 6 seeds/ft. The taller cultivars are supported by trellises; and where double rows are used, they may be separated by only a matter of inches on either side of the trellis. Trellises are separated by 2–5 ft to allow access. Pods can be picked as they reach the proper harvest stage, and multiple harvests are not only possible but may be desirable to prolong the harvest season.

Field plantings on large acreages are seeded in rows about 6–7 in. apart with about 6–9 seeds/ft of row, to give plant populations of 450,000–700,000 plants/acre. No artificial support is provided, but intertwining of tendrils and plants provides mutual support. Because the entire field is harvested at one time, it is desirable to have as many pods as possible with peas at the optimum harvest stage.

Because peas very rapidly pass through the optimum harvest stage, several cultivars and planting dates are used to secure a continuous supply of fresh succulent peas. For commercial pea production, plantings are scheduled according to temperatures, or ''heat units.'' The base temperature used in computing heat units is 40°F because this is about the minimum temperature for pea germination and growth. Temperatures above this base result in a ''heat unit accumulation.'' For example, if the average temperature for a day is 62°F, the heat unit accumulation for that day is $62° - 40° = 22$ heat units. Heat units thus accumulate slowly during the cool spring and rapidly as the season advances.

The number of heat units needed for a cultivar to reach the processing stage is fairly constant. This makes it possible, by referring to temperature records of past years, to predict the date on which the cultivar will be ready to harvest. Since records show the average heat unit accumulation just before harvest, the number of heat units that must separate planting dates is known. Because of the large difference in spring and summer temperatures, several days separation at planting time may be required to separate harvests by only 1 or 2 days. The exact number of heat units needed for a cultivar to reach harvest may vary somewhat with season and locality, but the heat unit system still works very well because the differences between cultivars remain remarkably constant.

The current harvest procedure for most peas grown commercially for canning or freezing is to cut the entire plant at ground level and to lay them into a swath or windrow. The windrow is picked up by a large threshing machine, called a viner, which threshes the shelled peas from the pods while the pods remain attached to the plants (Fig. 8.8). The

FIGURE 8.8.   A pea viner picking up and threshing freshly cut and windrowed pea plants.

shelled peas are collected in a bin on the viner while the threshed pods and vines are returned to the field. Efficiency of this harvest operation is influenced by the ratio of the volume of the whole plant to the volume of shelled peas recovered. Thus, a reduction in amount of vine and/or an increase in shelled pea productivity will be beneficial. Recently, self-propelled viners that strip the pods from the plants without the need for cutting and windrowing have made their appearance (Fig. 8.9).

Improvements of Plant Habit and Pod Characteristics

Yields of peas for the garden have been improved through the development of cultivars that bear more pods per node. Except for the very earliest cultivars, most garden peas now bear at least two pods per node. Selection has also been effective for more consistent performance. Although height or vine length is less critical than for field-grown plants, it has been generally reduced; and many dwarf cultivars growing only 2–2.5 ft tall have been developed.

The recent introduction of an edible podded pea with thick walls has aroused a great deal of interest among home gardeners and may offer some possibilities for expanded commercial production and use. The pods and developing ovules of this pea remain palatable over a long period of time. Raw pods containing the ovules can be used as hors d'oeuvres. Pods and ovules can be cooked together and eaten, or the peas can be shelled from the pods and the pods and peas eaten separately. They can be preserved by freezing, but present cultivars do not maintain their integrity after canning.

The pods of the snap pea are thick and fleshy due to the recessive *n* gene. They are recessive for the *p* and *v* genes, which condition fiber development in the pod, as are pods of other edible podded cultivars.

A detailed list of pea cultivars grown in the United States accompanied by written and pictorial descriptions was published in 1928 (*28*). Many of these cultivars were very tall— 3 ft or more in length—and most bore predominantly one pod per node. Few were listed as having disease resistance. I am unaware of any such recent compilation, and since new cultivars are regularly replacing old ones, such a listing would rapidly become outdated. Present cultivars differ from those grown previously in being mostly on dwarf habit (2–3 ft in length of vine), two pods per node, having smaller sizes of shelled peas, and with various disease resistances. A partial listing of some currently available cultivars showing a range of characteristics is given in Table 8.1.

The home gardner will select the cultivars to be grown, but where the crop is grown for canning or freezing, the processor, not the grower, selects the cultivars. Processors base their selection on cultivars that can be profitably grown to supply peas meeting certain market characteristics.

Shelled peas increase in size as they mature; hence within a cultivar the smaller sizes are usually sweeter, more tender, and more succulent. This has given rise to consumer preference for smaller sizes of the shelled peas. Their size is referred to in terms of sieve size, which represents the maximum size of pea that will pass through round holes having the following diameters in inches: no. 1, $9/32$; no. 2, $10/32$; no. 3, $11/32$; no. 4, $12/32$; no. 5, $13/32$; no. 6, $14/32$; and no. 7, $15/32$. Because of consumer preference for smaller sizes, newer cultivars have smaller peas.

Development of Cultivars for Canning and Freezing

While detailed descriptions of all currently grown cultivars is not feasible, the main characteristics of certain groups or classifications can be presented. First, cultivars can be

FIGURE 8.9. A self-propelled pea thresher equipped with pod-stripping head. Peas are harvested directly without the need for cutting and windrowing.

TABLE 8.1. Classification of Selected Pea Cultivars[a]

Berry color	Vine type	Pod shape	Seed surface	Cotyledon color	Cultivar
Light green	Tall (indeterminate)	Blunt	Smooth (starchy)	Green	Aska Kriter
			Wrinkled (sweet)	Green	Deli Early Sweet 11 No. 4683
	Short (determinate)	Blunt	Wrinkled	Green	Champ Charger Dart Dawn Mini Nugget Rally Resistant Early Perfection 326 Sybo Trend Trojan
Dark green	Tall (indeterminate)	Blunt	Smooth	Green	Juneau
			Wrinkled	Green	Laxton 8
		Pointed	Wrinkled	Green	Alderman
	Short (determinate)	Blunt	Wrinkled	Green	Abador Beacon Bolero Coronet Dark Skin Perfection Freezer 626 Frisky Ivy Lotus Mars Rigo Spring Sprite Trifect Trumpet Venus
		Pointed	Wrinkled	Green	No. 40 Progress No. 9 Rondo

[a]Adapted from Asgrow (2).

TABLE 8.1. (*cont.*)

Heat units (°F)	First blossom node	Vine color	Average sieve size	Use
1170	10	Light	2.6	Early June canning, dry
1280	10	Light	2.0	Extra-early canning, dry
1210	10	Light	2.5	Early sweet canning
1310	11	Light	3.9	Early sweet canning
1190	9–10	Light	3.5	Early sweet canning
1410	12–14	Dark	3.5	Midseason perfection, canning
1430	11–12	Dark	3.4	Extra-early perfection, canning
1310	10–11	Dark	3.2	Extra-early perfection, canning
1090	9–10	Medium	3.5	Extra-early perfection, canning
1430	14	Medium	1.9	*Petit pois,* processing
1470	14	Dark	2.9	Small-sieve perfection, canning
1320	11	Dark	2.8	Extra-early perfection, canning
1510	15–16	Dark	3.6	Full-season canning
1530	14	Dark	2.8	Small-sieve perfection, canning
1300	10–11	Dark	2.9	Early perfection, canning
1460	14–15	Dark	4.1	Midseason perfection, canning
1160	9–10	Dark	2.6	Extra-early freezing, canning
1250	10–11	Medium	5.4	Garden
1640	17	Dark	5.0	Garden, market
1420	14–15	Dark	2.8	Midseason freezing
1420	14–15	Dark	3.3	Midseason freezing
1500	14–15	Dark	4.0	Full-season freezing
1300	10–11	Dark	3.7	Early freezing
1550	15–16	Dark	4.4	Full-season freezing
1460	14–15	Dark	4.1	Full-season freezing
1300	10	Dark	2.4	First-early freezing
1580	15–16	Dark	2.6	Full-season freezing
1350	11–12	Dark	4.6	Early freezing
1410	13–14	Dark	4.2	Midseason freezing
1490	14–15	Dark	4.4	Full-season freezing
1100	9–10	Dark	5.0	First-early freezing
1170	9–10	Dark	4.8	First-early freezing
1520	15–16	Dark	4.5	Full-season freezing
1200	9–11	Dark	4.8	First-early freezing
1320	10–11	Dark	4.9	Early freezing
1640	14–15	Dark	5.0	Shipping, market, garden
1200	9	Dark	5.0	Shipping, market, garden
1600	15	Dark	5.0	Shipping, market, garden

divided into two groups based on the color of the fresh shelled peas. Shelled peas of canning cultivars have a lighter green color and are resistant to leaching of chlorophyll into the sugar–salt liquid in the can. Freezers are characterized by a dark-green shelled pea (called "berry" by raw-product personnel of the processing industry). This distinction was quite marked in the past but is gradually diminishing as some dual-purpose cultivars have been developed, and more can be expected. This is particularly advantageous to processors who have both canning and freezing capabilities and so can use the cultivar in either process. The northeastern production region has both canning and freezing plants, and both types are grown there. The north-central region has a preponderance of canning plants; hence relatively few freezing cultivars are grown commercially. The northwestern area grows primarily freezing cultivars.

Second, cultivars are classified according to maturity. The earliest cultivars require about 54 days from planting to harvest and the latest ones about 75 days. The correlation between the number of nodes to first flower and the time to harvest stage is very high and so a good estimate of relative maturity is given by knowing the node of first flower. The 54-day cultivars, for instance, will bear their first flower at about the eighth or ninth node above the cotyledonary node, while the very late cultivars may not flower until about the fifteenth node. The corresponding heat unit requirement may be 1150–1200 for the early cultivar and 1550 for the late.

Canning cultivars may have round or wrinkled seed. Round-seeded cultivars possess the dominant alleles at the R and R_b loci (genotype $R/R,R_b/R_b$), and the starchy cotyledons do not shrink or wrinkle upon drying, although in some genetic backgrounds a "dimple" may develop. Starch grains in the cotyledons are phenotypically compound. Most round-seeded garden or processing cultivars grown in the United States are of the Alaska type, which has been grown for many years. It is a short-season type that is very hardy and is usually the first to be planted in the spring. It has a light-green to almost yellowish foliage. It is widely grown in home gardens and is canned commercially as the early June type. It is mostly single podded although some Alaska selections have some double pods. Because of its starchy cotyledons, it rapidly becomes unpalatable with maturity.

A canning pea closely resembling the Alaska is rapidly replacing the Alaska type. It is a day or two later in maturity and has slightly higher yields due to more double pods. This type, first known as Alsweet and now known as early sweet or Alsweet, is considered to be sweeter in taste than Alaska due to an altered amylose : amylopectin ratio. It differs from Alaska in being recessive at the R_b locus (genotype $R/R,r_b/r_b$). The mature seeds are wrinkled since the cotyledons shrivel upon drying. They have phenotypically simple starch grains.

The remaining canning cultivars are known as midseason or late sweets, based on their maturity. They are considered "sweets" since they have an altered amylose : amylopectin ratio compared to round-seeded cultivars of the Alaska type. In this type the peas are sweeter, the mature seeds wrinkled from shriveling of the cotyledons upon drying, and the starch grains phenotypically compound, due to recessiveness at the R locus (genotype $r/r,R_b/R_b$).

Most freezer cultivars are wrinkle seeded due to recessiveness at the R locus (genotype $r/r,R_b/R_b$). A range in maturity is available.

Resistance to diseases is needed by all peas regardless of how and where they are grown. One of the most important diseases is common root rot, incited by the fungus *Aphanomyces euteiches* f. sp. *pisi* (Figs. 8.10 and 8.11). While resistance has not yet been

FIGURE 8.10.   Pea plant loss in the field due to common root rot caused by the fungus *A. euteiches* f. sp. *pisi.*

FIGURE 8.11.   Aphanomyces root rot of pea.

found, lines more tolerant of the disease have been developed and have improved stability of production. Fusarium wilt, caused by the fungus *Fusarium oxysporum* f. sp. *pisi* race 1, has been brought under control by incorporation of the recessive gene *fw*. Most pea cultivars now carry this resistance, noted in garden catalogs as "wilt resistant." A similiar disease called fusarium near wilt and incited by *F. oxysporum* f. sp. *pisi* race 2 (Fig. 8.12) has been controlled through use of the recessive gene *fnw*. Many cultivars have this resistance. Powdery mildew is a disease characterized by a white powdery, dustlike coating of leaves, stems, and pods by mycelium of the fungus *Erysiphe polygoni*. The disease is more prevalent on late-planted or late-maturing peas, where it can reduce yields by 50% or more and lower quality (20). At least two different genes have been shown to influence resistance (29), but in the United States satisfactory resistance has been conferred by the recessive gene *er-1*. Many later maturing cultivars are now resistant. Pea enation (Fig. 8.13) is a virus disease common in pea-growing areas of the Northeast and Northwest. Resistance can be mandatory for satisfactory production and has been achieved through use of the recessive gene *en*. One of the few virus diseases to

FIGURE 8.12.   Fusarium near wilt of pea. The one-sided wilting is characteristic of this disease.

FIGURE 8.13. Pea enation mosaic virus. The characteristic symptom is the blisterlike outgrowths on the bottom of the leaves and on the pods.

be seed transmitted, the pea seedborne mosaic virus (Fig. 8.14) has become a problem during about the last 10 years. Resistance has been found, its inheritance has been determined (*26*), and it is being incorporated into cultivars.

An exciting change in pea foliage morphology is possible (Fig. 8.15) and offers promise of easier and more efficient field harvest, more uniform shelled-pea color, and possibly other advantages (*92*). Possibly the most promising of several types available is that called afila. Plants homozygous for the recessive *af* gene have leaves that have leaflets replaced by a proliferation of tendrils. The intertwining of all these tendrils give the plants improved standing ability,which makes it possible to cut them with a swather from any direction. Badly lodged plants with conventional foliage can be cut from only one direction. The improved standing ability of the afila type also results in more uniform illumination of the pod and thus the shelled peas are more uniform in chlorophyll development and color. A few afila cultivars have been released, and more can be anticipated.

CURRENT GOALS OF BREEDING PROGRAMS

One of the major goals of any plant improvement program is to increase profitability of the crop. Since yield is a major component of profitability, improvement in yield is a major concern in developing new pea cultivars. At the same time, quality of the shelled peas must be maintained or improved to encourage consumption. Production of two or more pods per node is necessary for high yields, and these pods should be well filled with a high number of peas. Because peas rapidly pass through the optimum harvest stage (within 24 hr during hot, dry weather), the pods from only a limited number of nodes contribute to yield. In the field, generally only three to four nodes have peas that contribute to the yield, since peas at the upper nodes will not have developed sufficiently to be acceptable. Quality in the sense of color and taste is measured subjectively by panels, and new cultivars must have acceptable to superior quality compared to the cultivars they are to replace.

The thick-walled edible-podded cv. Sugar Snap pea has been such a successful new cultivar that many breeders have initiated programs to develop similar cultivars. One of the primary objectives of such programs is to eliminate the strings from the pods, such as was done to change the "string" bean to the "snap" bean. Further refinements are to have a dwarf vine, bearing double pods, and resistant to the prevalent diseases, especially powdery mildew. Of course, all this must be accomplished while maintaining the excellent flavor that helped make this cultivar successful.

The fasciated trait (gene symbol *fa*) has been utilized by some breeders in an attempt at greater uniformity of the shelled peas at harvest. This gene is variable in penetrance and expressivity, but where expression is pronounced it causes the angular stem to thicken and

FIGURE 8.14. Pea seedborne mosaic virus. In addition to a mosaic pattern in the leaflets, the stunted and malformed plants show a characteristic curling of tendrils into a tight ball.

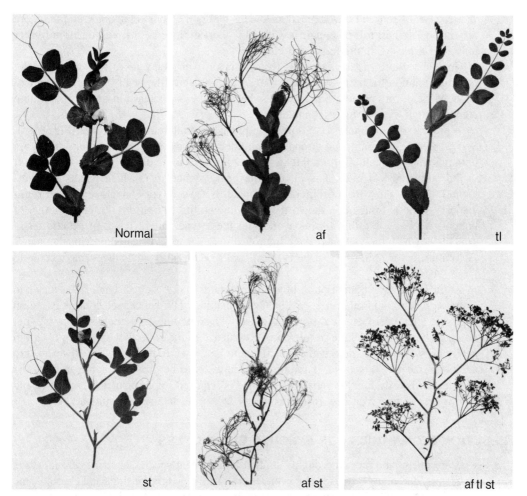

FIGURE 8.15. Some pea foliage types. Normal has the genotype *Af/Af, Tl/Tl, St/St*. The homozygous recessive gene(s) responsible for each different type is (are) shown below each type.

take on a ribbonlike appearance and the upper nodes to telescope so that flowers are produced in a more or less umbel at the top of the stem. Flowering time of fasciated plants is more concentrated than that of normal plants. The idea in incorporating this trait is that more shelled peas would reach the optimum harvest stage simultaneously, resulting in greater yield of uniform peas. The difficulty in utilizing the trait has been a tendency of fasciated plants to form pods with fewer peas in them, and to pass through the harvest stage more rapidly than normal peas, where a buffering of maturity is provided by the less mature peas at the upper nodes. Faciated plants are also more vulnerable to unfavorable weather if it should coincide with their brief period of flowering.

Just as developing two pods per node increased yields over one pod per node, there is interest in triple-podded cultivars. A few have been developed and are available, but in most trials they have not performed significantly better than the best double-podded cultivars. Under stress such as heat or drought, the triple-podded cultivars tend to have shorter pods or reduced development of ovules within the pods (blanking), and therefore may not reach their full potential. It is possible that continued effort will result in further improved triple- or multiple-podded cultivars.

Genetic variation is available for five or six or even many more flowers per node. Genetic variation for berry size exists; but since the size of the pea is one of the components of yield, any reduction in size must be offset by an increase in one or more other components if yield is to be maintained.

A present area of interest is in breeding plants with altered foliage types (Fig. 8.15). Some breeding programs are strongly committed to a new type, while many others are putting in only a small effort until they find out if one or more different types are superior to the normal. Most interest is in the afila gene (symbol *af*), which results in plants without leaflets but with a proliferation of tendrils. These types have been called semi-leafless. Various trials have shown their yielding ability to be from 90 to 100% that of normal. It is possible that further work with these types might result in genetic backgrounds in which the yields are even improved over normal. As stated previously, standing ability, color uniformity of berry, pest control, and other advantages may accrue to these types.

An important component of all breeding programs is disease resistance. Resistance to common root rot and other root rots behaves quantitatively; hence, selection is based on degree of tolerance. Genetic differences exist, with some lines extremely susceptible, but no lines show a high degree of tolerance. Resistance to fusarium wilt and often fusarium near wilt is especially important for cultivars to be grown in the Midwest. Resistance to additional races of *Fusarium* is required for cultivars to be grown in some areas of the Pacific Northwest. Enation mosaic resistance is required for most of the Northeast and Northwest. Powdery mildew resistance is needed for all late-season cultivars.

SELECTION TECHNIQUES FOR SPECIFIC CHARACTERS

Peas are one of the most thoroughly studied higher plant species in terms of qualitative characters, and many of these have been tested for linkage relationships and chromosome association (7, 102). The genetic map shown in Fig. 8.16 is quite helpful to breeders as it provides information about linkages, which may be either favorable or unfavorable depending upon the constitution of the material being hybridized. A partial listing of genes useful in the breeding program is given in Table 8.2.

Lines often differ at the R locus (round versus wrinkled seed), especially when one wishes to incorporate a trait from a plant introduction. Since dominance is expressed in cotyledons of the first filial generation, all F_1 seeds between contrasting types will be round. However, selection can be practiced for this trait in the F_2 seeds produced on F_1 plants. Classification is usually straightforward though in some genetic backgrounds discrimination can be difficult. Crosses involving the R_b locus behave similarly.

Another trait that often differs in lines being crossed is cotyledon color as determined by alleles at the I locus. Dominance results in yellow cotyledons and recessiveness in green. Almost all garden and canning or freezing cultivars possess green cotyledons. Since this trait is expressed in the cotyledons, selection can be practiced among F_2 seeds produced on F_1 plants. Because green seeds may bleach in strong light, prolonged exposure of mature seeds may make classification difficult. The seed coat can mask cotyledon color and so it is usually necessary to chip away a portion to expose a section of the cotyledon when scoring.

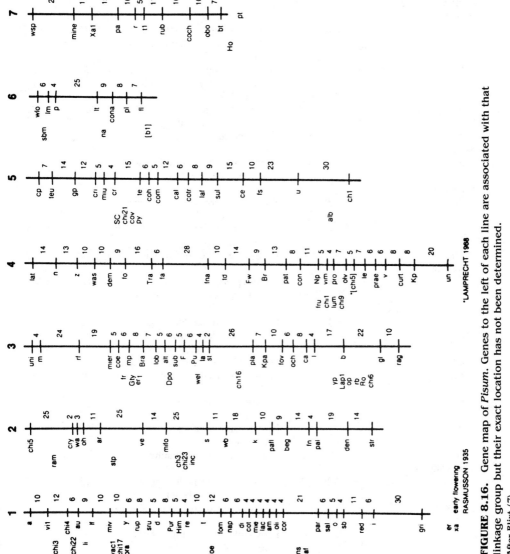

FIGURE 8.16. Gene map of *Pisum*. Genes to the left of each line are associated with that linkage group but their exact location has not been determined. After Blixt (7).

305

TABLE 8.2. Some Simply Inherited Characters of Pea and Their Gene Symbols and Chromosome Associations

Symbol	Chromosome number	Character description	Reference Symbol	Linkage
Plant Characters				
Height				
cry	2	Influences length of internodes and plant height along with la and le	81	35,71
la	3	Influences length of internodes and plant height along with cry and le	24	36,71
le	4	Influences length of internodes and plant height along with cry and la	100	43
Root				
age		Main root grows upward to soil surface and then horizontally	6	
Branching				
fr	3	With fru determines the number of basal branches	44	4
fru	4	With fr determines the number of basal branches	44	4
ram	2	Increases number of branches	74	73
Fasciation				
fa	4	With fas casus stem fasciation	100	47
fas	3	With fa causes stem fasciation	47	9
Leaves and Stipules				
af	1	Leaflets converted to tendrils	15	31
cri	5	Leaves, stipules, flowers, and pods folded and crisp	37	37
lat	4	Doubles leaflet and stipule area	38	38
st	3	Stipules reduced to straplike structures	93	43
tac		Tendrils present on acacia leaves	83	
tl	7	Leaves with extra leaflets and no tendrils	100	43
un	4	Leaflet margins undulating	58	58
Wax (bloom)				
wa	2	Without wax on pods, upper and lower stipule surfaces, and underside of leaflets	95	53
was	4	Reduced wax as with wa	38	38
wb	2	Pods without wax; little wax on rest of plant	95	43
wel	3	Wax absent from all parts of the plant	67	67
wex		Extra wax on all aerial plant parts	67	
wlo	6	Upper surface of leaflets without wax; other plant parts with normal wax	77	43

TABLE 8.2. (*cont.*)

Symbol	Chromosome number	Character description	Reference Symbol	Reference Linkage
wsp	7	Plants without wax except upper suface of leaflets almost normally waxy	*41*	*50*
Color				
a	1	Absence of anthocyanin; dominant allele required for anthocyanin production in plant, flower, and seed	*100*	*43*
alb	5	Plants without chorophyll; albina; lethal	*61*	*3*
alt	3	Plants deficient in chlorophyll from fifth node upward; lethal	*52*	*59*
au	1	Seedling golden yellow; lethal	*45*	*45*
ch-1	5	Plant light yellowish green	*62*	*62*
d	1	Green leaf axil; *D* dependent on *A* for manifestation of color	*100*	*43*
fl	6	No spotting or flecking of leaves or stipules	*85*	*43*
pa	7	Dark-green immature seed and foliage	*101*	*43*
vim	4	Effect similar to *pa*	*54*	*56*
xa-1	7	Seedlings pale yellow and dying upon exhaustion of cotyledon food reserve	*51*	*51*
Inflorescence				
Number of flowers				
fn	2	With *fna* determines number of flowers on the inflorescence; greatly influenced by environment	*100*	*46*
fna	4	With *fn* determines number of flowers on the inflorescence; greatly influenced by environment	*42*	*64*
Morphology				
coch	7	Flowers irregular and open; reduced fertility; stipules spatula shaped	*98*	*98*
k	2	Wings about same size and shape as keel and closely appressed to it	*80*	*43*
nap	1	Keel petals broad and winglike; more or less open	*49*	*65*
Pollen color				
yp	3	Yellow pollen	*75*	*75*
Color				
b	3	Flower pink; dependent on *A* for manifestion of color	*100*	*43*

(*continued*)

TABLE 8.2. (*cont.*)

Symbol	Chromosome number	Character description	Reference Symbol	Linkage
ce	5	Flower rose; dependent on *A* for manifestion of color	*97*	*97*
Seeds				
Form				
com	5	Sides of seeds flattened	*60*	*60*
di	6	Small dimpled depressions in seed; observable only with *a R* seeds	*96*	*96*
r	7	Seed cotyledons wrinkled	*100*	*43*
r_b	3	Seed cotyledons wrinkled	*33*	*17*
Surface				
gty	3	Surface texture gritty	*68*	*69*
Color				
F	3	Bluish-violet spots on seedcoat; dependent on *A Z Mp B* for manifestation; duplicate to *Fs*	*100*	*43*
fs	5	Bluish-violet spots on seedcoat; dependent on *A Z Mp B* for manifestation; duplicate to *F*	*101*	*43*
i	1	Green cotyledons; *I* produces yellow cotyledons	*100*	*43*
M	3	Brown marbling of testa with *A Z Mp*	*100*	*43*
Pl	6	Hilum black with *A Ar B*	*100*	*43*
Tra	4	Tragacanth on inside of seedcoat causing cotyledons and testa to stick together in spots; best observed with *R* or visible as flakes under a binocular	*55*	*55*
U	5	Testa black to brown violet with *A z*	*100*	*43*
Z	4	Basic gene with *A* for coloring of testa	*30*	*43*
Adherence to each other				
s	2	Dry seeds in pod stuck together with tragacanth on exterior of testa; expression varies with size of seeds and distance between them	*100*	*43*
Protein electrophoretic migration				
Lap	3	*Lap*[f] (fast moving) and *Lap*[s]: (slow moving) band of leucine aminopeptidase with starch gel electrophoresis	*1*	*1*
Pods				
Breadth				
lt	6	Increases pod width 25%	*57*	*57*

TABLE 8.2. (*cont.*)

Symbol	Chromosome number	Character description	Reference Symbol	Linkage
Form				
Bt	7	Apex of pods blunt	100	43
Con	4	Affects curvature of pod	39	63
Wall thickness				
n	4	Pod wall thick	94	43
Fiber				
Dpo	3	Pods tough and leathery; readily dehisce at maturity	7	69
p	6	Reduces or eliminates sclerenchymatous membrane on inner pod walls	100	43
v	4	Reduces or eliminates sclerenchymatous membrane on inner pod walls	100	43
Color				
gp	5	Young pod yellow	100	43
Pu	3	With A Pur pod color purple	48	63
Pur	1	With A Pu pod color purple	40	43
External growths				
Np	4	Proliferation of epidermal pod cells (neoplasm) associated with low light intensity	78	14
Disease Resistance				
Enation mosaic virus				
En		Resistant to enation mosaic virus	82	22
Fusarium wilt and near wilt				
Fnw	4	Resistant to *F. oxysporum* race 2	99	
Fw	4	Resistant to *F. oxysporum* race 1	90	90
Pea seedborne mosaic virus				
sbm	6	Resistant to pea seedborne mosaic virus	25	21
Powdery mildew				
er-1	3	Resistant to *E. polygoni*	27	69
		Moisture Stress		
wil	3	Wilts quickly under moisture stress	70	70

Except for some edible-podded cultivars, most garden and processing peas have white flowers. Field peas often have colored flowers, as do many plant introductions. The presence or absence of color is dependent upon a basic anthocyanin gene at the *A* locus. Dominance results in color, while homozygous recessive plants have white flowers. Alleles at several other loci determine the shade of color in the flower, pod color, color in leaf axils, and coloration of the seed coat.

Disease Resistance

Screening for common root rot caused by *A. euteiches* is generally done in the greenhouse and in field disease nurseries. In the greenhouse, roots of seedlings are usually inoculated with a mycelial–zoospore suspension of the fungus. The seedlings are then replanted in the growth medium and held at about 25°C. The lines are scored approximately 2 weeks after inoculation (*84*). Tests are highly variable with many escapes; hence the test works best where several plants of a line are available for testing. Usually, a line known to be especially susceptible and one or more lines that possess some degree of tolerance are included in the test as reference checks. Field evaluation employs a disease nursery in which the titer of the fungus has been brought to a high level by successive crops of peas. Seeds protected against seed and seedling rots by a fungicide such as Captan are planted when the soil has warmed to about 15°C. Adequate moisture provided by rainfall or irrigation is needed for good disease development. Plots are rated as the disease develops, with susceptible and tolerant checks included as references. Seed may be saved from plants surviving to maturity.

Screening for resistance to the different races of *Fusarium* is usually done by root inoculation of seedlings grown in the greenhouse (*84*). Field nurseries may also be employed, and such nurseries are commonly used in the Northwest.

Resistance to powdery mildew can usually be determined by planting late in the field, exposing the plants to high spore concentrations during midsummer. The disease will also develop under greenhouse conditions, and screening of limited numbers of plants may take place in this way. Infestations may be initiated by moving an infected plant into the house.

Both enation mosaic virus resistance and pea seedborne mosaic virus resistance may be determined by spraying sap expressed from infected plant tissue onto leaf areas where the cells have been ruptured by rubbing with carborundum powder (*82, 26*).

DESIGN OF BREEDING PROGRAM

The most common breeding strategies employed in pea improvement programs are backcross, pedigree, single-seed descent and bulk, and various modifications of them.

Because successful cultivars are a rare combination of desirable quality attributes and superior agronomic characteristics, it may be desirable to maintain that unique genetic constitution while incorporating one or a few simply inherited improvements. The backcross procedure is well suited to this need. The desirable line is used as the recurrent parent while the nonrecurrent parent serves as the donor of the desired gene. For instance, disease resistance may be transferred to an otherwise satisfactory cultivar, thereby prolonging its usefulness for years. Another advantage of the backcross system is that selection need be practiced only for the trait under transfer, often making it possible to advance generations through the greenhouse or winter nursery, where expression of the recurrent line might not be typical of its normal field performance. Five backcrosses in 2 years are easily accomplished for a dominant character because the F_1 progeny carrying the dominant allele can be easily identified. To transfer a recessive character, one can make a simultaneous testcross of each F_1 plant used to make the next backcross (Fig. 8.17). The heterozygotes carrying the desired recessive allele can thereby be detected and their backcross F_1 progeny selected as parents for the next backcross. The one generation of selfing to obtain true breeding lines with the desirable trait will establish the improved line.

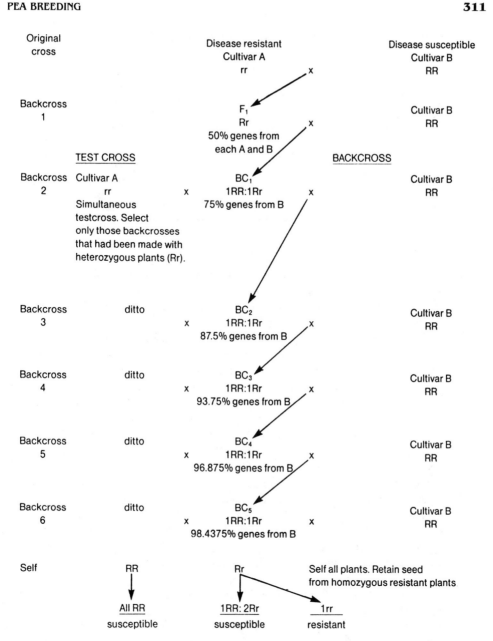

Original
cross

Disease resistant
Cultivar A
rr   x

Disease susceptible
Cultivar B
RR

Backcross
1

F₁
Rr   x
50% genes from
each A and B

Cultivar B
RR

TEST CROSS

BACKCROSS

Backcross
2

Cultivar A
rr   x
Simultaneous
testcross. Select
only those backcrosses
that had been made with
heterozygous plants (Rr).

BC₁
1RR:1Rr   x
75% genes from B

Cultivar B
RR

Backcross
3

ditto
          x

BC₂
1RR:1Rr   x
87.5% genes from B

Cultivar B
RR

Backcross
4

ditto
          x

BC₃
1RR:1Rr   x
93.75% genes from B

Cultivar B
RR

Backcross
5

ditto
          x

BC₄
1RR:1Rr   x
96.875% genes from B

Cultivar B
RR

Backcross
6

ditto
          x

BC₅
1RR:1Rr   x
98.4375% genes from B

Cultivar B
RR

Self

RR
↓
All RR
susceptible

Rr
↓
1RR: 2Rr
susceptible

Self all plants. Retain seed
from homozygous resistant plants

1rr
resistant

FIGURE 8.17. Procedure for transferring a recessive gene for disease resistance into a successful cultivar. The procedure shown incorporates the trait in the fastest possible time using a testcross to identify backcross plants carrying the desired gene. An alternative procedure, which requires additional generations, is to employ a generation of selfing after each backcross to allow the homozygous resistant types to come to expression.

TABLE 8.3. Pedigree–Bulk or Pedigree–Single-Seed Descent Breeding Scheme

Operation	Location	Season
Make cross	Greenhouse	Fall
Grow F_1 plants	Greenhouse	Spring
Grow F_2 plants and select desired types	Field	Summer
Grow F_3 plants to advance generations	Greenhouse	Fall
Grow F_4 plants to advance generations	Greenhouse	Spring
Grow F_5 plants and select desired types	Field	Summer
Disease test F_6 plants	Greenhouse	Fall
Increase seed of desired F_6 plants	Greenhouse	Spring
Grow single-plant selections on trellises	Field	Summer

The pedigree system or a modification of it has long been employed as the main pea improvement technique. A breeding scheme using a combination of pedigree–bulk or pedigree–single-seed descent and the way in which it may be used in pea breeding is shown in Table 8.3. It provides for new combinations of genes that can lead to progeny superior to any of the parents involved in the cross. It is common to advance generations in the greenhouse or winter nursery with minimum or no selection while practicing rigorous selection during the field season under environmental conditions such as the derived cultivar will encounter in production. Because uniformity is essential for canning and freezing, the final single-plant selection will not be made until F_6 or later. An extensive program of testing will be required to evaluate the worth of the selections. It has been found through experience that superior lines are much more likely to be extracted from hybridizations among successful cultivars than from crosses involving one or more unadapted or inferior lines. Where the inferior lines have some characteristics that are desired in a new cultivar, one or two backcrosses to a desirable parent may be used to increase the proportion of desirable genes, followed by typical pedigree selection.

The newest technique to receive widespread use is that of single-seed descent as proposed by Brim (*10*) in soybeans. Again, this has been most successful where highly selected desirable lines are intercrossed. Several F_1 plants are grown to produce 500 or more F_2 seeds. One seed only is harvested from each F_2 plant and all seeds are bulked and used to plant the F_3 generation. This procedure is followed to about the F_6 generation, at which time inferior progeny are discarded and phenotypically superior plants are selected for future evaluation. Lines obviously inferior will have been discarded along the way, and lines that fail to propagate themselves due to inviability of seed or unthrifty plants are usually genetically inferior and not worthy of continued propagation. A great advantage of this system is that it can be conducted with a minimum of time and effort on the part of the breeder, and can proceed in the greenhouse and field without regard to environmental effects on the phenotype. Several populations can be carried along simultaneously with other efforts such as backcrossing, pedigree, and evaluation of selections.

A limited effort has been put into mutation breeding, mostly for resistance to root rot, but without great success to date.

Recurrent selection has been practiced, primarily for root rot resistance, and has resulted in the release of two breeding lines possessing more tolerance to common root rot than currently available cultivars (*13, 32*).

Though the pea is a normally self-fertilizing species, hybrid vigor for many traits has long been noted. Dry-seed yield of F_1 plants compared to their midparent values averaged 56% higher in a study by Krarup and Davis (*34*), who found midparent heterosis values for pods/plant, peas/pod, and weight/seed to be 32, 10, and 7%, respectively. Gritton (*18*) reported 55% midparent heterosis and 28% high parent heterosis for dry-seed yield. Corresponding values for pods/plant, seeds/pod, and weight/seed were 31, 8, and 1%, respectively. In spite of the heterosis shown, most studies have indicated that for most traits, the genetic variance is predominantly additive. A range of narrow sense heritability estimates from 0 to 41% has been reported for dry-seed yield (*79*). All programs are designed for the development and release of pure lines.

TRIALS OF ADVANCED LINES

Seed of single-plant selections made in the F_6 to F_8 generation will usually number from 10 to 50. A few seeds are generally held back as reserve in case the planting is lost, while the remainder are planted in the field nursery where the plants will be grown in one unreplicated plot along trellises (Fig. 8.18). Spacing between seeds will be about 2–6 in. whether they are planted along only one side or both sides of the trellis. Trellises will be spaced 3–10 ft apart. This relatively low plant population will permit the plants to develop and produce to their maximum potential. It will also permit the breeder to make detailed observations and measurements of individual plants. The node at which the first flower is borne will be recorded because this is remarkably constant for a line and gives an indication of a line's maturity. Flowering date will be recorded when 50% or more of the plants in the plot have at least one open flower. Counts are made of the number of flowers per node and later the number of pods set per node as an indication of plant productivity. The number of peas per pod is counted or estimated, and the degree of "blanking," spaces in the pod where seeds failed to develop, will be noted. Size of the shelled peas (sieve size) will be recorded. Vine length will be measured.

FIGURE 8.18. Part of a pea field nursery. Plants are being grown along widely spaced wire trellises.

The number of plants to be sampled for yield of fresh shelled peas, number of pods per plant, number of peas per pod, and weight per ovule for given degrees of precision have been estimated (*19*). With a cost ratio of replication to plant of 18 : 1, 16 plants is indicated as the optimum number to sample per replicate for weight of shelled fresh peas, with three, five, and eight replications required for coefficients of variation of 13, 10, and 8%, respectively. Since yield of fresh shelled peas was the most variable trait, followed in order by number of pods per plant, number of peas per pod, and finally weight per ovule, it would be estimated with greater precision. Experience, though, has indicated this would be a minimum number of plants and replications, since pea performance is greatly influenced by the environment.

General thriftiness as well as any susceptibility to diseases will be recorded. Since these observations are made under conditions similar to those encountered in the garden, the evaluations should be directly applicable. Such observations are valid for highly heritable traits where the plants are to be grown under high densities in the field, but will not be sufficient to characterize the line for all traits (*16*).

Often some measurements of individual plants are made when the plants are mature. This avoids the problem of trying to evaluate each line at exactly the proper green pea harvest stage, and provides seed for maintaining and increasing the line. Again, highly heritable traits can be evaluated at maturity, but care must be exercised in interpretation of such observations (*16*).

The second step in evaluating a line is usually to plant 100–500 seeds along a trellis. These may be replicated or not. The same observations that were made on the single-plant plot are repeated here, and in addition a portion of the plot may be harvested at the proper green pea stage. This sample will be used to determine yield, sieve size, color, and other quality characteristics.

FIGURE 8.19. Long, narrow plots of peas planted with very limited amounts of seed early in the evaluation process. Only the center rows will be harvested for yield.

FIGURE 8.20. A pea field evaluation trial planted with a grain drill. The entire plot can be harvested by machine or the center area only can be pulled by hand.

The third evaluation step will usually involve growing plots under conditions that simulate those used in commercial canning or freezing pea production (Fig. 8.19). In a study of optimum plot size and shape estimates for pea yield trials, Zuhlke and Gritton (*104*) found that the optimum size for unguarded plots was 36 ft^2, while for bordered plots the estimate was 33 ft^2. Differences among plot shapes were not significant except that variances of some larger plots were lower for long, narrow shapes. The randomized complete block design is often used in these evaluations; however, Zuhlke and Gritton (*105*) found that lattices were always equal to or more efficient than randomized complete blocks, ranging from 100% with some short, wide plots in smaller block sizes to 236% for a long, narrow plot with a block size of seven.

Plants from small block sizes may be pulled by hand. This requires considerable hand labor and is not duplicative of the way fields are harvested; hence many investigators prefer to use a large block size (Fig. 8.20) and harvest it by machine. This may involve a plot 7–10 ft wide or whatever width is planted by the grain drill used, and will probably be 10–50 ft long. Sometimes winter wheat or other cereal is planted in the outside rows to provide competition and to separate plots of peas. In other instances, plots are separated by clean-tilled soil. Here, the assumption is made that border effect will affect all entries in the test similarly. The plots are cut and windrowed in the conventional manner, and threshed with a mobile combine. Detailed data on plant characteristics such as yield and sieve size are collected, and the shelled peas will usually be canned or frozen to evaluate their quality.

Those entries deemed superior in all trials in the above series will be offered to commercial canners and freezers for them to grow in small plots under their own conditions in their own area of production. Since they are the customers, they make the final decisions on which breeding lines will be accepted.

REFERENCES

1. Almgard, G., and Ohlund, K. 1970. Inheritance and location of a biochemical character in *Pisum*. Pisum Newsl. *2*, 9.
2. Anon. 1980. Seed for Today. A Descriptive Catalog of Vegetable Varieties, No. 23. Asgrow Seed Co., Kalamazoo, MI.
3. Blixt, S. 1966. Linkage studies in *Pisum*. V. The mutant albina$_{R2}$. Agri Hort. Genet. *24*, 168–172.
4. Blixt, S. 1968. Linkage studies in *Pisum*. XII. Linkage relations of the genes *Fr* and *Fru*, determining ramification. Agri Hort. Genet. *26*, 136–141.
5. Blixt, S. 1970. *Pisum*. *In* Genetic Resources in Plants—Their Exploration and Conservation. O. H. Frankel and E. Bennett (Editors), Int. Biol. Programme, pp. 321–326. Blackwell Scientific Publications, Oxford.
6. Blixt, S. 1970. The *ageotropum* mutant. Pisum Newsl. *2*, 11–12.
7. Blixt, S. 1972. Mutation genetics in *Pisum*. Agri Hort. Genet. *30*, 1–293.
8. Blixt, S. 1975. The pea. *In* Handbook of Genetics. R. S. King (Editor), pp. 181–221. Plenum Press, New York.
9. Blixt, S. 1976. Linkage studies in *Pisum*. XV. Establishing the *rms*-gene and the linkage of *rms* and *fas* in chromosome III. Agri Hort. Genet. *34*, 83–87.
10. Brim, C. A. 1966. A modified pedigree method of selection in soybeans. Crop Sci. *6*, 220.
11. Cooper, D. C. 1938. Embryology of *Pisum sativum*. Bot. Gaz. (Chicago) *100*, 123–132.
12. Darlington, C. D., and Janaki, Ammal, E. K. 1945. Chromosome Atlas of Cultivated Plants. George Allen & Unwin, London.
13. Davis, D. W., Shehata, M. A., and Bisonnette, H. L. 1976. Minnesota 108 pea breeding line. HortScience *11*, 434.
14. Dodds, K. S., and Matthews, P. 1966. Neoplastic pod in the pea. J. Hered. *57*, 83–85.
15. Goldenberg, J. B. 1965. "Afila," a new mutation in pea (*Pisum sativum* L.). Bol. Genet. *1*, 27–31.
16. Gritton, E. T. 1969. Comparison of planting arrangements and time of evaluation for peas (*Pisum sativum* L.). Crop Sci. *9*, 276–279.
17. Gritton, E. T. 1971. A case of apparent linkage between the gene *rb* for wrinkled seed and *st* for reduced stipule. Pisum News. *3*, 15.
18. Gritton, E. T. 1975. Heterosis and combining ability in a diallel cross of peas. Crop Sci. *15*, 453–467.
19. Gritton, E. T., and Chi, P. Y. 1972. Sampling procedures and optimum sample size for estimating yield components in peas (*Pisum sativum* L.). J. Am. Soc. Hortic. Sci. *97*, 451–453.
20. Gritton, E. T., and Ebert, R. D. 1975. Interaction of planting date and powdery mildew on pea plant performance. J. Am. Soc. Hortic. Sci. *100*, 137–142.
21. Gritton, E. T., and Hagedorn, D. J. 1975. Linkage of the genes *sbm* and *wlo* in peas. Crop Sci. *15*, 447–448.
22. Gritton, E. T., and Hagedorn, D. J. 1980. Linkage of the *en* and *st* genes in peas. Pisum News. *12*, 26–27.
23. Gritton, E. T., and Wierzbicka, B. 1975. An embryological study of a *Pisum sativum* × *Vicia faba* cross. Euphytica *24*, 277–284.
24. Haan, H. de 1927. Length factors in *Pisum*. Genetica *9*, 481–498.
25. Hagedorn, D. J., and Gritton, E. T. 1971. Inheritance and linkage of resistance to the pea seed borne mosaic virus. Pisum Newsl. *3*, 16.
26. Hagedorn, D. J., and Gritton, E. T. 1973. Inheritance of resistance to the pea seed-borne mosaic virus. Phytopathology *63*, 1130–1133.
27. Harland, S. C. 1948. Inheritance of immunity to mildew in Peruvian forms of *Pisum sativum*. Heredity *2*, 263–269.

28. Hedrick, V. P. 1928. Peas of New York, Vol. 1, Part 1. New York State Agric. Exp. Stn., Geneva, NY.

29. Heringa, R. J., van Norel, A., and Tazelaar, M. F. 1969. Resistance to powdery mildew (*Erysiphe polygoni* D. C.) in peas (*Pisum sativum* L.). Euphytica *18*, 163–169.

30. Kajanus, B. 1923. Genetische Studien an *Pisum*. Z. Pflanzenzuecht. *9*, 1–22.

31. Khangildin, W. V. 1966. A new gene, *leaf*, inducing the absence of leaflets in peas. Interaction between the genes *leaf* and *Tl^w*. Genetika *6*, 88–96.

32. King, T. H., Davis, D. W., Shehata, M. A., and Pfleger, F. L. 1981. Minnesota 494-All pea germplasm. HortScience *16*, 100.

33. Kooistra, E. 1962. On the differences between smooth and three types of wrinkled peas. Euphytica *11*, 357–373.

34. Krarup, A., and Davis, D. W. 1970. Inheritance of seed yield and its components in a six-parent diallel cross in peas. J. Am. Soc. Hortic. Sci. *95*, 795–797.

35. Lamm, R. 1937. Length factors in dwarf peas. Hereditas *23*, 38–48.

36. Lamm, R. 1947. Studies on linkage relations of the *Cy* factors in *Pisum*. Hereditas *33*, 405–419.

37. Lamm, R. 1949. Contributions to the genetics of the *Gp*-chromosome of *Pisum*. Hereditas *35*, 203–214.

38. Lamm, R. 1957. Three new genes in *Pisum*. Hereditas *43*, 541–548.

39. Lamprecht, H. 1936. Genstudien an *Pisum sativum*. I. Über den Effekt der Genpaare *Con-con* und *S-s*. Hereditas *22*, 336–360.

40. Lamprecht, H. 1938. Über Hülseneigen-schaften bei *Pisum*, ihre Vererbung und ihr züchterischer Wert. Züchter *10*, 150–157.

41. Lamprecht, H. 1939. Genstudien an *Pisum sativum*. IV. Über Vererbung von Waschslosigkeit und ein neues Gen für lokale Ausbildung von Wachs, *Wsp*. Hereditas *25*, 459–471.

42. Lamprecht, H. 1947. En-, två-och treblommighetens praktiska betydelse vid växtförädlingsarbete med ärter. Agri Hort. Genet. *4*, 79–98.

43. Lamprecht, H. 1948. The variation of linkage and the course of crossingover. Agri Hort. Genet. *6*, 10–48.

44. Lamprecht, H. 1950. The degree of ramification in *Pisum* caused by polymeric genes. Agri Hort. Genet. *8*, 1–6.

45. Lamprecht, H. 1952. Über Chlorophyllmutanten bei *Pisum* und die Verenbung einer neuen, goldgelben Mutante. Agri Hort. Genet. *10*, 1–18.

46. Lamprecht, H. 1952. Weitere Koppelung-studien an *Pisum sativum*, insbesondere im Chromosom II (*Ar*). Agri Hort. Genet. *10*, 51–74.

47. Lamprecht, H. 1952. Polymere Gene und Chromosomenstruktur bei *Pisum*. Agri Hort. Genet. *10*, 158–168.

48. Lamprecht, H. 1953. New and hitherto known polymeric genes of *Pisum*. Agri Hort. Genet. *11*, 40–54.

49. Lamprecht, H. 1953. Bisher bekannte und neue Gene für die Morphologie der *Pisum* Blüte. Agri Hort. Genet. *11*, 122–132.

50. Lamprecht, H. 1954. Die Koppelung des Gens *Wsp* und die Genenkarte von Chromosom VII von *Pisum*. Agri Hort. Genet. *12*, 115–120.

51. Lamprecht, H. 1955. Die Koppelung des Chlorophyllgens *Xa 1* von *Pisum*. Agri. Hort. Genet. *13*, 115–120.

52. Lamprecht, H. 1955. Die Vererbung der Chlorophyllmutante *albinaterminalis* von *Pisum* sowie Allgemeines zum Verhalten von Chlorophyll- und andere Genen. Agri Hort. Genet. *13*, 103–114.

53. Lamprecht, H. 1955. Studien zur Genenkarte von Chromosom II von *Pisum*. Die Koppelung des gens *Wa* und die Wirkung der übrigen Wachsgene. Agri Hort. Genet. *13*, 154–172.

54. Lamprecht, H. 1955. Zur Kenntnis der Genenkarte von Chromosom VII von *Pisum* sowie die Wirkung der Gene *Tram* und *Vim*. Agri Hort. Genet. *13*, 214–229.

55. Lamprecht, H. 1956. Über Wirkung und Koppelung des Gens *Tram* von *Pisum*. Agri Hort. Genet. *14*, 45–53.

56. Lamprecht, H. 1957. Die Lage der Gene *Ve* und *Vim* in den Chromosomen II besw. IV von *Pisum* sowie weitere Koppelungstudien. Agri Hort. Genet. *15*, 1–11.

57. Lamprecht, H. 1957. Über die Vererbung der Hülsenbreite bei *Pisum*. Agri Hort. Genet. *15*, 105–114.

58. Lamprecht, H. 1958. Gekräuselte Blättchen bei *Pisum* and ihre Vererbung. Agri Hort. Genet. *16*, 1–8.

59. Lamprecht, H. 1959. Über Wirkung and Koppelung des Gens *Alt* von *Pisum*. Agri Hort. Genet. *17*, 15–25.

60. Lamprecht, H. 1960. Zur Vererbung der Samenformen bei *Pisum* sowie über zwei neue, diese beeinflussende Gene. Agri Hort. Genet. *18*, 1–22.

61. Lamprecht, H. 1960. Über Blattfarben von Fanerogamen. Klassifikation, Terminologie und Gensymbole von Chlorophyllund anderen Farbmutanten. Agri Hort. Genet. *18*, 135–168.

62. Lamprecht, H. 1960. Zwei neue Chlorophyllmutanten von *Pisum*, *chlorina–virescens* und *chlorina–virescens–chlorotica–terminalis*, sowie zur Koppelung des *Ch* - gens. Agri Hort. Genet. *18*, 169–180.

63. Lamprecht, H. 1961. Die Genenkarte von *Pisum* bei normaler Struktur der Chromosomen. Agri Hort. Genet. *19*, 360–401.

64. Lamprecht, H. 1968. Die neue Genenkarte von *Pisum* und warum Mendel in seinen Erbsenkreuzungen keine Genenkoppelung gefunden hat. Arbeiten aus der Steiermärkischen Landesbibliothek am Joanneum in Graz, No. 10. Graz, Austria.

65. Lamprecht, H. 1974. Monograph of the genus *Pisum*. *In* Steiermarkische Landesdruckerei. K. Mecenovic (Editor), 655 pp. Graz, Austria (in German).

66. Layne, R. E. C., and Hagedorn, D. J. 1963. Effect of vacuum-drying, freeze-drying, and storage environment on the viability of pea pollen. Crop Sci. *3*, 433–436.

67. Marx, G. A. 1969. Two additional genes conditioning wax formation. Pisum Newsl. *1*, 10–11.

68. Marx, G. A. 1969. A new seed gene. Pisum Newsl. *1*, 11–12.

69. Marx, G. A. 1971. New linkage relations for chromosome III of *Pisum*. Pisum Newsl. *3*, 18–19.

70. Marx, G. A. 1976. "Wilty": A new gene of *Pisum*. Pisum Newsl. *8*, 40–41.

71. Marx, G. A. 1978. Map positions of *La* and *Cry*. Pisum Newsl. *10*, 41–42.

72. McComb, J. A. 1975. Is intergeneric hybridization in the Leguminosae possible? Euphytica *24*, 497–502.

73. Monti, L. M. 1970. Linkage studies in four induced mutants of peas. Pisum Newsl. *2*, 21.

74. Monti, L. M., and Scarascia-Mugnozza, G. T. 1967. Mutazioni per precocita e ramosita indotte in pisello. Genet. Agrar. *21*, 301–312.

75. Murfet, I. C. 1967. Yellow pollen—a new gene in *Pisum*. Heredity *22*, 602–607.

76. Myers, J. R. 1984. Male sterility, structural sterility and 2*n* pollen in the pea (*Pisum sativum* L.) Ph.D. Thesis Univ. Wisconsin, Madison. 194 pp.

77. Nilsson, E. 1933. Erblichkeitsversuche mit *Pisum*. VI–VIII. Hereditas *17*, 197–222.

78. Nuttal, V. W., and Lyall, L. H. 1964. Inheritance of neoplastic pod in the pea. J. Hered. *55*, 184–186.

79. Pandey, S., and Gritton, E. T. 1976. Observed and predicted response to selection for protein and yield in peas. Crop Sci. *16*, 289–292.

80. Pellew, C., and Sverdrup, A. 1923. New observations on the genetics of peas (*Pisum sativum*). J. Genet. *13*, 125–131.

81. Rasmusson, J. 1928. Genetically changed linkage-values in *Pisum*. Hereditas *10*, 1–152.

82. Schroeder, W. T., and Barton, D. W. 1958. The nature and inheritance of resistance to the pea enation mosaic virus in garden pea *Pisum sativum* L. Phytopathology *48*, 628–632.

83. Sharma, B. 1972. "Tendrilled Acacia," a new mutation controlling tendril formation in *Pisum sativum*. Pisum Newsl. *4*, 50.

84. Shehata, M., Grau, C. R., Davis, D. W., and Pfleger, F. L. 1976. A technique for concurrently evaluating resistance in peas to *Fusarium oxysporum* f. sp. *pisi* and *Aphanomyces euteiches*. Plant Dis. Rep. *60*, 1024–1026.

85. Tedin, H., and Tedin, O. 1926. Contributions to the genetics of *Pisum*. IV. Leaf axil colour and grey spottings on the leaves. Hereditas *7*, 102–108.

86. U.S. Department of Agriculture 1979. Agricultural Statistics. U.S. Government Printing Office, Washington, DC.

87. U.S. Department of Agriculture 1981. Crop Production. Annual Summary; Acreage, Yield, Production. Crop Reporting Board, Statistical Reporting Service, USDA, Washington, DC.

88. U.S. Department of Agriculture 1981. Vegetables. Annual Summary; Acreage, Yield, Production and Value. Crop Reporting Board, Statistical Reporting Service, USDA, Washington, DC.

89. Vavilov, N. I. 1926. Studies on the origin of cultivated plants. Bull. Appl. Bot. Plant Breed. *16*, 139–248.

90. Wade, B. L. 1929. The inheritance of *Fusarium* wilt resistance in canning peas. Res. Bull.— Wis., Agric. Exp. Stn. *97*.

91. Warnock, S. J., and Hagedorn, D. J. 1954. Stigma receptivity in peas (*Pisum sativum* L.). Agron. J. *46*, 274–277.

92. Wehner, T. C., and Gritton, E. T. 1981. Horticultural evaluation of eight foliage types of peas near-isogenic for the genes *af*, *tl* and *st*. J. Am. Soc. Hortic. Sci. *106*, 272–278.

93. Wellensiek, S. J. 1925. Genetic monograph on *Pisum*. Bibl. Genet. *2*, 343–476.

94. Wellensiek, S. J. 1925. *Pisum*—crosses. I. Genetica *7*, 1–64.

95. Wellensiek, S. J. 1928. Preliminary note on the genetics of wax in *Pisum*. Am. Nat. *62*, 94–96.

96. Wellensiek, S. J. 1943. *Pisum*—crosses. VI. Seed-surface. Genetica *23*, 77–92.

97. Wellensiek, S. J. 1951. *Pisum*—crosses. IX. The new flower colour "cerise." Genetica (The Hague) *25*, 525–529.

98. Wellensiek, S. J. 1962. The linkage relations of the *cochleata-mutant in Pisum*. Genetica (The Hague), *33*, 145–153.

99. Wells, D. J., Walker, J. C., and Hare, W. W. 1949. A study of linkage between factors for resistance to wilt and near wilt of garden peas. Phytopathology *39*, 907–912.

100. White, O. E. 1917. Studies of inheritance in *Pisum*. II. The present state of knowledge of heredity and variation in peas. Proc. Am. Philos. Soc. *56*, 487–588.

101. Winge, Ö. 1936. Linkage in *Pisum*. C. R. Trav. Lab. Carlsberg, Ser. Physiol. *21*, 271–393.

102. Yarnell, S. H. 1962. Cytogenetics of the vegetable crops. III. Legumes. A. Garden peas, *Pisum sativum* L. Bot. Rev. *28*, 465–537.

103. Zohary, D., and Hopf, M. 1973. Domestication of pulses in the old world. Science *182*, 887–894.

104. Zuhlke, T. A., and Gritton, E. T. 1969. Optimum plot size and shape estimates for pea yield trials. Agron. J. *61*, 905–908.

105. Zuhlke, T. A., and Gritton, E. T. 1970. Relative precision of different experimental designs and number of replications in pea yield trials. Agron. J. *62*, 61–64.

9
Carrot
Breeding

C. E. PETERSON
P. W. SIMON

The carrot is widely grown for use both fresh and processed. It provides an excellent source of vitamin A and fiber in the diet. Annual per capita consumption in the United States is estimated at about 7.5 lb; approximately 0.5 lb is frozen and 1.0 lb canned (69). These data for consumption probably are short of actual use because production is nearly 10 lb per capita. The estimates for processed consumption do not include a substantial tonnage of both processing and market types used in soups, mixed vegetables, and mixed juices. Probably the total production of 10 lb is divided more nearly 6 lb fresh and 4 lb processed.

Economically the carrot ranks among the ten most important vegetables, exceeded by potato, lettuce, tomato, onion, celery and sweet corn. The average annual farm value for the years 1978–1980, was approximately $162 million. California produced 44.1% of the U.S. carrot tonnage and 49.6% of the dollar value (Table 9.1). The three leading states, California, Texas, and Florida, produced approximately 65% by volume and 73% of farm value. In those states most of the production is for fresh market. In Washington, Wiscon-

321

TABLE 9.1. Acreage, Production, and Farm Value of Carrots by State
(1978–1980)[a]

State	Acres	Tons	%	Value ($1000)	%
California	36,300	494,850	44.1	80,364	49.6
Texas	15,800	122,200	10.9	19,807	12.2
Florida	11,500	115,000	10.3	18,000	11.1
Michigan	6,100	78,000	7.0	14,910	9.2
Washington	4,600	100,000	8.9	8,227	5.1
Wisconsin	4,000	75,500	6.7	5,524	3.4
Minnesota	1,600	30,560	2.7	3,514	2.2
Oregon	1,400	29,400	2.6	2,524	1.6
Arizona	2,000	12,850	1.2	2,398	1.5
Colorado	1,000	12,650	1.1	2,357	1.5
New York	1,100	17,400	1.6	2,095	1.3
Illinois	300	4,300	0.4	698	0.4
Other states	1,460	28,333	2.5	1,582	1.0
Total	87,200	1,121,100		162,000	

[a]Data from USDA statistics for 1981, with added estimates of Florida production, which
are not included in 1981 USDA figures.

sin, Minnesota, and Oregon, where the crop is produced mainly for processing, the higher
yields from processing cultivars are offset by lower prices, with the result that the dollar
return per acre is slightly lower for processing than for market carrots.

ORIGIN AND GENERAL BOTANY

The genus *Daucus,* which includes carrot, has many wild forms that grow mostly in the
Mediterranean region and southwest Asia. Fewer representatives are found in Africa,
Australia, and North America. For *Daucus carota* L. it is generally agreed that Afghani-
stan is the primary center of genetic diversity and therefore the primary source for
dissemination. There are more than 400 plant introductions currently available in the
United States.

In two comprehensive studies based on ancient writings and paintings Banga (8, 9) has
provided evidence that the purple (anthocyanin) carrot together with a yellow variant
spread from Afghanistan to the Mediterranean area as early as the tenth or eleventh
century. It was known in western Europe in the fourteenth and fifteenth centuries, in
China by the fourteenth century, and in Japan in the seventeenth century. The white and
orange (carotene) carrots are probably mutations of the yellow form. Orange carrots were
first cultivated in the Netherlands and probably in adjacent areas in the seventeenth
century. The existence of orange and yellow–orange carrots is proved by the fact that they
appear in seventeenth-century paintings exhibited in Netherlands museums (9).

The domestic carrot is closely related to and readily crosses with the highly diverse and
widely adapted wild carrot known as Queen Anne's Lace. In nature the wild carrot has an
annual or winter annual life cycle. Usually the seedling plants form a rosette in late fall,
which undergoes cold induction during winter and completes its reproductive cycle the
following summer. Wild accessions moved from semitropical to temperate zones will
generally behave as annuals, with bolting following promptly after a brief exposure of
seedling plants to low temperatures and to the longer photoperiod of northern latitudes.

Domestic cultivars have been selected for non-bolting and therefore behave as biennials or winter annuals. In practice they can be propagated on an annual basis by storing small roots (stecklings) for 6–8 weeks at 2°–5°C before replanting. In areas with relatively mild winters and an early snow cover, seed is produced from plants vernalized in the field during the winter. They will promptly bolt and mature a seed crop in a total of 12–13 months.

FLORAL BIOLOGY AND CONTROLLED POLLINATION

Floral Characteristics

The carrot umbel is a compound inflorescence (Fig. 9.1). The development of the umbel begins with a broadening of the floral axis and internode elongation (*18*). A primary umbel can contain over 1000 flowers at maturity, whereas secondary, tertiary, and quaternary umbels bear successively fewer flowers. Usually several floral stalks develop from a single plant.

In a given carrot flower, stamens, petals, and sepals develop simultaneously followed by carpel development. Flowers are arranged in umbellets and umbellets into an umbel. Within umbellets and umbels, floral development is centripetal and arrangement is spiral. Thus the first mature flowers are those on the outer edges of the outer umbellets.

FIGURE 9.1. Carrot umbel consisting of umbellets.

The primary umbel consists mainly of bisexual flowers, but male flowers can occur frequently (between the edge and center of an umbellet) in subsequent umbels (*19*). The pollen from flowers at the center of an umbellet is larger and more frequently fertile than that from peripheral flowers (*50*).

Carrots are protandrous. The petals separate and the filaments begin unrolling to release the anthers at anthesis, although the anthers in one flower may not extrude simultaneously. After the filament is straightened, the pollen is shed and the stamen is quickly abscissed. The petals then open fully and the style elongates. The carrot has a split style, which separates when the flower is receptive to pollination. The petals of male-fertile plants fall soon after the split stigma is receptive. It is interesting that the petals of petaloid, but not brown anther, male-sterile plants are persistent until the seed ripens.

Carrot flowers are epigynous with five small sepals, five petals, five stamens, and two carpels (Fig. 9.2). The mature flower and consequent mature fruit are approximately 2 mm long. Early in development, each carpel bears two ovule primordia, but only the lower one continues to grow. The carrot embryo sac is monosporic (developing from the chalazal macrospore) and 8-nucleate (*17*). Nectaries on the ovary wall are important in insect attraction for pollination (*33*). Pollen longevity in storage has not been evaluated.

The carrot fruit is a bilocular schizocarp, which dries and splits upon maturity to yield two mericarps (achenes) with one seed each. The mericarps are covered with spines, which must be removed for ease of handling (Fig. 9.3).

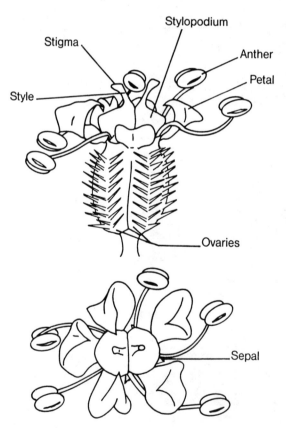

FIGURE 9.2. Carrot flower.

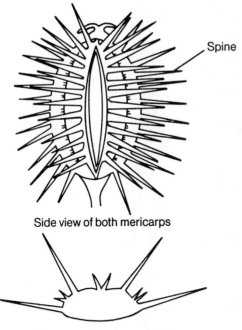

Side view of both mericarps

Cross section of a mericarp

FIGURE 9.3.   Carrot fruit.

Cytoplasmic Male Sterility (CMS)

There are two distinct types of male-sterile flowers, depending upon the source of ster-ility-inducing (S) cytoplasm. The brown anther type (Fig. 9.4), in which the anthers degenerate and shrivel before anthesis, is expressed in domestic cytoplasm from cv. Tendersweet, as reported by Welch and Grimball (72), and in cytoplasm from various commercial cultivars, as reported by Banga et al. (10). The other type of male-sterile flower is petaloid, in which the stamens are replaced by five petals (Fig. 9.5). Petaloid steriles exhibit a range of morphological structures, some of which have been described by Eisa and Wallace (30). In hybrid development petaloid steriles are more widely used than the brown anther type. If genetic and environmentally stable brown anther steriles were available, they would be preferred over petaloids because of their higher seed-yielding potential.

The 1947 report of Welch and Grimball (72) was followed in 1953 by the discovery of a sterile wild carrot (petaloid type) by H. M. Munger of Cornell University. From a detailed study of this wild carrot along with the brown anther material from Welch and three additional brown anther sources from Gabelman, Thompson developed a complex model for inheritance of pollen sterility (66). He concluded that there are at least two and probably three duplicate dominant maintainer genes and an epistatic restorer operating in the cytoplasm of both Tendersweet and the Cornell wild carrot. The useful maintainer line would therefore have to be free of the restorer and homozygous dominant at one of the *Ms* loci.

Hansche and Gabelman (37) reported digenic control of male sterility with either of two genes, dominant (*Ms-4*) or a recessive (*ms-5*), producing sterility interacting with

FIGURE 9.4. Brown-anther male-sterile carrot flowers.

cytoplasm from the cv. Tendersweet. They suggested that the recessive (*ms-5*) might have been introduced into Tendersweet from a plant introduction (PI 169486) known to be segregating for male sterility. It has not been determined if their *Ms-4* is allelic to any of the dominant maintainer genes postulated by Thompson (*66*). The model proposed by Hansche and Gabelman was supported by the extensive genetic analysis of Banga *et al.* (*10*). They found evidence for two additional complementary dominant restorer genes, which produced fertility if each locus had one dominant. It was suggested that one or both of these restorers may have been absent from the material studied by Hansche and Gabelman and that their data were not inconsistent with the hypothesis presented by Thompson.

Both the S cytoplasm and interacting nuclear genes that maintain sterility are widely distributed in wild and domestic forms of *D. carota*. In addition to the Cornell source, two other wild sources have been reported. McCollum (*44*) described petaloid sterility in a wild carrot accession from Sweden, and Morelock (*48*) reported the discovery of a similar wild carrot petaloid in Wisconsin in 1970. Banga *et al.* (*10*) searched seed fields for new sources and promptly found male-sterile plants of the brown anther type in Amsterdam Forcing, Flakee, Nantes, Vertau, Grelot, Parisienne, and some high-carotene selections. Four of these were used in their studies of the genetics of cytoplasmic sterility. From this experience it is expected that a systematic search would lead to discovery of cytoplasmic sterility in our domestic cultivars. No comparisons have been made to determine differences among sources of domestic cytoplasm, and so it is not known if any of those discovered in domestic cultivars in the Netherlands are superior to the Tendersweet source, which has been used to a limited extent for hybrid development in the United States.

A prolonged search in the wild carrot population around East Lansing, Michigan, failed to reveal any male-sterile plants. The petaloid sterile reported by Morelock (48) was found in a small isolated colony of wild carrot near Madison, Wisconsin. From populations segregating in this new cytoplasm, he concluded that the system of maintainer genes in the Wisconsin wild source was the same as that in Cornell wild cytoplasm and their morphological types were similar. He also demonstrated the decisive role of cytoplasm in determining the morphological type of sterility. By segregating identical genotypes from crosses between a petaloid maintainer (MSU 1558M) and a brown anther maintainer (W93M) it was found that only the brown anther type was expressed in domestic (Tendersweet) cytoplasm and only the petaloid type in the wild (Cornell) cytoplasm. His data supported the hypothesis that petaloid sterility in wild carrot cytoplasm is controlled by

FIGURE 9.5. Petaloid male-sterile carrot flowers. Note two whorls of petals.

two dominant nuclear genes (15 fertile : 1 sterile in F_2). The brown anther type in F_2 produced 15 sterile : 1 fertile in domestic cytoplasm to suggest control by two recessives.

The method used in these studies may have application in genetic studies involving other characteristics. First it is necessary to produce the F_1 from both sterile × fertile and fertile × fertile crosses, in this case W93S × 1558M and 1558S × W93M for sterile × fertile, and W93M × 1558M or 1558M × W93M for fertile × fertile. The segregating generations in the two types of cytoplasm were then produced as follows: [(W93S × 1558M) × (W93M × 1558M)], [(W93S × 1558M) × 1558M], [(W93S × 1558M) × W93M] to produce the appropriate segregation in domestic (W93S) cytoplasm, and similarly [(1558S × W93M) × (1558M × W93M)], [(1558S × W93M) × W93M], and [(1558S × W93M) × 1558M] to provide the same assortment of genotypes segregating in the Cornell wild (1558S) cytoplasm. If inbreeding is sufficient to establish homozygosity for characters in both the sterile and maintainer components of the inbred parents, then it is possible to produce the large populations needed for genetic studies without resort to laborious and often imprecise hand emasculation to produce backcross progenies.

TABLE 9.2. Simply Inherited Characters in Carrot and Their Gene Symobls

Gene symbol[a]	Character description	Gene source	Reference
A	α-*Carotene* synthesis (may be identical to *Io* or *O*)	Kintoki cv.	68
(Ce)*	*Cercospora* leaf spot resistance	WCR-1 Wisconsin inbred	3
(Cr)	*Cracking* roots (dominant to non-cracking)	Touchon cv.	27
Eh	Downy mildew (*Erysiphe heraclei*) resistance	D. carota ssp. dentatus	16
g	*Green* petiole	Tendersweet cv.	4
gls	*Glabrous* seedstalk	W-93 Wisconsin inbred	49
Io	*Intense orange* xylem	Miscellaneous	Kust (cited in 21)
L	*Lycopene* synthesis	Kintoki cv.	68
Ms-1, Ms-2, Ms-3	Maintenance of *male sterility*	Tendersweet cv.	66
Ms-4, ms-5		Tendersweet cv., Imperator 58 cv., PI 169486	37
O	*Orange* xylem	Miscellaneous	Kust (cited in 21)
(P-1), (P-2)	*Purple* root	Miscellaneous	42
Rs	*Reducing sugar* in root	Miscellaneous	36
y	*Yellow* xylem	Miscellaneous	Kust (cited in 21)
Y-1	Differential xylem / phloem carotene levels	Miscellaneous	Kust (cited in 21)
Y-2	Differential xylem / phloem carotene levels	Miscellaneous	Kust (cited in 21)

[a]Loci enclosed in parentheses were not named previously; suggested symbol.

Genetics and Cytogenetics

Only 20 genetic loci have been described in carrots (Table 9.2) and no linkage groups have been defined. A wide range of amino acid and nucleic acid analog resistant cell lines have been isolated in carrot tissue culture, but definitive genetic analyses have not been performed on plants regenerated from these variants. The carrot has nine pairs of chromosomes, and little variation in length exists between chromosomes (*12, 13, 47, 73*). Four chromosome pairs are metacentric, four are submetacentric, and one is satellited. There are 22 recognized species of *Daucus,* most with 11 or 10 chromosome pairs and only two others with nine pairs. One interspecific hybrid between *Daucus carota* and *Daucus capillifolius* has been synthesized (*45*).

Controlled Pollination

Emasculation to accomplish controlled crosses between male-fertile plants is laborious and time consuming, particularly for genetic studies that require relatively large backcross populations. Anthers are removed from the early-opening outer flowers in the outer whorl of umbellets until enough have been emasculated to ensure the needed supply of seed. Emasculation is accomplished before any stigmas become receptive. Unopened central florets in the emasculated umbellets and all late-flowering umbellets are removed, leaving the female parent inflorescence with only emasculated flowers. This umbel is then isolated under a small cloth cage (Fig. 9.6) with a pollen-bearing umbel from the selected

FIGURE 9.6. Small cloth cage for isolating one to five carrot plants.

male parent. Live house flies and pupae are introduced to ensure a continuing supply of active pollinators during the full period of stigma receptivity.

To produce fertile F_1 hybrids for breeding purposes it is more efficient and just as effective to make fertile × fertile crosses simply by putting under cloth cages one or two flowering umbels from each selected parent along with a supply of flies. Seed from each parent is then sown in adjacent field rows. In most cases each row will contain a mixture of hybrid and parent phenotypes that can be easily identified. If the cross involves inbred lines, hybrid plants are distinguished by their vigor, i.e., the selfs from the inbreds will produce much smaller roots. To ensure identification of hybrid plants, the stand should be uniform and relatively thin to permit expression of hybrid vigor.

BREEDING HISTORY

Until the early 1960s nearly all carrot cultivars in use were derived by selection in open-pollinated material. Because of the rapid loss of vigor, little effort had been made to achieve uniformity by inbreeding. With the discovery of cytoplasmic male sterility it became possible to follow the classic system of establishing pollen-sterile and -fertile inbreds for use as parents in F_1 hybrids that exhibit restored vigor and uniform horticultural quality. There was widespread interest in hybrid carrot development in the 1950s. Active projects were underway in several private seed and processing firms and at the Wisconsin, Idaho, California, New York, Oregon, and Michigan Agriculture Experiment Stations (AES) and the USDA at Beltsville, Maryland. As a result of this activity, hybrid cultivars began to appear in the early 1960s, and the frequency of new hybrid releases has continued to increase since then. By 1983, at least 50 hybrid cultivars had been introduced in the United States alone. The state breeding programs that provided parent inbreds for a substantial share of these hybrids were discontinued in the 1970s. Idaho, California, New York, Oregon, and Michigan AES had abandoned their carrot breeding projects by 1980. The increase in efforts by major vegetable seed firms has not replaced those of the terminated projects. The long-term effects of this decline in breeding activity and the concurrent shift of responsibility to a few private breeders cannot be estimated. A probable result will be to eliminate from competition the small seed firms that are unable to support independent breeding programs. High-risk, long-term objectives and development of unique types, for which there will be limited demand for seed, probably will be neglected.

CURRENT GOALS

Objectives have been to eliminate or minimize deficiencies in accepted open-pollinated cultivars and to exploit unique characteristics derived from introduced accessions. The main objectives of U.S. breeders are improvement of yield, visible characteristics such as color, shape, smoothness, and freedom from defects, resistance to a few common diseases, and non-bolting. During the past 5 years increasing attention has been given to the less obvious qualities of flavor, texture, and nutritional value.

Uniformity

Market carrots, which comprise nearly 80% of the U.S. production, are immature when harvested. For this reason, total yielding capacity is not as important as uniformity. A field population of nearly identical genotypes will yield a high percentage of marketable product. Efforts made by growers to achieve what they term ''high pack-out'' include seed coating, precision planting, irrigation, and fertilization to provide a uniform growing environment. Genetic uniformity contributes substantially to the success of refined cultural practices.

Even for processing carrots, uniformity and quality are currently so important that maximum tonnage has not yet become a major objective. It is expected that when improved quality is available in a number of hybrids, yield will become the major goal.

Appearance

Root shape requirements are determined by the preferences of growers, consumers, and processors. The diverse shapes in breeding populations and open-pollinated cultivars provide breeders with all the genetic variability needed to meet established demands for popular shapes (Fig. 9.7). The smooth exterior demanded for both fresh-market and processing types is important for ease of cleaning and reduction of paring losses.

Exterior and interior color are important characteristics receiving attention in all breeding projects. It is possible to achieve a deep red–orange color in both xylem and phloem and to eliminate green color, both exterior and interior, from the crown area.

In addition to shape, smoothness, and color, genetic differences in certain root defects are evident and these defects are being minimized by appropriate breeding methods. By producing breeding material under conditions where defects develop it is possible to achieve effective selection.

Disease Resistance

Resistance to alternaria leaf blight, *Alternaria dauci*, present in all U.S. carrot-growing areas, is a major objective. Most currently popular open-pollinated market cultivars are

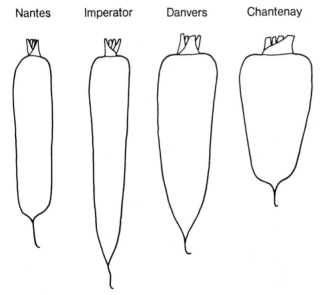

Nantes   Imperator   Danvers   Chantenay

FIGURE 9.7.   Typical shapes of popular U.S. carrot cultivar classes.

susceptible, and under severe conditions they require a regular fungicide program to avoid reduction in yield. The most serious damage is often the loss of so much foliage that roots cannot be lifted by mechanical harvesters. Processing types generally are more tolerant than market types. A high level of resistance has not yet been identified. With the available tolerance the size of petiole lesions is restricted and the plants generally retain enough foliage for mechanical harvest. In 1970, 241 plant introductions and 90 inbred lines were exposed to severe natural infestation at Belle Glade, Florida. Only nine retained enough foliage to be classified as resistant. The best were Japanese cvs. Kokubu (PI 261648), San Nai (PI 226043), and Imperial Long Scarlet (*64*). Resistance from these sources was incorporated in a breeding population released by the USDA and Florida AES in 1983.

Cercospora leaf blight *Cercospora carotae* is another common foliage disease with symptoms similar to those of alternaria leaf blight, but normally less severe. Resistance has been observed in certain lines. In areas where the disease is prevalent, selection for resistance under field conditions can be effective. Some use has been made of greenhouse screening for cercospora resistance (*3*).

One of the most devastating diseases in the northern United States is aster yellows, a mycoplasma transmitted to a wide range of hosts by the six-spotted leaf hopper *Macrosteles fascifrons*. Vector preference and probable plant tolerance in field exposure tests by the University of Wisconsin were reported in 1973 by Schultz (*54*). Since that time, continued field exposure and selection from many diverse cultivars and breeding materials by the University of Wisconsin have produced clear levels of plant resistance as well as vector preference. In recent replicated trials in which entries were planted with alternating rows of lettuce to ensure uniform exposure, the Wisconsin workers observed levels of infection low enough to suggest that eventually it might be possible for growers to produce resistant genotypes without pesticides. With these new sources of resistant germplasm, it is likely that carrot breeders will assume the major responsibility for long-term control of aster yellows.

Carrot motley dwarf, first described by Stubbs (*65*) is more common in the United Kingdom and Australia than it is in the United States. It is induced by two viruses, carrot mottle and carrot red leaf, transmitted by the willow aphid *Cavariella aegopodii*. The occasional outbreaks in California and the increasing prevalence of the disease in the Pacific Northwest justify including resistance as a minor breeding objective. Resistant germplasm is being developed in order to have useful material on hand in case the disease spreads to other producing areas.

Of many soilborne pathogens that affect carrots, only the *Pythium* species that have been implicated in a disease called brown root in the United States and rusty root in Canada have received attention. Selection has been for roots that do not develop forking in wet soils. Seedling resistance has been tested by Howard and Williams (*40*) using inoculated rooting media under controlled temperature. Field symptoms in muck-grown carrots are described by Mildenhall *et al.* (*46*).

Non-bolting

Premature development of seed stalks is a common cause of losses in yield and quality. In some genotypes there is good tolerance to low induction temperatures. To minimize losses from bolting, breeders are exposing segregating populations to induction environments and imposing continuing selection pressure for a high vernalization requirement.

Quality

Until recently, the opportunity to enhance eating quality has been neglected. This may be attributed to the fact that carrots have been included in diet recommendations as a source of provitamin A. Many consumers probably purchase carrots for their vitamins rather than taste. In open-pollinated cultivars the extremes of flavor, from harsh bitterness to sweet succulence, provide the opportunity for significant improvement; and increasing attention is being given to culinary quality and nutritive value. With improvements now becoming available, and the increasing interest in nutrition and food quality, these characteristics must be included as essential objectives.

Genetic variation has been reported for many quality attributes of carrot including carotenoids (22), fiber (52), texture (6, 41, 43), sugars (36), flavor (60), minerals (7) and toxicants (74). Reviews of genetic variation for eating quality have been prepared by Aubert and Bonnet (5) and by Simon et al. (62).

Inheritance of several of these quality attributes has been analyzed. Genetic variation for carotenoid type and amount in carrot roots, ranging from white to orange color, is controlled by at least three genes, with many more accounting for variation within the orange category (21, 42, 68). Consideration of carotenoid quantity is very important in carrots because they are the most important vegetable source of provitamin A in the U.S. diet, providing 14% of the total (56). An "average" carrot contains 90 ppm total carotenoids (71). Of this, approximately 20% is α-carotene, 50% is β-carotene, 0–20% is ζ-carotene, 0–2% is lycopene, and 0–10% γ-carotene (67, 68). Darker-orange carrots tend to contain a higher percentage of α-carotene, up to 50% of the total (42).

SELECTION TECHNIQUES

Color

Selection for color should be accomplished in uniform light. Under full daylight or incandescent light, roots may appear much darker and a more desirable deep orange than under fluorescent light. If light conditions are not uniform, it is easy to choose those exposed to full light and discard those illuminated by fluorescent light.

The desired bright-orange color should extend down to the tap root and up to the crown. The best color distribution can be achieved in populations segregating for interior color by selecting roots with the best exterior color and with color extending well down the tap root. The roots with color in the tap roots are referred to as "red tails." To observe interior color, a horizontal cut can be made approximately 1 in. from the bottom of the red tail selections. The color of interior tissues, cambium and the adjacent phloem and xylem, may differ greatly in a given root (Fig. 9.8). Roots with the best uniformity (match) between the color of xylem and phloem tissue and those with the least distinct cambium zone are then examined for color in the crown area. This selection is made by a transverse cut just into the xylem, one to two petioles deep, and extending outward and 1 or 2 in. down from the crown (Fig. 9.9). Interior color can then be selected easily by retaining those roots with little or no green in the cambium area or upper xylem, with the best color match between phloem and xylem, and with an indistinct cambium zone.

Improvements in the uniform dark-orange color that have been made by visual selection have been accompanied by increases in total carotene content. Some recently introduced hybrid cultivars with improved color have total carotene ranging from 120 to 150 ppm fresh weight, while the open-pollinated sources have 80–100 ppm.

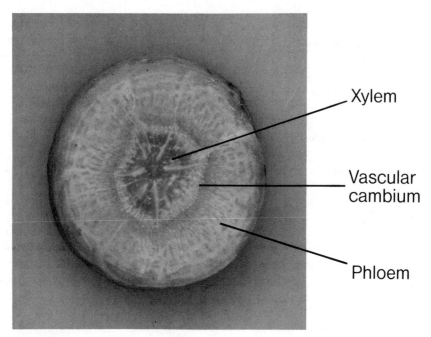

FIGURE 9.8. Cross section of carrot root.

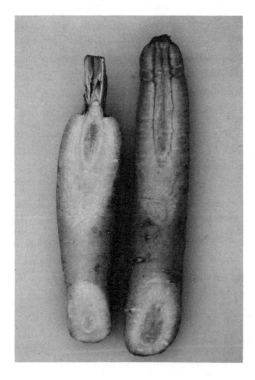

FIGURE 9.9. Carrot roots cut to select for interior quality. Note undesirable pigmented cambium in the crown of root on the right.

Visual selection for differences in carotenoid content is feasible if alleles for pale-orange, yellow, or white color (less than 50 ppm carotenoids) occur in the population being considered (42). The presence of high lycopene in the root is visually detectable by characteristic red color even in dark-orange populations (100–140 ppm carotenoids), except in the presence of high α-carotene content. Visual selection is adequate for improving carotene content up to 120 ppm total carotenoids. Above this level it is more difficult to make visual assessments of carotenoid content even though roots classified as "dark orange" may range from 130 to over 200 ppm carotenoids. Consequently, selection for total carotenoid content must be based upon laboratory analyses to assure continued improvement.

Total carotenoids are determined analytically by comparing spectrophotometric transmission of 450-nm light through a hexane extract of root tissue to the percentage transmission through a pure β-carotene standard solution. The hexane extract can be obtained by either of two methods: (a) blending frozen raw carrot slices in a mixture of hexane and acetone, followed by removal of the more polar acetone (along with water from the carrot) in a separatory funnel (67) or b) lyophilizing raw carrot slices and blending directly in hexane (60). The latter method is somewhat more rapid. Individual carotenes can be quantified by thin-layer chromatography but the procedure is time consuming (21, 22). High-performance liquid chromatography may be well suited for rapid analysis.

Interior Quality

Interior qualities other than color can be observed for selection purposes with the same cuts that expose interior root tissues for color selection. Among the interior defects that may be exposed are cottony xylem, hollow heart, and spongy phloem. These defects are most likely to appear under stress and in mature roots; therefore, it is prudent to provide an environment that will induce interior defects so that alleles conditioning their expression can be eliminated. To select against hollow heart and cottony xylem, the plants should be allowed to mature and remain in the soil for at least as long as the crop is held in the field under commercial practices. Of course, this applies only to processing carrots, which often are held (unharvested) for 4 or more weeks after they have attained nearly maximum size.

The spongy-phloem defect develops in winter carrots subjected to low temperatures in moist soils. It is a common defect in winter areas like the Imperial Valley of California, where the crop may be planted in September or October and harvested in February or March, a schedule that entails exposure to 1 or 2 months at relatively low soil temperatures. Selection must span several seasons to increase the probability of exposure to the environment that will induce spongy phloem. The defect has not been observed in our plots in midwest organic soils, and genetic susceptibility cannot be detected in such environments. Some otherwise good lines, carefully selected in Michigan, Wisconsin, and New York, were unacceptable because the spongy-phloem defect appeared upon exposure to the winter environments of California and Arizona. This experience illustrates the importance of early-generation selection of inbred lines and the testing of their hybrids in the environment where they are likely to be produced.

Growth cracks may occur under many growing environments, but appear to be most severe in processing cultivars when harvest is delayed after roots are mature. This field storage during wet autumn months results in longitudinal splitting, which is followed by healing and slow, prolonged growth. Such roots are unacceptable to the trade. This defect occurs to varying degrees in all U.S. growing areas. It is particularly severe in the Pacific

Northwest, where mild fall weather permits prolonged field storage. There are genetic differences in the incidence of growth cracks (14). Some of our lines that have undergone crack selection in the Puget Sound area of Washington have shown a significant reduction in the incidence of growth cracks compared with unselected populations.

A related defect, of some importance in both processing and market carrots, has been termed longitudinal cracking (27) and is generally called "harvest cracks" or "shatter cracks" by growers and shippers. It occurs when turgid roots are subjected to mechanical abuse at harvest. Genotypes differ in their tolerance to this type of cracking. Selection can be accomplished and inbred lines evaluated by pricking freshly dug roots with a pen knife. There is little experimental evidence to support the general belief that roots with the best eating quality are most subject to harvest cracking. The cv. Nantes, one of the best for eating quality, is vulnerable to longitudinal cracking when handled mechanically. If this relationship holds true generally, breeders may need to compromise quality in order to achieve the durability demanded for mechanical harvest and packing. It is possible that machinery will be designed and practices modified to minimize mechanical damage. This will permit breeders to develop genotypes with the highest possible table quality without the risk of rejection because of excessive damage in mechanized handling.

Sugar and Flavor

A simple taste test can be used to identify good flavor and texture. A thin cross-sectional slice is cut from the color-selected roots for tasting; those with harsh flavors are eliminated (62). The ability to detect some volatile components that contribute to bitter or harsh flavors varies among individual tasters. Therefore, if eliminating roots with undesirable flavor is to be effective, the selector must be able to detect flavor differences.

The concentration of stored sugars contributes to sweetness and is an important component of overall preference (60). Total sugars can be selected by refractometer estimates of total dissolved solids (TDS) (24). This can be accomplished by using a drop of juice from a freshly cut or frozen root sample for refractometer reading to estimate concentration. The small (5 g) frozen sample, sealed in a polyethylene bag and held at $-10°C$, is thawed and macerated within the sealed bag. A needle puncture will release the drops of juice needed for refractometer reading. This simple selection procedure is effective for increasing total sugars and total dry matter. Selection for high soluble solids and dry matter can be accomplished by specific gravity (11). High dry matter is important to processors of whole carrot products to meet drained-weight specifications.

Sugar quantity, TDS, and percentage dry weight exhibit quantitative inheritance patterns (53). Total sugar quantity and either TDS or percentage dry weight have correlation coefficients (r) of .75–.95, whereas TDS and percentage dry weight are more highly correlated ($r = .85–.95$). The latter correlation may be higher because soluble solids often account for over 90% of the dry weight, whereas total sugars account for only 60–75% of the soluble solids and 40–60% of the dry weight (24). As a point of reference, the "average" raw carrot is 88% water, 6–8% sugar, 1–2% fiber, 0.7–1.2% protein, 1% ash, 0–0.3% fat, and is starch free (51, 71).

Percentage dry weight, a standard variable in much biological research, can be determined quickly by simply weighing a fresh sample, drying, and reweighing. The measurement of total sugar quantity is more time consuming. Free sugars have been measured colorimetrically, but the advent of high-performance liquid chromatography has allowed for more rapid sugar analysis (36). With this technique three to five samples can be analyzed per hour by one worker, including all operations involved. Total dissolved solids or dry weight analysis can be determined in ⅓ to ¹⁄₁₀ the time.

The main goal in breeding for altered sugar quantity, TDS, and percentage dry weight has been to increase sugar content and thereby improve carrot flavor. The correlation between rapidly measurable TDS or dry weight and total sugars suggests that the use of either rapid technique effectively alters sugar levels.

Volatile terpenoids and other potent compounds, other than sugars, have important effects on carrot flavor. Carrot roots must be taste tested to achieve the correct balance between volatile terpenoids and sugar levels. The lack of consistent flavor improvement by selecting for increased TDS (53) and the failure to improve the flavor of high-volatile terpenoid carrots significantly by dipping slices in 30% fructose lead to the conclusion that volatiles play a decisive role in overall flavor. Genetic differences for sweetness found in raw carrots are also reflected in the processed carrots (59).

Total sugar is the sum of glucose, fructose, and sucrose since these sugars account for 98–100% of the free sugars in carrots. Even though complete sugar analysis is a time-consuming procedure for selection purposes, it has elucidated an interesting, simply inherited trait. The reducing sugar (Rs) locus controls the ratio of reducing sugars (glucose and fructose) to sucrose, independent of total sugar (36). Dominance is for high reducing sugars and low sucrose. The quality of deep-fried carrot chips may be damaged by the presence of high reducing sugars (Rs/–). Conversely, high reducing sugar may be desirable for improving flavor of raw carrots since the average sweetness of glucose plus fructose is 20% greater than that of sucrose (57). The potential for improvement with high reducing sugars remains to be tested since the sweetness levels cited are for weak, aqueous solutions of sugars, not solid matrices like carrot roots.

Selection for reduced levels of volatile terpenoids must be the prime concern in the improvement of raw carrot flavor. Carrots low in volatile terpenoids are bland but edible, whereas those high in volatile terpenoids are inedible. Terpenoid quantity and associated harsh flavor exhibit great genetic variation. An approximate range is from 5 to 200 ppm with surprisingly little environment (soil, climate) or maturity effect (39, 63). Volatile terpenoid levels of 20–50 ppm are desirable. Carrots with higher levels have a harsh flavor, while those with lower levels lack a desirable, typical carrot flavor.

Harsh flavor is common in available carrot cultivars (61), but rapid improvement of flavor is made feasible by the dominance for mild flavor in hybrids that have one mild parent (63). Since the inheritance of harshness is multigenic, selection for mildness should be initiated early in inbred development and applied in every generation. Laboratory assessment of terpenoid types and amount is useful in selecting for extreme individual roots to be used in establishing elite, experimental populations, but the use of a trained tester to ascertain flavor is more practical for routine selection. Volatile terpenoids and harshness are quite stable in cold-stored carrots, but processing reduces volatile terpenoid level, harshness, and overall raw carrot flavor (59).

Isocoumarin, which may contribute to bitter flavor, also demonstrates quantitative genetic variation. Hybrids tend to be comparable to the more bitter parent (23). Since isocoumarin bitterness is the consequence of extended cold storage, particularly in the presence of ethylene, it is not a primary concern in present carrot breeding programs. The role of isocoumarin in bitter flavor is not fully known (58).

Non-bolting

In northern latitudes, cold weather often occurs during early growth of seedlings and results in premature seeding (bolting) of a significant percentage of the population. Bolting can cause serious losses in yield and quality. It is important to impose continuing selection pressure for non-bolting in order to avoid a genetic shift toward easy bolting.

The first precaution is to provide a cold-induction treatment of stecklings adequate to bolt 100% of the breeding population. This procedure avoids selection against bolting-resistant genotypes. It has been established that 8–10 weeks at a temperature below 40 °F (5°C) are required for complete vernalization of stecklings harvested in Michigan in late August and intended for seed production in the greenhouse (28). The second recommended procedure in northern latitudes is planting as early as possible in the spring so that seedling plants are exposed to as much natural cold as possible. A cold-unit system for determining exposure needed for vernalization of seedling plants was worked out by Dickson et al. (29). They found that a good screening system required 650 hr below the base temperature of 50°F (10°C) during the first 2 months after planting, followed by 3 months of growth to allow all potential bolters to develop. Seasonal differences that do not provide the necessary cold may make it impossible to select non-bolters every year in spring-planted carrots. Relying on early planting and natural cold treatment further imposes a tight schedule on the breeder who must produce stecklings in time to provide for 2-month induction and 4–5 months to mature and process seed for the next early spring planting. In a breeding program based on an annual cycle, the best schedule is 8–10 weeks of root storage below 40°F (5°C), followed by planting in the greenhouse using a night temperature of 55°F (13°C) until seed stalk development begins. Then 70°F (21°C) nights and 80°F (27°C) days are used to hasten anthesis and seed maturity (29).

Another practical procedure for selection is to utilize winter cold induction in areas where winters normally do not produce 100% vernalization of fall-planted carrots and to select non-bolting or slow-bolting plants. Cultivars or breeding lines with known bolting tendency must be included as standards in such plantings. These locations may be used to index bolting tendency in established cultivars and breeding lines.

Disease Resistance

Selection for resistance to alternaria leaf blight has been accomplished by exposing breeding materials to natural infestation in areas where the disease is generally severe. Susceptible cultivars, planted between rows of breeding material, are inoculated to ensure the presence and uniform spread of the disease. Most years in central Florida these procedures have been effective in producing severe defoliation of susceptible cultivars. Resistance is not complete. The presence of vigorous tops is due in part to the ability of certain genotypes to produce new leaves rapidly to replace those lost to the disease. In addition to inherent top vigor, the resistant genotypes show a restriction in area or depth of lesions on leaf blades and petioles that reduces defoliation. Roots selected on the basis of top vigor and lesion type are used in developing inbred parent lines or breeding populations.

For resistance to cercospora leaf blight the same procedures can be used. Since both diseases may develop in the same season it is possible to make selections by using field exposure. With Cercospora a method for greenhouse inoculation and incubation was described by Angell and Gabelman (3). However, laboratory screening procedures for resistance to leaf blight diseases have had only limited use. Precisely controlled, repeatable screening methods are needed for more rapid and efficient development of resistant genotypes.

Field exposure has been used in the University of Wisconsin program to classify and select plants for resistance to aster yellows. The incidence and severity of the disease in northern carrot areas varies from year to year, depending upon the number and levels of pathogen infestation in migrating leaf hoppers and upon the timing of their movement from southern areas. Lettuce may be planted in alternate rows with carrot breeding material to encourage increase of the vector population and ensure uniform spread of the disease. Greenhouse screening techniques that might eliminate the effects of weather and vector preference have not been applied.

A similar field method has been adopted in our preliminary efforts to select for resistance to the virus complex that causes motley dwarf. At Corvallis, Oregon, motley dwarf is almost an annual occurrence, especially in carrots planted early enough to be exposed to the first generations of the aphid vector. It is difficult to grow virus-infected plants through the seed cycle even when foliar symptoms are not severe. It has been noted that plants with symptoms of carrot mottle may be quite vigorous with little loss in yields. Some genotypes, apparently tolerant to mottle, either escape or resist the red leaf component. It appears that there is resistance to one or the other of the viruses implicated in motely dwarf as well as resistance to both. The best procedure available at this time is early planting to ensure infection, then production of seed on surviving plants. Because of the uncertainty of seed production on surviving plants, it is necessary to produce the same populations in a virus-free area and make selections for self-pollination on the basis of data from the motley dwarf plots. This index system is slow and costly if the disease plots are remote from other breeding operations. It is more efficient to breed for resistance in an area where the disease is prevalent.

Inbred Vigor

Selection for tolerance to inbreeding has been accomplished without much effort to evaluate inbreeding depression. The loss of vigor associated with inbreeding is dramatic and especially noticeable in early-generation carrot lines extracted from open-pollinated cultivars. In early hybrid development, many inbreds simply failed to survive or were discarded when they appeared to be too weak to reproduce. Lines that survived to S_3 or S_4 and possessed improved horticultural qualities were tested in experimental single-cross (sterile × fertile) hybrids that frequently were substantial improvements over their source cultivars. The serious lack of vigor in inbred parents was partially solved by resort to three-way hybrids for female parents in order to secure acceptable yields of seed.

As surviving lines accumulated, systematic recycling of inbreds was initiated by making crosses between individual fertile plants selected from maintainer inbreds. Progenies from these crosses were inbred in turn to S_3 or S_4 before they were selected for an additional cycle of fertile × fertile crosses, followed by inbreeding and selection. The level of inbreeding has been dictated by survival of lines and the degree of uniformity demanded by the market. Some of our most promising inbreds trace back through four generations of such matings to the original open-pollinated source. The net result has been elimination of many recessive genes that contribute to inbreeding depression and the accumulation of those that contribute to color and other qualities for which selection pressure has been imposed. This strategy has enabled us to achieve intense inbreeding (as high as S_{10}) to establish the homozygosity required for genetic studies.

As this process continues, it is expected that increasing tolerance to inbreeding will be achieved and that inbred lines with enough vigor for direct use in single-cross hybrids will be developed. A similar long process in the development of inbred parents for hybrid corn (*Zea mays*) has resulted in gradual improvement of inbred vigor to a level that has permitted increasing use of inbred lines as parents in single-cross hybrids. Achievement of a comparable level of inbred vigor in carrot will take longer because of its long life cycle and because there are fewer breeding programs in which inbreds are being selected for vigor.

GROWING THE SEED CROP

Various types of cages are used to accomplish the isolation necessary for advancing selected plants (roots) through their reproductive cycle. A root selected for producing the next generation represents a substantial investment, not just in growing it to the root stage, but in the years of breeding involved in bringing it to a state of genetic improvement. Therefore, every precaution must be taken to ensure its successful propagation. The following procedures, evolved in our program, will not have universal application. They represent the type of measures that must be adapted and modified to ensure successful seed production wherever the project is carried on.

Storage and Vernalization

Stored carrot roots are vulnerable to a number of rotting organisms that are favored by a film of water on the root surface. An effective means of maintaining the high humidity necessary to prevent desiccation and still avoiding condensation of free water on root surfaces involves the use of polyethylene bags, paper bags, and dry wood shavings. The tops are cut to about 1 in. Roots are then dipped in a fungicide mix appropriate for the organisms to be controlled. Concentration is not critical because drying will leave a full-strength deposit of the chemicals on the root surface. After dipping, the roots must be dried thoroughly before packing. It is important that no water remains on the root surface. The dry roots are then placed in paper bags with a volume of dry wood-shavings approximately equal to that of the carrots. The paper packages are then placed in polyethylene bags, tied, and placed in a refrigerated storage at 38°F (3°C). After about 1–2 weeks, droplets of water will condense on the inner surface of the polyethylene bags and will be blotted up by the paper bags. At that time the polyethylene bags should be punctured to allow excess moisture to escape. By this method the roots remain dry while their microenvironment is maintained near the desired 100% relative humidity. Storage for 8 weeks will accomplish the vernalization necessary for prompt bolting in the field or greenhouse. Roots harvested in the fall can be stored for spring planting without losses from decay.

Planting

Care must be taken in planting to protect against a variety of hazards that will vary with location. Cuts made in the process of color selection leave the roots especially vulnerable to drought. They must be planted with crowns about 1 in. below the soil surface. Enough water is poured around each root to ensure good contact with the soil. They then receive a local soil application of a systemic insecticide to control the leafhopper vector of aster yellows. After the insecticide treatment, some adjacent soil is used to cover the crown to the desired depth and to provide a mulch that will delay drying.

Pest Control

In addition to the systemic pesticide applied at planting, a regular schedule of spray treatment, mainly for leafhopper control, is necessary until cages are installed. Effective control before flowering is important because of the difficulty of controlling pests after pollinating insects are introduced. To control aphids, lady bird beetles are introduced into isolation cages after insecticide applications are discontinued. In areas where rabbits and other carrot-loving animals are common, it is necessary to provide protection. In our Wisconsin plots where rabbits are the common pest, a fine (1 in.) mesh chicken wire fence, approximately 2 ft high, is placed around the area devoted to breeding plots and cage isolations. After 2–3 weeks of growth, the plants are no longer attractive and the fence can be removed.

Since breeding areas are small, appropriate herbicide treatments can be applied with hand-propelled or knapsack sprayers. The same cages are often used for one or two other insect-pollinated species along with carrots, and so care must be taken to use compatible herbicides or to direct applications only to the target species. Much hand weeding can be avoided by following a timely herbicide schedule.

Disease Control

Alternaria leaf blight, aster yellows, and bacterial blight (*Xanthomonas carotae*) are the most common diseases encountered in outdoor breeding plots and cages. Alternaria can be controlled by a weekly spray schedule with an effective fungicide. It is important to select a fungicide that is not toxic to pollinating insects. Spraying must be continued during pollination in order to prevent alternaria damage to the flowers.

In addition to the systemic insecticide applied at planting for control of the aster yellows vector, it is necessary to begin an insecticide schedule as soon as top growth begins and to continue until pollinators are introduced. A motorized backpack sprayer is effective in projecting a fine spray mist through the screens after the cages are installed.

Control of bacterial blight must be accomplished by hot-water treatment of the seed from which roots are produced. If the disease is introduced with systemically infected roots, some spread can be expected and no effective control is available. The disease is especially damaging in the environment provided under screen isolators where plants are shaded and air movement is restricted.

Isolation and Pollination

The cages used in controlling pollination vary from small cylindrical cloth cages to 6-ft-high screen enclosures that range from 3.5 × 6 ft to 12 × 80 ft (Fig. 9.10). The large cages, 12 × 24 ft and larger, are used to increase finished lines and to produce experimental hybrid combinations for trial. Under one screen, several female parents may be isolated with a single pollinator. For line increases it is preferable to isolate only the male-sterile and its companion maintainer in order to minimize chances for contamination. Small screen cages (6 × 6 ft and 3.5 × 6 ft) are used for advancing, on a mass basis, the early-generation inbred pairs, which usually are at the BC_2 or BC_3 generation in S cytoplasm.

Design of the large isolation cages will vary widely and can be made to conform to the breeder's preference. For our program, we use 6 × 6 ft panels constructed of heavy-duty electrical conduit as basic units. These panels are secured to supporting steel fence posts by means of heavy-grade filament tape. Screen covers, constructed with heavy-duty zippers for access, are installed before plants begin to flower. Multiples of the basic 6 × 6 ft panel can be used to fit the full range of screen cover sizes.

The small cloth-covered, cylindrical cages used to isolate umbels from one to five

FIGURE 9.10. Large screen cages for increasing parent inbreds and producing experimental carrot hybrids.

plants are constructed in a way that will permit easy introduction of house fly pupae (Fig. 9.6). At the top, the cloth cages are fastened around a short section of garden hose. These small isolators are used for selfing, making testcrosses, and for single-plant backcrosses in the early stages of inbred development. They are used for isolating two selected fertile plants to make fertile × fertile crosses from which new inbred lines are extracted. Umbels for isolation in these small cages should represent two or three stages of maturity. The late-flowering umbels will provide pollen during the time that stigmas on early-flowering umbels are receptive.

House flies (*Musca domestica*) are reared following a standard procedure with batches of pupae being produced at 3- or 4-day intervals to ensure a continuing supply. The pupae are dropped into the cages through the tube at the top and closed with a cork (Fig. 9.6). In hot weather the adult house fly is short lived. Mature pupae must be introduced two or three times per week to avoid interruption of the pollinating process. They should not be introduced until nearly mature. If immature pupae are used, they are likely to desiccate and succumb to high temperatures before they can emerge. Unlike honey bees, flies do not discriminate between lines on the basis of nectar quality and do not display preference for certain sources. Therefore, we introduce flies into the large cages to supplement bee activity and to ensure the necessary transfer of pollen from male-fertile to male-sterile plants. Nucleus hives of bees, placed in the large isolation cages at the beginning of bloom, are complete with queens and brood. The normal social organization needs to be complete to make sure that nectar-foraging bees will be available and that they will function. It must be remembered that nectar foragers are the ones that accomplish most of the necessary transfer of pollen from fertile to male-sterile parents. Pollen foragers generally collect only pollen and will not visit male-sterile plants. While nucleus hives are confined in screen cages, they must be provided with water and food; and their activity must be closely monitored so that replacements can be provided if the necessary activity does not occur.

Processing and Storing Seed

To ensure maximum yields, seed from parent lines in the screen cages should be harvested two or three times as it matures. Risk of mixture can be minimized if harvest is accomplished before umbels begin to shatter. After thorough drying the seed may be separated and despined by hand rubbing or with a motorized scarifier, a device used only to despine and not to scarify. A small blower designed for experimental samples is used to separate seed from chaff.

To accomplish the prescribed hot water treatment for control of bacterial blight (*X. carotae*), all seed lots (even down to a few grams) from single-plant isolations are placed, with an identifying number, in fine-mesh bags and submerged in a water bath. The range of acceptable combinations of time and temperature is narrow and must be controlled precisely. Too little heat will result in some survival of the pathogen and too much will cause injury to the seed. A temperature of 51°C for 12½ min is effective for carrots. After heat treatment the seed lots are removed from the water bath and plunged into cool water, then drained, and spread out to dry. When thoroughly dried they are packaged in new seed envelopes and transferred to a refrigerated seed storage. Because carrot seed is short-lived, it should be stored in a cool, dry room. The maximum should not exceed 50°F (10°C) and 50% RH. Longevity will be improved with lower humidity and temperature.

Commercial Seed Production

Inbred parent lines and hybrids will not serve their intended purpose unless profitable seed crops can be produced. The breeder should be familiar with essential production practices and with seed problems that relate to characteristics of inbred parents. In addition to acceptable yielding capacity of the female parent, there must be an adequate and timely supply of pollen, effective insect pollinator activity, and appropriate cultural procedures.

Seed is grown either from roots (stecklings) harvested in the fall and held in cold storage for spring planting or, in areas with relatively mild winters, from roots harvested in early winter and replanted immediately. Both schedules are defined as ''root-to-seed'' production. In the first case, cold induction (vernalization) occurs in storage; in the second it occurs in the field after stecklings are transplanted from root beds to production fields. The alternative to a root-to-seed system is a seed-to-seed schedule in which plants from seed sown in late summer remain in the field through the winter. Having received more than adequate natural cold induction, the plants promptly bolt and mature a seed crop the following summer.

Root-to-seed allows for inspection of harvested roots and elimination (roguing) of off-types that result from outcrosses or mixtures. Another obvious advantage is that less stock seed is required, approximately ¼ lb per acre compared with 1–2 lb for seed-to-seed. The disadvantage is the high cost of root harvest, storage, and replanting. Carrot roots are not easy to store and excessive losses frequently occur. Some lines are susceptible to storage diseases. If root storage is to be involved, attention must be given to the inherent storage quality of inbred parents.

The characteristically low seed yields from many cytosterile inbreds used as female parents in single-cross hybrids have encouraged breeders and seedmen to adopt three-way hybrids, in which male-sterile F_1 hybrids are used for seed parents. In the breeding program, it is necessary to produce and evaluate single-cross, male-sterile F_1 hybrids for use as seed parents. This must be done on a continuing basis as new sterile and maintainer lines become available. An obvious but essential requirement for hybrid seed parents is that roots of the inbred components must be as nearly identical as possible in visible characteristics such as shape, length, and color in order to minimize segregation in the three-way hybrid.

As a result of the high cost of root harvest, storage (including losses in storage), and replanting, most commercial production of hybrid seed is on a seed-to-seed schedule. In order to maintain genetic uniformity, increases of parent stocks should be made on a root-to-seed schedule. Elite stock seed must be increased from carefully selected roots under screen isolation. This cage-grown seed is adequate for outdoor root-to-seed production of the stock seed for commercial seed fields. One hundred plants grown under screen from carefully selected roots will produce about 1 lb of seed, enough for at least 4 acres of root-to-seed increase. Only one outdoor root-to-seed cycle should be used for increasing the components of a hybrid cultivar. To ensure uniformity, the roots for this one cycle should be rogued for off-types that result from mixtures or outcrosses.

Seed is produced in the arid west, mainly in Idaho, Oregon, Washington, and central California, where winters are cold enough to accomplish vernalization, yet mild enough for survival of overwintering plants. Areas like southern California occasionally have mild winters in which only part of the plants are vernalized. This kind of natural selection for easy induction will result in an increase of bolters in the hybrid crop. The risk is greater if stock seed is subjected to this unintended selection. Stecklings from cold storage can be planted for seed production in any of the seed-to-seed areas, a practice that is often followed to ensure some production in the event of winter-killing of the direct seeded crop. The late stages of hybrid development must include pilot production of candidate hybrids to permit evaluation of parents in their reproductive cycle and to provide enough seed for commercial trials.

In hybrid seed production fields, a common ♀ : ♂ ratio is 4 : 1, usually in an 8 : 2 arrangement with four, two-row beds of female alternating with single two-row beds of the pollen parent (Fig. 9.11). Some lines selected as good male parents by performance of their prototype hybrids produced under screen will be unsatisfactory seed producers under field conditions. For good field production an abundant supply of pollen is needed during peak flowering of the female. Some evaluation for pollen production can be accomplished in cage isolations. For conclusive evaluation of performance, they must be observed under field conditions. The female parents with their specific male parents also need to be evaluated for seed production potential in field isolations. Petaloid steriles generally set a lower percentage of their potential than do brown anther types. Brown anther steriles have been disappointing mainly because of their environmental instability (38). Lines that survived inbreeding and selection for brown anther CMS through many generations have later developed unacceptably high percentages of pollen-bearing plants in production fields. This experience, together with the fact that brown anther steriles are difficult to distinguish from fertiles in the roguing process, has resulted in a diminished interest and

FIGURE 9.11. Hybrid carrot seed production field.
Courtesy of Alf. Christiansen Seed Company, © 1982.

very little use of brown anther steriles in hybrid seed production. However, their superior seed set justifies continued efforts to establish stable lines.

Pollen transfer from a specific male to a specific female must be accomplished for successful production in hybrid seed fields. Whether or not this will occur can be determined only by experience with the parent lines and insect pollinators involved. Therefore, pilot seed production is an essential step in identifying hybrid combinations that can be reproduced economically. The domestic honey bee will discriminate between nectar sources, temporarily develop fidelity to a specific nectar, and return for frequent visits, while other lines are neglected. Differences in seed yield have been correlated with foraging preferences (34). Field testing is the only means now available for determining if a specific pair of parents will satisfy all requirements for nectar quality and pollen supply, to ensure adequate pollen transfer. Some of the many wild insects that visit carrot flowers are effective pollinators (15). By selecting sites where wild pollinators are prevalent, the domestic honey bee can be supplemented or replaced.

During the process of growing breeding lines through their reproductive cycle, some selection for seed-producing capacity can be accomplished. Inbreeding depression and the resulting low seed yields produced on inbred steriles have dictated the almost exclusive use of F_1 seed parents. Another means of ensuring reasonable yields is to use BC_1 females. In hybrid trials, single-cross F_1s are the most uniform and generally receive the highest marks for appearance. When an outstanding single-cross hybrid is identified, a near duplicate can be reproduced by using the BC_1 generation for the seed parent rather than the more intensely inbred seed parent, which usually is at BC_4 or more. In many cases, the small reduction in seed yield compared with an F_1 parent is offset by greater uniformity. This strategy can approximate the uniformity of inbred × inbred without the low seed yields generally produced on inbred parents. The BC_1 system has been used successfully in the recently introduced hybrid Orlando Gold. Its prototype was first observed as the single-cross B3640 × F524. The nonrecurrent component of the BC_1 parent B4367S was selected for its similarity to the recurrent line B3640M. To produce a near duplicate of the single cross we used [(B4367S × B3640M) × B3640M] × F524. The use of this BC_1 plan may not find favor with seedsmen who are reluctant to complicate their production program with the extra cycle necessary to produce the BC_1 parent. An obvious advantage is high yields of seed from F_1 plants that bear the BC_1 seed for field production.

BREEDING PLAN

At northern latitudes carrot breeding can be accomplished by growing roots of source material (open-pollinated cultivars, breeding populations, or segregating generations from fertile × fertile crosses) in the field and then producing seed from selected plants in the greenhouse. To produce seed in the greenhouse in time for spring planting it is necessary to replant roots from field selection before mid-November. To provide time for adequate vernalization, the crop must be harvested in early September. This inflexible schedule means that seed maturing in the greenhouse often must be rushed to the field for planting in order to complete the life cycle in 1 year. An additional disadvantage is the cost of greenhouse production and the limited quantities of seed of established parent lines and experimental hybrids that can be produced under glass. The obvious advantages of greenhouse production are the environmental controls and protection from weather. The relatively immature roots used in the greenhouse schedule permit more precise color discrimination. Xanthomonas blight does not damage greenhouse plants, making the hot water treatment unnecessary unless field seed production is anticipated.

If production sites are available at both southern and northern locations, it is possible to produce and select breeding material in typical carrot production areas during the winter months and then to grow seed from selected roots in the north. Under this schedule, seed is sown in central Florida, southern California or Texas in early October and roots are harvested in February or early March. Adequate cold induction can be accomplished by late April when the roots are planted for seed production in breeding plots and cages. Seed from these summer isolations is mature early in September in time for processing and planting at the southern sites in October, completing the life cycle in one year. Because summer seed production is difficult in areas most useful for winter root production, the breeder needs a site for seed production adapted to the root production schedule. Whether the seed cycle is produced in the greenhouse or field, it is possible to complete the life cycle of this normally biennial crop on an annual basis and thus hasten the achievement of breeding goals. The following annual operations are typical and represent the sequence and time required to complete development of a hybrid cultivar.

Year	Period	Step	Cycle[a]	Operation
1	Oct.–Mar.	a	V	In areas for which the eventual cultivar is intended, grow and select roots from a diverse collection of open-pollinated cultivars and breeding populations
2	May–Sept.	a	R	Self-pollinate and cross to a cytosterile tester as many as possible of the plants grown from these roots, taking care to include selections from a range of acceptable sources to maintain diversity
	Oct.–Mar.	b	V	Select in the production areas 5–10 roots of the best S_1 lines with their companion F_1 testcrosses
3	May–Sept.	a	R	Pair the 5–10 individual roots from each selected line with its F_1 under small cage isolators to produce S_2 and BC_1 seed
	Oct.–Mar.	b	V	From the best families of 5–10 sister S_1 lines, select 10 or more roots from only one of these sister lines in order to avoid narrowing the genetic base; the BC_1 roots should be selected to resemble their companion S_2 as nearly as possible; store and vernalize 40–50 BC_1 roots to use in progeny tests for maintainer genotypes
4	May–Sept.	a	R	Pair 5–10 single S_2 plants under small isolators, each with a plant from its BC_1 (making sure that the BC_1 plant is pollen sterile before caging) to produce S_3 and BC_2 seed
		b	R	Classify the 40–50 BC_1 plants (as male fertile or sterile) in the progeny tests in order to identify maintainer lines
	Oct.–Mar.	c	V	Follow the same procedure for the vegetative cycle as in 3b, again selecting roots from only one of the 5–10 sister S_3 lines; 40 or more roots are needed for isolation under increase and crossing cages.
5	May–Sept.	a	R	Mass under a large isolation cage 15 or more plants grown from S_3 roots of the maintainer (M) line and 5–10 phenotypically similar selections from the companion BC_2 (S) line
		b	R	For a preliminary test of combining ability, isolate an additional 10–20 S_3 plants with one or more established pollen-sterile (S) lines or F_1 hybrid seed parents to produce experimental hybrid combinations
	Oct.–Mar.	c	V	At as many locations as possible evaluate the prototype hybrids from step 5b to identify the lines that show promise as parents

(continued)

Year	Period	Step	Cycle	Operation
		d	V	At the same locations produce roots from the BC_3 and S_3 mass$_1$ seed produced under screen in step 5a; select for re-production 20–40 roots of the candidate lines identified in the hybrid trials (5c)
6	May–Sept.	a	R	Isolate under large cages the roots from 5d to produce BC_4 sterile (S) and S_3M_2 maintainer companion lines of the best inbreds
		b	R	Plant samples of 5–10 BC_3 roots from 5d under screen with selected pollen parents to test new lines as potential seed parents
		c	R	Isolate maintainer with selected female parents to reproduce the best hybrids in preceding trials 5c and to produce additional experimental combinations; selected single-cross steriles may be included to produce experimental three-way hybrids
	Oct.–Mar.	d	V	Test new hybrid combinations in observation plots; advance the best from preliminary trials 5c to replicated trials and expand to additional locations
		e	V	Produce and select roots of candidate lines for maintenance and increase
7	May–Sept.	a	R	Increase (BC_5 of the S and S_3M_3 of the maintainer) lines identified in replicated trials as potential hybrid parents
	Oct.–Mar.	b	V	Repeat in replicated trials the most promising entries of those tested in 6d
		c	V	Produce roots from 7a in numbers large enough for pilot seed production of candidate hybrids on a root-to-seed schedule
8	May–Sept.	a	R	Establish outdoor root-to-seed isolation in seed-producing area to produce hybrid seed for commercial trials; determine if seed yield is acceptable
		b	R	Make additional cage increases of parent inbreds from 7c roots to provide for seed-to-seed pilot production
	Oct.–Mar.	c	V	Distribute seed produced in 8a for commercial trials (strip plantings) in production areas using all standard production, harvesting, processing, packing, and distribution procedures
9	Sept.–Aug.	a	R	Plant the parent lines from cage increases (8b) on seed-to-seed schedule in a seed-producing area to provide hybrid seed for commercial trials and to evaluate seed-yielding potential
10		a		Distribute seed from 8a and 9a for second year and more extensive commercial trials in all producing areas where the

Year	Period	Step	Cycle	Operation
11		a		candidate hybrid is likely to be produced Release a hybrid cultivar and its inbred parent components if warranted by performance in trials for yield of roots, quality, and seed yield

[a]V, vegetative cycle (roots); R, reproductive cycle (seeds).

Notes on the Breeding Plan

At year 4 step (b) the progeny test may not be necessary if lines have been extracted from sources with a high incidence of maintainer genotypes. If pollen-fertile plants are observed, all of the BC_1 progeny should be discarded and the companion S_2 plants isolated to produce S_3 lines that may then be evaluated as pollen parents. Pollen-fertile plants in the BC_1 progeny indicate the presence of restorer genes in the recurrent line. Rarely is such a line worth the effort necessary to establish it as a maintainer. Hybrids in which it is the pollen parent will be fertile or will segregate fertile plants. This precludes its use in F_1 seed parents, but it may serve as a pollen parent for production of commercial hybrid seed. There is no disadvantage in having male-fertile individuals in the commercial root crop, unless the originator wishes to prevent the extraction of parent lines from the proprietary hybrid.

Also, at year 4 the most uniform S_2 lines may be massed rather than selfed in order to minimize inbreeding depression. This decision must be made on a line-by-line basis. Some lines may be massed while a few selected plants are selfed as well. The criterion in this case is to arrest the inbreeding process as soon as the level of uniformity dictated by the market is achieved. Massing at this stage can avoid the unacceptably low seed yields that result from inbreeding depression.

At year 5, step (a), a few S_3 lines that show good tolerance to inbreeding should be selected for additional generations of selfing. Homozygosity for selected horticultural qualities may thus be established. The inbred lines advanced to S_4 and beyond will provide parents for fertile × fertile crosses between lines of diverse origin. This type of recycling is a most productive source of improved inbred lines.

At year 5, step (c), some single-cross hybrids with similar but unrelated parents should be selected for testing as F_1 (sterile-hybrid) seed parents to be used in three-way crosses.

At year 6, step (b), there may be an assortment of BC_3 lines originating about the same time, but from roots of diverse origin. One or more of these male-sterile inbreds (BC_3) may be isolated with a proven pollinator to produce experimental F_1 hybrids. This provides an early test of combining ability of new sterile inbred lines.

At year 6, step (d), sufficient seed should be available for distribution to testers in all important producing areas. This schedule includes observation and replicated trials only at southern locations where the breeding lines are grown in the winter for selection and increase. If possible, these same hybrids should be tested in summer areas where the root production season corresponds to that used for the reproductive cycle in the above schedule.

In the crossing cages where sterile inbreds and single-cross hybrids are isolated with selected pollen parents to produce early-stage experimental hybrids, it is useful to record seed yields per plant. Such data provide some indication of seed production potential for candidate hybrids. Further data on seed yield are secured from the first root-to-seed and seed-to-seed pilot production.

All of the stages listed above probably will be proceeding at the same time. New and diverse material should be used each year in fertile × fertile crosses and for inbreeding to establish new lines. The number will be limited by resources available. Early-stage inbreds should be generated at a rate sufficient to replace those that are discarded.

Beginning at about year 5, step (a), when S_3 roots are in hand, serious consideration should be given to establishing breeding populations from which new breeding lines may be extracted. The method we have used involves isolating, under large screen cages, a single F_1 root from each of several crosses between plants selected from maintainer lines. The parent lines at S_3 or more are selected from diverse origins and possess at least one superior characteristic. Seed is harvested separately from each F_1 and grown in separate rows. To ensure an approximately equal contribution from each of the original parents the best four or five roots from each single-plant progeny are then planted for seed production under screen. The resulting population can be subjected to selection for several characteristics and recycled through mass selection as long as substantial genetic gains are achieved. One population has been selected through four generations for color and flavor and resistance to alternaria leaf blight. From the diversity of its original components it may be assumed that the population will yield a wide range of sugar and pigment levels when laboratory selection is applied.

TRIALS OF ADVANCED LINES

Because market carrots are harvested at a premature size that best meets market-grade specification, it is impossible to secure precise quantitative data in terms of total yields. The important performance characteristic of market carrots is marketable yield, which is likely to be influenced by stand density and time of harvest. The uniformity in size and shape and lack of serious grade defects, which depend to a large degree on genotype, become the important cultivar differences that need to be defined and evaluated in market carrot performance trials. In the case of processing carrots, one of the most important variables is total yielding capacity. Yields determined at maturity for processing carrots are more reliable and easier to secure than are those for market types.

The relatively wide range of maturity (root growth rates) represented by an array of entries in market carrot trials makes it impossible to select a harvest date optimum for all entries. Early-maturing entries may have a high percentage of oversized, low-value roots, while late entries may have undersized roots that do not make minimum marketable grade.

In our observation trials we have relied on subjective evaluation of new or early-stage experimental hybrids to identify specific combinations for advancement to replicated or commerical trial. These "beauty contests" have been judged by breeders, seedsmen, processors, growers, and shippers. The entries have been displayed and quality rated at several locations in carrot-growing areas. The numerical values for quality (1, unacceptable, to 5, excellent) have been reliable in identifying the entires preferred by a diverse group of judges. The scores have been useful for interpreting changes or trends in preference. During 10 years of such trials, it was apparent that exterior and interior color became increasingly important and that the long taper of cv. Imperator became less acceptable than those with slight taper and intermediate blunt tips. Recent trials have revealed an increasing preference for the nearly cylindrical blunt-rooted type. These subjective evaluations are important in providing data upon which selection of inbred lines and cultivars can be based, and they define future market trends that permit timely adjustment in selection criteria.

From the preliminary observation trials, entries with consistently high scores are selected for replicated and commercial trials. In most cases, remnant seed from cage production is sufficient to provide for replicated trials. Whenever possible, additional production under screen isolation is undertaken immediately to provide seed for the expanded testing program.

Replicated trials vary according to the type of information desired and the preference of the operator. Because of the difficulty experienced in establishing uniform stands it is usually necessary to overplant and thin to achieve the uniform stands needed to secure reliable yield data. For market carrots we have successfully used miniplots of single rows 1 m in length with nine replications. An estimated 60 seeds/plot are hand planted. After the stand is established, these plots are thinned to 30 plants/plot. At harvest we record the total number and weight of roots and the number and weight of marketable roots. The major types of culls are classified and weighed. In addition to these quantitative data, the roots are described from the standpoint of physical type and their quality factors are recorded (length, shape, smoothness, interior and exterior color). Recently, attempts have been made to conduct a taste panel evaluation of the best entries or those considered as candidates for release.

From small-plot data it is impossible to predict the performance of a carrot cultivar under commercial cultural and marketing practices. Therefore it has proved necessary to provide enough seed for planting and harvesting with the equipment used for the commercial crop and enough product to follow a candidate hybrid through harvesting, packing, and distribution. It is in commercial production and handling of the crop that potential deficiencies become evident. Among the defects that may emerge under commercial culture are (1) insufficient top length or strength for mechanical harvesters, (2) fragile roots that suffer mechanical damage in harvesting and packing, and (3) in processing carrots, an unacceptable canned or frozen product. In the latter case preliminary processing trials may avoid advancing an unacceptable candidate hybrid to commercial processing trials.

FUTURE GOALS

Hybrid carrot development through the 1970s was concerned with the most urgently needed and easily attained improvements. Until recently almost any combination of selected inbreds resulted in hybrids that were superior to open-pollinated cultivars. The apparent ease with which breeders were able to produce acceptable hybrids resulted in a proliferation of named releases. In many cases serious seed production problems emerged that made it difficult or impossible for seedsmen to provide a reliable, continuing supply of high-quality seed. As the number of available hybrids is narrowed by greater discrimination on the part of growers and by improved seed production of a few successful hybrids, priority will be given to objectives that have been receiving little attention.

One high-priority goal for future work is to improve seed-yielding capacity. This might involve the following: (1) establishing reliable brown anther lines, (2) early-generation evaluation of seed production potential of female lines under simulated commercial production, and (3) developing parent lines attractive to pollinating insects.

Other opportunities for future improvement lie in adapting to unusual uses. These include high-yielding, high-sugar lines for producing alcohol or sugar. Without the stringent quality requirements imposed on edible carrots, it should be possible to combine 10% or more fermentable carbohydrates with yields of more than 40 tons/acre. Such genotypes should be established so that material is on hand for prompt development in the event of a critical need for fuel alcohol. A winter carrot crop in the southwest sugar beet areas might provide a biomass or sugar source at a season of the year when sugar beet processing plants are idle. A non-bolting, high-sugar, white carrot would be suitable.

Breeding efforts probably will be directed toward other minor but potentially important uses. In some cases future product development will depend upon the imagination and initiative of breeders in creating raw products designed for special uses. Carrots can be fermented and processed like cucumbers to produce a pickled product of excellent eating quality. The carrot designed for this purpose will need to be low in fermentable carbohydrates, low in volatile terpenoids, high in carotene, and of a uniform interior color. These characteristics are also needed for juice or for blending with other vegetable or fruit juices. The special genetic characteristics required for producing quality deep-fried carrot chips are available. The most important requirement is for a low level of the reducing sugars, which char at high temperatures and result in an unattractive brown product. High-sucrose carrots produce attractive light-colored chips. The appearance and nutritive value of carrot chips can be improved by using high-carotene roots with uniform color distribution for chipping. Low-volatile compounds may not be a priority goal in products subjected to high temperatures in processing. The needs for an improved raw product for various kinds of dehydration should receive attention.

Resistance to additional serious or potentially serious diseases, insects, and nematodes will become high-priority objectives as the more urgent problems are solved and as improved germplasm becomes available. More comprehensive plant exploration and refinement of screening techniques are needed to identify sources of resistance and provide for efficient selection.

Powdery mildew, *Erysiphe polygoni*, first reported in California in 1976 (*1*), has been observed with increasing frequency since that time. The fact that in California resistance has been observed in some advanced breeding lines and hybrids suggests that progress can be achieved promptly. In France, Bonnet (*16*) found dominant resistance to a powdery mildew incited by a different species (*E. heraclei*) in *D. carota* ssp. *dentatus*. He suggested backcrossing and using his early inoculation method to transfer resistance to selected inbreds.

Work on resistance to root-damaging soilborne pathogens and pests has been limited to observations of levels of resistance in collections of introduced accessions and domestic cultivars or breeding material.

Apparent genetic differences in response to a type of root defect described as cavity spot have been observed. The cause or causes of cavity spot are not well established. Anderson *et al.* (*2*) proposed the names cavity spot for the disease caused by *Clostridium* spp. and rhizoctonia canker for that caused by *Rhizoctonia solani*. They classified as tolerant or resistant only 10 out of 125 cultivars and breeding lines tested against *R. solani*. In 1981 DeKock (*26*) reported that cavity spot was a typical calcium deficiency symptom that may be induced by overfertilization with potassium. Typical cavity spot symptoms described by Fawzi and Kelly (*35*) in 1982 resulted from feeding of the fungus gnat *Bradysia impatiens* (Joh.). They achieved complete control with a systemic insecticide and were unable to produce cavity spot symptoms under calcium deficient culture. Whatever its cause, cavity spot often results in serious losses; and there is enough evidence of genetic resistance to justify including it in future long-term goals.

One of the most serious soilborne pests on carrots is root-knot nematode *Meloidogyne hapla*, controlled by costly soil fumigation. Reports of genetic tolerance (*20, 25, 70*) give rise to hopes that eventually a level of resistance can be achieved that will minimize damage or contribute to reducing populations and to improving effectiveness of control measures. There are no active breeding programs now using controlled screening for nematode resistance.

In a 1977 report, Scott (*55*) suggested that differences in mortality of lygus bugs *Lygus hesperus* Knight and *Lygus elisus* van Duzee on inbred lines from the open-pollinated cv. Imperida were great enough to make breeding for resistance feasible. Lygus bugs on the seed crop cause serious reduction in seed quality. Immature seeds attacked by lygus bugs do not develop embryos and they cannot be separated from normal seeds in the milling process. Eventually, the possibility of exploiting genetic resistance will need to be explored.

Another potential pest is the carrot fly *Psila rosae*, reported as a common root feeder in Europe and Canada, but not yet a serious problem in U.S. carrot-producing areas. Ellis *et al.* (*31, 32*) recently found that strains of the cv. Nantes were damaged less than some processing types. Whether or not the observed level of resistance will provide adequate protection is not known.

Some widely publicized techniques for rapidly modifying the genetic constitution of higher plants have been neglected by carrot breeders, who generally have been inclined to concentrate on conventional breeding methods. New techniques classified as biotechnology or genetic engineering probably will not soon become major sources of genetic variability needed for carrot improvement. Most urgently needed genetic characteristics are available in domestic cultivars, breeding populations, and exotic accessions. Genes controlling needed characteristics are easily transferred by classical techniques of hybridization and selection. It is likely that conventional approaches will continue to occupy the attention of carrot breeders. However, the carrot is a good organism for exploring the potential application of advances in biotechnology. In cell culture it readily produces callus and protoplasts and provides a model for plant regeneration. Improvements that may be incorporated through biotechnology and that are not attainable by conventional means have not been identified. When they are identified and incorporated, conventional procedures probably will be necessary to establish stable genotypes that can be produced on the farm. Despite the general reluctance of breeders to adopt new techniques, biotechnology offers promise for future advances and should be exploited. One possible application is rapid development of inbreds by means of anther or pollen culture to obtain haploids. Another is selection for resistance in cell cultures exposed to disease toxins or herbicides.

It is impossible to predict future goals accurately, but we can be confident that unexpected problems will be encountered. New needs certainly will emerge and novel uses for carrots probably will be developed. Fortunately, most of the genetic diversity necessary to meet new demands is available. The breeder must be alert to develop germplasm and apply procedures for solving future as well as present problems.

REFERENCES

1. Abercrombie, K., and Finch, H. C. 1976. Powdery mildew of carrot in California. Plant Dis. Rep. *60*, 780–781.
2. Anderson, N. A., Davis, D. W., and Shehata, M. A. 1982. Screening carrots for resistance to cankers caused by *Rhizoctonia solani*. HortScience *17*, 254–256.

3. Angell, F. F., and Gabelman, W. H. 1968. Inheritance of resistance in carrot, *Daucus carota* var. *sativa*, to the leaf spot fungus, *Cercospora carotae*. Proc. Am. Soc. Hortic. Sci. *93*, 434–437.

4. Angell, F. F., and Gabelman, W. H. 1970. Inheritance of purple petiole in carrot, *Daucus carota* var. *sativa*. HortScience *5*, 175.

5. Aubert, S., and Bonnet, A. 1978. Nutritional value of the Umbelliferae with the example of the carrot (*Daucus carota*). Ombelliferes: Contrib. Pluridiscip. Syst., Actes Symp. Int. 2nd, 1977 pp. 809–822 (in French).

6. Aubert, S., Bonnet, A., and Szot, B. 1979. Rheological indices of texture in relation to biochemical characteristics of carrot (*Daucus carota* L.). Ann. Technol. Agric. *28*, 397–422 (in French).

7. Bajaj, K. L., Kaur, G., and Sukhija, B. S. 1980. Chemical composition and some plant characteristics in relation to quality to some promising cultivars of carrot (*Daucus carota* L.). Qual. Plant.—Plant Foods Hum. Nutr. *30*, 97–107.

8. Banga, O. 1957. Origin of the European cultivated carrot. Euphytica *6*, 54–63.

9. Banga, O. 1963. Origin and distribution of the western cultivated carrot Genet. Agrar. *17*, 357–370.

10. Banga, O., Petiet, J., and van Bennekom, J. L. 1964. Genetical analysis of male sterility in carrots. Euphytica *13*, 75–93.

11. Bassett, M. J. 1974. Screening carrot roots for high soluble solids by specific gravity. HortScience *93*, 232–233.

12. Bayliss, M. W. 1975. The effects of growth *in vitro* on the chromosome complement of *Daucus carota* (L.) suspension cultures. Chromosoma *51*, 401–411.

13. Bell, C. R., and Constance, L. 1960. Chromosome numbers in Umbelliferae. II. Am. J. Bot. *47*, 24–32.

14. Bienz, D. R. 1968. Evidence for carrot splitting as an inherited tendency. Proc. Am. Soc. Hortic. Sci. *93*, 429–433.

15. Bohart, G. E., and Nye, W. P. 1960. Insect pollination of carrots in Utah. Bull.—Utah Agric. Exp. Stn. *419*, 1–16.

16. Bonnet, A. 1983. Source of resistance to powdery mildew for breeding cultivated carrots. Agronomie *3*, 33–37.

17. Borthwick, H. A. 1931. Development of the macrogametophyte and embryo of *Daucus carota*. Bot. Gaz. (Chicago) *92*, 23–44.

18. Borthwick, H. A., Phillips, M., and Robbins, W. W. 1931. Floral development in *Daucus carota*. Am. J. Bot. *18*, 784–797.

19. Braak, J. P., and Kho, Y. O. 1958. Some observations on the floral biology of the carrot (*Daucus carota* L.). Euphytica *7*, 131–139.

20. Brzeski, M. W. 1974. The reaction of carrot cultivars to *Meloidogyne hapla* Chitw. infestation. Zesz. Probl. Postepow Nauk Roln. *154*, 173–181.

21. Buishand, J. G., and Gabelman, W. H. 1979. Investigations of color and carotenoid content in phloem and xylem of carrot roots (*Daucus carota* L.). Euphytica *28*, 611–632.

22. Buishand, J. G., and Gabelman, W. H. 1980. Studies on the inheritance of root color and carotenoid content in red × yellow and red × white crosses of the carrot, *Daucus carota* L. Euphytica *29*, 241–260.

23. Carlton, B. C. 1960. Breeding and physiological studies of a bitter compound in carrots. Ph.D. Thesis. Michigan State University, East Lansing.

24. Carlton, B. C., and Peterson, C. E. 1963. Breeding carrots for sugar and dry matter content. Proc. Am. Soc. Hortic. Sci. *82*, 332–340.

25. Clark, R. L. 1969. Resistance to northern root-knot nematode (*Meloidogyne hapla*) in plant introductions. FAO Plant Prot. Bull. *17*, 136–137.

26. De Kock, P. C., Hall, A., and Inkson, R. H. E. 1981. Cavity spot of carrots. An. Edafol. Agrobiol. *40*, 307–316.

27. Dickson, M. H. 1966. The inheritance of longitudinal cracking in carrots. Euphytica *15*, 99–101.

28. Dickson, M. H., and Peterson, C. E. 1958. Hastening greenhouse seed production for carrot breeding. Proc. Am. Soc. Hortic. Sci. *71*, 412–415.

29. Dickson, M. H., Rieger, B., and Peterson, C. E. 1961. A cold unit system to evaluate bolting resistance in carrots. Proc. Am. Soc. Hortic. Sci. *77*, 401–405.

30. Eisa, H. M., and Wallace, D. H. 1969. Morphological and anatomical aspects of petaloidy in the carrot, *Daucus carota* L. J. Am. Soc. Hortic. Sci. *94*, 545–548.

31. Ellis, P. R., Hardman, J. A., Jackson, J. C., and Dowker, B. D. 1980. Screening of carrots for their susceptibility to carrot fly attack. J. Natl. Inst. Agric. Bot. (G.B.), *15*, 294–302.

32. Ellis, P. R., and Hardman, J. A. 1981. The consistency of the resistance to carrot fly attack at several centers in Europe. Ann. Appl. Biol. *98*, 491–497.

33. Erickson, E. H., Garment, M. B., and Peterson, C. E. 1982. Structure of cytoplasmic male-sterile and fertile carrot flowers. J. Am. Soc. Hortic. Sci. *107*, 698–706.

34. Erickson, E. H., Peterson, C. E., and Werner, P. 1979. Honeybee foraging and resultant seed set among male-fertile and cytoplasmic male-sterile carrot inbreds and hybrid seed parents. J. Am. Soc. Hortic. Sci. *104*, 635–638.

35. Fawzi, T. H., and Kelly, W. C. 1982. Cavity spot of carrots caused by feeding of fungus gnat larvae. J. Am. Soc. Hortic. Sci. *107*, 1177–1181.

36. Freeman, R. E., and Simon, P. W. 1983. Evidence for simple genetic control of sugar type in carrot (*Daucus carota* L.). J. Am. Soc. Hortic. Sci. *108*, 50–54.

37. Hansche, P. E., and Gabelman, W. H. 1963. Digenic control of male sterility in carrots, *Daucus carota* L. Crop Sci. *3*, 383–386.

38. Hansche, P. E., and Gabelman, W. H. 1963. Phenotypic stability of pollen sterile carrots, *Daucus carota* L. Proc. Am. Soc. Hortic. Sci. *82*, 341–350.

39. Heatherbell, D. A., and Wrolstad, R. E. 1971. Carrot volatiles. 2. Influence of variety, maturity and storage. J. Food Sci. *36*, 225–227.

40. Howard, R. J., and Williams, P. H. 1976. Methods for detecting resistance to Pythium and Rhizoctonia root diseases in seedling carrots. Plant Dis. Rep. *60*, 151–156.

41. Kirtschev, N. A., and Kratchanov, C. G. 1980. On the pectins of certain varieties of *Daucus carota*. Z. Lebensm.-Unters. -Forsch. *170*, 31–33.

42. Laferriere, L., and Gabelman, W. H. 1968. Inheritance of color, total carotenoids, alpha-carotene, and beta-carotene in carrots, *Daucus carota* L. Proc. Am. Soc. Hortic. Sci. *93*, 408–418.

43. Martens, M., Fjeldsenden, B., and Russwurm, H., Jr. 1979. Evaluation of sensory and chemical quality criteria of carrots and swedes. Acta Hortic. *93*, 21–27.

44. McCollum, G. D. 1966. Occurrence of petaloid stamens in wild carrot (*Daucus carota*) from Sweden. Econ. Bot. *20*, 361–367.

45. McCollum, G. D. 1975. Interspecific hybrid of *Daucus carota* × *D. capillifolius*. Bot. Gaz. (Chicago) *136*, 201–206.

46. Mildenhall, J. P., Pratt, R. G., Williams, P. W., and Mitchell, J. E. 1971. Pythium brown root and forking of muck-grown carrots. Plant Dis. Rep. *55*, 536–540.

47. Moore, D. M. 1971. Chromosome studies in the Umbelliferae. *In* The Biology and Chemistry of the Umbelliferae. V. H. Heywood (Editor), pp. 233–256. Academic Press, New York.

48. Morelock, T. E. 1974. Influence of cytoplasmic source on expression of male sterility in carrot, *Daucus carota* L. Ph.D. Thesis. University of Wisconsin, Madison.

49. Morelock, T. E., and Hosfield, G. L. 1976. Glabrous seedstalk in carrot: Inheritance and use as a genetic marker. HortScience *11*, 144.

50. Nair, P. K. K., and Kapoor, S. K. 1973. Pollen morphology production of *Daucus carota* L. J. Palynol. *9*, 152–159.

51. Paul, A. A., and Southgate, D. A. T. 1978. The Composition of Foods. American Elsevier Publishing Co., New York.

52. Robertson, J. A., Eastwood, M. A., and Yeoman, M. M. 1979. An investigation into the dietary fibre content of named varieties of carrot at different developmental ages. J. Sci. Food Agric. *30*, 388–394.

53. Scheerens, J. S., and Hosfield, G. L. 1976. The feasibility of improving eating quality of table carrots by selecting for total soluble solids. J. Am. Soc. Hortic. Sci. *101*, 705–709.

54. Schultz, G. A. 1973. Plant resistance to aster yellows. Proc. North Cent. Branch Entomol. Soc. Am. *28*, 93–99.

55. Scott, D. R. 1977. Selection for Lygus bug resistance in carrot. HortScience *12*, 452.

56. Senti, F. R., and Rizek, R. L. 1975. Nutrient levels in horticultural crops. HortScience *10*, 243–246.

57. Shallenberger, R. S., and Birch, G. G. 1975. Sugar Chemistry. AVI Publishing Co., Westport, CT.

58. Simon, P. W. 1985. Effect of genotype, growing conditions, storage, and processing on carrot (*Daucus carota* L.) flavor. *In* Evaluation of Quality of Fruits and Vegetables. H. E. Pattee (Editor), Chapter 9. AVI Publishing Co., Westport, CT.

59. Simon, P. W., and Lindsay, R. C. 1983. Effects of processing upon objective and sensory variables of carrots. J. Am. Soc. Hortic. Sci. *108*, 928–931.

60. Simon, P. W., Peterson, C. E., and Lindsay, R. C. 1980A. Correlations between sensory and objective parameters of carrot flavor. J. Agric. Food Chem. *28*, 559–562.

61. Simon, P. W., Peterson, C. E., and Lindsay, R. C. 1980B. Genetic and environmental influences on carrot flavor. J. Am. Soc. Hortic. Sci. *105*, 416–420.

62. Simon, P. W., Peterson, C. E., and Lindsay, R. C. 1981. The improving of flavor in a program of carrot genetics and breeding. ACS Symp. Ser. *170*, 109–118.

63. Simon, P. W., Peterson, C. E., and Lindsay, R. C. 1982. Genotype, soil, and climate effects on sensory and objective components of carrot flavor. J. Am. Soc. Hortic. Sci. *107*, 644–648.

64. Strandberg, J. O., Bassett, M. J., Peterson, C. E., and Berger, R. D. 1972. Sources of resistance to *Alternaria dauci*. HortScience *7*, 345.

65. Stubbs, L. L. 1948. A new virus disease of carrots: Its transmission, host range and control. Aust. J. Sci. Res., Ser. B *1*, 303–332.

66. Thompson, D. J. 1962. Studies on the inheritance of male-sterility in the carrot, *Daucus carota* L. Proc. Am. Soc. Hortic. Sci. *78*, 332–338.

67. Umiel, N., and Gabelman, W. H. 1971. Analytical procedures for detecting carotenoids in carrot (*Daucus carota* L.) roots and tomato (*Lycopersicum esculentum*) fruits. J. Am. Soc. Hortic. Sci. *96*, 702–704.

68. Umiel, N., and Gabelman, W. H. 1972. Inheritance of root color and carotenoid synthesis in carrot, *Daucus carota* L.: Orange vs. red. J. Am. Soc. Hortic. Sci. *97*, 453–460.

69. U.S. Department of Agriculture 1981. Agricultural Statistics. U.S. Government Printing Office, Washington, DC.

70. Vrain, T. C., and Baker, L. R. 1980. Reaction of hybrid carrot cultivars to *Meloidogyne hapla*. Can. J. Plant Pathol. *2*, 163–168.

71. Watt, B. K., and Merrill, A. C. 1963. Composition of foods—raw, processed, prepared. U.S. Dep. Agric., Agric. Handb. *8*.

72. Welch, J. E., and Grimball, E. L. 1947. Male sterility in carrot. Science *106*, 594.

73. Whitaker, T. W. 1949. A note on the cytology and systematic relationships of the carrot. J. Am. Soc. Hortic. Sci. *53*, 305–308.

74. Yates, S. G., and England, R. E. 1982. Isolation and analysis of carrot constitutents: Myristicin, falcarinol, and falcarindiol. J. Agric. Food Chem. *30*, 317–320.

10
Onion Breeding

LEONARD M. PIKE

Onions are an important vegetable crop worldwide. There seems to be no limit to their use by any nationality. Even though their daily per capita consumption is small, the overall yearly per capita use was reported to be 10.1 lb in 1981 in the United States. This consumption was in the forms of fresh, frozen, and dehydrated bulbs and green bunching onions (*33*). Consumption in some countries is greater than in the United States, but accurate worldwide figures are not available. In addition, many closely related species are important food items such as garlic, leek, chives, and Welsh onions. Together, the onionlike plants are one of the most important horticultural crops in the world, considering the economic importance of the edible crop and its seed.

The 1981 U.S. acreage for dry onions was 111,630 with a production of over 35 million cwt and a farm value of approximately $472 million (*34*). The imports for the same period were reported to be 1.4 million cwt and the exports to be 2.6 million cwt (*33*).

World production in 1981 was reported to be 3.9 million acres with a production of 19.7 million MT. The leading countries in onion production are shown in Table 10.1 (*8*), and the major onion producing states in the United States are shown in Table 10.2 (*34*).

ORIGIN AND GENERAL BOTANY

Unlike most domesticated food crops, the origin of the onion *Allium cepa* L. is still somewhat a mystery. Linnaeus (*21*), Don (*5*), and Regel (*30*), men known for being

357

TABLE 10.1.   The Leading 20 Countries for Production of
              Dry Onions in 1981[a]

Country	Area (1000 acres)	Yield (cwt/acre)	Production (1000 tons)
China	534	111	2956
India	543	67	1815
United States	111	316	1759
USSR	420	76	1595
Japan	69	339	1174
Spain	79	291	1150
Turkey	178	124	1100
Brazil	183	94	855
Italy	54	219	583
Egypt	52	222	580
Pakistan	104	92	479
Poland	62	144	440
Romania	96	77	374
Yugoslavia	103	65	341
Iran	118	50	292
Indonesia	99	46	227
Bangladesh	79	39	154
Thailand	62	51	154
Viet Nam	106	27	147
Burma	49	46	116

[a]Adapted from FAO Production Yearbook (8).

monographers, could not pinpoint its origin. Vvedensky (36) listed A. cepa only as a
cultivated plant, and in a review Hooker (10) said its origin was not known. Most
botanists doubt that A. cepa exists today as a wild plant. Vavilov (35) suggested that the
onion originated in the area of Pakistan. Others have suggested Pakistan, Iran, and the
mountainous areas to the north (15).

One fact is certain: the onion has been around in its present edible form for thousands
of years. Tackholm and Drar (32) listed the use of onions recorded in tombs as early as
3200 B.C. It was mentioned as a food in the Bible and in the Koran with reference thought

TABLE 10.2.   Leading States for
              Commercial Dry
              Onion Production
              in 1981[a]

State	Area (acres)	Production (1000 cwt)
California	28,600	9,731
Texas	24,200	4,188
New York	14,300	3,933
Colorado	9,000	2,925
Michigan	7,300	2,446

[a]After U.S Department of Agriculture
(33).

to be around 1500 B.C. In addition to being a food, the onion was considered by people during ancient times to be a medicinal plant having certain healing powers. Some early drawings indicate that onions were used in spiritual offerings.

Common Onion, Multipliers, and Shallots (*Allium cepa* L.)

This group includes the common bulb-type onion, potato or multiplier onions, ever-ready onions, and shallots. The common bulb-type onion is by far the most important in commercial trade. It generally can be described as a bulbing onion under long- or short-day conditions, depending on its adaptation. It produces a single bulb and has an umbel-type inflorescence, producing seed and no (or rarely a few) bulbils. It is reproduced typically from true seed. It may be of several bulb shapes and colors. Bulbing, controlled by a combination of day length and temperature, varies depending on where the specific cultivars were developed. It may also vary in pungency from mild to very pungent and vary in keeping quality with storage life ranging from a few weeks after harvest to almost 1 year. Soluble solids may vary from 4 to 25%. These characteristics are described in detail by Magruder *et al.* (*24*) and Jones and Mann (*15*).

The potato or multiplier onions are generally thought of as garden onions. They usually have smaller bulbs, may or may not flower, and generally do not produce seed. They continually produce new bulbs by dividing and are propagated by bulb division.

The ever-ready onion is similar to the multiplier onion, but usually has smaller bulbs. It was described by Stearn (*31*) as being a perennial home garden cultivar giving rise to numerous bulbs each year. Ever-ready onions rarely flower and are always reproduced by division.

The shallot is of commercial importance in Europe and in the United States. Shallots differ from multiplier onions and ever-ready onions in two ways: they form into single bulbs following division and their tops die down, indicating maturity. They go into a state of rest or dormancy similar to the common onion. Some cultivars flower freely and produce seed. However, due to their prolific bulb division, they are propagated asexually. This group has had some taxonomic description problems; but since the shallot crosses freely with the common onion and produces fertile progenies, it will be treated here as *A. cepa* L.

A. cepa L., generally know as a bulbing vegetable, has several closely related species: *Allium ampeloprasum* L. (great-headed garlic, leek, and kurrat), *Allium chinense* G. Don (rakkyo), *Allium fistulosum* L. (Japanese bunching or Welsh onion), *Allium sativum* L. (garlic), *Allium schoenoprasum* L. (chives), and *Allium tuberosum* Rottler ex Sprengel (chinese chives). A brief description of these closely related species follows. Detailed descriptions can be read in the text "Onions and Their Allies," by Jones and Mann (*15*).

Great-Headed Garlic, Leek, and Kurrat (*Allium ampeloprasum* L.)

This species is extremely variable, with many cultivars described in detail by Feinbrun (*9*). Generally the leaves are very flat, typical of leek. Bulbs range from well defined, such as produced in a type known as great-headed garlic, to non-bulb-forming types. The bulbs are actually bladeless storage leaves known as cloves. The clove is very similar to garlic. The scape (or seedstem) is round and solid. The umbels produce seed, but this species rarely produces bulbils, which are common in true garlic. Sterility is common. Leek produces almost no bulbing structure and so the elongated foliage base is eaten. It is propagated by seed. Kurrat is a leeklike form of small stature. It is raised for its edible tops and is also propagated by seed.

Rakkyo (*Allium chinense* G. Don)

This vegetable onion is native to China and eastern Asia and is grown by home gardeners mainly of Chinese and Japanese background. It is a plant similar to chives except that it bulbs and divides and looks similar to small dividing onions. When mature, the leaves die down. Seed stems are solid, but leaf blades are hollow as in common onions, which is unique among cultivated alliums. Flowers are purple in color but do not set seed, because they are tetraploids, as reported by Kurita (*19*).

Japanese Bunching or Welsh Onion (*Allium fistulosum* L.)

Japanese bunching or Welsh onions have been popular garden onions of China and Japan. It is not clearly understood where the name Welsh onion originated. The plants are vigorous and have hollow leaves and seedstems. Non-flowering plants look very similar to *A. cepa*. The plants never produce enlarged bulbs. One of the most distinguishing characteristics between *A. cepa* and *A. fistulosum* is the shape of the seedstem. *A. cepa* is characterized by a swelling in its scale midway in its length, whereas *A. fistulosum* has a straight uniform scape. *A. fistulosum* flowers tend to open fully at anthesis and are somewhat unique in that they open in regular order from central flowers outward to the umbel base. Hybrids have been successfully made with *A. cepa* resulting in bunching types such as Beltsville Bunching. The hybrids are generally sterile.

Garlic (*Allium sativum* L.)

The garlic plant has flat, longitudinally folded leaf blades very similar to *A. ampeloprasum,* but does not produce bulblets around the main bulb. In flowering plants, the inflorescences produce mainly small bulbs instead of flowers. Some flowers may partially develop but produce no seed. The cape is round, smooth, and solid for the entire length and may be somewhat coiled. The bulb is a composite of several cloves. Garlic is propagated by planting individual cloves. Large bulbs are always made up of several cloves, which also helps distinguish garlic from *A. ampeloprasm* (great-headed garlic) which has large single bulbs with smaller bulbs or cloves attached.

Chives (*Allium schoenoprasum*)

The chive is probably the most variable in type and range. It is known as a wild plant in North America, Europe, and Asia. Its great variation has created difficulty for classification. Domestication has brought few changes to the species. Despite its variability, the chive is easily separated from other alliums by its morphology. Its flowers, usually purple in color, open first at the top of the umbel then successively toward the base, which is opposite the flowering habit from all other alliums except the Japanese bunching onion (*A. fistulosum*). It is also free flowering, has slender leaves, and undergoes very rapid vegetative multiplication, thereby forming dense clumps.

Chinese Chives (*Allium tuberosum*)

This species is grown widely in China and Japan for its edible leaves and young flowers. The name chinese chive probably occurred since only the leaves are eaten, similar to the common chive. It does produce enlarged rhizomes but they are not used for food. It is propagated by division of the clumps of rhizomes. The foliage consists of grasslike leaves that are solid in cross section. The scape is solid and has two or more sharp angles running the entire length of the stem. The flowers are white and borne on flat-topped umbels, and generally there are few flowers per umbel.

FLORAL BIOLOGY

As described in the introduction, there are several important *Allium* species; but the floral biology, pollination control, and breeding aspects will be limited to *A. cepa* L.

Flowering of the onion is initiated by environmental factors. The primary inductive factor is cool temperature, with day length playing no role as with bulb development. Other environmental factors that slow growth of the plant seem to interact with cool temperatures to cause flowering, but data to support this are not available. Temperatures of 40°F or below for 1 week will generally induce flower formation in bulbs or growing plants with four or more leaves. However, temperature prior to and following the 40°F week can alter flower induction. Very small seedlings do not normally respond to cool temperatures. The larger the plant, generally the more easily it can be induced to initiate flower development. When the onion plant is induced to flower, the shoot apex ceases to produce leaf primordia and initiates the inflorescence. The inflorescence may consist of a few to more than 2000 flowers per umbel. The flower stalk (scape or seedstem), which bears the umbel consisting of the spathe and the flowers, is actually a one-internode extension of the stem. The stalk is initially a solid structure, but with growth it becomes hollow as it develops. The number of seedstems produced per plant depends on the number of lateral buds contained on the stem, which is the compact base plate on the bottom of the bulb (Fig. 10.1).

Plants grown from seed usually produce only one seedstem if induced to flower. Plants grown from bulbs may produce six or more seedstems since several lateral buds may be

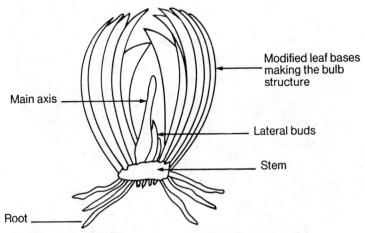

FIGURE 10.1. The stem of an onion is very compact and gener- ally not seen by the casual observer. The leaf bases enlarge upon bulb initiation and form the bulb. Upon flower initiation, seed- stems form in the apex of the leaf axis and elongate up through the bulb.

present that formed during development of the bulb. It should be noted that it is common for plants to produce bulbs and seedstems when grown during the winter and into the spring. This is due to the fact that one or more buds remain vegetative and produce leaves that form the bulb, while a lateral bud is initiated to form a seedstem. The plant then has both a bulb and a seedstem present at the same time.

The flowering structure is called an umbel (Fig. 10.2). It is an aggregate of many small inflorescences (cymes) of 5–10 flowers, each of which opens in a definite order, causing flowering to be irregular and to last for 2 or more weeks. If the plant produces two or more seedstems, the flowering sequence may actually occur for over a month (Fig. 10.3).

Each individual flower is made up of six stamens, three carpels united with one pistil, and six perianth segments (Fig. 10.4). The pistil contains three locules, each of which contains two ovules. The flowers also contain nectaries, which secrete nectar to attract insects. The anthers shed pollen over a period of 3 or 4 days prior to the time when the full length of the style is attained. The stigma becomes receptive at this time; and as a result of delayed female maturity (protandry), cross-pollination is favored. After pollination, the seeds develop; and as they mature, the capsules dry and split from the apex and down the center of each locule, which allows the seeds to fall free upon maturity.

The normal flower in onions is perfect, as previously described; but genetic and cytoplasmic sterility variations were discovered and reported by Jones and Emsweller (*14*) in a single-plant segregant of the cultivar Italian Red. Male-sterile plants developed from this original plant produced normal flowers except that the pollen did not develop into a viable stage. The inheritance was determined by Jones and Clarke (*12*) to be conditioned by a single recessive nuclear gene *ms/ms*, and a cytoplasmic factor, where one cytoplasm

FIGURE 10.2. A pair of healthy umbels showing good seed set. Note the difference in age of individual flowers. Some have seed development showing while others still show flower parts.

FIGURE 10.3. Typical seedstem and umbel formation in *A. cepa* L.

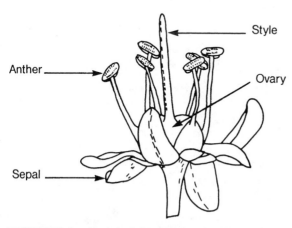

FIGURE 10.4. Each individual onion flower within the umbel is complete, having six stamens, three carpels united with one pistil, and six perianth segments. The pistil contains three locules each of which contains two ovules.

is considered normal (N) and the other sterile (S). To be made sterile, the onion plant must have the genetic and cytoplasmic condition S*ms/ms*.

The discovery, propagation, and techniques of maintaining male sterility in the onion have provided an excellent method for producing hybrid seed and will be discussed later in detail.

POLLINATION CONTROL

Pollination control in a cross-pollinating species such as onions is extremely difficult, considering each umbel has several hundred tiny individual perfect flowers. Therefore, it is important to understand onion flowering habits and the inheritance of as many characteristics as possible to be efficient in breeding the crop. Such information will help the student understand how to handle pollination so as to prevent many mistakes and the concomittant waste of time and money.

First, I shall discuss pollination control in a normal open-pollinated cultivar or breeding line. Selfing can be done only on a limited basis because inbreeding depression begins showing in the second (S_2) generation. To make the initial cross between two selections, the breeder has two choices. One is to hand emasculate stamens from one line, which is extremely difficult. The fact that flowers open on an umbel for 2 or more weeks adds to the problem of making the cross. The second choice is to make what is called a fertile × fertile cross, where two selections are caged together and then pollinated by hand or with

FIGURE 10.5. Small individual cages are used to obtain selfs on individual onion bulb selections. Flies are generally used to obtain pollination.

FIGURE 10.6. Small 2 × 2 ft nylon mesh cages used for making a three- to five-bulb mass following two generations of selection and selfing.

insects such as common house flies, blow flies, or bees. This method can efficiently be utilized if the two selections are different enough so that the F_1 can be differentiated from the two parents in the bulb stage. If they are different in that respect, seed should be saved separately from the two plants and planted in separate progeny rows. The hybrid bulbs can then be identified and distinguished from bulbs resulting from selfs of the two original parents that were caged together. The F_1 bulbs are harvested and then caged together to produce an F_2 progeny, or they can be used in a backcross program.

Once the selection process has begun in a segregating population, selfing is necessary for two generations to determine which progeny lines are desirable. To obtain the selfed populations, small individual cages are used to prevent outcrossing (Fig. 10.5). Flies are most commonly used to make these pollinations. Once desired progeny lines are selected, the breeder must begin making three to five bulb selections from the progeny rows and mass them in small cages. These are usually 2 × 2 ft, made of a nylon screen material, and fitted on a frame (Fig. 10.6). Flies are again the best choice to pollinate the small mass lots. Once a good line has been selected for commercial testing, a seed increase must be made in a large cage or small field isolation. Cages may be used up to 100 ft long by 12

FIGURE 10.7. Large screen cages are used to increase seed for small commercial testing. Bees are used to obtain pollination.

ft wide (Figs. 10.7 and 10.8). Bees are used to pollinate the onions in large cage increases. Small field isolations and commercial seed production must be separated from other flowering onions by one or more miles to ensure production of noncontaminated seed. Greater distances should be used, if possible, for breeding lines being produced without screen cages.

Male sterility is another factor in pollination control for the onion, which is normally a highly outcrossing species. The male-sterile mechanism involves both genetic and cytoplasmic factors as mentioned earlier and will be discussed as a means of pollination control in the onion breeding section.

BREEDING ONIONS

After this general presentation of flowering and pollination control in onions, we can proceed into breeding of the crop. Onions fall into two major types, known as short-day and long-day onions. However, a third group should be recognized as intermediate day length types, which bulb somewhere between the two major groups.

My observations suggest that some cultivars may actually be day length insensitive, but at present no actual data have been produced. Considering the normal long-day and short-day onions, those that bulb when the day length exceeds approximately 11.5 hr fall into the short-day group, and onions that require 14 hr or more to bulb fall into the long-day group. These are general groups and may vary 1 hr either way since temperature also alters bulbing. Therefore, onion breeders must be aware of the requirements of a particular location to select the correct germplasm to begin their program.

The life cycles of short- and long-day onions differ considerably and must be understood by the breeder. Short-day types are planted in the United States as seed in the fall, generally between October 15 and November 15, and are grown to maturity during April and May. This may vary by as much as 1 month either way depending on location. Bulbing requires approximately 180 days when grown as a seed-to-bulb crop. Bulbs are harvested and stored until September 15 to October 1, then reselected and planted to produce seed. Therefore, storage time is approximately 150 days. The planted bulbs begin growing immediately and will flower during May of the following spring with seed maturing in July.

Long-day onion seed is planted generally during late March and April to produce bulbs that mature in early September. Maturity requires approximately 150 days. Bulbs are harvested and stored until April, and then planted to produce seed. Flowering occurs in late June with seed maturing in August. Should the breeder decide to utilize seed-to-seed practices, seed must be planted in mid-August to produce a large plant to overwinter so that it will flower the following spring. Seed-to-seed production is a method of forcing an immature plant into flowering without producing bulbs. Plants must have produced four or more leaves and attained a diameter of approximately ⅓ in. before being subjected to temperatures below 45°F for flowering to be initiated. No definite size of plant or number of hours below 45°F can be presented as cultivars respond so differently. Most onions that are large enough to be initiated for flowering will flower if subjected to 7–10 consecutive

FIGURE 10.8. Large screen cages allow for isolation of several breeding lines of onions in close proximity of each other rather than providing one or more miles of separation for small field increases.

days of temperatures below 45°F or for longer periods of 50°F. Some cultivars are much more resistant to bolting and require more cold induction to cause flowering. It seems that several factors influence flowering. For example, if warm temperatures follow cold weather, bolting will be less than if followed by cool weather. Plant age and certainly plant size are important. The older and larger the plant, the easier it can be induced to flower. Therefore, when using seed-to-seed practices, it is extremely important to provide conditions that will cause all plants to flower, thus preventing selection pressure favoring easy bolters.

In addition to long- and short-day bulbing types, there are onions for different uses within those two categories. The three main types include fresh market, storage, and dehydration. In general, long-day types are primarily storage onions and short-day types are primarily fresh-market onions. Dehydrators may be either short- or long-day types. However, I recently developed two short-day onion cultivars that store for long periods of time, which shows that it is possible to develop short-day storage types (29), a development contrary to the conventional wisdom. Most long-day onions are much more pungent than short-day types. Again however, some long-day types are fairly mild and may be used for fresh market.

Major Breeding Achievements

The onion has been greatly improved in characteristics such as quality, yield, uniformity, and breeding practices. Breeders developed the crop during the years 1925–1940 to the point that it might be considered a classic example of crop improvement. The work done during that period, primarily by Henry Jones and co-workers, provided an enormous amount of information on the crop dealing with genetics and breeding methods. Jones worked on onions for over 50 years. Unfortunately, little new information has been published on the genetics and breeding of onions during recent years, mainly because Jones did such an excellent job of research and also because of the long-term efforts needed to study the crop. However, excellent cultivar improvement has been achieved by present-day onion breeders in relation to yield and quality.

The common onion, *A. cepa,* has a basic chromosome number of $n = 8$ and is known only as a diploid ($2n=16$). Extensive studies on meiosis and mitosis of the crop and related species have been done by Cochran (3), Emsweller and Jones (7), Levan (20), and Maeda (23). The reader is referred to these papers for detailed description.

Cytoplasmic Male Sterility

The most notable achievement with onion breeding began in 1925 when Jones and Emsweller (14) discovered a male-sterile plant in the cv. Italian Red growing in their breeding nursery at the University of California at Davis. The breeding entry was identified as 13–53 and is considered the historical pedigree number of male sterility. The plant produced no seed but did produce several bulbils among the florets, which saved it from extinction. The bulbils were saved and replanted to produce new plants. They were crossed with other onions, and the resulting progenies were studied to determine the inheritance of the male-sterile character. The results proved that the character was conditioned by an interaction between a recessive nuclear gene and a cytoplasmic factor. The cytoplasmic factor was designated N for normal fertile cytoplasm and S for the sterile condition. The nuclear genetic condition was designated as *Ms/*– for the normal fertile condition and *ms/ms* for the sterile condition. There are various combinations of genetic

TABLE 10.3. Progenies Resulting from Various Genetic and Cytoplasmic Combinations Crossed onto a Male-Sterile Onion Line[a]

Male-sterile line (A)	Male-fertile line (B)	Progenies
	Nms/ms	Sms/ms
	NMs/ms	SMs/ms
		Sms/ms
Sms/ms ×	NMs/Ms	SMs/ms
	SMs/Ms	SMs/ms
	SMs/ms	SMs/ms
		Sms/ms

[a]Only the Sms/ms condition is male sterile.

and cytoplasmic factors that provide male sterility and fertility. Table 10.3 illustrates why male sterility can only be maintained by having two lines with the following condition. The female line must be homozygous for Sms/ms and the male line must be homozygous for Nms/ms. Keep in mind that the fertile cytoplasmic factor does not transfer to the female during the cross. Therefore, the N cytoplasm provides for normal fertility in the male line (known as the maintainer or B line), yet does not restore fertility in the progeny when crossed onto the female line (known as the A line). Any other combination of genetic and cytoplasmic factors in the male line will give either fertile or segregating progenies.

It is important for the breeder to understand fully the information presented in Table 10.3 so the system can be handled in the breeding program. By using any known male-sterile line and any breeding selection that proves to a B line (identified by making testcrosses), the breeder can easily develop new A line–B line pairs in which the genotype of the A line will, through backcrossing, become identical to that of the B line.

The system shown in Table 10.4 shows that within six or more backcrosses, the A line

TABLE 10.4. The Method of Developing New A Lines in Onions Using the Backcross Method with Identified B Line Selections

Male-sterile line (101A)		Identified B line (101B)
Sms/ms	×	Nms/ms
Sms/ms	×	Nms/ms (BC$_1$)
(50% 101A, 50% 101B)		
Sms/ms	×	Nms/ms (BC$_2$)
(25% 101A, 75% 101B)		
Sms/ms	×	Nms/ms (BC$_3$)
(12.5% 101A, 87.5% 101B)		
Sms/ms	×	Nms/ms (BC$_4$)
(06.25% 101A, 93.75% 101B)		
Sms/ms	×	Nms/ms (BC$_5$)
(03.12% 101A, 96.87% 101B)		
Sms/ms	×	Nms/ms (BC$_6$)
(01.56% 101A, 98.44% 101B)		

originally used has become essentially identical to the recurrent parent B line for all other
genetic traits except fertility.

Genetics of Bulb Color

In addition to male sterility, studies on the inheritance of onion bulb color have been
major contributions. The inheritance of color is very complex and creates major problems
in both breeding the crop and producing commercial seed. A study by Clarke, Jones, and
Little (2) provided the major work. Jones and Peterson (17) reported on the inheritance of
a condition where yellow crossed by yellow gave all pink bulbs. El-Shafie and Davis (6)
reported data suggesting that two additional genes, G and L, were responsible for variations in color of onion bulbs. More research is needed to explain more completely the
many total color patterns that continually present themselves but do not fit the simple
explanations as reported.

When the I/I condition exists, the bulbs will be white regardless of the other color
genes (Table 10.5). The I/I factor is known as a color inhibitor gene and acts as a
dominant gene in the homozygous condition. In the heterozygous condition I/i this gene
gives an off-white or buff-colored bulb. When the genotype is i/i, the other color genes
$C/-$ and $R/-$ are expressed and condition the color of the onion bulb. It should be noted
that a recessive white condition occurs when the c/c genotype exists regardless of the
genotype at the $R/-$ locus. In addition, the fact that pink color occurs when crossing two

**TABLE 10.5. The Genetics of Several Traits in the Onion
Summarized[a]**

Onion traits	Genetic condition
Albino seedling	a/a
Yellow seedling linked with glossy	$y1/y1$
Yellow seedling not linked with glossy	$y2/y2$
Pale green seedling	pg/pg
Virescent seedling	v/v
Glossy foliage	gl/gl
Exposed anther	ea/ea
Yellow anther	ya/ya
Pink root resistance	pr/pr
Male sterility[b]	ms/ms
Bulb color	
Homozygous red	$i/i,C/C,R/R$
Heterozygous red	$i/i,C/c,R/R$
Heterozygous red	$i/i,C/C,R/r$
Heterozygous red	$i/i,C/c,R/r$
Homozygous yellow	$i/i,C/C,r/r$
Heterozygous yellow	$i/i,C/c,r/r$
Homozygous recessive white	$i/i,c/c,R/R$
Homozygous recessive white	$i/i,c/c,R/r$
Homozygous recessive white	$i/i,c/c,r/r$
Homozygous dominant white	$I/I,-,-$
Heterozygous dominant white (buff)	$I/i,-,-$

[a]The studies were conducted by Clarke, Jones, and Little (2) and
Stevenson (13).

[b]Male sterility is expressed as a recessive when in the presence of
sterile cytoplasmic factor S.

homozygous yellow bulbs creates a complex problem. This indicates the existence of complementary genes that modify bulb color. I have also observed variation of color intensity within most colors such as were reported by El-Shafie and Davis (6). For example, color may vary from deep purple to pink, dark yellow to very pale yellow, and even differences in brightness in the white bulbs. There are possibly many complementary and additive genes involved in onion bulb color. Selection in my program has led to development of bright-yellow cultivars and very deep red breeding lines, which have pigment going to the center of the bulbs.

Inheritance of Morphological and Other Characters

In addition to male sterility and color, most other characteristics seem to be conditioned by multiple genes or additive action. McCollum (26) reported heritability to be low for bulb weight and diameter, intermediate for height, and high for solids. Bulb shape is an important selection criterion, and the desired shape depends on the market preference. Genetic variability for shape ranges from extremely flat to oblong. Crosses between extreme shapes tend to produce F_1 bulbs intermediate between the two, but more toward the flat parent. Unless constant selection pressure is applied, the drift in shape moves toward flat. Single-center onions are important for producing nice, round-shaped onions with complete rings for onion ring processing and better storage. This trait is probably conditioned by additive gene effects and tends to drift to types with multiple centers unless constant selection pressure for single centers is applied. Ease of bolting is another important trait that acts the same way. Bolt resistance is important to prevent loss of yield to seedstem formation in bulb production. Unless constant selection pressure is applied for bolt resistance in a breeding program, the drift is toward easy bolting. Long storage and high percentage dry matter fall into the same categories. The trend of drift is toward poor storage quality and lower soluble solids. Foliage color, foliage morphology, disease resistance, and insect resistance are generally controlled by multiple genes. Jones and Perry (16) reported the inheritance of pink root disease resistance under field conditions to be controlled by a single recessive gene, which was later confirmed by Nichols et al. (26a) using a laboratory screening test. However, this resistant gene does not hold up to all strains of the fungus. My observations suggest a more complex inheritance because various levels of resistance have been observed. Several plant type characters under simple single-recessive gene control have been studied and are shown in Table 10.5. Thrip resistance was shown to be linked with the glossy-foliage character by Jones, Bailey, and Emsweller (11) and Peterson (27).

I have made several crosses between long-day and short-day types in my breeding program and observed that bulbing response is somewhat intermediate between the two parents but tends to be closer to the short-day parent. I have also observed that date of flowering correlates closely to bulb maturity. In other words, earlier maturing breeding lines or cultivars tend to flower earlier. This is important to know when planning to make hybrids, as flowering dates must be close enough to achieve making the cross in commercial hybrid seed production.

Goals and Objectives

Plant breeders must have established objectives within their programs before they begin breeding onions. They must know if they need to be working with long- or short-day types and whether the industry they intend to serve has a storage market, fresh market, or dehydration market. They must also be familiar with the disease and insect problems of the intended production areas and the onion bulb color preference of the intended market. In most instances, the size and shape of the bulbs are important. Pungency, lack of pungency, and percentage dry matter are important depending on the intended use of the onion. Examples include sweet mild onions for fresh-market use, firm long-storage types for processing and exporting, and a high-percentage dry matter with white color for dehydration. Onions used in salad bars should be mild, sweet, and red for good flavor and have an attractive appearance in the fresh salad. The various combinations of requirements are too many to list, but knowing the needs of the industry is very important in establishing the breeding objectives. These facts are mentioned because breeding onions is a long-term, complex, and expensive project.

In general, however, the resulting cultivar or hybrid should be of the type desired by the particular industry, including shape and color. It should be uniform for all characteristics including maturity. It should be high yielding, resistant to as many diseases and insects as possible, free of other defects such as short storage life, and easy bolting. It must be adapted to the area where it is to be grown, taking into consideration its sensitivity to day length, temperature, and other climatic conditions. Therefore, the breeding program should be in the location where the crop is to be grown; or at least the progeny selections must be made in that area because onions are not widely adapted to different growing conditions.

Selection Techniques

Selection techniques used in improvement of onion cultivars differ depending on whether the breeder is developing open-pollinated cultivars or hybrids. It will also differ between onion breeders. Some breeders practice more selfing than others; some believe only in developing hybrids. I shall discuss breeding techniques that have proven productive in my program for development of open-pollinated cultivars and shall present techniques used by the majority of onion breeders for improvement of hybrids.

The trend has been toward development of hybrids by commercial seed companies for two reasons: (1) hybrids are generally more uniform, and (2) this system prevents production of that hybrid by other seed companies. However, after passage of the Plant Variety Protection Act in 1970, some work is being done on development of improved open-pollinated cultivars. I feel that open-pollinated cultivars continue to have a place because they do not have such a narrow genetic base as hybrids and may prove to be better adapted to the environmental stresses that occur in many areas of production.

Several important qualities in onions are not visible to the eye, and so they need specialized selection methods. These will be briefly discussed here.

Selection for Dry Matter

The dehydration industry requires a much higher solids onion ($\geq 20\%$) than found in fresh-market or other processing types. The breeder cannot visually distinguish between low and high solids, and so instruments must be relied upon to assist in the selecting process. Most breeders use a refractometer to determine soluble solids, which provides a good estimate of dry matter. A good-quality hand-held refractometer is adequate for this selection process. Only one drop of juice is required from each bulb for determining its soluble solids. This can be obtained by taking a small plug of bulb tissue with a cork hole cutter or knife, and then squeezing the juice onto the glass plate of the refractometer.

Bulbs with high solids are saved for replanting. This method of selection has proved to be highly successful by breeders who put strong continual selection on dehydrator types.

Selection for Storage Quality

Storage quality is extremely important in onions as they are held for the most part in noncontrolled storage conditions. Many unknown factors seem to determine shelf life, and so the only method of screening is with storage trials. Long-day types should be stored under similar conditions (cool and dry) as the commercial crop. Short-day types should be stored under warm, humid conditions similar to those which occur during harvest of the crop. Good ventilation should be provided in storage for both short- and long-day types. The reason for suggesting storage of short-day onions under warm, humid conditions is that a high percentage of the short-day production is in hot, humid areas of the world. Therefore, they must be adapted to those poor natural handling and storage conditions. Selection is made in each generation for bulbs that do not sprout or rot in storage. Excellent progress can be made for better storage quality in three generations. There seem to be two major contributing factors involved in storage quality: in my observations, dormancy seems to be the primary factor and disease resistance a major secondary factor. If either one is weak, sprouting and or rotting occurs soon after harvest.

Selection for Flavor

Onion flavor is a factor that has no real guidelines. To one group of consumers, pungency is desired, while to another a lack of pungency is preferred. Generally, pungency is important for cooking onions while mildness is favored in fresh-market types. Pungency, lack of pungency, and sweetness are difficult to measure and actually vary within a cultivar depending on what time of year and where the onions are grown. Extreme differences also occur between cultivars and can be incorporated into new cultivars or hybrids through selection. The method for selection is difficult for two reasons: (1) variability as a result of growing conditions, and (2) the lack of a good measuring technique. At present, taste tests are the best selection method, with most of the effort going into the short-day fresh-market types, where sweetness and mildness are extremely important.

Development of Open-Pollinated Cultivars

It must be understood that open-pollinated cultivars are extremely heterozygous for many genetic traits. They may look uniform when observed in a production field; however, after selfing, the resulting progenies will look very different.

First Year

Breeders should begin by growing out several open-pollinated cultivars or plant introductions that they feel will contribute to achieving their objectives. Selection of 200 bulbs from each line should be made during the growing season in the area of commercial production, and then stored until planting time to produce the seed crop. For long-day types, the selection is in the fall and bulbs are stored at above-freezing temperatures until early spring. They are induced to bolt in storage and will flower following growth when planted in the field the next spring. Short-day types are selected in the spring, as they are grown during the winter months and bulb in early spring. They are stored during the summer at ambient temperatures in covered buildings and then replanted in the fall after

FIGURE 10.9. Onion bulbs stored in ventilated wood
pallet boxes. Good ventilation is essential for
long storage.

their dormancy is broken (Fig. 10.9). They sprout and grow during the winter and flower
the following spring. These are the main differences in handling mother bulbs of the two
types. Selection pressure should be applied for storage and possibly single centers, but
little else, because much segregation will result from the first selfing.There should be at
least 100 bulbs left for planting from each cultivar or as many as can be handled in cages
and progeny rows the following year.

Second Year

A minimum of 100 bulb selections from each cultivar should be planted for selfing. At
the same time I suggest making several fertile × fertile crosses to bring in additional
germplasm that might not be achieved by selfing within a single cultivar. If breeders
intend also to work simultaneously on hybrids, they should pair a known sterile bulb with
the selection they intend to self. This immediately provides a testcross progeny to determine if that particular selection is a maintainer (B line). This will be discussed more fully
in the section on hybrid development. The selected bulbs must be planted in a manner so
they can be caged when flowering occurs (Figs. 10.10 and 10.11). Several types of selfing
cages are used, ranging from paper bags covering the umbels to special designs of

FIGURE 10.10.   Bag showing field number and instructions to plant as a five-bulb mass. Bulbs and numbered field stake are in bag ready to plant before going to the field plot. The two stakes limit the area to be planted so that a cage can be placed over the flowering onions in the spring.

FIGURE 10.11.   Laying out small-lot selections of onions in a breeding-seed increase plot.

aluminum screen mesh or nylon cage covers. Once the cages are placed over the plants, pollination must be achieved. I use common house flies placed inside the cage as pupae, which hatch and pollinate the flowers as they feed on nectar. Other breeders use house flies or blow flies, and some even use honey bees or small brushes. Handling of bees is difficult as is using brushes, and obtaining blow flies in large quantities has been a problem. Whatever method of pollination works best for the breeder is undoubtedly the one that should be used. After selfing or the making of fertile × fertile crosses, seed should be allowed to mature. When black seed can be seen in a few capsules, harvest and dry the seed being careful not to mix lots. The seed should be quickly threshed and placed in a good seed storage room because onion seed loses viability within a few weeks if exposed to hot, humid conditions. Keep in mind that it takes 2 years to complete one generation of breeding onions: it takes 1 year to grow, make, and store bulb selections, and a second growing season to produce the seed. Many breeders, including myself, have attempted to obtain early flowering from selected bulbs but have not been successful. Seed to seed is possible in 1 year, but no selection for bulb characteristics is possible using that technique. However, seed to seed is practical in developing new A lines and will be discussed later.

Third Year

Plant S_1 seed from selfs or F_1 seed from crosses in the breeding plots during the same season, as commercial growers might be expected to do in their fields. I prefer to use a lower seeding rate than commercial rates, however, so that bulb evaluation will be easier to accomplish. For example, many growers plant four rows per bed or at high rates to obtain stands under poor weather conditions or to achieve maximum yields. I prefer two rows per bed so that bulbs will have the potential to develop size and express their true shape. Different cultural practices in different onion-growing areas will dictate breeder production practices. It is important that selection be practiced in the same area and under similar conditions as on commercial farms. Ten feet of progeny row should be sufficient for producing F_1 bulbs and 40 ft for S_1 lines. If seed is available, two separate plantings should be made to ensure against loss due to uncontrollable conditions. Thirty bulbs of F_1 and 100 bulbs of S_1 progenies should give the breeder sufficient observations for making selection decisions and providing material for advancement. Each progeny line should be kept separate to evaluate progeny lines at this time. Strong selection pressure for all traits in the objectives for each generation should be applied. For example, if resistance to pink root disease (*Pyrenochaeta terrestris*) is an objective, grow the plots on pink-root-infected land (Fig. 10.12). There is no reason to select bulbs that do not have the necessary genetic potential to make a new cultivar. The only exception is if the F_1 bulbs are being grown from a cross. In that case, the only interest in obtaining the F_1 bulbs to replant is to get the F_2 seed.

First, select progeny lines that show resistance to disease and the desired shape, color, etc. Discard poor progeny rows completely. Then make selections within the best progeny rows. I prefer to keep all the acceptable bulbs from the selected progeny rows, up to 100, because this is only the initial selection process. Many of these bulbs may be lost in storage, which is fine, as storage quality should be a major goal in all breeding programs. Also, extra bulbs allow for further selection such as for single centers, good interior color, and high dry matter. Then only the best 10 bulbs within each selection will be used for replanting to obtain the second selfing. I should note that occasionally, if a large number

FIGURE 10.12. A progeny resistant to the pink-root organism, *P. terrestris*. Note the healthy root system even at bulb maturity.

of selfs are made, progeny rows may be found that are extremely uniform and may be massed for immediate increase to release as an improvement of its parent cultivar. I have seen it occur only once out of several thousand selfs, but the possibility does exist.

In most instances, the new onion breeder will be surprised to see that most selfs will give very different progenies from what would be expected from the cultivar chosen as the parent source. For example, 100 selfs out of a given cultivar might provide 100 different phenotypes. Some will seem very poor as breeding lines, while others may show good uniformity, and occasionally one will appear outstanding. This is due to the fact that the onion is an outcrossing species and is in a heterozygous condition at many loci.

Fourth Year

Store the selected S_1 and F_1 bulbs until time for replanting, when further selection is made for other quality traits just prior to planting. Most breeders cut off the top one-third of the bulbs to check for color, single centers, ring thickness, etc., before planting. Selection pressures should be severe at this time to prevent ending up with thousands of progeny rows the next generation. The final selection for selfing should include 5–10 of the best bulbs from each of the selected progeny rows. The breeder should make a few three- to five-bulb masses from rows that were uniformly good, as the possibility exists for quick improvement of a cultivar in onions. Jones and Mann (*15*) reported success with this method also. If segregation occurs in the next generation, those progenies can be discarded or new selections can be made. Obtain seed from the second selfing as described for the first selection.

Fifth Year

Plant the S_2 and F_2 seed to produce bulbs. The progenies this year represent bulbs that have resulted from two selfings or F_2 progenies from original crosses, depending on which choice was made in the beginning of the program. Progeny lines resulting from selfing will generally be surprisingly uniform for most characteristics, depending on the selection criteria in the previous generation. Therefore, in this third selection process only very uniform progeny rows should be saved. The main exception is color if two different-colored onions were crossed in the beginning. If that was the case, study the inheritance of bulb color in order to make the right decisions regarding color selections. If not, the work to obtain the desired true-breeding color will take many years longer than necessary. Store bulbs as described earlier.

Sixth Year

Since onions suffer from inbreeding depression, it is not advisable to plan on a third selfing for development of open-pollinated cultivars.

Following the final selection process, and just prior to planting, select several 12- to 15-bulb groups that look identical for the traits desired. These are then planted to be massed in small cages. I use Saran screen cages on pipe frames 2 × 4 ft. The reason for several small masses selected from each progeny line is that these bulbs are still heterozygous at several loci, and the probability of getting a uniform cultivar is greater by using several cage masses rather than one. This procedure has worked very well for me, especially where bulb color, shape, and maturity are also objectives.

It should be mentioned that breeding procedures are the same in handling bulbs resulting from an original cross beginning with the F_3 progeny. In other words, F_3 bulbs should also be handled as small masses. In most instances, seed obtained from the small three- to five-bulb mass will produce uniform progenies.

Seventh Year

I prefer to plant seed obtained from the small masses of the S_2 or F_3 bulbs in observational trials, which are grown in the same manner as commercial onions. Select only from the best progenies. Harvest 200 or more bulbs and store for use in seed production.

Eighth Year

Selection should be made for storage quality, uniformity, single centers, solids, or other qualities, depending on program objectives as before. Approximately 100 of the best bulbs should be planted in a 9 × 10 ft cage to produce enough seed to begin replicated trial testing. Seed yields should be observed and recorded along with other plant characteristics that might be useful in obtaining plant variety protection or use as a future pollinator in hybrid seed production.

Ninth Year

Seed should be planted in replicated yield trials and, if possible, in small plantings on commercial onion farms. Several locations will help in deciding which lines are best. If lines prove to be superior, select enough bulbs to plant a 12 × 24 ft cage to obtain seed for several small grower evaluations.

Tenth Year

Plant the 12 × 24 ft cage to make the seed increase. If the line is not sufficiently uniform, cut the bulbs to observe for single centers, solids, and other quality characteristics. I prefer to continue to make such selections until the final release. Again, observe and record plant flowering habit, seed yields, and other plant characteristics.

Eleventh Year

Seed produced is entered into the yield trial and if possible put out with a few selected growers. The decision is made during that season to increase the line further. Shipping quality is observed and the final decision is made for release as a new cultivar. Note that onion breeding is certainly a long-term program, requiring 11–13 years to develop the new cultivar (Table 10.6). I stress using strong selection pressures and large numbers of progeny lines to ensure success.

Development of New A and B Lines

To utilize the male-sterile condition in hybrid development, the breeder must develop pairs of breeding lines known as A and B lines. These lines are developed as was shown in Table 10.4 by continual backcrossing of the B line (fertile) to the A line (male sterile). After five or more backcrosses, the A line has essentially the same nuclear genotype as its B line. The B line is used to maintain the seed supply of the male-sterile A line. Since the fertile cytoplasm factor is not transferred to the female, the A line remains male sterile when crossed with the B line, which differs only by its sterile cytoplasmic factor. Both factors, genetic sterility and cytoplasmic sterility, must be present for the plant to be male sterile.

TABLE 10.6. A Schematic System for Breeding Improved Open-Pollinated Cultivars.

Year	Procedure
1	Grow source lines and select 100 bulbs; store bulbs
2	Plant selected bulbs and self
3	Plant S_1 seed in progeny rows, select best bulbs from best progeny rows, discard poor progenies completely; store bulbs
4	Plant S_1 bulbs and self 5–10 selected bulbs from each progeny; also make a few 3–5 plant masses from same progeny rows
5	Plant S_2 seed and three-bulb mass seed obtained in fourth year; select lines that look best; if any 3- to 5-plant masses look good and are uniform, select those lines also; store bulbs
6	Plant S_2 bulbs, i.e., mass 12–15 S_2 bulbs from selected progeny rows
7	Plant seed to begin observation trial testing and early evaluation of bulbs; select superior progenies and discard others
8	Mass 100 bulbs and plant in a 9 × 10 ft cage for small seed increases of selected lines
9	Plant seed for yield trials, and in small plantings on commercial farms; several locations will help in deciding which lines are best; select stock bulbs for seed increase
10	Plant bulbs to make a 12 × 24 ft cage for seed increase; observe seed yields
11	Plant several commercial plantings to evaluate for all requirements such as shipping, storage, and processing, which was not possible during earlier testing
12	Release superior line as a new cultivar

The best way to develop new A lines is to make testcrosses of plants from open-pollinated cultivars with known A lines to identify any *ms/ms* genotypes within those populations (Table 10.7). From my experience and that of other onion breeders, one can expect to find approximately 5% B lines in any population of most cultivars (Fig. 10.13). It has been reported by Little *et al.* *(22)* and Davis *(4)* to be as high as 50% for the *Ms/ms* condition. Therefore, by pairing up single bulb selections from cultivars or breeding lines with known A-line bulbs, one can determine which pollinators were B lines by growing out the F_1 progenies. The F_1 progenies with 100% male-sterile plants indicate which selections were B lines. Seed of the selfed selection from each testcross must be saved so that once the B lines are identified, they can be used in a backcross program to develop the A line to the point where it is identical in genotype to the B line selection. The above procedure usually involves many pairs in the onion-breeding program because the development of a new A line does not ensure that it will make a good hybrid. Once several A lines are developed, they must be tested with other inbred B lines or C lines in hybrid combinations to determine if they will make superior hybrids.

Development of new A lines and their maintenance requires the same repetitive procedure. It is a continual backcrossing process, the only difference being cage size or field isolation. The A lines must continually be rogued for any fertile plants and the B lines should continually be rogued for the type desired. Extreme care must be taken to prevent mixing seeds or bulbs in the program. In all phases of bulb or seed storage, growing, and handling, the A and B lines should be kept separate. For example, A-line bulbs should be grown as a group separate from B- or C-line bulbs. Seed should also be cleaned separately, as either A- or B-line groups to prevent accidental mixing.

The procedure described is the easiest way to develop new A lines and the only way to maintain or increase A-line seed. However, it is possible to make crosses between known B and C lines to make new B lines, and then go into a backcrossing program with a known

TABLE 10.7. A Procedure to Develop New Male-Sterile Onion Breeding Lines (A Lines) Using the Testcross and Backcross Method

Year	Known A-line bulb	Fertile selection bulb
1	Plant bulb	Plant bulb; self and cross to A line
2	Plant F_1 seed	Plant S_1 seed
3	Plant F_1 bulbs following storage at 40°F or plant and overwinter; read at flowering and pair with B-line selection under a screen cage if all are sterile to make first backcross; if not 100% sterile, discard	Plant S_1 bulbs along side F_1 bulbs, using 5–10 bulbs to prevent severe inbreeding; pair with sterile F_1 plants; if F_1 was segregating or fertile, discard at this time; cage with the sterile F_1 to make 1st backcross; mass 3–5 bulbs of the S_1 (B line)
4	Plant BC_1 seed to produce BC_1 bulbs	Plant seed from small mass of B-line selections to produce S_2 B-line bulbs; make selections for type desired
5	Store at 40°F or plant and overwinter; observe flowers for male sterility; all should be sterile; cage with B line	Same induction procedure as A line; plant beside BC_1 A line; cage with new A line to make second backcross
	Continue until fifth backcross	Continue until fifth backcross
	New A line	B line, maintainer for A line

FIGURE 10.13. Small cages are used for making testcrosses where a known A line is paired with a selected bulb. Several hundred pairs are usually made, as about 5% of the selections are expected to be B lines from any given population.

A line. A third method of obtaining new A lines is possible if the breeder has the opportunity to observe flowering in large fields where seed is being produced. In several cultivars, a very low percentage of male-sterile plants has been observed. Peterson and Foskett (28), Kobabe (18), Banga and Petiet (1), Makarov (25), and Yen (37) all reported observing such male-sterile segregants in seed production fields. I have observed occasional steriles in breeding lines as well as cultivars. When male-sterile lines are found within a cultivar, they can be paired with a fertile in a testcross. If by chance the seed saved from the male sterile is 100% sterile, the breeder knows that the selection used in the testcross was a B line. There is thus, an immediate new A line with a maintainer B line without the necessity of going through all the backcrossing described above.

Development of Hybrids

The development of hybrids in the onion follows a completely different approach from development of open-pollinated cultivars. In cultivars, uniformity of certain desired characteristics is sought while at the same time maintain heterozygosity in order to keep plant vigor. In hybrids, it is possible to develop inbred lines having much more homozygosity and less vigor, because that vigor will be restored when the inbreds are crossed. In many ways, the development of hybrids is the easier of the two methods of cultivar improvement as it allows the breeder to work with more homozygosity. Better evaluations of the true genotypic value of desired selections can be made. The time required to develop improved hybrids and the effort needed to maintain inbred lines are much greater. Hybrids require development of A lines, their B-line maintainers, and C lines that are inbreds used as pollen parents in hybrids. If the breeder decides to make three- or four-way hybrids, then the time of development and testing is greater and the procedure more complex.

When planning a new hybrid onion-breeding program, one should not anticipate new hybrids before 15–20 years. If good A lines are available, new improved hybrid combinations can be expected in approximately the same length of time to develop new open-pollinated cultivars. As mentioned earlier, it is possible to work both programs simultaneously if when selfing, a sterile is included in the cage to make a testcross. However, I shall approach development of hybrids as a separate method and readers can decide which system to use in their programs.

First Year

Grow and select 50–100 desirable bulbs from each of the cultivars adapted to the area that show potential for use as a parent. At the same time, save an ample supply of A-line bulbs that have been grown from seed obtained from some other program. Sources include breeders within state Agricultural Experiment Stations, USDA, and commercial seed companies. Some hybrids are made up from A-line × B-line crosses and are therefore sterile. If necessary, use F_1 hybrids that were determined to be sterile. Store selected bulbs to allow them to go through dormancy as described earlier.

Second Year

Plant in two-bulb pairs: one will be a known A line and the second will be a selection. Cage the plants prior to flowering and leave a seedstem out on the sterile bulb if it has more than three. This allows for easier observation when checking for sterility. If it is sterile, simply break it off. If fertile, remove the entire plant and transplant another sterile in its place. Onions transplant fairly easily even when they are flowering. Be sure to pick off all open flowers on the fertile side of the pair if this is necessary, because some crossing may have occurred in the cage with the first plant, which was thought to be sterile. Begin placing fly pupae inside the cage and continue to do so twice a week for 3 weeks. Note that the plants must be free of insects such as thrips and spider mites prior to adding pollinators, because further spraying will also kill the flies.

After pollination is complete and seed set is observed, gently separate the umbels between the A line and its pair so that upon maturity, no possible mixing of seed can occur. The cage is removed at this time and it is advisable to stake the seedstems to prevent them from falling or getting broken by wind. Harvest the seed heads separately into small paper bags and number them as pairs so they can be maintained together throughout the remainder of the program. Also clean all seed from sterile parents separately from selfs. What is accomplished during the making of a testcross is identification of a possible B line from a chosen cultivar and selfing the fertile selection to start the inbreeding process. I prefer to number the pairs such as 40001 and 40002. The smaller odd number will always be the sterile pair for the next higher even number. The paper bags would read 40001 for the seed collected from the sterile and 40002 for the selfed fertile plant. Of course, other breeders use different systems and each breeder should use the system preferred. My numbering system always uses the year as the first digit and provides for 10,000 entries each year. This system provides numbers that do not show up again for 11 years. Note that entry numbers only need to be changed each generation, or every 2 years, because the bulbs grown and replanted are still the same generation as the seed planted to produce those bulbs. In other words, the new number assigned to the seed at planting will also be the same number assigned the bulbs when they are planted the following season.

I also begin another series of bulb selections the second year. By doing so, one does not have all bulb selections in the first year and all seed production in the second. The result is to have essentially twice as large a program with only half the required cages.

Third Year

The seed from the testcrosses and selfed bulbs are now ready to be planted. The seed should be assigned new numbers. It is important to separate all pairs into sterile and fertile groups and also to plant in that order. This prevents accidental mixing of sterile and fertile lines during planting and later during harvesting. At bulb maturity, select for desirable progeny rows within fertile lines obtained from the selfs. Discard all fertile lines not up to the standard of selection and also discard the corresponding sterile testcross progenies, because they serve no further purpose. Make selections within the selected fertile lines, a procedure similar to that described for improvement of open-pollinated cultivars. From the sterile testcross lines corresponding to the selected rows, select 25 bulbs that look healthy. No strong selection pressure is necessary at this time for the potential sterile line. Store selected paired progenies separately to ensure against accidental mixing.

Fourth Year

Plant 10 selected bulbs from the fertile S_1 progeny and the 25 taken from the testcross F_1 in paired rows (Table 10.8). Take extreme care not to mix bulbs during planting.

At flowering, carefully observe each F_1 plant for the presence of fertile umbels. If the total 25 plants are sterile, the probability is good that the paired self is a maintainer line having the cytoplasmic–genetic condition Nms/ms and can be classified as a B line. All open flowers on the 10 fertile plants and 10 of the sterile plants must be picked off to make the first backcross. The remaining 15 F_1 plants can be discarded. They should be caged as pairs to obtain the second selfing on the potential new B line and to make its backcross. Seed should be saved separately from each pair (self and cross) and handled as described earlier.

F_1 progeny lines that produced fertile umbels should be discarded along with their fertile pairs as they serve no purpose in the hybrid program. However, if you are also breeding open-pollinated cultivars, move the fertile line over into that program.

Fifth Year

Plant seed from both plants (S_2 and BC_1) as previously described in the nursery plots and again select the best progeny lines of the fertile side of the pair. Since this is now the second generation of selfing, the B lines will show considerable uniformity and the A-line pairs will begin to look similar to their maintainer. However, at this time much will

TABLE 10.8. A Schematic System for Breeding Improved Onion Hybrids

Year	Procedure
1	Grow and select 100 bulbs, store, and plant; also grow supply of male-sterile bulbs for use in testcrosses
2	Self selected bulbs and at the same time, testcross with known sterile
3	Grow out bulbs from self and F_1 testcrosses, select, discard poor progeny rows and their F_1 pair, store
4	Plant bulbs for seed production; observe sterility characteristics in F_1 lines; if 100% sterile, self selection and make backcross to F_1; discard pairs with fertile F_1 lines
5	Grow out A and B lines as pairs, continue to select in B line side, save best bulbs from A line for next backcross
6	Self B line of selected progenies and make backcross to sterile side of pair
7	Grow bulbs and make final selection on basis of B line side
8	Mass B line using 10–20 bulbs in cage while making the second backcross to the sterile side of the pair
9–12	At this point, begin going seed to seed and continue through the fifth backcross using the same procedure as in the eighth year; several A and B lines should have been developed; begin making hybrid combinations for testing

depend on how closely the original sterile resembled the fertile used in the cross. Select uniform bulbs from the most uniform progeny lines on the fertile side and, where possible, select A line bulbs that look the most like their maintainer. Although the A line will eventually become identical to its maintainer, some speeding up of the process can be made with selection. Harvest the selections and store until planting time.

Sixth Year

Plant bulbs (S_2 and BC_1) as for last cycle to obtain the third selfing and to make the second backcross. Not more than 10 sterile plants are needed, but one must always check for sterility when making the backcross. Harvest seed and handle as before.

Seventh Year

Plant the seed in progeny rows as before. We now have fertile lines that have been selfed three times and that should be very uniform and have much reduced vigor. The second backcross to the sterile side of the pair now has 87% of the genes of the maintainer and should be looking quite similar. At this time select only the one or two best progeny lines from the 10 under observation. From this time on, the next two generations are handled as masses of 20 bulbs for the B-line backcrosses. Further selfing would only lead to such loss of vigor that the B line could not be maintained in most cases and certainly would not contribute to any improvement of a new hybrid. The next two backcrosses will carry through the eleventh year.

The procedure for making new A and B lines to use in hybrid combinations has been explained, and now the procedure for making and testing hybrids will be discussed. With the A- and B-line genetic and cytoplasmic factors in onion, F_1 hybrids, three-way hybrids, and even four-way hybrids can be made. However, I feel that F_1 or three-way hybrids are superior and will limit the discussion to them.

F_1 hybrids are made by crossing A lines with unrelated B or C lines. Three-way hybrids are made by crossing A lines with unrelated B lines to produce the sterile F_1 and then by crossing that F_1 with either another B or C line. The main reason to use a three-way hybrid is to improve seed production, which is generally low when using inbred A lines. The F_1 seed parent will exhibit hybrid vigor and will produce greater seed yields.

The making of hybrids is essentially the same procedure as maintaining the A- and B-line pairs except that unrelated parents are used. The decision to make hybrids must be based on the objectives desired by the plant breeder. The most obvious reason for making hybrid onions is to obtain uniformity of the onions produced and to control the date of maturity. In some instances, seedling vigor and even yields may be improved. Other reasons might include adding disease, insect, or environmental stress resistance to the crop or producing a hybrid that will mature somewhere between that of the two parents.

Once the breeding objective is determined, the breeder must make testcrosses between various inbred A, B, or C lines using knowledge of these inbreds. Only small amounts of seed are needed for onion yield trials, and so the normal procedure to produce experimental hybrids is to place several A lines and one B or C line in a screen cage. Using a hive of bees for pollination, several hybrids can be produced in each cage. The same group of A lines is therefore placed in several cages, with each cage containing a different fertile inbred line.

After the pilot production of these experimental F_1 hybrids, seed is planted in yield trials in the various areas of onion production so they can be evaluated as potential new hybrids. It is extremely important that they be tested in the actual areas of production since onions are sensitive to day length and temperature. Yield trials should be replicated with each test row being 10 ft or longer. Rows of 10 ft seem to be sufficient if the soil is uniform in the field. I plant six replications, using one for field day observation and five for collecting data. It is advisable to repeat this planting in at least three locations.

The experimental hybrids must be critically evaluated for all characteristics important to onions in the particular growing area. Yield alone means little in today's market. Size, shape, color, date of maturity, storage or shipping quality, and disease resistance must all be observed in determining the value of the hybrid. Seed production is also just as critical in the value of a cultivar as bulb production, since good seed yields are necessary for the seed industry to be able to grow and sell the crop.

This evaluation procedure must be made for at least 3 years in each location to ensure the adaptation of the crop. One good year in one location has little value, since growing seasons vary so much from year to year. Three good years, and especially in several locations, give the breeder a good idea of the value of the new hybrid.

In my opinion, hybrids need more testing than open-pollinated cultivars since they have a narrow genetic base and have been observed to vary more in performance due to environmental stress. Once these hybrid combinations are tested and proven in trials, small commercial tests are the next step as with any crop. The remainder of testing would be the same as described in development of open-pollinated cultivars.

MANAGEMENT OF ONION SEED PRODUCTION

Management of onion breeding lines to obtain seed from selections is possibly the most important and critical part of a breeding program. Many good bulb selections are made by the breeder and lost for various reasons, thereby leaving the development of improved lines or cultivars incomplete. These include loss in storage, bulb mixtures, seed mixtures, or outcrossing occurring during seed increase. The handling and increasing of seed on A and B lines further complicates the management of the breeding program.

Assuming bulbs are at the stage for selection in the field, we shall proceed with steps to manage the physical operation of a program. Bulbs will be selected, based on objectives, from the progeny rows and laid in a group next to the field stake. Nonselected bulbs or progeny rows are discarded. In dealing with short-day onions, the bulbs are topped (foliage clipped leaving at least 1 in. of neck on the bulbs) and put into mesh onion sacks. The roots do not need clipping. The stake from the plot is placed inside the sack and a wire-strung shipping tag is labeled and attached to the sack. A knot should not be tied in the sack string because it may be necessary to open the sack several times during the selection process. The sacks of onions are left in the field for 2 or 3 days if the weather is dry so that the necks can dry down. If that is not possible, they must be placed in a dry place where fans force air through the sacks to dry them quickly. In the case of long-day storage onions, less drying is needed because they are generally allowed to become dry in the field before harvest.

Extreme care must be exercised at all times during handling and storage to keep progenies separate. It is even more important when working with A and B lines. Remember that they should always be planted separately from each other to prevent seed mixtures and bulb mixtures. The bulbs must then be stored until dormancy is broken. Short-day onions are harvested in the spring and long-day onions are harvested in the fall. In both cases, the bulbs must be kept dry and well ventilated.

Short-Day Onion Storage Procedures

Short-day onions are usually stored in open-air structures with only shade provided for temperature control. Small lots can be kept in any kind of container that is ventilated. I use wire milk carton crates, which each hold approximately 25 lb of bulbs and stack easily. They can usually be purchased from milk companies, which are currently switching to lighter, plastic crates. There are also several sources of ventilated plastic containers. Some commercial seed companies use wooden crates, and I have seen bulbs stored simply in mesh or burlap bags. However, ventilated crates are best (Fig. 10.14).

The bulbs are harvested in April, May, or June and stored until the latter part of August. At this time, they should be taken out and selections made for planting. Sprouts, rots, and bulbs showing doubling should be discarded. The decision based upon objectives should be made at this time. Bulbs should be labeled for selfing, making testcrosses, masses, and increases; then placed in paper or mesh bags with field numbers and instructions attached. For example, all selfs should be together for planting in small cages, three-to five-bulb masses next, 9 × 10 ft cages next, etc., so that cage plantings will end up in an organized manner (Fig. 10.15). Careful notes must be taken as selections are made and bagged for planting. If single centers are part of the objectives, cut off the top one-third of the bulb to be sure they are single centered. Other selection pressures are applied at this time. At the completion of this selecting period, which in a large program may last 2–3 weeks, plant the bulbs in the field in a way so that cages can be placed over them at flowering the next spring (Fig. 10.16). The bulbs should be pressed into loose soil or into a shallow furrow so that only the bottom half of the bulb is in the soil. Do not cover the entire bulb as wet weather can cause considerable loss due to rotting.

FIGURE 10.14. Short-day onion storage boxes should be placed under a covered shed in a dry area. Long-day varieties must be placed in an enclosed building to prevent freezing.

FIGURE 10.15. A view of an organized group of onion-breeding cages. The 2 × 2 ft cages begin in the foreground, and larger cages follow until the field is completed with 12 × 80 ft cages. Such an arrangement helps in physical handling and erecting, management of cultural practices, introduction of flies and handling bees, and seed harvesting.

Grow the onions as any other winter crop using good cultural practices. Insect, disease, and weed control is necessary to ensure good seed production. The bulbs will rapidly sprout and grow when irrigated. Lateral buds will grow and the resulting plant will usually have three to five growing points. The foliage will continue to grow, and flowering will be initiated during the winter months. Begin caging the plants as soon as bolting begins in the spring, because flowering will generally occur within 30–45 days once seedstems appear.

FIGURE 10.16. Bulbs being set with root plates down after being scattered in the breeding-cage areas. Soil should be pulled up to the bulbs but bulbs should not be covered.

Long-Day Onion Storage Procedures

Long-day bulb selection and harvest occurs in the fall. Following selection, the bulbs are stored in similar containers as for short-day bulbs. The bulbs are stored inside a building and kept dry and above freezing temperature. Air circulation is important as with short-day types. Sometime during midwinter, the bulbs are observed for sprouting and rotting. Generally, data are collected and weights are recorded again to determine storage quality. In most programs, bulb evaluations are made at this time to determine superior storage lines. Decisions are made in early spring as to what breeding steps need to be taken as with the short-day program. The bulbs are planted in a similar manner as discussed earlier, when the danger of hard freezes is past. The bulbs are ready to flower since they were stored under cool temperatures. Thus they produce fewer new leaves than short-day types and quickly send up seedstems.

Caging of Onions for Seed Production

When seedstems begin to show, cages should be taken out of storage and carried to the field. Frame parts and cages should be arranged by size and laid out in an organized manner to correspond to planted onions in the field. This organization is very important since up to several hundred cages must be erected in a 1-week period (Fig. 10.17). Cage frame legs, cross braces, length braces, etc., should be stacked separately as different-sized cages use pieces of different lengths. I color-code all pieces and keep them stored in separate bundles to simplify organization. Cages must be erected and covered prior to any flowering to prevent outcrossing (Fig. 10.18). It is advisable to use cross wiring on the frames tp provide additional bracing, as the screen covers catch wind and may be blown over (Fig. 10.19). Such precautions are well worth the cost and effort, since many years of work lie under the cages. It is very important to stake and tie up the seedstems on the

FIGURE 10.17. It is very important that frame parts and cages be arranged in an organized manner because several hundred cages will be erected during a one-week period.

FIGURE 10.18. Cages should be erected over the planted bulbs when seedstems begin to develop and prior to flowering. This allows for normal cultivation practices during plant growth, yet still gives enough time to put up cages before actual flowering occurs.

FIGURE 10.19. Additional bracing against wind damage is provided by cross wiring the cage frames.

outside rows of the cages to prevent them from touching the screen. If allowed to touch, the flowers may be pollinated through the screen with foreign pollen (Fig. 10.20). The same is true for the ends of all the rows. When flowering begins, flies or fly pupae need to be placed in the small cages and bees in the larger cages. Water must be provided for the bees at all times and the life of flies will also be extended if they are also provided with water (Fig. 10.21). Flies generally need to be added to the small cages twice each week for 3 to 4 weeks.

When the seed is set, the bees need to be removed, as excessive death loss occurs in the confinement of cages. Cage covers should be left on, however, until umbels are ready to harvest so as to provide protection from winds or hail. The umbels are ready to harvest when a few black seed coats can be seen as capsules begin to split. Remove the cages carefully and cut off any umbels that might have touched the screen.

When harvesting the seed, be sure that the bags or containers are correctly labeled and placed in the cage grouping. Small cages are harvested into paper bags and stapled closed. Larger cages are harvested into tightly woven burlap or cotton bags or pallet boxes. If the cage contains A and B lines, extreme care must be taken to prevent mixing of seed.

The cage frames are then taken down and organized by cage pieces in bundles. The cage covers are folded and everything is loaded on a trailer to be put into storage. Wire that was used to brace the frames should be picked up and disposed of to prevent future problems with equipment in the field. Seed should be dried quickly to prevent heating or molding. If the area is humid, artificial drying is needed to get the seed dry so that it can

FIGURE 10.20. The rows next to the side of the cage and plants on the row ends must be staked and tied to prevent umbels from touching the screen. Insects can pollinate flowers touching the screen and can cause contamination of the isolation.

FIGURE 10.21. A good seed set in a small massing
cage. String is used to tie umbels to prevent them from
touching the screen during flowering. String is also tied
around the perimeter of the cage to keep the wind from
pressing the screen into the umbels. Water is provided
for flies in small containers.

be threshed (Fig. 10.22). Seed can be dried at temperatures of 100°–110°F. When the
umbels are dry enough to feel brittle, they should be rubbed through ⅛-in. mesh hardware
cloth, which breaks the seed free from the capsules and causes little damage. The old
method was rubbing with a corrugated rubber mat in a box. However, some seed was
usually damaged and the procedure was slow.

For larger field isolations, I have a custom-built belt thresher, which consists of two
rubber belts moving in the same direction. The top belt moves 1.5 times as fast as the
bottom belt, which creates a gentle rubbing action as the seed moves through the machine.
The threshed seed is cleaned with an air column blower or an aspirator-type cleaner for
small lots. Larger lots are cleaned in a small, two screen seed cleaner, which has a blower
to remove light seed and trash.

If the seed lots are poor, washing may be necessary to remove light seed and trash.
Float off the unwanted material, collect the good seed, which sinks, and dry quickly by
using screens and forced dry air. Once seed is cleaned, it must be packaged, labeled, and

FIGURE 10.22. An onion seed drying bin. Seed is placed on the perforated sheet metal, which has holes too small for onion seed to fall through. Air is forced up through the seed, which may be stacked 4–5 ft high.

stored in a good seed storage room. An excellent storage condition consists of a temperature of approximately 40°F and a relative humidity of 40% or less.

Once breeders have the cleaned seed in the seed storage room, they can finally relax and feel good that they have been successful in moving one generation closer to the development of a new cultivar of onions.

REFERENCES

1. Banga, O., and Petiet, J. 1958. Breeding male sterile lines of Dutch onion varieties as a preliminary to the breeding of hybrid varieties. Euphytica 7, 21–30.
2. Clarke, A. E., Jones, H. A., and Little, T. M. 1944. Inheritance of bulb color in the onion. Genetics 29, 569–575.
3. Cochran, F. D. 1953. Cytogenetic studies of the species hybrid *Allium fistulosum* × *Allium ascalonicum* and its backcross progenies. La. State Univ. Stud, Biol. Sci. Ser. 2, 1–170.
4. Davis, E. W. 1957. The distribution of the male-sterility gene in onion. Proc. Am. Soc. Hortic. Sci. 70, 316–318.
5. Don, G. A. 1827. A monograph on the genus *Allium*. Mem. Wern. Nat. Hist. Soc. (Edinburgh) 6, 1–102.
6. El-Shafie, M. W., and Davis, G. N. 1967. Inheritance of bulb color in the *Allium cepa* L. Hilgardia 38, 607–622.
7. Emsweller, S. L., and Jones, H. A. 1935. An interspecific hybrid in *Allium*. Hilgardia 9, 277–294.
8, FAO Production Yearbook 1982. Crops. F.A.O. Stat. Ser. 36, 155–156.
9. Feinbrun, N. 1943. *Allium* sectio *Porrum* of Palestine and the neighbouring countries. Palest. J. Bot., Jerusalem Ser. 3, 1–21.

10. Hooker, J. D. 1947. The Alliums of British India (rev. and suppl. by W. T. Stearn). Herbertia *12*, 73–84.

11. Jones, H. A., Bailey, S. F., and Emsweller, S. L. 1934. Thrips resistance in the onion. Hilgardia *8*, 215–232.

12. Jones, H. A., and Clarke, A. E. 1942. A natural amphidiploid from an onion species hybrid. J. Hered. *33*, 25–32.

13. Jones, H. A., Clarke, A. E., and Stevenson, F. J. 1944. Studies in the genetics of the onion (*Allium cepa* L.). Proc. Am. Soc. Hortic. Sci. *44*, 479–484.

14. Jones, H. A., and Emsweller, S. L. 1936. A male-sterile onion. Proc. Am. Soc. Hortic. Sci. *34*, 582–585.

15. Jones, H. A., and Mann, L. K. 1963. Onions and Their Allies. Leonard Hill, London.

16. Jones, H. A., and Perry, B. A. 1956. Inheritance of resistance to pink root in the onion. J. Hered. *47*, 33–34.

17. Jones, H. A., and Peterson, C. E. 1952. Complementary factors for light-red bulb color in · onions. Proc. Am. Soc. Hortic. Sci. *59*, 457.

18. Kobabe, G. 1958. Entwicklungsgeschichtliche und genetische Untersuchungen an neuen mannlich sterilen Mutanten der Kuchenzwiebel (*Allium cepa* L.). Z. Pflanzenzuecht. *40*, 353–384.

19. Kurita, M. 1952. On the karyotypes of some *Allium*-species from Japan. Mem. Ehime Univ. Sect. 2 *1* (3), 179–188.

20. Levan, A. 1935. Cytological studies in *Allium*. VI. The chromosome morphology of some diploid species of *Allium*. Hereditas *20*, 289–330.

21. Linnaeus, C. 1753. Species Plantarum. (Facsimile) (Royal Society, London, 1957, Vol. I).

22. Little, T. M., Jones, H. A., and Clarke, A. E. 1946. The distribution of the male-sterility gene in varieties of onion. Herbertia *2*, 310–312.

23. Maeda, T. 1937. Chiasma studies in *Allium fistulosum*, *Allium cepa*, and their F_1, F_2 and backcross hybrids. Jpn. J. Genet. *13*, 146–159.

24. Magruder, R., Wester, R. E., Jones, H. A., Randall, T. E., Snyder, G. B., Brown, H. D., and Hawthorn, L. R. 1941. Storage quality of principal American varieties of onions. U.S., Dep. Agric., Circ. *618*, 1–48.

25. Makarov, A. A. 1960. [Cytoplasmic pollen sterility in the onion (in Russian).] News Timirjazev Agric. Acad. *1*, 209–216. abstract in Plant Breed, Abstr. *30*, 825–826 (1960).

26. McCollum, G. D. 1968. Heritability and genetic correlation of soluble solids, bulb size and shape in white sweet spanish onion. Can. J. Genet. Cytol. *10*, 508–514.

26a. Nichols, C. G., Gabelman, W. H., Larsen, R. H., and Walker, J. C. 1965. The expression and inheritance of resistance to pink root in onion seedlings. Phytopathology *55*, 752–756.

27. Peterson, C. E. 1947. Some factors related to the resistance of onion varieties to *Thrips tabaci* Lind. Ph.D. Thesis. Iowa State University Library, Ames.

28. Peterson, C. E., and Foskett, R. L. 1953. Occurrence of pollen sterility in seed fields of Scott County Globe onions. Proc. Am. Soc. Hortic. Sci. *62*, 443–448.

29. Pike, L. M., and Leeper, P. 1982. Five new short day onion varieties for an expanded production season in Texas. Tex. Agric. Exp. Stn. [Misc. Publ.] MP *MP-1514*, 1–5.

30. Regel, E. 1875. Alliorum adhuc cognitorum monographia. Acta Horti Petropol. *3*, 1–266.

31. Stearn, W. T. 1943. The Welsh onion and the ever-ready onion. Gard. Chron., Ser. *114*, 86–88.

32. Tackholm, V., and Drar, M. 1954. Flora of Egypt. Bull. Fac. Sci., Egypt. Univ. *3*, 1–644.

33. U.S. Department of Agriculture 1982. Agricultural Statistics. U.S. Government Printing Office, Washington, DC.

34. U.S. Department of Agriculture 1981. Vegetables—Annual Summary: Acreage, Yield, Production and Value. Economics and Statistics Service, Crop Reporting Board, USDA, Washington, DC.

35. Vavilov, N. I. 1951. The origin, variation, immunity and breeding of cultivated plants. Chron. Bot. *13*, 1–6.

36. Vvedensky, A. I. 1946. The genus *Allium* in the USSR. Herbertia *11* (1944), 65–218 (translation by H. K. Airy Shaw).

37. Yen, D. E. 1959. Pollen sterility in Pukekohe Longkeeper onions. N.Z. J. Agric. Res. *2*, 605–612.

11
Cabbage Breeding

MICHAEL H. DICKSON
D. H. WALLACE

In the United States, cabbage is the most economically important member of the genus *Brassica* (Table 11.1). However, less cabbage is being consumed in recent years, principally because consumers now have a wider range of vegetables from which to choose.

Per capita consumption of cabbage in this country decreased from 23 lb in 1940 to around 11 lb in the mid-1970s. About 9 lb were eaten as fresh cabbage, primarily as coleslaw, and approximately 2 lb were consumed as sauerkraut (*41*).

From 1974 to 1978 the major fresh-market cabbage production states (based on aver-

TABLE 11.1. Acreage, Yield Production, and Value of Cabbage Grown for Fresh Market and Processing in the United States[a]

Year	Harvested acres	Yield per acre (cwt)	Production (1000 cwt)	Value ($1000)
1920[b]	121,670	178	21,672	19,167
1930[b]	160,790	134	21,482	20,024
1940[b]	191,410	140	26,800	15,149
1950[b]	176,630	181	31,931	31,573
1960[b]	134,610	190	25,545	45,476
1970	107,930	220	23,744	81,973
1978	96,100	246	24,575	163,888

[a]After U.S. Department of Agriculture (40).
[b]After U.S. Department of Agriculture (41).

age harvested acres) were Texas, 17,920; Florida, 16,920; New York, 11,880; California, 8400; North Carolina, 7320; Wisconsin, 5280; New Jersey, 4360; Michigan, 4260; and Ohio, 3180 (see Table 11.2).

Processing cabbage (for sauerkraut) is primarily produced in New York, Wisconsin, and Ohio (40). Cabbage for coleslaw is produced in most parts of the country, but in the winter much of it comes from cabbage produced and stored in New York because it is whiter than winter-grown southern cabbage. It is also less juicy and therefore makes better coleslaw. Cabbage for fresh market is produced in norther states in the summer and in southern states in the winter. Texas and Florida are the major centers for winter production. Cabbage is also an important crop in the northern European countries, and in the USSR, Japan, China, and Australia. Cabbage breeding occurs to some extent in all these countries, but Holland and Japan are the major breeding centers.

ORIGIN AND GENERAL BOTANY

Historical evidence indicates that modern hard-head cabbage cultivars are descended from wild non-heading brassicas originating in the eastern Mediterranean and in Asia Minor. The ancient Greeks held these early forms of cabbage in high regard and believed that they

TABLE 11.2. Acreage (1974–1978) by Season and Production of States Producing over 3000 Acres of Cabbage Annually[a]

State	Harvest season				Total production (1000 cwt)
	Winter	Spring	Summer	Fall	
California	2640	2620	1480	1660	1907
Florida	9720	5480	—	1720	4167
Michigan	—	—	2480	1780	722
New Jersey	—	780	2300	1280	890
New York	—	—	3460	8420	4482
North Carolina	—	2100	3120	2100	1108
Ohio	—	440	1200	1540	950
Texas	8780	3560	—	5580	3887
Wisconsin	—	—	3040	2240	2009

[a]After U.S. Department of Agriculture (40).

were a gift from the gods. The Celts and, later, the Romans disseminated cabbage throughout Europe. In fact, the Latin name *Brassica* is derived from the Celtic word *bresic,* meaning cabbage. Over a period of centuries, hard-headed cabbage types evolved in northern Europe, while loose-heading, heat-resistant types developed further south (*38*).

Cabbage was first introduced into the Americas when the French explorer Jacques Cartier planted seed in Canada on his third voyage in 1541. Because cabbage was such a commonly grown vegetable throughout northern Europe, the earliest colonists brought seed to America, where the Indians also adopted the crop.

Most of the cultivars grown in the United States today are descended from types originally grown in Germany, Denmark, or the Low Countries. Round-headed types are older than flat or egg-shaped cultivars, which apparently did not evolve until as late as the seventeenth or eighteenth century (*38*). Because cabbage is easily stored for 2 or 3 months, it was an especially popular winter vegetable in the northern United States until well into the present century, when modern transportation and refrigeration made it possible to obtain other fresh vegetables the year around.

Cabbage, *Brassica oleracea* L. var. *capitata,* belongs to the Cruciferae or mustard family. Broccoli, cauliflower, Brussels sprouts, collards, kale, and kohlrabi are all readily intercrossed members of this species.

During the early growth and development of the cabbage plant, the first leaves expand and unfold, forming what is commonly referred to as the frame. Once the frame has been produced, the newly expanding leaves only partially unfold, forming the shell or outer skin of the head. Next, the growing point increases in size, while the core and stem enlarge in diameter and become a storage area for essential nutrients. Finally, the head is filled with a number of sessile fleshy leaves. Under favorable growing conditions, the inner leaves can exert sufficient pressure at maturity to cause the head to burst or crack open. Usually the crack occurs at the top or along one side of the head, but in the case of short-cored cultivars the split often occurs at the base of the head.

Taxonomy

Taxonomy of the *Brassica* is complicated. Figure 11.1 gives the interrelationship of the *Brassica* (*42*) genome designations and chromosome numbers. The following six *Brassica* species, plus *Raphanus sativus,* radish, $2n = 18$, have been intercrossed with difficulty (*17*), requiring embryo culture to obtain F_1 plants: *Brassica nigra* Koch, black mustard, $2n = 16$; *Brassica carinata* Braun, Ethiopian mustard, $2n = 34$; *Brassica juncea* L. Coss, brown mustard, $2n = 36$; *Brassica napus,* swedes, rape, rutabagas, $2n = 38$; *Brassica campestris,* turnip group and Chinese cabbage, $2n = 20$; *Brassica oleracea,* cole crops, $2n = 18$.

The usual objectives of interspecific crosses are to transfer disease resistance (*39*) or cytoplasmic male sterility (*17*). The amphidiploid species (Fig. 11.1) evidently originated in nature from crosses between the elementary species. In meiosis of the amphidiploids, pairing of homologous chromosome bivalents shows a secondary pairing, indicating lack of duplication of chromosomes in different species. In most cases *B. oleracea* has been successfully used only as the pollen parent.

Cytoplasm Terminology

Williams (*51*) has suggested use of the single capitalized letter representing the uncapitalized genome descriptor (Table 11.3) to designate the cytoplasm in which the

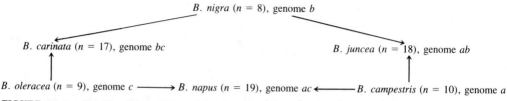

FIGURE 11.1. Relationships of *Brassica* taxa based on chromosome numbers.
After U. N. (39).

nuclear genes are functioning. Thus, for example, the *B. oleracea* genome is *cc* and its
cytoplasm is C. The *B. oleracea* genome in Ogura's radish cytoplasm is designated as
R*cc*, *B. campestris* in radish cytoplasm is designated as R*aa*, and *B. juncea* genome in *B.
campestris* cytoplasm is designated as A*ab*.

FLORAL BIOLOGY AND CONTROLLED POLLINATION

Flower Development

During differentiation of the flower, the successive development of four sepals, six
stamens, two carpels, and four petals occurs. The carpels form a superior ovary with a
"false" septum and two rows of campylotropous ovules. When the buds are about 5 mm
long, the megaspore in each ovule divides twice, producing four cells, one of which
becomes the embryo sac, while the other three abort. The nucellar tissue is largely
displaced by the remaining embryo sac; and when the buds open, the ovules mainly
consist of the two integuments and the ripe embryo sac.

The androecium is tetradynamous, i.e., there are two short and four long stamens.
When the anthers are a few millimeters in length, the pollen mother cells, after meiosis,
give rise to the tetrads. The pollen grains are 30–40 μm in diameter and have three
germination pores.

TABLE 11.3. Designation of Cytoplasmic and Nuclear Genomes of Agriculturally
Important *Brassica* and *Raphanus* Species (with ssp. or var.
of *B. campestris*)[a]

Species	ssp. or var.	Cytoplasm	2n genome descriptor	Common name
B. nigra	—	B	*bb*	Black mustard
B. oleracea	—	C	*cc*	Cole crops
B. campestris	—	A	*aa*	—
	chinensis		*aa.c*	Pak-choi
	nipposinica		*aa.n*	—
	oleifera		*aa.o*	Turnip rape
	parachinensis		*aa.pa*	Choy sum
	pekinensis		*aa.p*	Chinese cabbage, petsai
	rapifera		*aa.r*	Turnip
	trilocularis		*aa.t*	Sarson
B. carinata	—	BC	*bbcc*	Ethiopian mustard
B. juncea	—	AB	*aabb*	Mustard
B. napus	—	AC	*aacc*	Fodder and oil rape, swede
R. sativus	—	R	*rr*	Radish, daikon

[a]After Williams and Heyn (51).

The buds open under pressure of the rapidly growing petals. Opening starts in the afternoon, and usually the flowers become fully expanded during the following morning. The bright-yellow petals become 15–25 mm long and about 10 mm wide. In contrast to those of some other *Brassica* species, the sepals are erect. The anthers open a few hours later, the flowers being slightly protogynous. The flowers are pollinated by insects, particularly bees, which collect pollen and nectar. Nectar is secreted by two nectaries situated between bases of the short stamens and the ovary. Situated outside the bases of each of the two pairs of long stamens is one additional nectary, but these two nectaries are not active (Fig. 11.2).

The flowers are borne in racemes on the main stem and its axillary branches. The inflorescence may attain a length of 1–2 m. The slender pedicels are 1.5–2 cm long (Fig. 11.3).

Seed Development

After fertilization the endosperm develops rapidly, while embryo growth does not start for some days. The embryo is generally still small 2 weeks after pollination. It fills most of the seed coat after 3–5 weeks, by which time the endosperm has been almost completely absorbed. Nutrient reserves for germination are stored in the cotyledons, which are folded together with the embryo radicle lying between them.

Fruit

The fruits of cole crops are glabrous siliques, 4–5 mm wide and sometimes over 10 cm long, with two rows of seeds lying along the edges of the replum (false septum, an outgrowth of the placenta). A silique contains 10–30 seeds. Three to four weeks after the opening of its flower the silique reaches its full length and diameter. When it is ripe, the two valves dehisce. Separation begins at the attached base and works toward the unattached end, leaving the seeds attached to the placentas. Physical force ultimately separates the seeds, usually by the pushing of the dehisced siliques against other plant parts by the wind or by threshing operations (Fig. 11.4).

Male Sterility

A number of recessive mutations monogenic (*5,9,11,24,34*) for male sterility have been reported, but it was not until Pearson (*28*) crossed *B. nigra* with broccoli that cytoplasmic

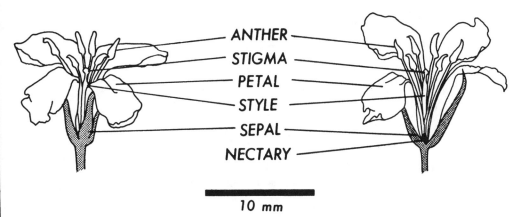

ANTHER
STIGMA
PETAL
STYLE
SEPAL
NECTARY

10 mm

FIGURE 11.2. Floral parts of a cabbage flower.

FIGURE 11.3.   Cabbage plant in bloom (left) and with developed pods (right).

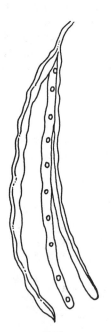

SILIQUE

FIGURE 11.4.   A dehisced silique
showing seeds along the
edges of the replum.

10 mm

male sterility, designated N*ps/ps,* was obtained in *B. oleracea* and developed in cabbage. Unfortunately, this system was complicated by petaloidy and lack of development of the nectaries, giving a male sterile that was unattractive to bees. These problems were being solved through backcrossing and selection, when an easier-to-use cytoplasmic male sterility was obtained. This occurred when Bannerot (*4*) crossed a cytoplasmic male-sterile radish (R cytoplasm) from Ogura (*26*) with cabbage. In the BC$_4$ generation he obtained normal plants with $2n = 18$ that were totally male sterile, the flowers having only empty pollen grains or vestigial anthers. Male-sterile plants had the problem of pale or white cotyledons and also of pale-yellow leaves during plant development, accentuated at low temperatures. These problems are cytoplasmically inherited, and various attempts to overcome them are being made via cell fusion, with the hope of retaining the factor for sterility and eliminating the factor for pale leaves and low temperature sensitivity. If these attempts are not successful, it is doubtful that R cytoplasm-based male sterility can be used successfully, at least in the more temperate climates. In contrast to the cytoplasmic steriles produced with *B. nigra* cytoplasm, many cultivars of which had fertility-restoring genes, all the fertile *B. oleracea, B. campestris,* and *B. napus* genotypes tested on R cytoplasm have carried only *ms* genes. All brassicas apparently have no fertility-restoring genes for the R cytoplasm type of sterility and act as maintainers. Thus, any *B. oleracea* plant can be combined with R cytoplasm to produce a sterile plant. Restorer genes are present in radish. Obtaining a male-sterile version of an inbred line only requires converting any source plant with R cytoplasm to that inbred genotype by backcrossing. The inbred line will always be a B line (fertile maintainer) for the male sterile. The recurrent inbred parent must be selected for self-compatibility, or a closely selected OP line heterozygous for several *S* alleles can be used as a B line. Figure 11.5 shows flowers of an R-cytoplasmic male-sterile A line and its maintainer B line. Racemes of these plants are shown in Fig. 11.6. Petals of the male-sterile flowers are smaller than those of fertile flowers. Fortunately, nectaries are usually normal in male-sterile flowers with R cytoplasm, and bees usually work these sterile flowers.

FIGURE 11.5. Male-fertile and -sterile flowers showing normal stamens (below) and vestigial stamens (above).

FIGURE 11.6. Racemes from a cytoplasmic male-sterile
A line (left) and the fertile maintainer B line (right).

The R cytoplasm of the male-sterile radish has also been combined with *B. campestris* and *B. napus* to obtain cytoplasmic male sterility in these species.

Self-Incompatibility

Cabbage flowers must be cross-pollinated. Depending on the cultivar, a few to most plants are self-incompatible. Few seeds will be set following self-pollination. Pollination in the field must be by insects because the sticky pollen is not windblown. In the greenhouse or with proper protection in the field, the pollination can be done by hand. The pollen is viable and will achieve fertilization and seed set in most cross-pollinations, excluding that small percentage for which the incompatibility specificity of the plant functioning as female is the same as the incompatibility specificity of the pollen. Maximum seed set does not occur in self-pollinations, because there are identical incompatibility specificities in both the stigma (female) and pollen (male). When the male and female *S*-allele specificities are identical, self-incompatibility acts to prevent the pollen from germinating on and growing into the stigma or style. By this mechanism, self-incompatibility prevents self-fertilization. The self-incompatibility phenomenon also prevents fertilization in crosses between plants of identical genotypes, and in crosses between plants of near identical genotype when dominance or codominance conditions the identical expression of incompatibility specificities. No such barrier inhibits pollen germination and penetration on stigmas of plants of nonidentical *S*-allele genotype, which therefore have different incompatibility specificities.

Incompatibility Specificities

The incompatibility specificities of cabbage are controlled by one locus, called the S gene. About 50 alleles, $S_1, S_2, S_3, \ldots, S_{50}$, each giving one specificity, have been identified. Thus, homozygous S-allele genotypes S_1S_1, S_2S_2, S_3S_3, etc., have incompatibility specificities of S_1, S_2, S_3, etc., for both their stigma (female) and pollen (male) reproductive organs. Most cabbage plants have heterozygous S-allele genotypes because of the cross-fertilization enforced by the self-incompatibility. The two S-alleles of heterozygous plants complicate the expression of incompatibility specificities. The incompatibility specificities for S-allele heterozygous cabbage plants are complicated because control of the pollen specificity is by the sporophyte, the whole (diploid) plant, rather than by the individual (haploid) pollen grain. These complications of the sporophytic control of incompatibility of pollen are more easily comprehended after understanding the simpler gametophytic control of pollen specificity. Therefore, the next section describes gametophytic incompatibility, as it occurs in *Petunia,* alfalfa, clovers, and many fruit trees (22). The subsequent section will return to the sporophytic incompatibility of cabbage.

Gametophytic Incompatibility

Self-incompatibility has gametophytic control when the haploid S-allele genotype of each pollen grain (male gamete) exactly indicates that gamete's expressed incompatibility specificity. For a heterozygous S_1S_2 plant, meiosis will give haploid male gametes of S_1 and S_2 genotypes in about a 1 : 1 proportion. Therefore, 50% of the pollen grains will have S_1 specificity and 50% will have S_2 specificity. Determination of whether the gametophytically controlled pollen grain will function compatibly or incompatibly in any given self- or cross-pollination requires an answer to only one question: Does the S allele (haploid genotype) conferring specificity on the pollen grain also occur in the female flower? If not, the pollination will be compatible; if so, whether the female plant is homozygous or heterozygous, the pollination will be incompatible. (These relationships for compatible and incompatible pollinations for gametophytic incompatibility are illustrated later in Table 11.5)

Sporophytic Incompatibility

For sporophytic incompatibility, we must ask: Does the S allele given to the pollen grain by meiosis also occur in the female flower of this pollination? However, this only gives a partial answer as to whether the pollination will be incompatible or compatible. If no, then the pollination will be compatible; if yes, the pollination may be compatible or incompatible, depending on answers to the following further questions that must be asked separately but jointly considered for the plant functioning as male and for the plant functioning as female:

1. Is the male plant homozygous for one or heterozygous for two S alleles? If heterozygous, which of the following four levels of interaction between the two S alleles characterizes the incompatibility specificity of the pollen: dominance, codominance, mutual weakening, or intermediate activity?
2. Is the female plant homozygous for one or heterozygous for two S alleles? If heterozygous, which of the following four levels of interaction between the two S alleles characterizes the incompatibility specificity of the female organ

**TABLE 11.4. S-Allele Interactions in
Heterozygous Genotypes with
Sporophytic Incompatibility**

Dominance	$S_1 < S_2$
Codominance	$S_1 = S_2$
Mutual weakening	No action by either allele
Intermediate gradations	0–100% activity by each allele

(stigma): dominance, codominance, mutual weakening, or intermediate activity (Table 11.4)?

Knowing in advance whether a pollination will be compatible or incompatible requires knowing whether either or both plants are homozygous, whether the two S alleles of a plant that is heterozygous interact with dominance, codominance, mutual weakening, or intermediate activities, and which of the alleles is dominant and which recessive (Table 11.4).

For a cabbage plant that is homozygous, only one S allele is designated for incompatibility specificity, but the intensity of expression may be weak, intermediate, or strong. When two plants that are homozygous are cross-pollinated, the question that must be asked is the same one as for the gametophytic system: Is the same S allele present in both plants? However, it is usually not known if either plant is homozygous. Therefore, determination of the incompatibility/compatibility of all cabbage pollinations—whether the plant is used as male or female, and whether one or both plants are homozygous— ultimately depends on asking all the questions that must be asked for all plants that are heterozygous for S alleles.

Assaying Self-incompatibility

The procedure first developed for quantifying self-incompatibility was to count the number of seeds that develop to maturity after each specific self- or cross-pollination. One disadvantage is the 60-day time duration between pollination and seed maturity. A second disadvantage is that subsequent to expressions of compatibility or of weak incompatibility, the numbers of seeds developing and ultimately reaching maturity may also be reduced by disease, water shortage, high temperature, or other stresses. Thus, seed counts at maturity often do not strictly reflect the intensity of expressed compatibility/incompatibility.

Ability of the fluorescent microscope to display readily those pollen tubes that have penetrated the style provides a direct measure of incompatibility that can be completed within 12–15 hr. However, it is adequate and more convenient in a large breeding program to pollinate on day 0. The pollinated flowers are then collected 16–30 hr later (day 1). On the same day, the excised ovaries are then softened in 60% NaOH and placed in aniline blue for staining. At about 48 hr (day 2) after pollination, the stigma and style are squashed on a microscope slide. The aniline blue stain accumulates in the pollen tubes and fluoresces when irradiated with ultraviolet light. Therefore, with appropriate light filters, under a fluorescent microscope, the tubes are visible, whereas the background of stylar tissues is largely unseen. Penetration of the style by none or a few tubes indicates incompatibility, penetration by many tubes indicates compatibility, and penetration by intermediate numbers indicates intermediate strength of the expressed incompatibility/compatibility (Figs. 11.7 and 11.8). In facilitating a breeding program, the

major advantage of the fluorescent microscope is availability of the incompatibility data
and attendant conclusions within 2 days, as compared to the 60 days required for seed
counts, or the 20–40 days if developing seeds are counted. With the assay completed and
conclusions available within 48 hr, additional pollinations can be specifically planned to
verify a conclusion, to test a seemingly erroneous conclusion, or to further other breeding
objectives directly by accepting the conclusions. This early acceptance facilitates moving
on to work with other populations, inbreds, or objectives. Using these procedures, the
breeder has knowledge about the expressed incompatibility while sufficient flowers remain to maximize use of the conclusions. Seed set data provide this information only after
most of the cabbage plants have finished flowering. With the use of the fluorescent
microscope plus improved understanding of the sporophytic incompatibility system (47),
homozygous S-allele genotypes can be identified in 1–3 weeks. This contrasts to the 3
years usually required when only seed set data have been used.

Identifying S-Allele Genotypes

Developing cabbage inbred lines that have all possible desirable horticultural and
disease resistance characteristics, plus homozygosity for a single S allele, is an essential
step in developing the inbreds to be used as parents in producing seed of hybrid cabbage
cultivars. As a first step, plants from open-pollinated cultivars or hybrid populations
should be individually selected for their disease resistance, head solidity, head size, short
core, leaf color, and other desirable horticultural characteristics. This directed selection,
when accompanied by the usual open- and cross-pollination, will most often result in
random selection with respect to the S alleles present within the source population.

FIGURE 11.7. Fluorescent-microscope view of pollen grains with very short pollen tubes
and strong fluorescence (callose formation) indicates incompatibility. The papillae are
very faintly fluorescent and therefore barely visible. Nonfluorescing stigma tissues
are not visible.

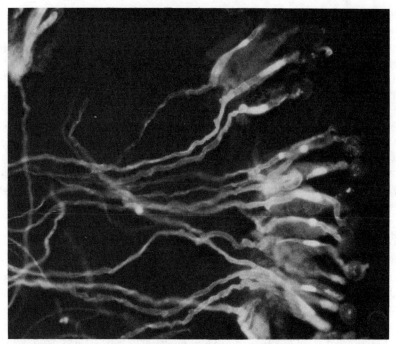

FIGURE 11.8. Fluorescent-microscope view of pollen grains on the papillae of a cabbage stigma, with very long pollen tubes extending beyond the papillae and downward into the style. The numerous long pollen tubes that fluoresce because they contain callose indicate compatibility. The papillae are faintly fluorescent, and nonfluorescing stigma and stylar tissues that are present are not visible.

Individual plants selected as described above will usually be heterozygous for two S alleles, S_1S_2, S_2S_{26}, or S_1S_{49}, for example. Each selected individual plant must next be selfed by bud pollination (self-pollination is discussed later in this chapter) for maintenance and seed increase. Simultaneously, open flowers on the plant should also be selfed. The resultant seed set or pollen tube penetration from the open flowers will be used to measure intensity of the self-incompatibility of the selected plant. If compatible or weakly compatible, all resultant seed for this plant may be discarded. For selfing, neither the buds nor flowers need be emasculated. Outcrosses with other flowers must be totally avoided since they would result in a false indication of compatibility for plants that are actually incompatible. Also, care should be taken to see that no outcrosses occur during the bud pollinations, because identification of S-allele genotypes is relatively easy and efficient in populations carrying precisely two S alleles that have segregated into all three possible genotypic combinations. Three or more S alleles in the population make the task nearly impossible. Therefore, the I_1 generation (first generation of inbreeding) from selfing an S-allele heterozygous plant is by far the most effective population for identifying S-allele genotypes. This population will contain the three genotypes, S_aS_a, S_aS_b, S_bS_b, for example, in a $1:2:1$ ratio. There will be no other genotypes if no outcrossing is permitted. (This genotype labeling is used because it will not be known if either of the two possible alleles in the I_1 population represents S_1, S_2, S_3, . . . , or S_{50}. S_a and S_b are tentative assignments to be used until a specific numerical S-allele designation is assigned.)

A group of 11 plants from an I_1 population provides a 95% probability of having at least one plant of each of the three genotypes (two homozygous and one heterozygous). Identifying a plant of each of these three S-allele genotypes requires incompatible/compatible interpretations from a series of reciprocal pollinations between subsets of two of the 11 I_1 sibling plants. The next several paragraphs show, in fact, that the efficient procedure is to begin by reciprocally crossing a series of individual I_1 plants to each of two other individual I_1 plants. This procedure is continued until the three different (all possible) genotypes of three plants become simultaneously evident. Thus, the three possible genotypes will each be represented by one of the three I_1 plants. On the assumption that each I_1 plant is self-incompatible, i.e., that mutual weakening does not apply, the incompatible/compatible expectations for these reciprocal pollinations are presented in Table 11.5. The expectations come from knowledge that dominance or codominance occurs in the pollen and also in the stigma (the two sexual organs), giving four sexual organ × S-allele interactions, designated types I, II, III, IV, as summarized in Table 11.5. It is assumed that each S-allele heterozygote will belong to one of these four types. This assumption will usually be valid and the interpretation is then straightforward from Table 11.5.

When the above assumption is not valid, the heterozygote will be of a type intermediate to the extreme types I, II, III, IV. Such intermediate types will lack strong dominance or codominance in the heterozygote. Intermediate activities in the heterozygote by both S alleles constitute mutual weakening. This weakening in the S-allele heterozygote is specific for given pairs of S alleles. Such mutual weakening is indicated when many of the reciprocal crosses between I_1 plants give intermediate seed sets or penetrations of pollen grains into the stigmas that indicate neither definite incompatibility nor definite compatibility. Rather, S-allele action intermediate to these extremes is evidence. Also, repeats of the identical pairs of reciprocal crosses will sometimes indicate near incompatibility and sometimes near compatibility. Thus, the compatible/incompatible interpretations are more variable than usual.

Intermediate S-allele interactions with mutual weakening of both S-allele activities in the heterozygote do not permit rapid and efficient identification of S-allele genotypes of the individual I_1 plants. Also, one or both S-allele homozygotes is likely to have weak intensity of self-incompatibility. More seriously, the mutual weakening will prevent use of the two S alleles in a moderately uniform single-cross F_1 as is essential for producing the strongly desired three- and four-way hybrid cultivars. For these reasons, an I_1 population should be discarded after determining that the S-allele interaction in the heterozygote involves mutual weakening. The only merit that could justify keeping any plants from such I_1 populations is the presence of one or more rare but valuable characteristics.

For types II and III, but not for type I or IV, a reciprocal difference is expected from any reciprocal crosses between the recessive S-allele homozygous genotype S_aS_a and the heterozygous genotype S_aS_b (from Table 11.5). This reciprocal difference is expected because S_b is dominant to S_a (symbolized by $S_a < S_b$) in either the pollen or stigma for types II and III, but it is not dominant in both sexual organs (Table 11.5). Therefore, observation of a reciprocal difference between any two plants of an I_1 population immediately indicates either type II or III. It also indicates that one of the two I_1 plants is the genotype S_aS_a (recessive) while the other is the heterozygote S_aS_b (with $S_a < S_b$ in either pollen or stigma). The S_aS_a and S_aS_b genotypes cannot be positively assigned to either of the two I_1 plants, until or unless some more reciprocal crosses to these two plants have already been made. Information is essential relative to the incompatibility/compatibility

TABLE 11.5. Expected Incompatible/Compatible Interpretations and Reciprocal Difference Interpretations[a] from Reciprocal Intercrosses among Plants of the Two Homozygous and One Heterozygous S-Allele Genotypes for Sexual Organ × S-Allele Interaction, Types I, II, III, and IV

	S-allele interaction		Genotype and phenotype[b]			
Type of interaction	Stigma	Pollen	Female	Male		
I	Dominance	Dominance		S_aS_a	$S_a < S_b$	S_bS_b
			S_aS_a	Inc	Com	Com
			$S_a < S_b$	Com	Inc	Inc
			S_bS_b	Com	Inc	Inc
II	Codominance	Dominance		S_aS_a	S_aS_b	S_bS_b
			S_aS_a	Inc	Com	Com
			$S_a = S_b$	Inc	Inc	Inc
			S_bS_b	Com	Inc	Inc
III	Dominance	Codominance		S_aS_a	$S_a = S_b$	S_bS_b
			S_aS_a	Inc	Inc	Com
			$S_a < S_b$	Com	Inc	Inc
			S_bS_b	Com	Inc	Inc
IV	Codominance	Codominance		S_aS_a	$S_a = S_b$	S_bS_b
			S_aS_a	Inc	Inc	Com
			$S_a = S_b$	Inc	Inc	Inc
			S_bS_b	Com	Inc	Inc

[a]Reciprocal crosses with a reciprocal difference are indicated by dashed arrows and reciprocal crosses without a reciprocal difference by solid arrows.

[b]The S-allele phenotype of heterozygotes is indicated by the symbols < and = where < specifies recessive vs. dominant and = indicates codominance. These phenotypes correspond with the described S-allele interactions in stigma and pollen. The phenotype for homozygotes always corresponds with the genotype.

of additional reciprocal crosses between the one of these two I_1 plants that is S_aS_a and one of the additional sibling I_1 plants that is S_bS_b. The already achieved tentative assignment of either S_aS_a or S_aS_b (with $S_a < S_b$) makes these two I_1 plants more efficient than any other I_1 plants for continued use toward ultimate identification of the three genotypes. Their use will result in positive genotype assignments after the fewest additional reciprocal crosses. Use of both is required because it is not known which of the two is S_aS_a. Therefore, each of the two I_1 plants should continue to be reciprocally crossed once to each additional I_1 plant if sufficient reciprocal crosses have not already been made. When a pair of reciprocal crosses between either of these two I_1 plants and another I_1 plant gives reciprocal incompatibility, this information is not immediately useful in ascertaining the genotypes. This information will nevertheless be useful for assigning the genotype of the fourth, fifth, or eighth, etc., I_1 plant, after the genotypes of I_1 plants 1, 2, and 3 have been assigned as follows. The S_aS_a and S_aS_b genotypes can be positively assigned to the two I_1 plants giving the reciprocal difference as soon as one of the two (plant 1) is found to have reciprocal compatibility with a third I_1 plant (plant 3). Plant 1 is S_aS_a since it was both reciprocally compatible with plant 3 (which must therefore be the homozygous dominant S_bS_b), and since it also had a reciprocal difference in crosses with plant 2. Plant 2 therefore must be the heterozygote S_aS_b (with $S_a < S_b$). It is type II if S_aS_b was the female parent in the incompatible cross and the male in the reciprocal but compatible cross (from Table 11.5). It is type III if S_aS_b was the male parent in the incompatible cross and the female in the reciprocal but compatible cross.

With all three of the possible genotypes now identified, the plant with S_aS_a genotype becomes the most efficient for identifying the genotype of all the additional plants. This known genotype S_aS_a will be reciprocally incompatible with an unknown S_aS_a; it will be reciprocally different with an unknown S_aS_b; and it will be reciprocally compatible with a plant of unknown genotype S_bS_b.

With type I there is a dominance for one of the two alleles of the heterozygote in both the pollen and stigma (Table 11.5). Type I, therefore, is indicated when a first I_1 plant is reciprocally compatible with one of a second and third I_1 plant that are reciprocally compatible with each other, while this first I_1 plant is on the contrary reciprocally incompatible with the other of the second and third reciprocally compatible I_1 plants (from Table 11.5). One of the second and third reciprocally compatible I_1 plants is the homozygous recessive genotype (S_aS_a). The second plant is either the heterozyous ($S_a < S_b$) or homozygous dominant (S_bS_b) genotype. The third plant is also either the heterozygous ($S_a < S_b$) or homozygous dominant (S_bS_b) genotype. Within the I_1 generation, there is no way of further differentiating among these three genotypes. In the I_2 generation all plants from the homozygous recessive and from the homozygous dominant I_1 genotype will breed true. Each I_2 family from each such I_1 plant will have the same S-allele specificity. Each such I_2 population will be reciprocally cross incompatible with all the plants of other populations of its own homozygous S-allele genotype, and will be reciprocally cross compatible with all plants of the other true-breeding homozygous S-allele genotype. In the I_2 generation any populations from an I_1 plant of heterozygous S-allele genotype will segregate and behave like the I_1 generation. The true-breeding dominant homozygous I_2 plants will be identified by their incompatibility with about three-fourths (the heterozygous $S_a < S_b$ plus homozygous dominant S_bS_b genotypes) of the plants of any I_2 populations that are segregating, and by simultaneous compatibility with the remaining smaller fraction (about 25%) of the plants of the segregating I_2 (the S_aS_a recessive genotypes). Simultaneously, the *recessive* homozygous true-breeding I_2 populations will be indicated by compatibility with the same larger part of the I_2 populations that are segregating and by incompatibility with the same smaller fraction of the plants.

In the procedures above for types II and III, no immediate use was possible for expression of reciprocal incompatibility between two I_1 plants. After one plant with each of the three I_1 genotypes has been identified, knowledge about incompatibility between a plant of known S-allele identity and other I_1 plants of unknown S-allele genotype will either identify, help to identify the unknown genotype, or will verify an already determined genotype. Therefore, all information about incompatibility between the I_1 sibs should be saved. These same relationships apply fully for type I and partially for type IV. Therefore, for all four types, it is expressed compatibility for one or both of the pair of reciprocal crosses between two I_1 plants that is the first information required in order to identify the S-allele genotype of any and then all I_1 plants.

The procedure described above assumed a reciprocal difference was observed for at least one of the first pairs of reciprocal crosses made between the I_1 sibling plants. That is, it accepted that compatibility (or incompatibility) was observed when a first I_1 plant was crossed as female parent to a second as male, but that the reverse (incompatibility or compatibility) resulted from the reciprocal cross for which the second plant was female while the first was male.

The reciprocal difference between two I_1 plants was the first of two compatibility relationships that each by itself indicated the same thing, as follows: One of the two reciprocally crossed I_1 plants must be the homozygous recessive genotype, while the second plant must have the dominant S-allele phenotype. Expression of dominant phenotype indicates the genotype to be either heterozygous $S_a < S_b$ or homozygous S_bS_b, but does not differentiate. The reciprocal difference simultaneously indicated the type to be II or III (Table 11.5), which only occurs for the crosses between the homozygous recessive genotype and a heterozygous genotype with the dominant S_b phenotype expressed by either male (type II) or female (type III), while simultaneously expressing both S_a and S_b (codominance) in the other sexual organ. Therefore, simultaneously, a reciprocal difference indicates three facts: (1) one of the I_1 plants is S_aS_a; (2) the second I_1 plant with its dominant S-allele phenotype is genotype $S_a < S_b$; (3) the second plant is not S_bS_b.

Reciprocal compatibility is the second and only other possible compatible relationship between two I_1 plants. By itself reciprocal compatibility also demonstrates, like reciprocal difference, that one of the two reciprocally crossed I_1 plants must be the recessive genotype S_aS_a, but cannot indicate which plant carries which of the two genotypes, and also leaves unresolved the alternative of heterozygote vs. homozygote for the plant with the dominant phenotype. The second I_1 plant, however, must have the phenotype of the dominant allele, but may be either the heterozygote $S_a < S_b$ (for type I) or the homozygote S_bS_b (for any of types I, II, III, or IV).

Neither a reciprocal difference nor reciprocal compatibility can by itself specify which of the two I_1 plants is the homozygous recessive genotype, and which is the I_1 plant with the dominant S allele(s) and consequent dominant phenotype. This differentiation is achieved as both of these two compatibility results have identified the same I_1 plant as being either the recessive S_aS_a or one of the two genotypes with dominant phenotype. By expressing reciprocal compatibility with one or more of its I_1 sibs and reciprocal difference with one or more other I_1 sibs, a plant is identified as the homozygous recessive genotype, since only the recessive S_aS_a can express both and only for types II and III. Since the reciprocal difference identifies the second involved I_1 plant as the $S_a < S_b$ genotype, and the reciprocal compatibility identifies the third involved I_1 plant as the homozygous S_bS_b genotype, differentiation is simultaneously achieved between all three genotypes: S_aS_a, S_aS_b, and S_bS_b.

The procedure described above for types II and III assumed that a reciprocal difference was observed prior to observing reciprocal compatibility. Purposely, thereafter, to efficiently identify that one of the two I_1 plants that was truly the recessive genotype, both of these two plants were reciprocally crossed to a third, fourth, etc., I_1 plant until reciprocal compatibility was found. If reciprocal compatibility is observed prior to finding a reciprocal difference, which must always occur for types I and IV because they cannot have a reciprocal difference, then the efficient second step toward identifying the genotypes of all I_1 plants is to cross the two reciprocally compatible plants with a third, fourth, etc., I_1 plant in search of a reciprocal difference (for types II and III), or to find a third I_1 plant that is reciprocally incompatible with both of the two reciprocally compatible I_1 plants (for type IV). The order in which reciprocal compatibility and then reciprocal difference is found is of no consequence; the essential ordering is to find expression of both by the same I_1 plant. Type I presents an additional problem because dominance of allele S_b over S_a in both stigma and pollen makes it impossible to differentiate between S_aS_b and S_bS_b except by a test to detect either segregation or homozygosity of progeny.

A reciprocal difference will occur for ⅛ of the pairs of reciprocal crosses among the sibling I_1 plants if, but only if, the type is II or III (Table 11.5). The ⅛ probability arises from the I_1 generation ratios of ¼ S_aS_a × ½ genotype S_aS_b = ⅛ probability from each pair of reciprocal crosses for a reciprocal difference. For each of types I, II, III and IV a reciprocally compatible result will occur for 1/16 of the crosses between two I_1 plants, from the ratios ¼ S_aS_a × ¼ S_bS_b. For type I, reciprocal crosses will occur with an additional frequency of ⅛, to give a total probability of 3/16, because the homozygous recessive ¼ of the plants of genotype S_aS_a will also be reciprocally compatible with the ½ of the I_1 plants that are the heterozygous genotype $S_a < S_b$. Another factor that influences the proportion of crosses between the I_1 plants that will give a reciprocal difference is that about half of the sexual organ × S-allele interactions are usually type II or III, leaving about ¼ for each of types I and IV.

Type IV has codominance in both pollen and stigma of the heterozygous S-allele genotype (from Table 11.5). Thus, the heterozygote has strong activity by both alleles in both its pollen and its stigma. Type IV is therefore indicated when an I_1 plant is reciprocally incompatible with both of two I_1 plants that are reciprocally compatible with each other. The I_1 plant that is reciprocally incompatible with both reciprocally compatible plants is the heterozygous genotype. The two reciprocally compatible I_1 plants must be one plant of each of the two homozygous genotypes. These two plants can be arbitrarily assigned tentative S_aS_a and S_bS_b identities, because the alleles are codominant and there is no recessive versus dominance S-allele interaction. That is, both alleles are simultaneously expressed with nearly equal intensity in the heterozygote. With the same criteria, the genotypes of all of the other I_1 plants can now be designated after each plant of unknown genotype has been reciprocally pollinated to two of the three known I_1 genotypes.

Permanent S-Allele Identities

Most permanent S-allele identities ($S_1, S_2, \ldots, S_{50}$) are assigned by the breeder, but the National Vegetable Research Station at Wellesbourne, England, has a collection of all known S alleleles, which constitutes the internationally accepted nomenclature.

A homozygous plant of tentative S_aS_a or S_bS_b genotype will be known to represent the breeder's S_3S_3 genotype when S_aS_a or S_bS_b is demonstrated to be reciprocally incompatible with the breeder's inbred of S_3 genotype. Similarly, it will be $S_{26}S_{26}$ of the international allele nomenclature when it is reciprocally incompatible with plants of known international S-allele genotype S_{26}. The S_aS_a and S_bS_b will be compatible with all plants having other S-allele genotypes.

Unknown alleles in plants of heterozygous S-allele genotype can usually be specifically identified by partial incompatibility with reciprocal crosses with the corresponding homozygous S-allele genotype. Thus, with dominance the S-allele activity will be strongly expressed only in the stigma (type IV from Table 11.5) or only in the pollen (type II) or strongly expressed in both (type I). Both alleles will be strongly expressed in both the stigma and pollen for type IV. Alternatively, with recessiveness the allele will be weakly expressed in the pollen but strongly in the stigma (type III), or weakly in the stigma but strongly in the pollen (type II), or weakly expressed in both stigma and pollen (type I). Because of these complexities, assignment of permanent S-allele designations is most easily done using plants known to be homozygous S_aS_a or homozygous S_bS_b.

Production of Hybrid Cultivars

The self-incompatibility character is used to enforce the cross-fertilization required in producing hybrid seed of cabbage, cauliflower, broccoli, Brussels sprouts, and kale. Sibling plants of inbred lines that have been selected for homozygosity of an S allele followed by selection for strong expression by this S allele of self-incompatibility will not cross-fertilize each other. As a result there will be little selfed (inbred) seed. On the other hand, such an inbred S_1S_1 when planted in rows alternating with another inbred S_2S_2 in every other or every second or third row, or with both inbreds in alternating blocks of three or four rows, will be readily cross-fertilized. The fertilization will be by pollen carried from one inbred to the other by pollinating insects (mostly bees). These insects must readily visit the flowers of both entries if high seed production is to be obtained. The cross-compatibility between inbreds S_1S_1 and S_2S_2 assures the production of F_1 hybrid seed. If both inbreds are incompatible, the seed produced on the S_1S_1 and S_2S_2 inbreds will be identical except for possible maternally conditioned and/or inherited characteristics. Thus, commercial seed of the same F_1 hybrid can be harvested from both inbreds. The inbred A (S_1S_1) × inbred B (S_2S_2) cross and its reciprocal will constitute a standard A × B single-cross F_1 cultivar.

To date, most of the cabbage hybrids produced in the United States are topcrosses. United States breeders have not fully advanced to single crosses. For topcrosses an open-pollinated cultivar with good horticultural characteristics is used as the pollen parent. It is used to pollinate a single self-incompatible inbred line that is the female parent. Topcross hybrids are numerous because U.S. seedsmen have not until recently had full understanding of efficient procedures for identifying S-allele genotypes (47). Japanese *Brassica* breeders, and a few U.S. and European seedsmen, have advanced beyond single-cross F_1s to three- and four-way crosses. For three- and four-way hybrid cultivars, a single-cross F_1 such as A × B derived above is used as the seed parent. The heterosis of this F_1 as seed parent compared to the vigor of an inbred used as the female parent gives increased seed yield. A three-way cross may use an inbred line as a pollen parent only; or if it is self-incompatible, the inbred may simultaneously be used as a seed parent. Also, an open-pollinated line may be used as the pollen parent, thus giving a three-way topcross hybrid cultivar.

When the third line is an incompatible inbred, seed can be harvested from all the plants. When a highly self-incompatible single-cross C × D is used in conjunction with A × B, the resultant four-way hybrid seed can also be harvested from every plant in the field. This plus the larger seed production per plant due to the hybrid vigor of both single-cross seed parents (which are both also pollen parents) will give the lowest production cost for the hybrid seed.

Successful seed production of hybrids requires much effort to select not only for horticultural characteristics but also for horticultural combining ability and for simultaneous flowering of the parents. Many excellent potential hybrids have not been successfully produced because the intended parents did not nick with respect to flowering time. Two inbreds developed from the same I_1 population, but selected for homozygosity for the opposite S alleles present, are often used as the two inbred parents of a single-cross F_1.

Their common origin combined with the selection for common horticultural characteristics (except for different S alleles) minimizes the genetic variability that the single-cross F_1 will introduce into the three- or four-way F_1 hybrid cultivar. Uniformity in all respects, and especially for time of harvest, is the major advantage of hybrid cultivars. Much selection and effort is required to achieve this and attendant goals.

Because uniformity is highly important, it may be beneficial to select and self the S-allele heterozygotes through the I_1 and then identify the S-allele genotypes in the I_2 generation. If the S_aS_a and S_bS_b counterparts are selected after two generations of selfing, they should give an S_aS_b single cross that will produce three- and four-way hybrids that are more uniform for desired horticultural characteristics. A disadvantage of two and especially more generations of selfing prior to selecting the S_aS_a and S_bS_b counterparts is that hybrid vigor of the S_aS_b single cross may be reduced.

The use of male sterility in hybrid seed production is expected to increase rapidly. The same requirements for successful seed production expected of inbreds using incompatibility will be required of the male-sterile inbred. Likewise, a three-way cross can be made using male sterility just as with incompatibility. In addition, the three-way cross using male sterility may be easier and more seed productive, since the F_1 hybrid is likely to be heterozygous for incompatibility factors, resulting in less problems from incompatibility when the third parent line is crossed with the F_1 hybrid.

Pollination by Hand

Self-Pollination

Self-pollination of cabbage can be obtained by brushing or shaking the open flowers if the plant is self-compatible. If the plant is self-incompatible, the buds can be opened 1–4 days before they will open naturally and be bud pollinated. As illustrated in Fig. 11.9, the largest unopen bud is probably too old for successful bud pollination as the incompatibility factor will be biosynthesized by that stage. A younger bud 3–4 days prior to natural flower opening will have the least self-incompatibility and still be large enough for bud pollination. Thus six to eight buds can be opened at one time with a pointed object such as a toothpick or forceps, and pollen from an older open flower can be transferred to the stigma and seed obtained. Pollen can be transferred with a brush or onto a thumbnail. Also, a fertile flower in full bloom is often used by brushing the anthers across the stigmatic surface to transfer the pollen. Williams (49) has suggested the use of bee sticks, using the hairy abdomen of bees mounted on a toothpick as pollen carrier.

Cross-Pollination

When the pollination is for a breeding objective such that self-incompatibility is not being considered, the bud should be emasculated using buds expected to open in 1 or 2 days if elimination of selfing is essential. The desire pollen is then transferred to the stigma in the same manner as for bud pollination. If a male-sterile or self-incompatible plant is used as a female parent, then an open flower with protruding pistil can be used to avoid the time and effort required to open the flower bud. However, only certain lines have these protruding pistils. When the cross-incompatibility between two plants is being assayed, the open flowers need not be emasculated. Cross-pollinating without emasculation also saves considerable time and labor and permits estimations of the porportion of cross-fertilization (hybrid) and selfed (inbred) seeds of the F_1 plants of that inbred.

FIGURE 11.9. Raceme showing flower buds at stage for bud pollination and toothpick used to open bud and transfer pollen.

Seed Increases

Hand pollination can best be performed in the greenhouse or in a large screened cage to eliminate insects. If crosses or self-pollination are desired in the field, cheesecloth bags can be used to enclose the blossom of one or two plants. More preferable is enclosure of several plants in a screen cage 6 ft high for a small increase. Bees are the best pollinators and are supplied to facilitate cross-pollination in a cage. If large-scale, outdoor increases are to be made, the minimum isolation distance between lots should be 400 yards. More isolation is needed if one lot is downwind from another. Asexual plant propagation can be obtained by cutting off the head and allowing the lateral buds to develop. The buds can be excised, rooted on a moist medium, vernalized, and allowed to flower in small 2- to 4-in. pots (Figs. 11.10 and 11.11). Buds from stored cabbage heads that have been previously cut from their supporting stems can be similarly excised, rooted, vernalized, and allowed to grow, bloom, and be pollinated.

MAJOR BREEDING ACHIEVEMENTS OF THE RECENT PAST

Two achievements stand out above all others in terms of the genetic improvement of cabbage cultivars. All important U.S. cultivars now have fusarium yellows resistance and most are hybrids. Many imported hybrids do not have yellows resistance but otherwise

FIGURE 11.10. Cabbage stump with large buds ready for removal and rooting.

FIGURE 11.11. Small plants from rooted and vernalized buds.

may be well adapted. The second achievement that has required major efforts at the basic and applied levels is the development of hybrids utilizing self-incompatibility. In 1960 essentially all cabbage cultivars were open-pollinated. By 1980 most were hybrid except for some old cultivars that still perform well in special locations.

Old cultivars were of varied shapes from pointed to flat and many had large frames. New cultivars are all essentially round and have small compact frames, allowing more plants per acre. The heads have become more solid and in some cases more resistant to splitting.

Thirty years ago bolting was a problem, but it is rarely so now. Twenty years ago the cultivars used for processing had large heads, but the heads were loose, the frames very large, and the dry matter low. The new cultivars have compact frames and solid heads, and are adapted to machine harvest. The heads can withstand being rolled or can be dropped without shattering. Tolerance to virus, mildew, and clubroot has been incorporated in some hybrids.

Twenty years ago cabbage was harvested by hand. Now almost all cabbage for processing as kraut or for immediate use in coleslaw is machine harvested. The harvester loads the crop directly into a truck. (At present all harvesters are made by Castle Harvester of Seneca Castle, New York.)

Recent advances in storage techniques, using refrigerated or even controlled-atmosphere (CA) storage, and improvement in cultivars have resulted in the ability to store cabbage for up to 6 months. This contrasts with a previous maximum of 2 to 3 months common storage. Common storage functions by inserting cool air into the storage during nights and colder days and by preventing entry of the warmer daytime air. The stored product, principally from New York, is excellent for coleslaw and is shipped around the United States for this purpose.

CURRENT GOALS OF BREEDING PROGRAMS

The incorporation of multiple-disease resistance [especially black rot and turnip mosaic virus (TuMV) resistance], insect resistance (especially lepidopterous worm and root maggot), and tipburn resistance is a major current breeding objective. To combine these characters in hybrid cultivars suitable for production of fresh-market, processing, and storage cabbage is the broad aim of almost all breeding programs. Selecting for incompatibility or incorporating male sterility is an additional major concern of broad programs because one of the two systems is essential for the now required hybrid seed production for new cultivars. The commercial cabbage seed market demands that all new cultivars be hybrids.

Black-rot-resistant hybrids and hybrids using cytoplasmic male sterility are just starting to appear on the market or for extensive trial. Some researchers and commercial breeders anticipate a rapid increase in the use of male sterility in new hybrids because the system is simpler to develop and less influenced by the environment than the incompatibility system. Whether or not this occurs depends to a large extent upon the ease with which compatibility can be incorporated in the most desirable parental lines.

In special areas such as New York, there is interest in developing better quality in cultivars that are adapted to long-term storage in refrigerated or CA storage. Selected late-maturing lines are harvested in late October or early November and placed in storage. They are evaluated two or three times during the winter for keeping ability. Heads will exhibit storage breakdown due to TuMV or bacterial speck if they are susceptible and infected with these diseases. This is a major problem in commercial storages. The heads will also be assayed for color and leaf retention.

For sauerkraut production, the following selection criteria are important: high yield, good solidity, small ribs, white internal head color, high-percentage dry matter, ease of removal of outer wrapper leaves, short core length (25% or less of head diameter), core not excessively tough and woody, and cracking resistance both in the field and following mechanical harvest. All selections are evaluated for these characteristics and final selections are made on the basis of recorded data.

For the Texas winter fresh market, plants that are resistant to bolting are essential because the long, cool growing season with widely fluctuating temperatures can result in premature bolting. Cultivars adapted to summer production in the North are usually not suitable. For fresh market, heads with dark-green, well-developed wrapper leaves are essential.

SELECTION TECHNIQUES FOR SPECIFIC CHARACTERS

Cabbage Head Shape

Cabbage head shape has changed from pointed, flat, or round to almost exclusively round heads within the past 20 years. Pointed head is dominant to round. However, it is generally agreed that many genetic factors determine head shape. Selection is made by cutting the head vertically through the core, allowing selection for head shape, internal solidity, leaf configuration, and core size and length. The cut core will heal and flower stalks will arise from leaf axials of the upper stem just below the head.

Heading vs. Non-heading

The distinguishing character of cabbage is the development of several wrapper leaves surrounding the terminal bud. These are tight enough to form a head or heart. It is generally agreed that heading is recessive to non-heading. The F_1 between a non-heading type and a heading cabbage will be intermediate. Depending how wide the cross, two genes ($n1$ and $n2$) or more are involved in heading (29). Likewise the loose head of a savoy cabbage is recessive to the hard head of a smooth-leaf cabbage.

Head Maturity and Annual vs. Biennial Habit

There have been several papers (10,46) on B. oleracea reporting that annual habit is dominant over biennial and that early maturity is dominant over late. There must be a dominance series of genes and/or alleles to account for the range of genetic variability in vernalization requirements and earliness. The genetics of variation of extremes has not been studied. However, there is a continuous range from Chinese kale and very early summer cauliflowers, such as Pusa Katki, which need no vernalization even when grown at 27°C, to late-maturing cabbage and winter hardy kales and Brussels sprouts, which require 12 weeks of vernalization below 10°C. Summer cabbages mature early and, if left in the field, will often produce seed stalks soon after splitting. Except for the strongly annual B. oleracea cultivars, all cultivars require some cold for veneralization and can only be vernalized after reaching a minimum size. In broccoli and summer cauliflower the maximum temperature for effecting vernalization may be as high as 20°–25°C, while for true biennials it is generally considered to be 10°C.

Head Leaves

In a cross between lines with few and those with many wrapper leaves, few is dominant. There are modifying factors. Evaluation and selection is made following vertical splitting of mature heads with a knife.

Size of Head

Head size is inherited quantitatively with hybrid vigor sometimes expressed. Cabbage is quite responsive to photoperiod. Large heads develop under the long summer days of the north and extremely large heads develop in Alaska, whereas much smaller heads develop in Texas and Florida under the short days of winter.

Plant Height

There is a major gene T for height (*30*). Most cultivars are recessive in this respect. In addition there are modifiers of this character (*31*). A tall, long stem is undesirable as it will fall from the weight of the cabbage head.

Core Width

Core width inheritance has not been studied, but wide core appears to be dominant to narrow core. A narrow core is desirable, but is less important than a short core.

Core Length

Core length is controlled by two incompletely dominant genes for short core (Fig. 11.12). A short core less than 25% of head diameter would be desirable (*13*).

Core Solidity

The solidity or toughness likewise has not been studied. A soft core is desirable over a tough core, especially for processing cabbage where the core is cut up and processed. Some hybrids, such as Roundup, have very tough cores that result in woodiness in the processed product.

Frame Size

Older cultivars generally had large frames and large basal leaves. In adapting cabbage to mechanical harvesting and in the effort to develop high-yielding cultivars, the trend has been toward smaller frames. Fresh-market cultivars generally have had smaller frames than processing types. However, an adequate frame is needed for photosynthesis, depending on the size of the head desired.

Head Splitting

Head splitting is controlled by three genes acting additively with partial dominance for early splitting. According to Chiang (*6*), narrow sense heritability for splitting was 47%. To evaluate cultivars for splitting, they all must be allowed to go to full maturity to assess the splitting tendency. Long-cored cultivars usually split at the top of the head, while short-cored cultivars tend to split at the base.

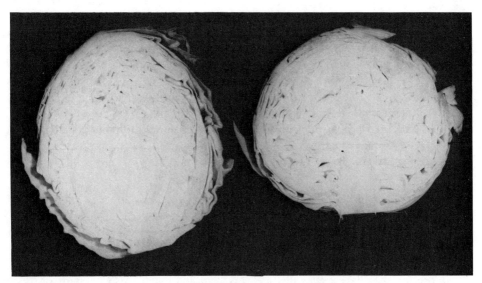

FIGURE 11.12. Cross section of cabbage heads with short and long cores.

Axillary Heading

Axillary heading occurs as buds or large sprouts below the head and as buds in the head. Both factors are undesirable since they make for a loose head and cause difficulty in harvesting the crop. Precocious axillary heading is recessive and controlled by one major gene plus modifiers (*1*). Heritability is low, making selection a difficult and slow process. Axillary heading is best assessed by splitting the head vertically through the core.

Red Coloration

The anthocyanin color of red cabbage is due to several factors and is quantitatively inherited (*18,36*). An F_1 plant of green × dark red will be pink. The gene *M* produces magenta (*20*) and with *S* gives purple on the upper side of the leaf. There have been a number of studies on anthocyanin inheritance without good agreement on the inheritance pattern.

Dry Matter

Dry matter or percentage soluble solids is inherited quantitatively (*15*) with low dry matter (6%) being recessive and present in early cultivars, whereas the highest dry matter (9–10%) occurs in late-winter storage types and savoy cabbage. Dry matter reaches a plateau when the cabbage is mature. Thus, if dry matter is to be measured, the head must be mature. If dry-matter readings are in agreement for two successive readings taken 2 weeks apart, then the cultivar is mature. Density can also be used in the same way as a criterion of maturity. Density is measured by placing a weighed head in water and measuring the water displaced. The denser the head, the closer to unity will be the ratio of head weight to weight of water displaced. Dry matter is measured by sampling a portion of the head (200 g), not including core tissue, and measuring the wet and dry weights of the samples to obtain percentage dry matter. Dry matter is correlated ($r = .78$) with soluble solids, but soluble-solid readings on cabbage juice are difficult to read on a hand refractometer. Usually analysis of dry matter of five individual heads will give a good assessment of the dry-matter potential of the cultivar. If the samples of a mature cultivar are not ±0.5% of the mean, the cultivar is either not uniform or not mature.

Dry matter is positively correlated with later maturity, and it will always be difficult to develop early, high-yielding, high-dry-matter cabbage cultivars. In effect, high dry matter is a concentration of the product of photosynthesis.

Individual heads are sampled for dry matter in a segregating population at the time when the head is split for internal character assessment. Plant stumps of heads having high dry matter can then be saved for seed production.

Storageability

Late, slow-growing cultivars are best suited for storing. Dry matter of the better storage types, such as the new Dutch hybrids, is higher than standard and early cultivars. The heads are firm and have think finely veined leaves. Intermediate-storing cultivars, such as Green Winter, are coarser. Eating quality of the very late storage cultivars is inferior because they have leaves that are very hard and tough. Also, in some of the very late cultivars the leaves wrap together very tightly. Heads of desirable types are placed in storage for extended periods, and then selected heads, which have maintained their color and quality, can be rooted and seed produced. The head will form roots at the base of its core, or alternatively the axillary buds at the base of each leaf can be excised and rooted (Figs. 11.10 and 11.11). The desired color depends on potential use. Cabbage to be used for coleslaw should be white, while if it is to be sold for fresh market, it should be green.

Winter Cultivars

Some cultivars, such as Round Dutch, Greenback, Rio Verde, and Superette, are adapted to overwintering and growth during midwinter in the South. These cultivars are bolting resistant, requiring a long cold period to induce flowering. They are often slightly savoyed or have blistered leaves.

Selection for bolting resistance can best be done in the South. Planting is done earlier than for the main crop, resulting in larger plants when cold weather occurs and greater susceptibility to vernalization. Plants not bolting rapidly under such conditions must be bolting resistant. Alternatively, in the North, seed can be planted in late February or early March and large seedlings transplanted to the field in early April. Plants that bolt instead of producing heads are eliminated.

Savoy

The savoy leaf texture is controlled by three or more genes. The yellow savoy cultivars are high yielding and considered by many to be better flavored and less gas producing than the smooth-leaved cultivars. The savoy cultivars are the highest in solids or dry matter of all cabbages.

SELECTION FOR PEST RESISTANCE

Procedures for screening for single- or multiple-disease resistance are described in detail by Williams (*49*).

Cabbage Yellows

Cabbage yellows is a soilborne fungus disease [*Fusarium oxysporum* f. sp. *conglutinens* (Wr) Snyder & Hansen]. Fusarium yellows is a soilborne vascular wilt favored by warm soil temperatures with the optimum at 28°C; the symptoms are progressive yellowing followed by brown necrosis of the plant starting with the lower leaves; frequently unilateral vascular browning occurs. The plant is stunted and premature leaf drop may occur.

There are two types of cabbage yellows resistance, designated type A and type B by Walker (*43*). Type A is determined by one dominant factor and is carried by most resistant cultivars. This resistance is not influenced by the temperature. Type B resistance is conditioned by several genes and breaks down at temperature above 22°C. Testing for type A resistance is done by dipping young seedlings in an inoculation suspension (*49*) and then growing them at 27°C. In 2 or 3 weeks susceptible plants will be dead. Recently, a pathotype of the fungus capable of breaking type A resistance in cabbage has been found. Fusarium yellows is a classic example of stable resistance controlled by a single gene.

Downy Mildew

Downy mildew is a fungus disease [*Peronospora parasitica* (Pers. ex Fr.)]. Downy mildew is soilborne and spreads when soil is transferred by wind and rain. It attacks the lower leaves first, beginning as a sparse to densely packed white mat on chlorotic to partially chlorotic lesions. Resistance has been found in cabbage and broccoli, but the genetics of resistance is complicated by the existence of numerous pathogenic races (*21*). Cabbage PI 245015 has dominant genes for resistance to races 1 and 2, each inherited independently (*21*).

Testing for resistance is done by spraying the cotyledons of young plants with a spore suspension and placing the treated plants in a saturated atmosphere at 16°C for 12–18 hr and 5 days later returning them to the chamber for 24 hr. Susceptible plants will have a heavy sporulation on the lower leaf surface.

Black Rot

Resistance to black rot bacteria [*Xanthomonas campestris* (Pam. Dows.)] was found in the cv. Early Fuji (*52*). The disease is seedborne and is also spread by wind and splashing water. The symptoms are vascular bacteriosis causing yellowing of the leaves and brown to black discoloration of the veins. Resistance is controlled by a major gene *f*, plus two modifier genes, one dominant and the other recessive.

For artificial inoculation a bacterial spore suspension is sprayed on well-developed plants early in the morning. This introduces bacteria into the guttation droplets. As the day warms, the bacteria will be drawn back into the leaves through the hydathodes. In 2–3 weeks susceptible plants will develop large lesions on the leaf margins and blackening through the veins of the leaf and stem. Resistant cultivars will only show slight necrotic infections at the leaf margins (Fig. 11.13).

Powdery Mildew

Resistance to powdery mildew fungus (*Erysiphe polygoni D.C.*) was found in Globelle, and the resistance in this cultivar is controlled by a single dominant gene, although modifying genes may influence its expression (*44*). Powdery mildew shows up in late fall,

FIGURE 11.13. Black rot in cabbage (note V-shaped lesions). (A) Resistant plant: arrows indicate hypersensitive reaction at hydathodes. (B) Moderate resistance. (C) Susceptible. (D) Highly susceptible.
Courtesy of P. W. Williams.

especially if it has been a dry season. The common symptom is a fine necrotic flecking on the exposed head and lower leaves, near the base of the large ribs. Natural field infection (if heavy) is the easiest method to use for selection for resistance.

Turnip and Cauliflower Mosaic Virus

These viruses are spread by infected aphids feeding on the plants. Wild plant hosts are the overwintering disease reservoir. The symptoms are mottling and nerotic spots or ringspots. There is no known resistance to cauliflower mosiac virus, but some cultivars, such as Globelle, exhibit quantitative resistance to races 1 and 2 of turnip mosaic virus (TuMV). Provvidenti (*32*) has shown that TuMV has at least four or more races, and there is complete resistance to all four races in Chinese cabbage. Resistance to each race is inherited independently. Screening for resistance can best be done in the greenhouse. Seedlings are inoculated, held for 7 days at 25°C to allow development of the virus, and then grown at 15°C to encourage expression of susceptibility. If plants are inoculated and planted in the field, the virus will express itself best when cool weather develops (Fig. 11.14).

Clubroot

Clubroot (*Plasmodiophora brassicae* Woron.) is a soilborne fungus disease. The symptoms are fusiform or spherical enlargements and malformations (galls) on the roots. Severely infected plants wilt in direct sunlight, and the plants will be more or less stunted. There are many races of clubroot that infect various species. Resistance in *B. oleracea* usually involves several genes for control of each race. Resistance has come primarily from kale and rutabaga. Interspecific hybridization has been extensively used by crucifer breeders, primarily for the purpose of transferring resistance to specific races (*7,8*). Screening is best performed by growing seedlings in heavily infested soil at temperatures of 20°–25°C and pH about 5–6. Alternatively, seedlings can be dipped in a suspension of spores for a few hours and then planted. Severe infection may be obtained after 5–6 weeks. Races 2, 3, 5, 6, and 7 are known in North America, and over 24 races have been identified around the world. Jersey Queen and Badger Shipper cabbage and Laurentian and Wilhelmburger rutabaga are used as international differential hosts (*48*). Badger Shipper, which has resistance derived from kale, is resistant to races 1, 3, 5, and 6. Chiang (*7*) found in *B. napus* resistance to races 1 and 3 controlled by a single dominant gene and resistance to race 5 controlled by two recessive genes. Strandberg (*37*) found a dominant gene for resistance to races 6 and 7 in Chinese cabbage.

Rhizoctonia or Bottom Rot

The bottom rot pathogen *Rhizoctonia solani* Kuhn. is soilborne. Early infection produces seed decay and pre- and postemergence damping-off. Later infection on the stem produces cankers and wire stem, and older plants develop a bottom rot of the lower leaves and head

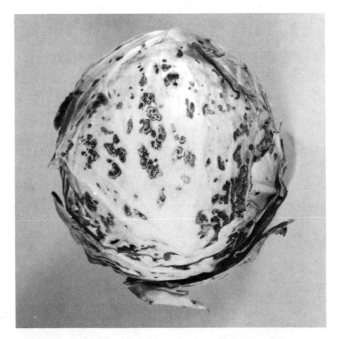

FIGURE 11.14. Cabbage head showing TuMV lesions.

rots. Williams and Walker (*53*) have shown that resistance is controlled by a single dominant gene that is present in the cv. Globelle.

Tipburn

Tipburn is a physiological disease of major importance worldwide (Fig. 11.15). It is associated with lack of calcium translocation to the tips of rapidly growing leaves, resulting in death of the cells and consequent browning and blackening of leaf tissue in the head. High nitrogen and rapid growth are associated with the disease. Palskill *et al.* (*27*) reported that factors that produced or increased root pressure deficits were associated with increased tipburn. Thus, for effective selection against tipburn, conditions that stress the plant are required in order to identify resistant plants. Selection efficiency can be meaningful by creating conditions for rapid vigorous growth and then stressing the plant by withholding water, root pruning, or reducing translocation by holding the plants in a mist chamber for several days. Walker *et al.* (*45*) reported that resistance was controlled by two or three recessive genes, whereas Dickson (*12*)—studying other selections—presented evidence for dominant resistance controlled by two or three genes. Many factors influence tipburn, making selection difficult. Repeated selection over several generations is necessary to be reasonably sure the breeder has developed a resistant line. Cultivars differ widely in susceptibility; for example, Green Boy and Rio Verde are very susceptible while Titanic, Roundup, and Superboy are much less so.

Lepidopterous Worm Resistance

There is a variability in susceptibility of cabbage cultivars to worms. Red cabbages are generally much more resistant than the green cabbages. This in part is due to vector preference for the green cabbage. Dickson and Eckenrode (*14*) found that PI 234599 cauliflower, which has glossy leaves, was immune in the field to the cabbage looper (*Trichoplusia ni* Hubner), imported cabbage worm (*Pieris rapae* L.), and to diamondback moth (*Plutella xylostella* L.). Resistance is recessive and quantitative, and it is partially linked to the glossy character. Cauliflowers with leaves having normal bloom

FIGURE 11.15. Cabbage heads showing internal tipburn.

and moderate resistance have been developed. Screening in the field, using natural pest populations developed by growing a susceptible genotype every third row, was preferable to screening in cages or greenhouses. Both high nitrogen and low light reduce resistance. Resistance is due to antibiosis, i.e., the larvae do not grow as healthily on the resistant lines. Chemical identity of the antibiotic substance is unknown, but resistant plants in feeding studies imparted no deleterious effects on rats.

Radcliffe and Chapman (*33*) have reported the differences in cultivar response by various cruciferous crops to attack by various insects in the field.

GENE LIST

The gene list (Table 11.6) was prepared in part from the lists published by Yarnell (*55*) and by Wills (*54*) in the *Eucarpia Cruciferae Newsletter*. People interested in crucifer genes should contact either A. B. Wills, Scottish Horticultural Institute, Mylnefield, Dundee, Scotland or Paul Williams, University of Wisconsin, Madison, for an up-to-date gene list.

DESIGN OF A COMPLETE BREEDING PROGRAM

Let us assume that a poor-quality plant introduction line (A) of cabbage carries downy mildew resistance, having a single dominant gene for resistance to two races. We want to incorporate the downy mildew resistance into a desirable inbred cabbage cultivar (B). Let us assume also that the A line has no resistance to fusarium yellows, but cultivar B does carries the A-type dominant resistance to fusarium yellows.

Generation 1

May 1. Plant 10 or more plants each of parents A and B.
June 5. Transplant A and B to the field.
September 1. Dig and pot several plants of both the A and B lines, removing the outer leaves. Place the plants in cold storage at 5°C or in a cold greenhouse. During storage remove any rotting leaf bases or tissue. If cabbage maggots were observed in the stem when the plants were dug, treat the plants with insecticides after potting.
November 15. Remove the plants from cold storage and move to a greenhouse at 15° ± 5°C. Water plants and fertilize as needed. Trim the outer part (leaves) of the heads of both A and B plants without damaging the core.
December 1. Remove dehisced leaves from plants. These will begin to rot and may cause the whole plant to rot if they are not removed.
January 1. Make crosses using B (the yellows-resistant parent of desirable horticultural phenotype) as the pistillate parent. Emasculate the buds before they are fully open to guarantee that no self-pollination occurs.
March 1. Harvest seed as pods turn pale or become dry. Do not let pods dry to the point that they dehisce.

Generation 2

About May 1, plant 20–30 F_1 seed in a row of a flat or in a single pot. When the seedlings are 2 weeks old and the cotyledons fully developed, screen them for mildew resistance to both races. All should be heterozygous at both loci and therefore resistant. The B parent should be susceptible and the A parent resistant. Transplant the F_1 survivors plus 10 plants

TABLE 11.6. Simply Inherited Characters of Cabbage and Their Gene Symbols[a]

Symbol	Character description	Reference
Foliage color		
c	Anthocyanin-less: recessive, epistatic to A; anthocyanin suppressed in all parts	18,36
A	Basic *anthocyanin* color factor: allelic series; intensifiers	18
A^rc	Colored lamina: color intensifier in red cabbage	18,36
B	Light red midrib alone: with A gives a dark-red violet	18
G, H	Complementary factors for deep purple: Gh, sun color; gH, gh, green	19
M	*Magenta* plant color	20
R-1, R-2	Duplicate factors for sun color: need reconciling with G and H, and also with S	19
S	*Sun* color: see G, H, R-1 and R-2	20
Leaf morphology and heading		
As	*Asparagoides*: bizarre protuberances from leaf midrib and veins; originally designated A	30
En	*Entire* vs. hyrate leaf: originally E	30
fc	*Fused* cotyledon: outer edges of cotyledon fused to form funnel	53
gl	*Glossy* foliage: both recessive and dominant; series of non-allelic genes with similar phenotype; wax inhibited; dominant types produce wax on stems	3
Hr-1	*Hairy* first leaf: hairs on margin; sometimes poor expression in heterozygote	35
sm	*Smooth* leaves: with wr; originally S	19
Pet	*Petiolate* sessile vs. leaf: originally P	30
W-1, W-2, W-3, W-4	Series of factors for frilled leaves	2
K	Dominant factor for heading	2

of cultivar B to the field. In September dig three F_1 plants and three of parent B. For transfer of the mildew resistance only one is needed, but three allow for possible loss to rots, genetic variability, or other contingencies. More than three should be used if the breeder is interested in breeding hybrids or open-pollinated cultivars. The larger number of S alleles will enhance the potential for the cross-pollination required. Pot and place these plants in cold storage for vernalization. Repeat the procedures for vernalization and transfer to the greenhouse for pollination.

Backcross one or more of the F_1 plants to parent B as female. Emasculation and pollination of about 50 parent B flowers should provide plenty of seed. The emasculation guarantees that there are no selfed plants in the backcross pollinations. Emasculation is not essential, since any selfs of parent B will be mildew susceptible. They should not remain after selection for this resistance, except when the disease incidence is low and escapes occur. Crossing without emasculation saves time during the crossing procedure. The use of all three F_1 and more than one B parent plant will maximize the number of S alleles occurring in the ultimately developed disease resistant population. Harvest the seed when ripe.

TABLE 11.6. (*cont.*)

Symbol	Character description	Reference
n-1, n-2	Recessive factors for heading; it is suggested that these might be substituted for *k-1, k-2, k-3*	29
W	*Wide* leaf	29
Plant habit		
Ax	*Axil* sprouts: originally *A*	1
dw	*Dwarf*: short internodes, round leaves	56
T	*Tall* vs. short plant	29
Flower color		
An	*Anther* spot: tip of anther purple spotted; suppressed by *C*	54
Wh	*White* petal	28
Flower morphology		
cp	*Crinkly petal*: allelic series	53
Genic male sterility		
ap	*Aborted pollen*: anthers appear normal, but pollen does not germinate	54
ms-1	Broccoli, Anstey's *M2*, linkage group 5	9
ms-2	Brussels sprouts	34
ms-4	Purple cauliflower	5
ms-5	Cauliflower	24
ms-6	Broccoli	11
Cytoplasmic male sterility		
ms	Cytoplasmic factor with R, radish cytoplasm	4,26
ms	Cytoplasmic factor with N, *nigra* cytoplasm	28
Self-incompatibility		
S	*Self-incompatibility*: multiallelic	25
Miscellaneous		
f	Major gene for resistance to black rot	52

[a]Adapted from Yarnell (55).

Generation 3

Plant 500–600 BC_1 seed, plus cultivar B. Screen for resistance to mildew; one-fourth should have resistance to both races. Transplant survivors and parent B in the field.

In September select for vigorous plants with moderate-sized basal leaves, no lateral buds sprouts, and round solid heads. Cut off the heads, and split them vertically through the core. Observe heads for internal tipburn, short core, head solidity, and lack of open spaces in the head, especially at its base. Select the 5–15 best plants and place them in cold storage. During storage, check the plants for rot of the cut surface. Remove any rot and dust the infected area with fungicide.

Generation 4

Backcross the selections (BC_1) onto cultivar B. Plant 500–600 BC_2 seed in the greenhouse. At 2 weeks of age test seedlings with fusarium yellows inoculum. Include some seedlings of A or other known susceptible such as Copenhagen Market or Pennstate Ballhead as a susceptible check. All seedlings should be resistant to yellows. Test 4-week-old seedlings in a mist chamber for resistance to both races of downy mildew (one-fourth should be resistant to both). Transplant the 100–150 survivors to the field. Repeat the selection procedures for type in September. If recovery for the type of cultivar B has been achieved, self-pollinate selections. Make a further backcross and repeat the cycle if the type of cultivar B has not been achieved.

If the type of cultivar B was achieved, select 5–10 or more plants; pot and vernalize them as described above. The larger the number of plants, the larger the number of S alleles that will be represented. At bloom, self-pollinate each plant separately. Bud selfing will probably give the maximum seed set, and open flower selfing will provide an assay of the plant's self-incompatibility. Make test crosses to adapted male sterile or self-incompatible inbreds (6–12 flowers). This will test each new selection for combining ability.

Generation 5

The objective is to verify the disease resistances and to identify the homozygous resistant plants of generation 4. Screen seedling progeny of each selfed BC_2–F_2 selection separately for yellows resistance and resistance to both races of mildew. Seven out of eight selections should be homozygous resistant to yellows, and one-fourth of the seedlings from each selection should be resistant to both races of mildew. Transplant the survivors, keeping all the progeny plants from each F_2 selection separate. Repeat the selection and self-pollination procedure. Repeat and make further testcrosses onto adapted male-sterile or self-incompatible inbreds to test for combining ability in terms of seed production and primarily for field performance.

If incompatibility is desired, it must be recognized that the F_2 generation is also the I_1 generation, which is the most effective population for use in identifying S-allele genotypes. Also, the observed number of seeds per pod from pollination of open flowers of the F_1 plants (generation 4 here) provides an assay of self-incompatibility/self-compatibility of the line.

Generation 6

For further verification of disease resistance and homozygosity, repeat the screening procedure as in generation 5. This will identify BC_2–F_3 lines that are homozygous resistant to fusarium yellows and mildew. For the development of self-incompatible inbreds with the yellows and mildew resistances, the BC_2–F_3 lines represent the I_2 generation. During this generation, homozygosity of the S-allele homozygous genotypes identified during the previous (BC–F_2, i.e., I_1) generation can be verified. Selection for or against strong self-incompatibility can be applied to this (generation 6) and subsequent generations. To obtain uniform inbred lines, selection and selfing through the F_3 and F_4 (I_2 and I_3) generations is usually required.

At this stage the lines will be ready for release as new inbreds that will give experimental hybrids with mildew resistance. Resistance in this one parent of the hybrid is adequate since resistance is dominant for both diseases.

TRIALS OF ADVANCED LINES

Testing of open-pollinated cultivars or potential hybrid cultivars can be done as follows. A satisfactory procedure is to have two concurrent trials. One should be an observational trial of new hybrids with only one replicate. The second is a replicated trial using three or four replications with 30–40 plants per replication, testing only those hybrids previously selected in the observational trial. Complete data will be taken on the replicated trial. The best and worst hybrids are noted in the observation trial. The most promising of these are advanced into the replicated trials of the next planting season. Following the advanced trials the best lines will be recommended to the grower for trial, at which time commercial small-scale evaluation of their potential for shipping, storage, or processing may be initiated.

Plant spacing within trials should be adjusted to 24-in. rows and in row spacing of 12 in. for fresh market. For processing and storage cabbage, use 30- to 36-in. rows and in row spacing of 15–18 in. The trials can be direct seeded and thinned or can be transplanted. Transplanting is usually simpler for a trial and also conserves scarce seed. Cabbage growth rates change with seasonal variations in moisture and temperature. Therefore, duration from seeding to harvest will fluctuate ±2–3 weeks from year to year. There is no precise way of knowing when a cabbage is mature, and so a sequential harvest at weekly or biweekly intervals provides the best evaluation. The first harvest should be made when the line is first considered mature or when the first head splits. If the cabbage is for processing or storage, constant week-to-week dry-matter or head density determinations will indicate if maturity is reached. Both density and dry matter will increase from week to week prior to maturity. Once maturity is reached, both values will remain constant until the head splits.

Heads should be evaluated for splitting, and for protective formation of wrapper leaves about the head. Heads must be cut open and evaluated for tipburn, head solidity or lack of open spaces, core length and width, internal head color, rib size, and leaf thickness. If thrips are a problem, this is also the time to record their presence or absence.

Cabbages for storage are harvested at early maturity and placed in storage for 3–6 months to evaluate their storage potential. Fully mature heads will not store as well. Enough heads must be placed in storage to allow sampling two or three times during the storage period.

Currently, two types of breeding lines are released to seed companies by public-sector breeding programs. One type is fully developed inbred lines for use in specific hybrid combinations, which the seedsmen may produce and market. They may also use the inbred line in combination with their own lines to produce private hybrids. Partially developed lines with special attributes are another common type of germplasm release. Seedsmen will further inbreed and refine these lines prior to using them in hybrid combination. Some public breeders do not select for S-allele homozygosity and self-incompatibility. In this case the seedsmen may find that additional selection for incompatibility is necessary before the inbred can be used as the female parent of a hybrid. They may also find it impossible to obtain adequately strong incompatibility if the breeder has not endeavored to maintain a relatively large frequency of S alleles. This is more likely to be true if the line was developed with some specific attribute in mind, such as disease or insect resistance.

Commercial Hybrid Seed Industry

Commercial cabbage seed is produced in Washington, Oregon, and California. The seed bed is planted in early September and transplanted to the field, with every fifth row being left for the male row if hybrid seed is to be produced. When both inbreds are self-incompatible so that seed can be harvested from both, then two rows of each parent usually are planted alternately. The plants will grow and develop until cool weather stops growth, and vernalization occurs during the cold months. When warm weather comes in the spring, the plants produce seed stalks, and bloom in May and June. The seed is harvested in July and August. Yields of hybrid seed will vary from 200 to 600 lb/acre.

Breeding programs for *Brassica* crops occur all over the world. In Europe, Japan, or various parts of the United States, the seed companies breeding cabbage have their own research facilities. Seed production in Washington is rather concentrated and the location of each field or plot is carefully planned to avoid contamination with foreign pollen.

In Washington, a very cold winter may sometimes kill many of the plants. Excess rain in the summer can also ruin the seed crop. Another serious concern is to prevent seed from being contaminated with disease, such as blackrot or phoma. Great care is taken to ensure that seed planted in the seed production area is free of disease.

Male sterility is just starting to be used in the commercial production of hybrid cabbage. However, it is expected that its use will increase quite rapidly as new hybrids are developed, since it is simpler to manipulate than the self-incompatibility system and less subject to environmental effects. Also it is impossible to have inbreds as contaminants in commercial seed if male sterility is used, whereas when using incompatibility in some crosses and some years a considerable percentage of inbreds can occur. The inbreds are smaller than the hybrids and are considered undesirable. A seed lot will be discarded if the inbred contamination is in excess of about 5%. However, along with development of male-sterile inbreds and their maintainers will be the need to select for self-compatibility.

Useful additional references are Niewhof (*23*), Eucarpia-Cruciferae Newsletter (*16*), and Williams (*50*).

REFERENCES

1. Akratanakul, W., and Baggett, J. R. 1977. Inheritance of axillary heading tendency in cabbage, *Brassica oleracea* L. J. Am. Soc. Hortic. Sci. *102*, 5–7.
2. Allgayer, H. 1928. Genetic investigations with garden cabbage by crossing trials by Richard Freedenberg. Z. Indukt. Abstamm. Vererbungsl. *47*, 191–260.
3. Anstey, T. H., and Moore, J. F. 1954. Inheritance of glossy foliage and cream petals in green sprouting broccoli. J. Hered. *5*, 39–41.
4. Bannerot, H., Loulidard, L., Cauderon, Y., and Tempe, T. 1974. Cytoplasmic made sterility transfer from *Raphanus* to *Brassica*. *In* Eucarpia—Cruciferae Conference, pp. 52–54. Scottish Horticulture Research Institute, Mylnefield, Dundee, Scotland.
5. Borchers, E. A. 1966. Characteristics of a male sterile mutant in purple cauliflower (*B. oleracea* L.). Proc. Am. Soc. Hortic. Sci. *88*, 406–410.
6. Chiang, M. S. 1972. Inheritance of head splitting in cabbage (*Brassica oleracea* L. var. *capitata* L.). Euphytica *21*, 507–509.
7. Chiang, M. S., Chiang, B. Y., and Grant, W. F. 1977. Clubroot resistance transferred to cabbage. Euphytica *26*, 319–336.
8. Chiang, M. S., and Crete, R. 1970. Inheritance of clubroot resistance in cabbage (*Brassica oleracea* L. var. *capitata*). Can. J. Genet. Cytol. *12*, 253–256.
9. Cole, K. 1959. Inheritance of male sterility in green sprouting broccoli. Can. J. Genet. Cytol. *1*, 203–207.
10. Detjen, L. R., and McCue, C. A. 1933. Cabbage characters and their heredity. Del., Agric. Exp. Stn., Bull. *180*.
11. Dickson, M. H. 1970. A temperature sensitive male sterile gene in broccoli, *Brassica oleracea* L. var. *italica*. J. Am. Soc. Hortic. Sci. *95*, 13–14.

12. Dickson, M. H. 1977. Inheritance of resistance to tipburn in cabbage. Euphytica 26, 811–815.
13. Dickson, M. H., and Carruth, A. F. 1967. The inheritance of core length in cabbage. J. Am. Soc. Hortic. Sci. 91, 321–324.
14. Dickson, M. H., and Eckenrode, C. J. 1980. Breeding for resistance in cabbage and cauliflower to cabbage looper, imported cabbage worm and diamond back moth. J. Am. Soc. Hortic. Sci. 105, 782–785.
15. Dickson, M. H., and Stamer, J. R. 1970. Breeding cabbage for high dry matter and soluble solids. J. Am. Soc. Hortic. Sci. 95, 720–723.
16. Eucarpia Cruciferae Newsletter. A. B. Wills (Editor). Scottish Horticulture Research Institute, Mylnefield, Dundee, Scotland.
17. Inomata, N. 1978. Production of interspecific hybrids in *Brassica campestris* × *B. oleracea* by culture in vitro of excised ovaries. I. Development of excised ovaries in the crosses of various cultivars. Jpn. J. Genet. 53, 161–173.
18. Kristofferson, K. B. 1924. Contributions to the genetics of *Brassica oleracea*. Hereditas 5, 297–364.
19. Kwan, C. C. 1934. Inheritance of some plant characters in cabbage *Brassica oleracea* var. capitata. J. Agric. Assoc. China 126, 81–127.
20. Magruder, R., and Myers, C. H. 1933. The inheritance of some plant colors in cabbage. J. Agric. Res. (Washington, D.C.) 47, 233–248.
21. Natti, J. J., Dickson, M. H., and Atkin, J. D. 1967. Resistance of *Brassica oleracea* varieties to downy mildew. Phytopathology 57, 144–147.
22. Nettancourt, D. 1977. Incompatibility in Angiosperms. Springer-Verlag, Berlin and New York.
23. Niewhof, M. 1969. Cole Crops, World Crop Ser. Leonard Hill, London.
24. Niewhof, M. 1961. Male sterility in cole crops. Euphytica 10, 351–356.
25. Odland, M. L., and Noll, C. J. 1950. The utilization of cross-compatibility and self-incompatibility in the production of F_1 hybrid cabbage. Proc. Am. Soc. Hortic. Sci. 55, 390–402.
26. Ogura, H. 1968. Studies on the new male sterility in Japanese radish with special reference to the utilization of this sterility towards the practical raising of hybrid seed. Mem. Fac. Agric., Kagoshima Univ. 6, 39–78.
27. Palskill, D. A., Tibbits, T. W., and Williams, P. H. 1976. Enhancement of calcium transport to inner leaves of cabbage for prevention of tipburn. J. Am. Soc. Hortic. Sci. 101, 645–648.
28. Pearson, H. J. 1972. Cytoplasmically inherited male sterility characters and flavor components from the species *Brassica nigra* (L.) Koch × *B. oleracea* L. J. Am. Soc. Hortic. Sci. 97, 397–402.
29. Pearson, O. H. 1929. A dominant white flower color in *B. oleracea* L. Am. Nat. 63, 561–565.
30. Pease, M. S. 1926. Genetic situation in *Brassica oleracea*. J. Genet. 16, 363–385.
31. Pelofske, P. J., and Baggett, J. R. 1980. Inheritance of internode length, plant form, and annual habit in a cross of cabbage and broccoli (*Brassica oleracea* var. *capitale* L. and var. *italica* Plenck). Euphytica 28, 189–197.
32. Provvidenti, J. J. 1980. Evaluation of Chinese cabbage cultivars from Japan and the People's Republic of China for resistance to turnip mosaic virus and cauliflower mosaic virus. J. Am. Soc. Hortic. Sci. 105, 571–573.
33. Radcliffe, E. B., and Chapman, R. K. 1966. Varietal response to insect attack on various criciferous crops. J. Econ. Entomol. 59, 120–125.
34. Sampson, D. R. 1966. Genetic analysis of *Brassica oleracea* using nine genes from sprouting broccoli. Can. J. Genet. Cytol. 8, 404–413.
35. Sampson, D. R. 1967. Linkage between genes for hairy first leaf and chlorophyll deficiency in *Brassica oleracea*. Euphytica 16, 29–32.
36. Sampson, D. R. 1967. New light on the complexities of anthocyanin inheritance in *Brassica oleracea*. Can. J. Genet. Cytol. 9, 352–358.

37. Strandberg, J. O., and Williams, P. H. 1967. Inheritance of clubroot resistance in Chinese cabbage. Phytopathology *57*, 330.
38. Sturtevant, E. L. 1919. Notes on edible plants. N.Y. State Agric. Exp. Stn., 27th Annu. Rep. *2*, 1–686.
39. U.N. 1935. Genome-analysis in *Brassica* with special reference to the experimental formation of *B. rapus* and peculiar mode of fertilization. Jpn. J. Bot. *7*, 389–452.
40. U.S. Department of Agriculture 1979. Annual Summaries—Acreage, Yield, Production and Value. Crop Reporting Board, USDA, Washington, DC.
41. U.S. Department of Agriculture 1979. Economic Research Service, Vegetable Section, Quarterly Reports. USDA, Washington, DC.
42. Vaughan, J. G. 1977. A multidisciplinary study of the taxonomy and origin of *Brassica* crops. BioScience *27*, 35–40.
43. Walker, J. C. 1930. Inheritance of fusarium resistance in cabbage. J. Agric. Res. (Washington, D.C.) *40*, 721–745.
44. Walker, J. C., and Williams, P. H. 1965. Inheritance of powdery mildew resistance in cabbage. Plant Dis. Rep. *49*, 198–201.
45. Walker, J. C., Williams, P. H., and Pound, G. S. 1965. Internal tipburn in cabbage; its control through breeding. Wis., Agric. Exp. Stn., Bull. *258*.
46. Walkof, C. 1963. A mutant annual cabbage. Euphytica *12*, 77–80.
47. Wallace, D. H. 1979. Procedures for identifying *S*-allele genotypes of *Brassica*. Theor. Appl. Genet. *54*, 249–265.
48. Williams, P. H. 1966. A system for the determination of races of *Plasmodiophora brasscia* that infect cabbage and rutabaga. Phytopathology *56*, 624–626.
49. Williams, P. H. 1981. Bee-sticks, an aide in pollinating Cruciferae. HortScience *15*, 802–803.
50. Williams, P. H. 1981. Screening Crucifers for Multiple Disease Resistance. Crucifer Workshop, September 1981. University of Wisconsin, Madison.
51. Williams, P. H., and Heyn, F. W. 1980. The origins and development of cytoplasmic male sterility in Chinese cabbage. *In* Chinese Cabbage, Proceedings of the First International Symposium, pp. 293–300. AVRDC, Taiwan.
52. Williams, P. H., Staub, J., and Sutton, J. C. 1972. Inheritance of resistance in cabbage to black rot. Phytopathology *62*, 247–252.
53. Williams, P. H., and Walker, J. C. 1966. Inheritance of Rhizoctonia bottom rot resistance in cabbage. Phytopathology *56*, 36.
54. Willis, A. B. 1977. A preliminary gene list in *Brassica oleracea* Crucifer Newsl. *2*, 22–24.
55. Yarnell, S. H. 1956. Cytogenetics of vegetable crops. II. Crucifers. Bot. Rev. *22*, 81–166.
56. Yeager, A. F. 1943. The characteristics of crosses between botanical varieties of cabbage *Brassica oleracea*. Proc. Am. Soc. Hortic. Sci. *43*, 199–200.

12
Lettuce Breeding

EDWARD J. RYDER

Lettuce is the most important vegetable crop produced for fresh market in the United States in terms of acreage, production, and market value. Most lettuce in the United States is of the crisphead or iceberg type. There are five other lettuce types: cos or romaine, butterhead, and leaf lettuces, which are grown extensively in the United States; Latin lettuce, a Mediterranean and South American type; and stem lettuce, which is grown in the Orient and in Egypt. About 90% of U.S. commercial production is of the crisphead type, the rest is of cos, butterhead, and leaf types. The last three are also most commonly found in home gardens (Fig. 12.1).

Crisphead lettuce in the United States is largely a western crop (Table 12.1). Approximately 87% is produced in two states, California and Arizona, and 73% in California alone. California production is concentrated in five districts: Salinas, Imperial, San Joaquin, Santa Maria, and Palo Verde valleys. In Arizona, the Yuma Valley is the largest district. Other important districts and states are the San Luis Valley in Colorado, the Lower Rio Grande Valley of Texas, the Everglades of Florida, the Las Cruces–Hatch area of New Mexico, Orange and Oswego Counties in New York, and Cumberland County in New Jersey.

In other lettuce-producing countries, crisphead lettuce is much less important. Butterhead lettuce is more highly favored in England, France, Holland, and other western

BREEDING VEGETABLE CROPS

FIGURE 12.1. Principal types of lettuce produced in the United States. (A) Crisphead, cv. Francisco, (B) butterhead, cv. Midas.

European countries. Around the Mediterranean, cos or romaine is the most common type. The important districts of western Europe are the Thames Valley, Lancashire, and Lincolnshire in Great Britain, the Rhine Valley of Holland, and Brittany and Roussillon in France.

In the United States lettuce is primarily a salad vegetable. It also is a common ingredient of sandwiches. Over 2 billion heads of lettuce are harvested each year in this country.

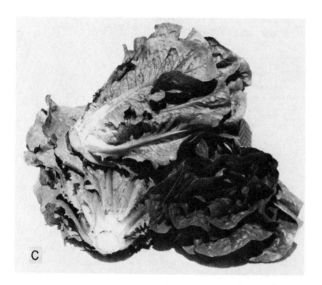

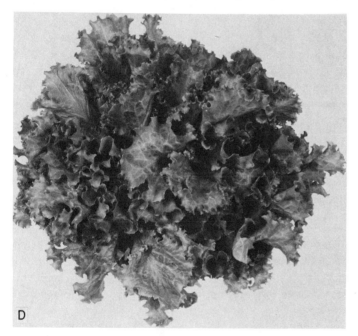

FIGURE 12.1. (*Cont'd.*) (C) Cos, cv. Parris Island, (D) leaf, cv. Grand Rapids.

Three recent trends in the consumption of lettuce should be noted. One is the appearance in restaurants of the salad bar, in which the customer creates a salad from among the lettuce and other salad vegetables and ingredients. Another is the use of lettuce shredded near the harvest site and shipped in commercial or consumer packages to stores and restaurants.

The third trend is on an international level and may be the most significant. An increasing proportion of people in England, Scandinavia, Germany, and other western

**TABLE 12.1. Area, Production, and Market Value of
Crisphead Lettuce in the United States
(1978)**[a]

State	Area (acres)	Production (1000 tons)	Value ($ million)
California	159,400	2,217	456.3
Arizona	39,500	429	96.4
Florida	10,500	103	41.0
Texas	5,400	59	11.1
New Mexico	5,000	53	11.0
New York	3,500	33	6.6
Colorado	5,400	58	5.8
New Jersey	2,600	25	5.6
Michigan	1,400	14	3.7
Wisconsin	1,200	13	2.0
Washington	1,200	10	1.6
Other	1,300	13	6.9
Total	236,400	3,026	647.5

[a]Adapted from U.S. Department of Agriculture (57).

European countries are eating and demanding crisphead lettuce. Some is locally grown, but a high proportion is shipped from Israel, Spain, and particularly the United States. As the demand for crisphead lettuce continues to increase, imports will continue to increase until new, more suitable cultivars are developed in Europe for the European production areas.

The food value of lettuce varies with type. Crisphead lettuce contributes moderate amounts of ascorbic acid (6 mg/100 g), vitamin A (330 IU/100 g), and calcium (20 mg/100 g). Butterhead lettuce has 8 mg of ascorbic acid, 970 IU of vitamin A, and 35 mg of calcium in 100 g of tissue, while cos and leaf lettuces, probably because of a greater proportion of green tissue, have 18 mg of ascorbic acid, 1900 IU of vitamin A, and 68 mg of calcium in 100 g. Differences among the various cultivars have not been measured (60).

Lettuce is twenty-sixth on the list of common fruits and vegetables arranged in order of nutritional value (10 vitamins and minerals) in unit terms (IU or mg/100 g of tissue). However, M. A. Stevens [in Rick (29)] weighted nutritional value in terms of actual consumption. Lettuce, a staple food, places fourth, behind tomatoes, oranges, and potatoes, in the rating scheme.

There are some less common uses for lettuce. A cigarette containing no nicotine is made from lettuce leaves. Seeds of a primitive form of lettuce in Egypt are a source of edible oil (16). Latex from a wild lettuce relative, *Lactuca virosa* L., is used to manufacture a sleep-inducing medicine.

ORIGIN AND GENERAL BOTANY

Lettuce is a highly self-pollinated crop that originated in the Mediterranean area (59). The earliest indication of its existence came about 4500 B.C. in tomb paintings in Egypt, which showed plants with narrow pointed leaves of the type that is now sometimes called asparagus lettuce. It apparently was similar to the modern form found in Egypt and known as *balady* (meaning local). Lettuce spread rapidly through the Mediterranean basin, particularly in the ancient Greek and Roman civilizations. It subsequently spread to western Europe and then to the New World, where it was mentioned as early as 1494 (*9,48*).

Cultivated lettuce prospers in a cool environment and can tolerate short periods of freezing or near-freezing temperatures. In addition, varieties have been selected for tolerance to high temperatures. The Mediterranean basin varies from mild to hot, from damp to dry. Egypt, the most likely center of origin, is relatively mild near the seacoast and very hot in the interior. It seems more likely that lettuce originated closer to the coast than in the interior. If so, and if it then spread to the south and later to colder areas in the north, certain adaptive changes may have occurred. In the hotter climates, seed stalk elongation, a high-temperature response, would have been accelerated, stimulating selection for slower bolting tendency to permit sufficient leaf development to maximize size and competitive ability.

On the other hand, movement to the colder northern climates would have led to development of types with the long-day flowering requirement of many modern lettuces. This would be necessary for the plant to avoid freezing during the reproductive phase.

Lettuce is in the family Asteraceae (Compositae), tribe Cichoreae, genus *Lactuca*. The number of species assigned to the genus varies, depending upon author. One author includes over 300 species (*17*). [There are seven chromosome levels in the genus: $2n =$ 10, 16, 18, 32, 34, 36, 48 (*5*).]

Botanically, lettuce (*Lactuca sativa* L.) is described as follows by Ferakova (*5*, p. 68):

Annual to biennial up to 120 cm high with a slender tap-root. Stem erect, 30–100 (most common, 30–70) cm, glabrous not setose. Basal leaves in rosette, undivided or runcinate–pinnatifid, shortly petiolate, cauline leaves simple, ovate to orbicular in outline, entire, cordate–amplexicaul, sessile, not held vertically. Apex blunt. Inflorescence a dense, corymbose, flat-topped panicle. Heads with 7–35 florets (most common, 7–15), ligules longer than involucre. Involucre of 3–4 rows of bracts, 12–15 mm long, erect when achenes ripe. Outer bracts ovate, inner ones lanceolate. Ligules yellow. Achenes 6–8 mm, obovate, narrowed and often finely muricate at the apex, greyish. Beak white, filiform, as long as body. Pappus white.

Interspecific Relationships

Of the many species of *Lactuca*, only four were found to form a useful breeding group by methods available in the period 1940–1960 (*20,56*). These are *L. sativa* L., *Lactuca serriola* L. (Fig. 12.2), *Lactuca saligna* L. (Fig. 12.3), and *L. virosa* L. (Fig. 12.4). Each has $2n = 2x = 18$ chromosomes. Despite morphological differences discussed below, *L. sativa* and *L. serriola* cross freely with each other and could well be classified as subsections of the same species. *L. saligna* and *L. virosa* are significantly different. All four are obligate self-fertilizing species.

L. saligna looks much like *L. serriola,* but they can be distinguished from each other by several morphological traits. The two best diagnostic characters are leaf shape and inflorescence type. *L. saligna* has very narrow, nearly linear stem leaves. Those of *L. serriola* are wider and are held vertically *L. saligna* has a spikelike panicle with sessile flowers. The *L. serriola* inflorescence is a pyramidal panicle (*5*). *L. saligna* can be crossed with *L. sativa* or *L. serriola* when *L. saligna* is used as the female parent. The

FIGURE 12.2. *Lactuca serriola,* a wild relative of lettuce. Common names: common wild lettuce, prickly lettuce.

hybrids are fertile (*20*). There is no published record of crosses made with *L. saligna* as the male parent.

    L. virosa may be crossed with *L. sativa* and *L. serriola* in both directions. The hybrids are sterile, however (*20*). A successful cross between *L. virosa* and *L. sativa* was obtained by Thompson and the hybrid was made fertile by doubling the chromosome number (*54*). Lindqvist was unable to obtain a cross between *L. saligna* and *L. virosa* in either direction (*20*).

    L. sativa can be distinguished from *L. serriola* by several character differences:

	L. serriola	*L. sativa*
Involucre bracts	Reflexed when mature	Erect when mature
Cauline leaves	Held vertically	Not held vertically
Inflorescence	Pyramidal panicle	Flat or rounded corymbose panicle

Also, *L. serriola* usually has spines, forms a rosette before bolting, and often has multiple stems. The achenes are black and usually smaller than those of *L. sativa,* which does not have spines. The rosette is relatively more upright and there is only one stem at bolting. Achenes may be black, white, yellow, or brown.

    Additionally, there is a group of forms with similarities to both species. These are referred to as primitive forms of *L. sativa* by Lindqvist (*22*). Several have been given

FIGURE 12.3. *Lactuca saligna,* a wild relative of lettuce.

FIGURE 12.4. *Lactuca virosa,* a wild relative of lettuce.

species names, such as *Lactuca altaica* and *Lactuca augustana*. They have large seeds and single stems, may or may not have spines, form a flat rosette, and have erect involucres when mature. Leaves may not be held vertically, and the panicles are usually corymbose and flat topped.

These primitive forms cross readily with each other and with *L. sativa* and *L. serriola*, and their existence lends some uncertainty to the relationship between *L. sativa* and *L. serriola*.

Alternative evolutionary paths may have been taken. *L. sativa* (and additionally the primitive forms) may have evolved from *L. serriola* (4,56). Lindqvist suggests also the possibility of hybridization between *L. serriola* and a third species leading to *L. sativa*, or between *L. sativa* and a third species leading to *L. serriola* (22). Whatever the evolutionary pathway, it is clear that *L. sativa*, *L. serriola*, and the primitive forms are very closely related—so closely that they could be considered as forms of one species.

There has been no success, using standard crossing techniques, in obtaining crosses between the species of the *L. sativa* group and other species. It is possible, however, that the new parasexual methods may be applied successfully in some cases (52,56,68).

FLORAL BIOLOGY AND CONTROLLED POLLINATION

Seedstalk Development

Lettuce growth occurs in two stages. The first is vegetative: the formation of a rosette of leaves and/or a head. This is followed by seedstalk elongation and flowering. Seedstalk formation in lettuce is an agriculturally undesirable event, signaling the end of the marketability of the crop. It is, of course, biologically necessary, and in plant breeding, it is essential when the object is to make crosses or to produce a seed crop (Figs. 12.5–12.7).

It is convenient in the greenhouse, when working with the head-forming types, to subvert the heading process by forcing the leaves outward and away from the growing point. This allows the main stem to elongate freely and to remain straight and unbroken. In the field, we allow head formation to take place. Then we either slash the top with a knife to allow the seedstalk through or break the head off at its base, encouraging the elongation of axillary stems.

FIGURE 12.5. Young stages of growth in lettuce. Cotyledon stage (left); rosette stage (right).

FIGURE 12.6. Cv. Climax in early heading stage.

FIGURE 12.7. Lettuce plants in the greenhouse. Foreground, plants not yet bolted; background, plants in flowering stage.

Flower Development

At the earliest stages of flower development, according to Jones (*13*) the corolla primordial whorl is first to appear, followed by those of the stamens and pappus, and the carpels. The microspores develop earlier than the megaspore. A single row of pollen mother cells is formed, each producing four microspores (pollen grains). Each pollen grain has two filamentous sperms and a tube nucleus.

The megaspore mother cell is embedded in a single layer of nucellar tissue. It divides to form four megaspores, three of which degenerate. The surviving megaspore, in a series of successive divisions, produces one egg, two polar nuclei, two synergids, and at most three antipodal cells.

The elongating seedstalk produces a single terminal capitulum (composite flower) and a series of branches with many capitula forming the panicle. Each capitulum is composed of an involucre of overlapping bracts surrounding several florets: as few as 10 to over 20. Each floret consists of a ligulate (strap-shaped) yellow corolla enclosing the sexual parts (Fig. 12.8). The stamens are fused to form a tube, and anthers shed pollen at the inside surface as the flower opens in the morning. The elongating style with two stigmatic lobes sweeps the pollen along as it emerges from the top of the stamen tube.

The capitulum begins to swell about 24 hr before anthesis as the florets enlarge. The top circle of bracts opens. The following morning the florets elongate and open, exposing the staminal column, through which the style and stigmas emerge. The stigmatic lobes

FIGURE 12.8. Close-up of a lettuce flower (6×).

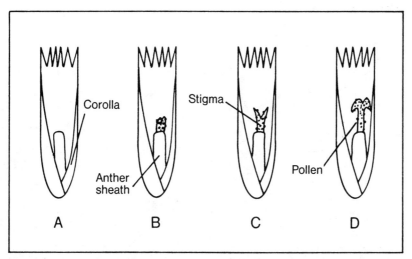

FIGURE 12.9. Structure and stages of anthesis of a lettuce floret. (A) Early, stigma not yet emerged from anther sheath. (B) Stigma emerging, covered with pollen. (C) Ideal stage for pollen removal. (D) Too late for crossing; selfing has occurred.

separate, forming a V, and then curl backwards, signaling the end of receptiveness (Fig. 12.9).

Lettuce flowers open only once. The process takes place in 1–2 hours. Higher temperature and increased light speed up the process.

Crossing Techniques

The structure of the lettuce flower obligates self-fertilization. The sequence of events in flower opening dictates the need for special crossing techniques. Specifically, the enclosing stamen tube is quite delicate and difficult to remove, although, with a fine pair of forceps, the tube can be removed early in the morning before the style has emerged above the top of the stamens. However, removal of several tubes in a single flower can be quite tedious and time consuming.

Other methods have been proposed. The most commonly used method was developed by Oliver (27). The flower is washed with a fine stream of water when the stigmatic lobes have begun to separate, but before they have started to curl back. Since anthesis has already occurred, some of the pollen grains have already germinated. Most can be washed off, however, and replaced with foreign pollen by simply touching another flower to the washed one (Fig. 12.10). Earlier or later washing is less effective in eliminating selfing pollen. However, even when the washing is done at the optimum time, some selfing will occur. It is necessary, therefore, to have some means of distinguishing selfed from crossed plants in the progeny. Most desirable is a seedling marker gene; if the male parent is dominant for this gene, hybrids will show the dominant trait, while selfs will show the recessive. In lettuce, genes for anthocyanin, shades of green, and leaf shape are useful as markers.

If specific seedling genes do not differentiate the parents, more subtle growth or color differences may be used. For example, crosses between butterhead and crisphead lettuces

FIGURE 12.10. Making a cross.
(A) Washing pollen off stigma. (B) Flower
from male parent is used to apply pollen
to washed flower. (C) Attaching tag
identifying cross.

are easily distinguished by degree of color, leaf shape, texture, and method of growth.
Both parents should always be grown for comparison.

Several disease-resistance genes, such as downy mildew resistance, are available as
markers; of course, inoculation with the organism is required. Finally, late growth stage
characters may be useful. These include flower color and seed color. These are less
desirable, since the plants have to be grown for a longer period, up to 5 months, before the
characters can be observed and the selfed plants eliminated.

It is possible to adapt Oliver's method in order to ease the problem of differentiation between selfs and crosses. The pollen may be washed off two or three times instead of only once, starting just as the stigmas begin to emerge. This allows for maximum pollen removal. A large number of flowers can be washed at once by placing a fine spray nozzle over the plants and washing the entire panicle intermittently or continuously during the stigma emergence period (*39*). We have used both methods and often achieve 100% crossing. This is particularly important in backcross experiments in which a few selfed plants can skew results. The original method of Oliver is adequate when only a few hybrids are needed and when the crosses can be easily distinguished from selfed plants.

Another method of crossing has been proposed by Pearson (*28*). If the flower is clipped quite early in the morning, it is possible to remove most of the corollas and most of the stamen tubes before the styles have fully elongated, leaving the latter intact. A blast of compressed air removes debris. The pollination can then be made. This method also permits virtually 100% crossing.

Crosses are most easily made in the greenhouse. They can also be made outdoors, although wind, rain, or insects may cause loss of the hybrid seed. There is little danger of contamination by unwanted foreign pollen. Lettuce pollen is sticky and not easily wind-blown. There is also a dearth of insects that work lettuce flowers. Research by Thompson *et al.* (*55*) showed an average of 1% natural cross-pollination between adjacent rows.

An additional factor in lettuce hybridization is male sterility. Male sterility in lettuce has been identified and investigated by Lindqvist (*21*) (three complementary recessive alleles) and by Ryder (*31,33,37*) (dominant–recessive epistasis, single recessive, single dominant). None have a cytoplasmic basis, and they would not be cost effective in a commercial hybridization program. However, the dominant–recessive and recessive types would be useful in a breeding program when used in female parents of crosses, thus eliminating the need for the tedious preparation for crossing. These genes have been transferred to a few other lines but have not been thoroughly exploited.

MAJOR BREEDING ACHIEVEMENTS OF THE RECENT PAST

Disease resistance, improved yield, and improved quality are standard breeding objectives. Significant accomplishments have been made for lettuce in these areas as well as in bolting resistance.

Disease Resistance

Breeders have released cultivars with resistance to brown blight, downy mildew, lettuce mosaic, and tipburn in California, to lettuce mosaic and Bidens mottle in Florida, to downy mildew in Texas, and to downy mildew in several European countries.

The first documented success was achieved by I. C. Jagger in 1926 with release of the first brown-blight-resistant Imperial strains (*10*). Brown blight was a disease of the cv. New York that was manifested as yellowing, necrosis, and death. (The cause of brown blight has not yet been discovered.) Nearly all plants in a field were affected. Jagger selected healthy survivors, which further tests showed to be highly resistant, and the first three Imperial cultivars were developed from them.

Crosses with the French cv. Blonde Lente á Monter led to a second series of Imperial strains that were highly resistant to both brown blight and downy mildew, incited by the fungus *Bremia lactucae* (*10*) (Fig. 12.11).

However, a new strain of the fungus, designated as race 5, appeared in 1932. All existing cultivars were susceptible. Further breeding led to new resistant cultivars: Imperi-

FIGURE 12.11. Imperial 44, one of the second series of Imperial cultivars, resistant to brown blight and downy mildew.

al 410 in 1945, Valverde in 1959, and Calmar in 1960. The development of Valverde by T. W. Whitaker and P. W. Leeper was credited with saving the Lower Rio Grande lettuce industry from virtual extinction by the disease (*16,18*).

Calmar (Fig. 12.12) quickly replaced the existing group of Great Lakes strains in the California coastal districts (*61*). In addition to its mildew resistance, it was also quite resistant to tipburn and moderately resistant to big vein. A group of cultivars derived from Calmar through selection or further crosses was also mildew resistant, until a new virulent form appeared in 1974.

Downy mildew is a relatively sporadic problem in most U.S. lettuce districts. In Western Europe, however, it is a continuing major problem. The organism produces new physiologic forms in response to the appearance of new resistant cultivars. The result has been a continuous cycle of new cultivars, new races, new cultivars, etc. In the 60 years since about 1925, a total of 120 cultivars of lettuce have been bred to offer resistance to the continuous appearance of strains of the fungus (*3*) (Fig. 12.13).

Lettuce mosaic, a virus disease vectored primarily by the green peach aphid (*Myzus persicae*), has been a chronic and often serious problem in most lettuce-growing areas of the world. In many districts in the United States it is controlled by mandatory use of virus-free seed (*7*). Different sources of resistance were identified in the United States and in Argentina. However, resistance is controlled at the same locus in both sources (*34–36,58*). Several cultivars with resistance were released in Europe and Vanguard 75 was the first resistant cultivar released in the United States. It is used primarily for late-winter production in the California and Arizona desert districts (*44*). Florida 1974, a cos type, is resistant to mosaic and also to Bidens mottle, a virus disease found in Florida (*8*).

FIGURE 12.12. Calmar, a major cultivar in California coastal districts in the 1960s.

Yield Improvement

Improvements in yield have also been made, although they are somewhat more difficult to document. Yield of lettuce is measurable in more than one way. In most U.S. lettuce districts the unit of yield is the number of cartons per acre. Plant spacing is standardized within each district so that the approximate upper limit of heads per acre is known. There are usually 24 heads in a carton. These figures can be used to calculate the percentage of

FIGURE 12.13. Sporulating lesions of downy mildew (*B. lactucae*) on cotyledon and young leaf.

heads harvested. The harvest percentage is affected by the state of the market. A strong market, in which demand is high, has an enhancing effect on the number of heads cut and a weak market has a depressing effect. The harvest percentage, however, tells nothing about head weight, which varies depending upon cultivar, season, and stage of maturity at harvest. USDA reports yields in terms of weight, usually in hundreds of pounds per acre.

At any rate, various sources indicate that there have been improvements in both measures of yield in recent years. At least four factors have contributed to yield improvement: disease control, increased head size, improved cultural practices, and greater uniformity and percentage cut.

In 1953, 22,857,000 cwt were harvested from 123,600 acres in California, an average yield of 18,500 lb/acre or the equivalent of 411 cartons, each weighing 45 lb. In 1977, 41,806,000 cwt were harvested from 156,700 acres. This is an average yield of 26,700 lb/acre or 534 cartons, each weighing 50 lb (*49*).

The increase in carton weight from 1953 to 1977 is primarily a cultivar difference. In 1953, lettuce in California consisted of Great Lakes and Imperial strains. These were smaller and lighter cultivars than those grown in the 1970s: Calmar and related cultivars in the coastal areas and Empire, Merit, Climax, and Vanguard, released by R. C. Thompson in the late 1950s, in the desert districts.

The change in the number of cartons harvested was due to two factors: (1) Within-row spacing of plants decreased from 14 to 10 in., thus increasing the number of plants per acre. (2) A higher percentage of plants was cut because new cultivars and improvements in cultural practices contributed to more uniform head development.

In the eastern United States, a group of cultivars developed by G. W. Raleigh in the late 1960s and early 1970s was successful because it brought together tipburn resistance and bolting resistance into desirable cultivars, resulting in increased yields. Fulton, Minetto, Ithaca, and Fairton soon replaced Great Lakes 659, Imperial 456, and Pennlake in the eastern and midwestern states and Florida.

Quality Improvement

The concept of quality is difficult to evaluate. It touches, however, on the single most remarkable breeding accomplishment of recent years.

Lactuca virosa can be crossed with lettuce, but the hybrid is sterile. R. C. Thompson crossed PI 125130 (*L. virosa*) with a line derived from a complex *L. sativa–L. serriola* cross and treated the F_1 with colchicine. The resulting amphidiploid was fertile. Subsequent backcrosses to domestic lettuces led back to the diploid state and finally to the cv. Vanguard, which was released in 1958 (Fig. 12.14). Vanguard has a number of qualities different from other crisphead lettuces: dull green color, smooth and relatively soft texture, flat ribs, an extensive root system, and less tendency toward bitterness and deterioration when mature. Some or all of these traits stem from the *L. virosa* parent. The texture, rib type, and superior flavor of Vanguard, together with its resistance to tipburn, give it a high-quality rating unmatched in earlier crisphead lettuces. These traits are also found in Winterhaven, Moranguard, and other selections from Vanguard, and in Vanguard 75 and Salinas, which were derived from further crosses. The Vanguard type is now the dominant one in the western United States (*43,44,54*).

An important quality trait is bolting resistance, which was a factor in the success of the

FIGURE 12.14. Vanguard, the principal cultivar grown for the spring season in California and Arizona desert districts in the late 1960s and 1970s.

eastern cultivars discussed above and also of Empire, which is planted during periods of high temperatures in the fall in the West.

Other Lettuces

The principal breeding efforts with butterhead lettuces have been made in Europe, with resistance to downy mildew as the main objective. Probably the most important accomplishment has been the development of a rational hypothesis for the relationship between the host and the pathogen, a fungus. For years the evidence accumulated that resistance to the various races was due to several dominant genes acting singly. Mildew breeding and genetics research was conducted in several countries, and races and resistances were identified independently. This resulted in much confusion, as it was not known which races were the same and which were different. The specific relationships between host and fungus were not known.

Crute and Johnson (2) postulated a gene-for-gene relationship between host and fungus, such that a dominant allele in the host conferred resistance to all races of the fungus except those with the same numbered virulence factor. Several host genes have been specifically identified from segregation data. The significance of this work is that the breeder can make specific crosses based upon specifically identified genes and gene sources. It is no longer useful to refer to races of the fungus, as the race designation itself tells nothing of the virulence factors present. These can be distinguished by testing against a set of cultivars known as differentials, whose downy mildew gene makeup is known. Table 12.2 shows several U.S. cultivars with identified genes.

TABLE 12.2. Identity of *R* Genes for Resistance to Downy Mildew in Certain U.S. Cultivars[a]

R gene	Cultivar	Protects against race with V factors:[b]	Infected by race with V factors:
0	Empire, Ithaca, White Boston	None	All
5	Valmaine	1–4, 6–11	5
6	Grand Rapids	1–5, 7–11	6
7	Vanguard 75, Mesa 659	1–6, 8–11	7
8	Valverde, Valrio, Valtemp	1–7, 9–11	8
7 + 8	Salinas, Calmar, Montemar, Cal K-60	1–6, 9–11	7 + 8

[a]Adapted from Crute and Johnson (2); I. R. Crute (personal communication).
[b]V factors, virulence factors.

CURRENT GOALS OF BREEDING PROGRAMS

The major goals of the various lettuce-breeding programs remain similar to those of the recent past and can most easily be generalized by stating those of our program at the U.S. Agricultural Research Station, Salinas, California. This is a very large comprehensive program. We stress four general goals under the broad objectives of cultivar and germplasm development, especially of crisphead lettuce:

1. Resistance. This includes primarily resistance to field diseases plus additional work on insect resistance, salt tolerance, and resistance to postharvest disorders.
2. Horticultural improvement. We select for favorable expression of many traits of appearance and quality, often on a subjective basis.
3. Uniformity. Selection is among lines for uniform head formation and maturation.
4. Adaptation to specific environments. Primary effort is for the coastal, interior, and desert valleys of the West, with secondary interest in other lettuce districts.

The major emphasis in our program has been on the incorporation of resistances in superior horticultural types.

Resistance

Some of the specific goals are described below:

Big Vein

This is a disorder characterized by vein clearing (Fig. 12.15) and apparent stiffening of leaves. The latter gives the plant a bushy appearance. These symptoms are unsightly and may also be associated with delay or prevention of heading. Big vein is caused by a viruslike agent introduced into the plant by the root-feeding fungus *Olpidium brassicae* (Wor.) Dang. Since the program began in 1957, we have attempted to bring together genes for resistance from diverse sources. The rationale for this is that low levels of resistance, possibly multigenic, exist in many cultivars. We are attempting to reach successive plateaus of resistance with each cycle of hybridization and selection. This work is carried on concurrently with genetic studies and a survey of all lettuce cultivars to assess the distribution of resistance genes.

FIGURE 12.15. Big vein. Above, healthy leaves. Below, leaves showing the typical vein-clearing symptom.

Two resistant cultivars, Sea Green and Thompson, and five resistant breeding lines were released to the seed trade in the spring of 1981 (46,47). These are compared with two standard cultivars in Table 12.3.

Sclerotinia Drop

This fungus disease is characterized by severe wilting of maturing plants, collar rot, and death (Fig. 12.16). It is usually incited by *Sclerotinia minor* (small sclerotial form), although in some areas *Sclerotinia sclerotiorum* (large form) may be the agent. Limited field and greenhouse testing has shown significant variability in percentage of drop. This

TABLE 12.3. Comparison of Big Vein (BV) Percentage and Percentage First Harvest (Harv.) for Two Cultivars Selected for Resistance in a Pedigree Program and Two Moderately Susceptible Cultivars[a]

	1978		1979		1980	
	BV (%)	Harv. (%)	BV (%)	Harv. (%)	BV (%)	Harv. (%)
Thompson	12a[b]	81a	36ab	54bc	21a	64ab
Sea Green	14a	62bc	55bc	70ab	20a	75a
Salinas	56b	58bc	71c	82a	76b	48b
Calmar	54b	30d	91d	75ab	67b	11c

[a]Adapted from Ryder (46,47).
[b]Mean separation by Duncan's multiple range, 1% level.

FIGURE 12.16. Sclerotinia drop. Typical wilting of the lower leaves prior to yellowing necrosis and death.

is probably genetically based, and as with big vein, we shall attempt to combine resistance from several sources.

Lettuce Mosaic

Lettuce mosaic is a virus disease. The virus is seedborne and may be transmitted at the rate of 1–3%. It is transmitted in the field by the green peach aphid. Early infection causes vein clearing, followed by mottling, recurving of the leaves, and increased marginal frilliness (Fig. 12.17). Maturing plants are yellowed and stunted and therefore not harvestable.

Resistance is conferred by a single recessive allele and has been bred into two U.S. cultivars and several European cultivars. Our present activity consists of continued breeding of resistance into new cultivars and germplasm, studies of linkage relationships, and studies of the nature of resistance.

Resistant plants become systematically infected but at an apparently lower level so that symptom expression is much milder. Growth and development are essentially normal and the plants can be harvested (Fig. 12.18). We are studying various environmental effects on symptom expression and seed transmission of the virus (*41*).

Tipburn

Tipburn is a physiological disorder of lettuce expressed as a browning of portions of margins of internal leaves of the head. It usually occurs at harvest time and is a function of the failure of calcium to reach marginal tissues. An inspector who finds more than minimal damage as a field is being harvested is empowered to stop the harvest. Therefore, tipburn can represent a serious economic loss to the grower.

Calmar and Salinas are highly resistant to tipburn. We use the latter as a parent in crosses and select in the field for resistance in the progenies. Tipburn may be induced on detached heads in a high-temperature chamber (*25*). Further refinement of this technique may enable breeders to select more effectively for resistance.

FIGURE 12.17. Lettuce mosaic. Typical mottling effect caused by patches of chlorotic tissue.

Cabbage Looper

Particularly in the desert districts, insects are serious pests on lettuce. The cabbage looper (*Trichoplusia ni* Hubner) is one of several species contributing to a serious worm problem.

Oviposition and larval non-preference (antixenosis) and antibiosis have been identified in an accession of *L. saligna* (*65*). Oviposition non-preference has also been found in *L. serriola* (*15*).

Oviposition non-preference has been transferred to *L. sativa* lettuces (*14*). However, field studies showed that this type of resistance was not effective because the insect laid eggs on both the preferred and non-preferred plants (*24*).

FIGURE 12.18. Lettuce mosaic. Left, typical restricted symptom on a leaf from a resistant plant. Right, typical mottling symptom on a leaf from a susceptible plant.

Salt Tolerance

In the desert production areas, salt accumulation in the soil may limit the production of lettuce. At 13.0 decisiemens or above, lettuce cannot be grown. Below this level, seedling growth and subsequent development may be restricted directly in proportion to the amount of salinity. Differences in salt tolerance have been demonstrated (50). These will be exploited in a crossing and selection program to develop lines that will grow and develop normally in soil that would ordinarily limit growth.

Horticultural Improvement

Two important concepts in lettuce breeding are type and quality. Type includes several characteristics: color, leaf texture and shape, head size and shape, and butt appearance. Together these create a type. In the modern history of lettuce breeding in the United States, there have been four types of crisphead lettuce:

1. Imperial. Light- or medium-green leaves, relatively soft texture, leaves serrated or wavy, varied in butt color and ribbiness.
2. Great Lakes. Bright-green leaves in various shades, very crisp, leaves serrated, butt whitish with prominent ribs.
3. Vanguard. Dull-green leaves, softer texture than Great Lakes, leaf margins wavy, butt green with flat ribs.
4. Empire. Light-green leaves, leaves deeply serrated, very crisp, heads often conical in shape, butt whitish with flat ribs.

There are other cultivars with characteristics of more than one of the above types. The New York group, for example, has the lighter green shades of the Imperials with the leaf texture of Great Lakes.

For periods of 10–20 years after the release of a cultivar with a new type, breeders have developed cultivars within that type until a break with the old is achieved by development of a new cultivar having new combinations of horticultural traits. It may be assumed that this will continue, and that, as before, type will be combined with improvements in specific resistances and other traits.

Uniformity

At one time it was taken for granted that lettuce matured nonuniformly, requiring three or more harvests. At present, economic needs dictate greater uniformity and a more efficient harvest. The ultimate goal is the so-called once-over harvest.

Steady progress has been made toward this goal in the western states, where multiple harvests have been the rule. In the smaller fields and more rapidly grown crops of the East and Midwest, it has been the usual practice to harvest once. In the West, in the early years, three and four harvests per field were common, decreasing to two or three in the 1950s and 1960s and to one, two, or three at present.

Unfortunately, there have been no adequate studies separating the effects of cultivar, cultural practices, and harvesting procedures in the achievement of greater yields and uniformity. The assumption is that all three have contributed.

In our program, it has been a conscious goal to select breeding lines with greater uniformity. The procedure is simple and somewhat subjective, consisting of estimating readiness by squeezing the heads to assess their firmness. It is also possible to assess

FIGURE 12.19. Four beds of Monterey lettuce in a field of Montemar lettuce, just after harvest. The greater uniformity of the Monterey is indicated by the greater proportion of heads harvested.

firmness or head density by passing X rays or gamma rays through the head (*6,19*). The greater the attenuation of the beam, the greater the density and more mature the head. This characteristic can be used to provide a more objective breeding tool than present methods.

Uniformity is a component of yield in lettuce. Yield of the first cutting is very important, since other factors such as market price may play a part in the shippers' decision to harvest a second cutting. The loss in yield may be substantial if the first cut is, say, between 70 and 80% and the decision is made not to cut a second time. A more uniform field, resulting in a higher percentage cut the first time, minimizes the effect of the negative decision (Fig. 12.19).

A large first cutting does not always indicate greater uniformity. Some shippers will cut when 25–30% of the field is mature, in order to maintain quality and obtain a higher price. Others will wait for a higher proportion of heads to mature before cutting. This sacrifices some quality, reduces harvesting costs, and lowers the price obtainable for the crop.

Adaptation to Specific Environments

Lettuce is grown all year round in the United States, moving from north to south and back with the seasons. There are wide variations in soil types, temperature, light, day length, water quality, and relative humidity, particularly among major areas but even within those areas. The Imperial Valley of California is a case in point. The first lettuce is planted about September 15–20. Plantings are made through October and early November. Harvesting starts in mid-December and continues through mid-March. The earliest lettuce thus is planted during hot weather when day length is about 12 hr. Germination and emergence is hampered and some insect problems are severe. Day length decreases to a minimum and temperatures become lower as the plants mature. October plantings are made in cooler weather. Aphid numbers increase. Temperatures become lower, reaching the coldest levels, with chances of frost at harvest. Day length decreases and then increases. November plantings are made during still cooler weather and still shorter days. Early growth is made during the coldest period of the year and during the shortest days. The plantings mature as day length increases and as temperatures become quite warm.

Susceptibility to bolting, resistance to different diseases and insects, and adaptation to various temperature and day length sequences thus change in importance through the season. Present cultivars reflect these differences: Empire, Merit, Climax, Winterhaven, and Vanguard types are grown in that order. Each performs well in its "slot" and poorly during other time periods.

Similar differences occur within other districts, particularly those in which the growing season is long.

Therefore, our fourth general goal is the development of cultivars adapted to specific seasons and locations. This requires starting selection in early segregating generations (F_2 and/or F_3) at each location and season. Certain crosses are made with a specific problem and location in mind. Others are more general in nature, and we try to produce sufficient F_2 seed to enable us to split the lot among two or more locations.

The alternative to our procedure is to breed for wide adaptive ability. This requires the testing and evaluation of all selected materials in each location of interest in each generation. If the number of plantings over all districts of interest is large, the amount of seed needed each generation may also be very large. Few cultivars of wide adaptive ability have been developed; Great Lakes 659 is probably the best example. It has been grown in all districts in the United States in most time periods.

As a part of our procedure, the most promising lines in each sub-program are tested, starting in F_5 or F_6, in many or all locations, to determine if they have broader adaptive ability. Thus broad adaptation is considered a bonus benefit in our program.

Other Breeding Programs

Breeding programs in other parts of the United States and in other countries will, of course, vary depending upon needs as dictated by soil type, climatic conditions, various pest and stress problems, and local industry requirements.

For example, most lettuce in the Northeast and Midwest is grown on muck soils under relatively high temperatures conducive to rapid, uniform growth and maturity. Selection for uniformity is a less critical problem than in the West. Conversely, bolting resistance is more critical because of the high temperatures.

Lettuce breeding projects in northern Europe have a rather different set of goals. Most lettuce is of the butterhead type. The climate is different from that of the lettuce-growing districts of the United States. Lettuce in Great Britain, Holland, and France is grown farther north than any in the United States; therefore, summer days are longer and winter days are shorter than in corresponding districts in the United States. Also, the frequency of rains and extent of cloud cover is greater, and temperatures are lower in Europe. Several consequences result. The downy mildew organism thrives and is a far greater problem than in the United States. Breeding for mildew resistance is a major feature of nearly all programs and is complicated by a cyclical epidemiology: development of resistant cultivars that protect against existing strains of the pathogen, replacement of existing strains with new virulent strains, search for new resistances, breeding of new resistant cultivars, etc. (2).

The nature of this problem has led to a questioning of the present breeding strategy of incorporation of major genes conferring specific resistances, one at a time, into new cultivars. At least two replacement strategies have been proposed.

One is to have a series of cultivars with different resistance genes, replacing one with another when the first is attacked by the matching virulent form. Theoretically this would maintain virulent forms at low levels and prevent extensive damage from severe epiphytotics (2).

A second strategy makes use of horizontal or nonspecific resistance. In theory, this is based on the presence of multiple genes of small individual effect conferring a moderate level of resistance to the organism regardless of the specific virulent forms present. Although this type of resistance has been shown to exist for downy mildew of lettuce, it is not known whether it can be successfully exploited (3).

A second concern for butterhead lettuce breeding in Europe stems from the fact that lettuce is grown in greenhouses during the winter to compete with imported lettuce from the Mediterranean basin and the United States. A primary breeding goal is the development of cultivars with sufficient vigor to grow to adequate size under the adverse conditions of the winter season. These include short days, poor light because of the extensive cloud cover, and increasing need to reduce temperatures in order to save on high fuel costs.

Potential Goals

In addition to adding specific goals under the broad categories listed above, there is some need and potential for breeding in other broad areas. The following are three potential areas for research.

1. *Nutritional.* Variation exists among the different lettuce groups for content of several nutrients. It may therefore be possible to increase the nutritional value of crisphead lettuce through crosses with the other kinds. Little is known about varietal differences as no broad survey has ever been done. A breeding program must start with an assessment of the existing variation. To do this, and to enable the breeder to select in segregating generations, appropriate testing procedures must be developed.

2. *Energy efficiency.* Lettuce requires a large energy input for a relatively low output. Using present farming methods in the Salinas Valley, total energy input is 8 million kcal/acre. This includes machinery, fuel, fertilizer, irrigation, and pesticides. Output is only 1.7 million kcal/acre (45).

The narrowing of this difference may become possible with the development of a larger and more efficient root system, with improved plant architecture, or with selection for improved biochemical pathways.

3. *Cultivar development among non-crisphead lettuces.* Lettuce use in the United States is heavily oriented toward crisphead lettuce. The consumer's increasing interest in the other types should stimulate the breeding of cultivars with resistances, adaptability, shipping ability, and other improvements already bred into the crispheads.

TABLE 12.4. Simply Inherited Characters in Lettuce and Their Gene Symbols[a]

Preferred symbol	Character description	Reference
Seed, leaf, and bolting characteristics		
C, G	Anthocyanin formation; controls presence of red color[b]	51
R, R[s], R[bs], R[t]	Anthocyanin distribution (pattern) controlled by an allelic series[c]	21,51
Sc	*Scallop* leaf margin	32
t	Slow seedstalk formation; bolting resistance	1
u, u[o]	Leaf lobing; allelic series[d]	4,64
w	*White* seed; W confers black seed color	4
y	*Yellow* seed; w epistatic to y	53
Male sterility		
ms-4, Ms-5	*Male sterile*, genetic[e]	31
ms-6	*Male sterile*, genetic	33
Disease resistances		
bi	*Bidens* mottle virus resistance	71
Dm-X	*Downy mildew* resistance[f]	11,12,69
mo	Lettuce *mosaic* virus resistance	36
Nl-1, Nl-2	*Non-lethal* reaction to virulent lettuce mosaic strain	70
Pm	*Powdery mildew* resistance	66
Tu	*Turnip* mosaic virus resistance	69

[a]Adapted from Robinson *et al.* (*30*).
[b]Complementary factors; both present in dominant form (C/–,G/–) results in anthocyanin formation. G formerly designated as T.
[c]R, full red; R[s], spotted; R[bs], red–brown spotting; R[t], tinged with red.
[d]u, lobed; u[o], oak leaf; U, non-lobed.
[e]Act together to give 13 : 3 ratio in F_2.
[f]X = 1, 2, 3, 4, 5, 6, or 10; identified alleles for resistance.

These goals, as well as the more traditional ones, can be achieved partly through greater use of introduced germplasm. Much of this is already available, but largely uncatalogued for the genes that might be useful.

At this writing 59 genes have been identified in lettuce and closely related species (*30*). Of these, several have either been used or have potential use for breeding (Table 12.4).

Plant Genetic Engineering

A large group of techniques have been or are being developed that may have application in lettuce breeding. These include cell and tissue culture, protoplast fusion, microspore culture and other forms of haploid creation, recombinant DNA, organelle transfer, and various forms of mutagenesis. Some of these are new and some relatively old.

These techniques may be expected to provide (1) variability, some of which may be useful in breeding programs, (2) new means of rapid plant population increase, and (3) means for transfer of existing variability from wild *Lactuca* species or other composites into domestic lettuce.

SELECTION TECHNIQUES

Selection techniques in lettuce vary considerably, depending upon the character or combination of characters being selected. Single-gene characters, including certain kinds of resistance, lend themselves to selection under closely controlled conditions and limited space as in the greenhouse, laboratory, or growth chamber. More complex characters may require more complex techniques, such as alternate or concurrent greenhouse and field evaluation. Other complex characters and character combinations, such as head conformation, can only be selected in the field.

Single-Gene Traits

Examples of selection procedures for lettuce traits in this category are discussed in the following subsections.

Lettuce Mosaic

Once the source of resistance was identified and a description of the difference between the resistant and susceptible phenotypes obtained, crosses were made between the resistant wild plant introduction and appropriate adapted susceptible crisphead cultivars.

In classic backcross technique, the F_1 is crossed to the recurrent parent. The progeny from this cross is then crossed again to the recurrent parent and repeated in subsequent generations. If a dominant allele from the donor parent is to be preserved, an appropriate parent can easily be chosen from each progeny. If the desired allele is recessive, half the progeny carries the allele, and the backcross can be made blindly using several plants to ensure transfer of the allele. A safer procefure is to grow the F_2 and select the desired parent plant from among the segregating progeny.

The latter was done with the transfer of mosaic resistance. This required that we inoculate the segregating progeny and select resistant plants for the subsequent cross.

Early in the program, repeated backcrosses were made, until BC_6 or BC_7, with no attempt to select for horticultural type. However, we found that too many lines were developed that had off-type traits rendering the lines useless, as the use of F_2 segregates tended to keep undesirable genes in the population longer. Thereafter, we interrupted the backcross procedure after two and four backcrosses to grow a generation of progenies from several resistant segregates in the field to select for acceptable horticultural type. This was particularly important to eliminate non- or poorly heading plants, excessive suckering, and undesirable leaf shapes.

We also carried on parallel programs of pedigree selection. We selected resistant plants in an F_2 population or in BC_1F_2 or BC_2F_2 populations, planted progenies in the field, and selected for horticultural improvement. As resistance is recessive, the resistance allele was fixed early and it was not necessary to use the screening procedure to maintain it each generation.

In resistance breeding, the screening technique is an integral part of the breeding program. There were two reasons why field screening for mosaic resistance could not be done. First, one method of disease control in the Salinas Valley is a mandatory program, under which only indexed seed free of virus may be planted in order to eliminate primary inoculum as a source for secondary spread by *M. persicae* Sulz., the green peach aphid. While special dispensation could be obtained for experimental materials, we felt that the continuous use of infected materials in the field would be contrary to the goals of the indexing program.

Second, inoculation in the field is a tedious and unreliable procedure, and symptoms in the field are often difficult to detect.

Therefore, a greenhouse screening technique was developed. Mechanical inoculation of large numbers of plants proved unreliable and time consuming, and an aphid inoculation procedure was soon adopted (*34,36*). Green peach aphids colonized on White Icicle radish leaves were fed on infected lettuce plants and then transferred to healthy seedlings. The aphid is a nonpersistent host, and virus particles on the mouth parts are transferred mechanically to the healthy seedlings upon feeding.

Symptoms appear on susceptible plants after 10–14 days. The remaining plants have either escaped or are resistant. The latter are identified positively by appearance of a mild expression of the virus that we refer to as the "resistant symptom."

J. E. Welch of the University of California approached the mosaic resistance problem in another manner. He crossed susceptible cultivars with non-seed-transmitting materials, including lines of *L. serriola*. Segregating progenies were subjected to hard selection for zero transmission under at least two environments. This work is continuing with emphasis on elimination of certain wild characters.

Downy Mildew

Several features distinguish the downy mildew resistance programs:

1. Resistance is race specific and dominant. Resistant phenotypes are easily identified, unlike those of lettuce mosaic resistance.
2. In California, downy mildew is relatively sporadic, because the pathogen requires an environment of relatively low temperatures and high humidity, and these conditions do not occur often there. Consequently the inoculation procedure requires control of temperature and humidity. This has been obtained in the greenhouse during the winter, or more efficiently in growth rooms at any time (2).
3. In northern Europe particularly, downy mildew has been a chronic, often serious problem. Breeders have reacted until recently by developing cultivars resistant to the "races" of the organism existing at a specific time and location, only to discover that the resistance was overcome with the appearance of a new race. The work of Crute, Johnson, and others explained what may have been happening in genetic terms, but this did not eliminate the breeding difficulty. The aggressive, sexually reproducing nature of the fungus allows it to meet each new challenge by the breeder, i.e., a resistant cultivar, by producing a new genetically different form with virulence against the resistant cultivar.

A strategy of breeding for broad, race-nonspecific resistance may produce a more stable resistance than presently obtained. This would require field conditions providing continued exposure to the organism (26). In the United States, this technique might be used in Texas, where winter growing conditions are conducive to epiphytotics, but it would be less useful in California, because the environment is favorable only sporadically.

Complex Traits

When the genetic basis for a character is unknown and/or obviously complex, selection procedures may be less precise and require more steps and more time.

Big Vein

Big vein is incited by big vein agent (BVA), so called because its nature is unknown, although it has viruslike characteristics. It is introduced into the lettuce plant by a root-feeding fungus *O. brassicae*. Although the fungus is ubiquitous in western U.S. soils and occurs in other areas as well, the expression of big vein varies, particularly in the proportion of plants showing symptoms. The highest proportion and most severe expression occur when air temperature is low and when the plants are in heavy-textured, wet soils (*62,63*).

Certain strategies seemed appropriate as we began the work in 1957, at which time some of the above, meager facts were known intuitively or not at all. There was some variation among cultivars and we assumed that the differences were inherited at least in part.

We found that progenies could be distinguished from one another by the proportion of plants exhibiting the typical vein-clearing symptom at or near maturity. We also found that, in the field, these differences were most apparent in the cool spring on certain soils later designated as big vein prone. We also developed a winter greenhouse technique, using an unheated section, to confirm field results. This technique also allowed us to follow the progress of the disease and to conduct inheritance studies. In both greenhouse and field, inoculation occurs naturally if the soil is not sterilized or fumigated.

As the percentage of big vein varies considerably with the environment and as big vein is only one of several active factors affecting yield, we have employed a technique of repeated trials to evaluate both big vein and yield. We usually plant three or four trials each spring over a period of 3–5 years. Lines are ranked according to both big vein and yield. The best lines overall become candidates for release as cultivars or germplasm.

A more precise inoculation technique developed in studies of big vein by R. N. Campbell, F. V. Westerlund, and others encourages much earlier symptom expression and will permit more precision for genetic studies (*62*). This technique employs soilless culture, temperature control, and inoculation with measured amounts of zoospores of *O. brassicae*.

Insect Resistance

Breeding for insect resistance introduces another complicating factor. Unlike disease organisms, insects are capable of independent movement and may choose egg-laying and feeding sites. Resistance based on oviposition or larval non-preference (antixenosis) assumes that the insect will forego one or the other if the preferred host is absent. Unfortunately, many insects will choose the non-preferred host when the preferred one is not present and perhaps even when it is present.

As part of a program for breeding for resistance to cabbage looper, the following techniques were developed by A. N. Kishaba for screening for larval non-preference in *L. saligna* and oviposition non-preference in *L. serriola*.

Larval Non-preference   Ten first instar larvae were confined in a 6.3-cm diameter cage containing one intact leaf each of Prizehead, the susceptible check, and plants of *L. saligna*. Larvae were counted after 5 days. Plants that supported no more than one-fifth as many larvae as Prizehead were selected as potentially resistant and retested. Those supporting no more than one-fifth as many as Prizehead in the second test were selected for breeding.

Oviposition Non-preference   Plants of *L. serriola,* with Prizehead and Great Lakes as checks, were arranged randomly in three-row blocks in a screen house. In the evening, 210 pairs of previously mated moths were released. On the first and second mornings after release, the blocks and rows within blocks were rerandomized to prevent differences due to location. Eggs were counted on the third morning. Plants with 30% or less of the mean number of eggs on the checks were selected for breeding.

Resistant plants were crossed with adapted cultivars. Resistant F_1 plants were allowed to produce F_2 seed to fix resistance, followed by backcrossing to return to types with good horticultural traits.

Field-Evaluated Traits

The most important of these traits is head type. This is not a single character but a group of characters, some of which can be defined and measured precisely and others that are ill defined and difficult to measure.

The characters of interest may include most or all of the following: exterior color, interior color, size and number of frame leaves, roundness of head, leaf texture, leaf margin type, firmness, arrangement of leaves on top of head, length of stem, diameter of stem, midrib prominence, leaf base color, arrangement of interior leaves, leaf shape.

Some of these traits are in turn divisible into subtraits. Leaf shape, for example, is a function of length, width, and marginal folding and undulation. Exterior color can be various shades of green, with or without anthocyanin.

Relative weight of emphasis in making selections will depend upon personal idiosyncracies of the breeder, growers' and buyers' preferences, packing and shipping constraints, or other criteria.

Most of these traits are evaluated subjectively by the breeder. Color, general head conformation, and shape and size of frame (lower), wrapper (leaves surrounding the head), and head leaves are all gauged at a glance, and a first acceptance or rejection is made.

If accepted, we assess firmness and confirm shape by squeezing the head near the top and around all sides. Acceptance at this point is rewarded by marking the head with a colored stake.

Further acceptance requires examination of the interior of the head. It would perhaps be useful at this point to pause and describe the head that we look for in our program.

The desired crisphead lettuce type varies, depending upon the following:

1. Country or region. These reflect the needs of the industry in terms of size, color, shape, and various leaf appearance characters.
2. Time. Needs and tastes change over the years.
3. Breeder. The breeder's personal tastes are part of the input. This may sometimes have profound effect if new and different colors, sizes, or other traits become available for the breeder's choice.

At this writing, the so-called Vanguard-type lettuce has become popular in California and Arizona as the result of pioneering work by R. C. Thompson, refinement by the author, and acceptance by the western industry groups (49).

The salient features, and therefore selection criteria, of this group of cultivars are dark, dull-green exterior color; creamy to yellow interior color; spherical head shape; relatively soft pliable leaves; flat ribs; green color of the outer leaves extending very close to the stem; wavy, not serrated, leaf margins; and firm, but not hard, head.

Returning to selection of heads in the field, the next step after marking the head is inspection of its interior. This is accomplished by cutting twice at right angles across the top of the head to a depth of about 3 in. The quarters are then pulled apart and inspected for interior color, tipburn lesions, and other blemishes. Also, leaves should be arranged in a regular overlapping manner. Then, all leaves, except a few around the growing point, are stripped off in preparation for digging the plants up to be transplanted into pots in the greenhouse for seed production. In warmer areas, seed may be produced in the field, but it is too cool and windy in our part of the Salinas Valley and the plants must be moved into the greenhouse.

In the F_3 and following generations, the characteristics of the row become increasingly important in relation to those of the individual plant. These are principally uniformity characteristics and proportions of plants showing disorders such as tipburn and big vein.

Uniformity also may be thought of in terms of proportions. The breeder is looking for the rows with the highest proportions of (1) firm heads at a given date, (2) well-shaped heads, (3) heads with acceptable frame, wrapper, and head leaf distribution, or collectively, the rows with the highest proportions of heads with similar appearance and development.

Within the most uniform rows, individual plants are selected on the basis of all characters listed above, until the F_6 or F_7 generation. At this stage, we identify those rows that look most alike, are most uniform, and have the best appearance. We either mass the remnant seed for field trials the following season or mass select in those rows. (We also may mass portions of seed from single selections as early as F_4 for field trials, while continuing further single-plant selection.)

Greenhouse Production

Certain traits are particularly important for the requirements of greenhouse production. Greenhouse lettuce is grown in the late fall, winter, and early spring in areas or countries where lettuce cannot be grown outdoors during these periods. In the United States the bulk of the greenhouse industry is in several states in the Northeast and Midwest. In Europe it is in Holland, France, Belgium, United Kingdom, and other northern areas.

Two specific environmental characteristics are critical: temperature and light. In the winter, additional heat and supplemental light are often required to maintain adequate growth and development. However, both of these requirements are costly energy items, and may make the total cost of raising greenhouse lettuce prohibitive.

The alternative is for the breeder to breed genotypes with lower temperature and light (intensity and day length) requirements. This is a goal, for example, of the lettuce breeding work at the Glasshouse Crops Research Institute (GCRI), Littlehampton, England. Recent cultivars released from GCRI have had these attributes (23).

THE OVERALL BREEDING PROGRAM

For many years, the USDA had two lettuce breeding programs. One was located in the Imperial Valley and was initiated by I. C. Jagger in 1922. The work was later carried on by T. W. Whitaker and G. W. Bohn. The other was initiated by R. C. Thompson in 1938 in Beltsville, Maryland, moved to Salinas in 1956, and joined by the author in 1957. J. D. McCreight took over the desert program in 1977 and moved to Salinas in 1979.

The present program gives primary emphasis to the needs of year-round Western lettuce production but with national and international responsibility as well.

The basic materials for the two programs in the early years were old cultivars from this country and from Europe and accessions of *L. sativa* and the closely related *L. serriola, L.*

virosa, and *L. saligna.* From these, a large array of breeding lines was developed from which most the cultivars used in the United States since the late 1920s originated.

The desert program produced most of the Imperial cultivars and the original Great Lakes cultivar from which all others of this type were selected. The Thompson program produced an array of diverse cultivars, of which the Vanguard type is most notable because of its origin from *L. virosa* and its still increasing popularity.

In 1957 several cultivars had just been released or were close to release. The bulk of the remaining material was in various stages of inbreeding development. One line was released shortly afterwards, in 1962, as Francisco. The others were used as foundation germplasm as we began to introduce other materials enabling us to combine horticultural improvement with disease resistance. In 1957, the big vein program began with a cross between Merit, a moderately resistant cultivar, and 2741, a slightly resistant breeding line. In 1959 we began the search for lettuce mosaic resistance. In 1971, we found a line highly resistant to tipburn, which, with further refining, became Salinas (*43*). In 1977, we identified lines moderately resistant to sclerotinia drop and began a breeding program. In 1979, we added work on downy mildew resistance; in 1980, on corky root rot; and in 1981, on root aphid and infectious yellows.

The output from these efforts has so far been seven cultivars, 14 breeding lines, and the identification and naming of 21 genes and three linkage groups. The major events of the program are shown in Table 12.5.

BREEDING FOR MOSAIC RESISTANCE

A description of the mosaic resistance breeding program will illustrate the procedures, pathways, and complexities of a major breeding effort (Fig. 12.20).

In the 1950s, lettuce mosaic was a serious, worldwide disease problem. In 1958, on the basis of research at the University of California, Davis, by Grogan, Zink, and others (*7*), a local ordinance mandated that only seed with $\leq 0.1\%$ seedborne mosaic virus could be planted in the Salinas Valley lettuce district. This was the first step in attempted control of the spread of the virus. Later, the standard became zero infected seeds in 30,000 tested. Indexing of seed by the seed companies and later by growers' associations became standard practice in many districts and has been successful in controlling mosaic.

We decided in 1959 to begin a mosaic resistance breeding program. At that time, the potential for success of the indexing program was not known. Several years later, however, it was apparent that the indexing program was successful. Nevertheless, the resistance program was continued for several reasons. Occasional serious losses still occurred; there was room for human or sampling error in the indexing procedure; there was still a potential reservoir of inoculum among related weed species; and finally, it seemed desirable to offer more than one kind of protection to the lettuce grower.

The first step was to search for sources of resistance. This required a screening procedure, which is described earlier as well as in previous publications (*34,36*). We screened 192 cultivars, 205 Plant Introduction lines, 290 breeding lines, and 1711 segregating lines from seeds treated with X rays, thermal neutrons, or chemical mutagens.

The material treated with mutagens yielded many variants, but no resistant forms. In 1962, we found three related Plant Introduction accessions from Egypt: PI 251245, PI 251246, and PI 251247, which were resistant. PI 251245 and PI 251246 were each crossed with a crisphead breeding line. Progenies from the latter cross were pedigree selected until F_3 and discarded. Crisphead × PI 251245 progenies were selected to F_6.

TABLE 12.5. Summary of Overall Lettuce-Breeding Program at U.S. Agricultural Research Station, Salinas, California

Year	Program beginning[a]	Cultivar output	Genes and linkages identified[b]
1957	Big vein, horticultural improvement, uniformity		
1958	Genetics	Vanguard, Climax, Golden State A, B, C, D	
1959	Lettuce mosaic		
1960			
1961			
1962		Francisco	
1963	Linkage		ms-4, Ms-5, sh
1964			
1965			Sc, fr, sn, ct, cr
1966	Specific adaptation		
1967			ms-6
1968			
1969			
1970	Quantitative inheritance		mo
1971	Tipburn		Ms-7, go, vi, pa, ab, pl
1972		Monterey	
1973			
1974		Portage	
1975		Salinas, Vanguard 75	fr/w, fr/ms-6, vi/w, en, cd, al-1, al-2, al-3
1976			
1977	Sclerotinia		
1978			
1979	Downy mildew		
1980	Corky root rot		
1981	Root aphid, infectious yellows	Thompson, Sea Green	

[a]Material input: R. C. Thompson materials followed by continuous input of cultivars, PI lines, breeding lines.

[b]See Robinson et al. (30).

None of this material had both resistance and good heading characteristics, probably because there were too few favorable heading genes introduced from the crisphead parent. Therefore, the F_1 and resistant segregates from F_2, F_4, F_5, and F_6 generations of this cross were backcrossed to three crisphead cultivars: Great Lakes 118, Calmar, and Vanguard.

Several routes were pursued with the backcross material. Six backcrosses were made to Vanguard. There were no intervening generations for selection or field observation. Each successive backcross was made with resistant segregates from the F_2 of the previous backcross. After BC_6, two mass selections were made and BC_6F_3 seed was released as Vanguard 75 (Fig. 12.21).

The Great Lakes and Calmar materials were carried to BC_6F_2 and BC_7F_2, respectively, but have not been released as they do not have the desirable horticultural type of the Vanguard–Salinas group.

Some of the resistant segregates from these backcrosses were crossed with other cultivars, followed by backcrossing or pedigree selection. The latter procedure is now

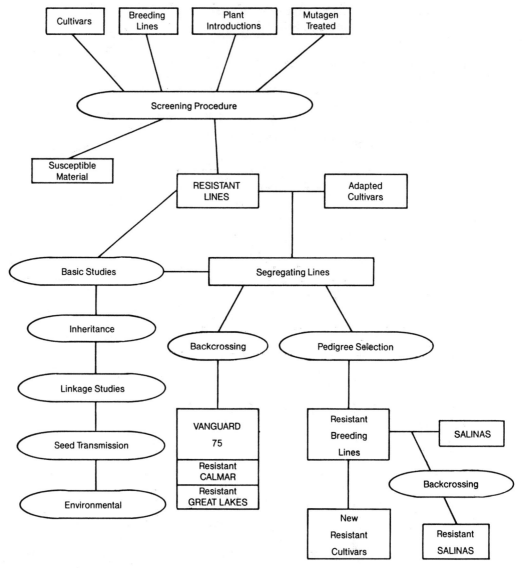

FIGURE 12.20. Flowchart showing progress of breeding program for lettuce mosaic resistance. Rounded boxes show procedures; rectangles show materials used.

focused primarily on Salinas- and Vanguard-type selections. Our aim with these materials is to create an array of Salinas–Vanguard horticultural lines with mosaic resistance to replace all existing cultivars with appropriate adapted resistant cultivars.

Additionally, we have segregating material from crosses of Vanguard 75 with big-vein-resistant lines and with downy-mildew-resistant lines to develop cultivars with combined resistances.

Beginning in 1966, we began to extract lines from the greenhouse mosaic program and plant them in the field for pedigree selection for horticultural improvement. At intervals, some of these lines were in turn used as resistant parents in crosses with other mosaic-susceptible promising lines that were prospective cultivars. Two of the latter were later

FIGURE 12.21. Vanguard 75, the first crisphead cultivar with lettuce mosaic resistance.

released: Monterey in 1972 and Salinas in 1975 (Fig. 12.22). Monterey became moderately popular, but was replaced along with others of the Calmar type by Salinas. Therefore, the backcrossing program with Monterey as the recurrent parent has been discontinued. Backcrossing to Salinas continues and eventually we plan to replace the present cultivar with a mosaic-resistant version.

Simultaneously with the breeding program we began a series of fundamental studies on

FIGURE 12.22. Salinas, the principal cultivar grown in California coastal districts.

inheritance (*35*), linkage (*40*), seed transmission of the virus (*38*), and effects of the environment on resistance (*41*).

Finally, we have made available both genetic and breeding materials to interested lettuce breeders and other scientists. These materials include officially released germ-plasm as well as small amounts of seed of breeding lines sent to other researchers at various times. The cos cv. Florida 1974 was a product of this type of material (*8*).

FIELD TECHNIQUES AND PRACTICES

Breeding material field plantings are often made under conditions paralleling commercial practices. Deviations are dictated by experimental needs or equipment availability. In California, lettuce is commercially grown on double-row raised beds on 40-in. centers. Seed is planted by machine; it is space planted or sowed and is usually coated. Lettuce is an irrigated crop. Most plantings are sprinkled until thinning and then furrow irrigated. Nitrogen and phosphorus are required but rarely potassium. For greater detail on cultural practices see Whitaker *et al.* (*67*) or Ryder (*42*).

Our breeding trials are planted either on single-row beds spaced 28 in. apart or on double-row beds. Within-row spacing is usually 10 in.; this is also the spacing most commonly used in commercial plantings. The single-row planting procedure is dictated by the availability of equipment fitted for planting of other crops. The plants are more easily inspected in single rows, which is of some advantage in looking at early-generation materials. However, the advantages of being able to evaluate plant size, development, and appearance under conditions similar to commercial conditions are great enough so that we may eventually convert completely to a two-row system.

All of our trials are planted with a Planet, Jr. Jiffy Planter, in which seed is distributed from a small hopper by a vibrating notched plate. We obtain seed separation by mixing dead and live seed in varying proportions. This technique is also useful in extending the small amounts of seed available in a breeding program.

Our standard planting contains 180 rows, 100 ft. in length, arranged in six blocks of 30 rows each. This may vary depending on available material. This size of planting enables us to observe and select in all materials within a few days. Rapid evaluation is extremely important in lettuce because the optimum maturity period is very short. Undermature and overmature lettuce heads cannot be properly evaluated for color, size, head shape, or tipburn reaction.

Plots are treated with an herbicide for weed control, appropriate insecticides for aphids, cabbage loopers, wireworms, or other insects, and occasionally with a fungicide for sporadic disease problems.

The principal pests in the greenhouse are spider mites, greenhouse whiteflies, and aphids. These are treated in a weekly spray program.

TRIALS OF ADVANCED LINES

The overall philosophy of our breeding program is that cultivars are to be developed for environments of limited scope. Adaptation on a larger scale is an unsought but acceptable, benefit.

Although our general responsibilities are national in scope, our primary geographic emphasis is the Far West. The lettuce districts in this area fall into three categories:

1. Coastal. The Salinas and the Santa Maria districts are the two most important districts. They provide lettuce from March through November, with heaviest volume from May through mid-October.
2. Desert. The three main districts are the Imperial Valley and Palo Verde Valley of California and the Yuma Valley of Arizona. These are winter districts, producing lettuce from December through April.
3. Interior. Two sections of the west side of the San Joaquin Valley, Central and Southern, produce lettuce in the spring (March and April) and in the fall (September through December).

Our breeding efforts are directed toward each type of district. Our major effort is in the Salinas district, which actually encompasses four valleys in four counties: the Salinas Valley in Monterey County, the Pajaro Valley in Santa Cruz County, the San Benito Valley in San Benito County, and the Santa Clara Valley in Santa Clara County.

As the first step in the development of lines for specific environments, we plant our breeding materials in three plantings, for spring selections (early May), summer selections (early July), and late-summer selection (mid-September). After the F_5, F_6, or F_7 generation, promising lines from the spring planting are put in spring commercial field trials the following year, and promising lines from the summer and late-summer plantings in summer and late-summer trials, respectively.

In the Salinas Valley, there is a movement of plantings from the south to the north and back to the south through the season. We try to match this pattern with our trial distribution. Our trials are planted once a month from December to August. The number of lines in each trial varies from as few as four to as many as 48. The larger numbers are planted in early stages of testing, and the actual number of lines depends upon the materials that we are currently testing. In later stages, the number is reduced to a core of lines that we then subject to repeated testing over a larger range of environments.

The number of replications varies from one to four, depending inversely upon the number of entries. Plot lengths are usually 50 ft., but occasionally are 100 ft. Overall length of the plot does not usually exceed 600 ft, to ensure not exceeding the length of the field itself and to minimize variation.

Most of our trials are planted in widths of four two-row beds, as the field planters plant four beds at a time and we therefore need interrupt the planting cycle only once to plant the trial with our own planter. Each replication thus consists of one or more four-bed blocks.

The purposes of the field trials are to measure yield, evaluate horticultural traits, and assess adaptive ability. The assessment of yielding ability is a bit complicated. Plants are either harvestable or nonharvestable. Nonharvestable plants include (1) diseased plants, (2) off-type plants, (3) late-germinating plants that will not form heads, and (4) physically damaged plants. Harvestable plants include (1) overmature plants that will be discarded, (2) mature, well-shaped plants, (3) undermature plants requiring several more days of growth, and (4) poorly shaped plants that are relatively undesirable. The percentage of plants in the harvestable category that are actually harvested is the measure of yield.

The first-cut percentage is therefore a measure of earliness and uniformity. The trials are harvested by the shipper and are as affected by the vagaries of the market situation as the field itself. The shippers vary in their harvesting policy. Some try to maximize the first-cut harvest, taking heads with substantial variation in size, shape, and maturity. Others will cut only those heads at optimum maturity. Also, if the market is poor, the shipper may be more selective in harvesting than when the market is good.

Thus, there is a certain variation in lettuce yields that is dependent upon factors other than biological and environmental ones, unlike the yields of crops like wheat or tomatoes. To compensate for this variation, our trials are planted in sets of three or four, or more. Our conclusions are drawn from the performance of lines in relation to each other and to the checks and field cultivar rather than on the percentages per se.

Also, we rely upon horticultural traits for making comparisons among lines. Broadly speaking, these traits include size; color; earliness; head shape and covering; butt characters such as ribbiness, core width, and flatness of the butt; and internal characters including stem length and interior leaf color.

Adaptation is the reaction of the line to the overall environment. A line may be poorly adapted if it is too small in a cool environment or too large in a warm environment, if it fails to form a head, if it is subject to bolting or tipburn in hot weather, or if for any other reason it fails to produce an acceptable number of harvestable heads in a given environment.

Trials of advanced lines may also be for more specific purposes. For example, we test big-vein-resistant materials under field conditions. Resistance is measured as the percentage of plants with vein clearing, the principal symptom. Big vein plantings are made in the spring each year. They are on a regular basis because natural infections can be relied upon to occur each season. Other disease resistance trials are also planted, but on a more sporadic basis.

Desert field trials are planted principally in the Imperial Valley. They are planted at intervals to mature throughout the season, and are cooperative trials consisting of lines from the USDA coastal and desert programs and from the University of California breeding program.

The purpose of advanced-line field trials is, of course, to decide which lines, if any, should be released as new cultivars. Lines that survive early field trials will remain in trial for about 5 years. Those that appear to have wide adaptation will be tested repeatedly over all the environments that we sample. This enables us to predict an environment or environments under which each candidate line is likely to perform best. The progress of development, testing, and release of a new cultivar is illustrated by the history of the Salinas cultivar (Table 12.6).

The final step before release is a seed increase. Part of the seed is used for large-scale trials and part for release to eligible seed companies. We also have the opportunity to assess whether there will be any difficulty in seed production, although this problem is unlikely to occur with lettuce (Fig. 12.23).

USDA policy is to release, without fee, lettuce seed of new cultivars on a pro rata basis to all bona fide primary lettuce seed producers based in the United States. USDA cultivars are not protected under the Plant Variety Protection Act of 1970.

This Act, which permits sexually reproduced cultivars to be protected by patent, may have profound effects on the future course of lettuce breeding. Nearly all the cultivars of lettuce currently in use were developed and released by public institutions or were selected from existing cultivars by seed companies. In general, the publicly developed cultivars are innovative in the sense that they incorporate new forms of disease resistance, are derived from wide crosses, including interspecific ones, or otherwise show a strong departure from the older cultivars. The privately developed cultivars have either been single-plant selections from existing cultivars or selections in segregating progenies from crosses between cultivars. Rarely have they been significantly different from existing cultivars.

The Act has stimulated greater interest in cultivar breeding among seed companies.

TABLE 12.6. Breeding, Selection, and Trial Sequence Leading to Release of Salinas

Phase	Year	Parents and selections	Comment
Breeding	1962	Calmar ──────── 8830	8830: *L. virosa*-derived breeding line
	1966	66-93	F_4, selected 10 plants
	1967	67-345 (66-93-8)	F_5, selected 10 plants
Selection and testing	1968	66-93-2,-8,-9	Trial of parent lines
	1969	66-93-2,-8,-9	Trial of parent lines
	1970	66-93-2,-8,-9; 67-345-1, 2,-4,-5,-6,-7,-8,-9,-10	Three trials; parent lines and selections; selection among 67-345 lines
Testing	1971	67-345-7	Found to be tipburn free in mechanical-harvest trial
		67-345-7 sublines	Summer trial to test for tipburn resistance
	1972	67-345-7 sublines	Seven trials, evaluated for tipburn and yield
	1973	67-345-7 sublines	Nine trials
	1974	67-345-7-6	Twelve trials plus eight trials by grower
Release	1975	Salinas	

FIGURE 12.23. Production of lettuce seed in the greenhouse.

Most cultivars that have been patented are still little different from existing cultivars, as the cultivar has only to demonstrate an easily identifiable difference from earlier cultivars, and this difference may be a minor one such as a few days earliness or a difference in seed color. However, as the seed companies build up staffs of trained personnel, we may expect future developments to show more radical departures from the current cultivar array.

REFERENCES

1. Bremer, A. H., and Grana, J. 1935. Genetic research with lettuce. II. Gartenbauwissenschaft *9*, 231–245.
2. Crute, I. R., and Johnson, A. G. 1976. The genetic relationship between races of *Bremia lactucae* and cultivars of *Lactuca sativa*. Ann Appl. Biol. *83*, 125–137.
3. Crute, I. R., Norwood, J. M., and Johnson, A. G. 1980. Investigations into field resistance to *Bremia lactucae* in lettuce. *In* Proceedings of the Eucarpia Meeting on Leafy Vegetables 11–14 March 1980, pp. 119–125. Glasshouse Crops Res. Inst., Littlehampton, U.K.
4. Durst, C. E. 1930. Inheritance in lettuce. Ill., Agric. Exp. Stn., Bull. *356*.
5. Ferakova, V. 1977. The Genus *Lactuca* L. In Europe p. 68. Univerzita Komenskeho, Bratislave.
6. Garrett, R. E., and Talley, W. K. 1970. Use of gamma ray transmission in selecting lettuce for harvest. Trans. ASAE *13*, 820–823.
7. Grogan, R. G. 1980. Control of lettuce mosaic with virus free seed. Plant Dis. Rep. *64*, 446–449.
8. Guzman, V. L., and Zitter, T. A. 1977. Florida 1974, cos-type lettuce breeding line. Hort Science *12*, 168.
9. Hedrick, U. P. 1972. Sturtevant's Edible Plants of the World. Dover Press, New York.
10. Jagger, I. C., Whitaker, T. W., Uselman, J. J., and Owen, W. M. 1941. The Imperial strains of lettuce. U.S., Dep. Agric., Circ. *596*.
11. Johnson, A. G., Crute, I. R., and Gordon, P. L. 1977. The genetics of race specific resistance in lettuce to downy mildew (*Bremia lactucae*). Ann. Appl. Biol. *86*, 87–103.
12. Johnson, A. G., Laxton, S. A., Crute, I. R., Gordon, P. L., and Norwood, J. M. 1978. Further work on the genetics of race specific resistance in lettuce (*Lactuca sativa*) to downy mildew (*Bremia lactucae*). Ann. Appl. Biol. *89*, 257–264.
13. Jones, H. A. 1927. Pollination and life history studies of lettuce (*Lactuca sativa* L.). Hilgardia *2*, 425–479.
14. Kishaba, A. N., McCreight, J. D., Coudriet, D. L., Whitaker, T. W., and Pesho, G. R. 1980. Studies of ovipositional preference of cabbage looper on progenies from a cross between cultivated lettuce and prickly lettuce. J. Am. Soc. Hortic. Sci. *105*, 890–892.
15. Kishaba, A. N., Whitaker, T. W., Vail, P. V., and Toba, H. H. 1973. Differential oviposition of cabbage loopers on lettuce. J. Am. Soc. Hortic. Sci. *98*, 367–370.
16. Knowles, P. W. 1978. Personal communication. University of California, Davis.
17. Koster, J. T. 1976. The Compositae of New Guinea. V. Blumea *23*, 163–175.
18. Leeper, P. W., Whitaker, T. W., and Bohn, G. W. 1959. Lettuce mildew resistant variety. Am. Veg. Grow. *7*(9), 18.
19. Lenker, D. H., and Adrian, P. A. 1971. Use of X-rays for selecting mature lettuce heads. Trans. ASAE *14*, 894–898.
20. Lindqvist, K. 1960. Cytogenetic studies in the serriola group of *Lactuca*. Hereditas *46*, 75–151.
21. Lindqvist, K. 1960. Inheritance studies in lettuce. Hereditas *46*, 387–470.
22. Lindqvist, K. 1960. On the origin of cultivated lettuce. Hereditas *46*, 319–350.
23. Maxon-Smith, J. W. 1979. Personal communication. Glasshouse Crops Res. Inst., Littlehampton, England.

24. McCreight, J. D. 1980. Personal communication. USDA-ARS, Salinas, CA.

25. Misaghi, I. J., and Grogan, R. G. 1978. Effect of temperature on tipburn development in head lettuce. Phytopathology *68*, 1738–1743.

26. Norwood, J. M., and Crute, I. R. 1981. Evaluation of lettuce cultivars for field resistance to downy mildew. Ann. Appl Biol. *97*, Suppl., 7–9.

27. Oliver, G. W. 1910. New methods of plant breeding. U.S. Bur. Plant. Ind. Bull. *167*.

28. Pearson, O. H. 1962. A simplified method for emasculating lettuce flowers. Veg. Imp. Newsl. *4*, 6.

29. Rick, C. M. 1978. The tomato. Sci. Am. *239*, (2), 66–77.

30. Robinson, R. W., McCreight, J. D., and Ryder, E. J. 1983. The genes of lettuce and closely related species. Plant Breed. Rev. *1*, 267–294.

31. Ryder, E. J. 1963. An epistatically controlled pollen sterile in lettuce (*Lactuca sativa* L.). Proc. Am. Soc. Hortic. Sci. *83*, 585–589.

32. Ryder, E. J. 1965. The inheritance of five leaf characters in lettuce (*Lactuca sativa* L.). Proc. Am. Soc. Hortic. Sci. *86*, 457–461.

33. Ryder, E. J. 1967. A recessive male sterility gene in lettuce (*Lactuca sativa* L.). Proc. Am. Soc. Hortic. Sci. *91*, 366–368.

34. Ryder, E. J. 1968. Evaluation of lettuce varieties and breeding lines for resistance to common lettuce mosaic. U.S., Dep. Agric., Tech. Bull. *1391*.

35. Ryder, E. J. 1970. Inheritance of resistance to common lettuce mosaic. J. Am. Soc. Hortic. Sci. *95*, 378–379.

36. Ryder, E. J. 1970. Screening for resistance to lettuce mosaic. HortScience *5*, 47–48.

37. Ryder, E. J. 1971. Genetic studies in lettuce (*Lactuca sativa* L.). J. Am. Soc. Hortic. Sci. *96*, 826–828.

38. Ryder, E. J. 1973. Seed transmission of lettuce mosaic virus in mosaic resistant lettuce. J. Am. Soc. Hortic. Sci. *98*, 610–614.

39. Ryder, E. J., and Johnson, A. S. 1974. Mist depollination of lettuce flowers. HortScience *9*, 584.

40. Ryder, E. J. 1975. Linkage and inheritance in lettuce (*Lactuca sativa* L.). J. Am. Soc. Hortic. Sci. *100*, 346–349.

41. Ryder, E. J. 1976. The nature of resistance to lettuce mosaic. *In* Proceedings of the Eucarpia Meeting on Leafy Vegetables, Wageningen, Holland, 15–18 March 1976, pp. 110–118. Inst. Hortic. Plant Breeding.

42. Ryder, E. J. 1979. Leafy Salad Vegetables. AVI Publishing Co., Westport, CT.

43. Ryder, E. J. 1979. 'Salinas' lettuce. HortScience *14*, 283.

44. Ryder, E. J. 1979. 'Vanguard 75' lettuce. HortScience *14*, 284–285.

45. Ryder, E. J. 1980. Lettuce. *In* Handbook of Energy Utilization in Agriculture. D. Pimentel (Editor), pp. 191–194. CRC Press, Boca Raton, FL.

46. Ryder, E. J. 1981. 'Sea Green' lettuce. HortScience *16*, 571–572.

47. Ryder, E. J. 1981. 'Thompson' lettuce. HortScience *16*, 687–688.

48. Ryder, E. J., and Whitaker, T. W. 1976. Lettuce. *Lactuca sativa* (Compositae). *In* Evolution of Crop Plants. N. W. Simmonds (Editor), pp. 39–41. Longman, London.

49. Ryder, E. J., and Whitaker, T. W. 1980. The lettuce industry in California: A quarter century of change, 1954–1979. Hortic. Rev. 2, 164–207.

50. Shannon, M. C. 1980. Differences in salt tolerances within 'Empire' lettuce. J. Am. Soc. Hortic. Sci. *105*, 944–947.

51. Thompson, R. C. 1938. Genetic relations of some color factors in lettuce. U.S., Dep. Agric., Tech. Bull. *620*.

52. Thompson, R. C. 1943. Further studies on interspecific genetic relationships in *Lactuca*. J. Agric. Res. (Washington, D.C.) *66*, 41–48.

53. Thompson, R. C. 1943. Inheritance of seed color in *Lactuca sativa*. J. Agric. Res. (Washington, D.C.) *66*, 441–446.

54. Thompson, R. C., and Ryder, E. J. 1961. Description and pedigrees of nine varieties of lettuce. U.S., Dep. Agric., Tech, Bull. *1244*.

55. Thompson, R. C., Whitaker, T. W., Bohn, G. W., and Van Horn, C. W. 1958. Natural cross-pollination in lettuce. Proc. Am. Soc. Hortic. Sci. *72*, 403–409.

56. Thompson, R. C., Whitaker, T. W., and Kosar, W. F. 1941. Interspecific genetic relationships in *Lactuca*. J. Agric. Res. (Washington, D.C.) *63*, 91–107.

57. U.S. Department of Agriculture 1979. Agricultural Statistics, 1979. U.S. Government Printing Office, Washington, DC.

58. Van der Pahlen, A., and Crnko, J. 1965. Lettuce mosaic virus (*Marmor lactucae* Holmes) in Mendoza and Buenos Aires. Rev. Invest. Agropecu., Ser. 5 *2*, 25–31.

59. Vavilov, N. I. 1935. The origin, variation, immunity, and breeding of cultivated plants. Chron. Bot. *13* (1/6), 1–366.

60. Watt, B. K., and Merrill, A. L. 1963. Composition of food. U.S. Dep., Agric., Agric. Handb. *8*.

61. Welch, J. E., Grogan, R. G., Zink, F. W., Kihara, G. M., and Kimble, K. A. 1965. Calmar. Calif. Agric. *19*, 3–4.

62. Westerlund, F. V., Campbell, R. N., and Grogan, R. G. 1978. Effect of temperature on transmission, translocation and persistence of the lettuce big-vein agent and big-vein symptom expression. Phytopathology *68*, 921–926.

63. Westerlund, F. V., Campbell, R. N., Grogan, R. G., and Duniway, J. M. 1978. Soil factors affecting the reproduction and survival of *Olpidium brassicae* and its transmission of big vein agent to lettuce. Phytopathology *68*, 927–935.

64. Whitaker, T. W. 1950. The genetics of leaf form in cultivated lettuce. I. The inheritance of lobing. Proc. Am. Soc. Hortic. Sci. *56*, 389–394.

65. Whitaker, T. W., Kishaba, A. N., and Toba, H. H. 1974. Host–parasite interrelations of *Lactuca saligna* L. and the cabbage looper, *Trichoplusia ni* (Hubner). J. Am. Soc. Hortic. Sci. *99*, 74–78.

66. Whitaker, T. W., and Pryor, D. E. 1941. The inheritance of resistance to powdery mildew (*Erysiphe cichoracearum*) in lettuce. Phytopathology *31*, 534–540.

67. Whitaker, T. W., Ryder, E. J., Rubatzky, V. E., and Vail, P. V. 1974. Lettuce production in the United States. U.S., Dep. Agric., Agric. Handb. *221* (rev.).

68. Whitaker, T. W., and Thompson, R. C. 1941. Cytological studies in *Lactuca*. Bull. Torrey Bot. Club *68*, 388–394.

69. Zink, F. W., and Duffus, J. E. 1970. Linkage of turnip mosaic virus susceptibility and downy mildew, *Bremia lactucae*, resistance in lettuce. J. Am. Soc. Hortic. Sci. *95*, 420–422.

70. Zink, F. W., Duffus, J. E., and Kimble, K. A. 1973. Relationship of a non-lethal reaction to a virulent isolate of lettuce mosaic virus and turnip mosaic susceptibility in lettuce. J. Am. Soc. Hortic. Sci. *98*, 41–45.

71. Zitter, T. A., and Guzman, V. L. 1977. Evaluation of cos lettuce crosses, endive cultivars and *Cichorium* introductions for resistance to Bidens mottle virus. Plant Dis. Rep. *61*, 767–770.

13
Sweet Corn Breeding

KARL KAUKIS
DAVID W. DAVIS

As a crop plant for human and animal uses, maize (*Zea mays* L.) has been important in the Western Hemisphere for centuries. Local peoples of Central and South America have long used maize in various forms both as a vegetable and as a less-perishable staple. Its development was interwoven with that of civilization in the hemisphere.

Beginning with a mutation of chromosome 4 at the *Su* locus of cultivated corn, humans gradually tailored the crop now known as ''sweet corn.'' Its uniqueness is manifested in accumulation of sugars and water-soluble polysaccharides in endosperm tissue that becomes translucent and brittle by the completion of maturation. Sweet corn is of relatively recent origin and is produced primarily in North America, but foreign consumption, primarily as a processed product, has increased over the last two decades.

475

BREEDING VEGETABLE CROPS

Copyright © 1986 by AVI Publishing Co.
All rights of reproduction in any form reserved
ISBN 0-87055-499-9

Because sweet corn is a genetic rather than a taxonomic entity, the literature pertaining to field corn botany, origin, development, breeding techniques, and breeding methodology can be readily applied to sweet corn. However, the breeding objectives and product evaluation procedures of sweet corn differ drastically from those of field corn with respect to product use and quality. These differences require specific evaluation techniques and approaches as emphasized in this review.

Information about the relationship between the sweet corn plant and its environment was summarized by Huelsen, who observed that little was known concerning the ecological aspects of sweet corn growing and cautioned against conclusions about climatic requirements for sweet corn based on information dealing with field corn (87).

In recent years, research work in sweet corn has become more broadly based to include the development of new endosperm types with altered chemical composition, unique taste properties, and utility for human consumption. A body of basic studies on genetic controls and pathways leading to starch synthesis in corn has had significant practical use (37,69,104,106,134).

Production information about the commercial sweet corn crop is available from various USDA summaries. Among the vegetable crops, sweet corn ranks second and fourth in farm value for processing and fresh market, respectively. The total farm value normally exceeds 260 million dollars annually from the more than 550,000 acres grown. Although only about 30% of this acreage is for the fresh market, this sector accounts for more than half of the total farm value. However, the added value that occurs through processing expands the total farm value of corn grown for canning and freezing by 300–400%.

The principal area of sweet corn production in the United States coincides with the northern limits of the corn belt. Of the acres planted for processing, Minnesota and Wisconsin each contribute about 25%. However, for the fresh market Florida is the leading state with 30% of the production, primarily contributing to fall, winter, and spring markets. Considerable acreage is grown for fresh market as a summer crop by states surrounding large population centers: New York, Pennsylvania, Ohio, Michigan, and California. In almost all states sweet corn is an important home and market garden crop.

Significant yield increases have occurred over the years. From an average yield of 4.6 tons/acre in 1970, yields for processing have gradually increased by more than 25% to exceed 5.7 tons/acre today. The corresponding increase for the fresh market has been 13%, from 69 to 79 cwt.

TABLE 13.1.   U.S. Sweet Corn Consumption and Processed Pack, 1970–1980[a]

	Canned sweet corn		Frozen			Fresh, per capita consumption (on-cob basis) (lb)
Year	Cases (millions of 303s)	Per capita consumption (lb)	Cut (millions of lb)	Corn on cob (cut basis) (millions of lb)	Per capita consumption (cut basis) (lb)	
1970	47.0	5.9	216.1	80.9	1.6	7.9
1972	53.0	6.5	273.8	133.1	1.5	7.9
1974	46.4	6.0	298.2	152.3	1.6	7.7
1976	54.7	5.7	282.4	187.6	1.5	8.2
1978	57.9	6.0	303.2	307.1	1.8	7.3
1980	50.6	5.8	270.5	258.6	1.9	7.2

[a]Adapted from U.S. Department of Agriculture (157a).

TABLE 13.2.   U.S. Sweet Corn Seed Production, 1978–1980[a]

	Production			Acres harvested		
	1978	1979	1980	1978	1979	1980
		(million of lb)			(1000)	
Sweet, hybrids	15.4	17.1	16.4	9.5	10.1	9.3
Sweet, open pollinated	0.4	0.2	0.2	0.2	0.1	0.1
Non-sweet[b]	0.5	0.4	0.3	0.2	0.2	0.1

[a]Adapted from U.S. Department of Agriculture (157a).
[b]Basically field corn (Su) types used for fresh market.

Per capita consumption of processed sweet corn in the United States has approximated 7 lb, with the strongest recent increases occurring in the frozen corn-on-the-cob sector (Table 13.1). Additional consumption of fresh-market sweet corn probably exceeds 7 lb of raw snapped product.

Single-cross hybrids account for virtually all of the sweet corn grown (Table 13.2), and more than 90% of the seed is produced in Idaho. Japan and Canada import most of the more than 3,000,000 lb of seed exported annually from the United States.

ORIGIN AND GENERAL BOTANY

Origin

The origin of sweet corn cannot be associated with a definite point in time or a single locale. Recessive mutations at what is known now as the *Su* locus on chromosome 4 may have occurred many times and in different races of corn. The adaptive characteristics and utility of such mutations determined the further spread of a type variously designated as sweet corn, maiz dulce (1), *Zea saccharata,* "shriveled corn, which is obviously sweet" (51), and others (87).

The literature on archaeological sweet corn was summarized by Carter (25). Papoon, the first sweet corn of Colonial times, apparently was obtained from Iroquois Indians in 1779 (59). The first listing of sweet corn in seed catalogs appeared in 1828, but as early as 1821 a flattering statement appeared in print: "Maize of the kind called sweet corn is the most delicious vegetable of any known in this country" (87). This indicates that sweet corn as a specific crop must have appeared by 1820 and was sufficiently popular for listing by a seedsman.

Taxonomy

Corn belongs to the family Graminaeae, subfamily Panicoideae, and the tribe Maydeae. Of interest to the breeder may be genera of *Maydae* (teosinte) and *Zea* (maize) (81). Additional information on taxonomy is available in Galinat's recent discourse on the origin of corn (64). Numerous earlier treatises are available in selected references (22,89,162).

Structure

An understanding of the structural elements of corn and their functional interrelations facilitates the work of the breeder. The literature bearing on the subject is voluminous

(*14,87,99,133*). Kiesselbach (*99*) and Sass (*133*), in particular, summarize investigations concerning the structure and reproductive processes of interest to breeding and genetic studies. The potential usefulness of our knowledge about the evolutionary change in the structure of the maize ear can be seen in work by Galinat on string cob and vestigial glume (*60,61*).

Sweet Corn and Its Relatives

We pointed out that sweet corn began as a single gene mutation in cultivated maize ($2n = 20$ chromosomes). Therefore, we can draw upon the research effort directed toward one of the most extensively studied of cultivated crops (Fig. 13.1) (*31,65*). Reviewing the origin of corn, Galinat concludes that two alternatives remain as viable options: (a) present-day teosinte is the wild ancestor of corn and (b) an extinct form of pod corn was the ancestor of corn with teosinte a mutant from this pod corn (*66*). More recently Beadle has concluded that there is no strong evidence that a wild corn other than teosinte ever existed (*10,11*).

Considerable attention has been given to perennial teosinte, *Zea diploperenis* ($2n = 20$), for long-term experimentation (*89*). Guatemalan teosinte has been suggested as a relic link in the evolution of corn (*62*). An extensive review of the races of corn was made by Brown and Goodman (*22*).

Floral Biology

The development and structure of the reproductive organs of maize have been discussed by many workers (*14,99,133*). As a monoecious plant, corn possesses staminate flowers in the tassel and pistillate flowers on the ear shoots (*99*). Anthesis in the tassel begins near the center of the spike and proceeds upward and downward, with the upper flower of the upper spikelet commencing the process first. The germinating pollen grain sends out through the germ pore a pollen tube that continues to grow down the entire length of the silk and eventually enters the embryo sac. On entering the embryo sac the end of the pollen tube ruptures, setting free the two sperm nuclei. The nucleus of one sperm fuses with the egg nucleus, forming the zygote and restoring the diploid number of 20 chromosomes to the new sporophyte. The other sperm nucleus fuses with one of the two polar nuclei and this in turn fuses with the other polar nucleus, thus establishing the primary $3n$ endosperm nucleus with 30 chromosomes (*99*). The mature kernel is a dry, single-seeded, indehiscent fruit known as a caryopsis and is composed of an embryo and an endosperm with a specialized outer layer, the aleurone. All of these tissues are enclosed in the outer layer or pericarp, a maternal tissue varying in thickness from two cell layers in teosinte, to five or more in sweet corn, and to 20–30 or more in some popcorns.

Controlled Pollination

Sweet corn is both self- and cross-pollinated. The development of inbred lines requires controlled self-fertilization to obtain the desired degree of homozygosity. The technique for controlled pollination involves (1) covering the tassel of the male to collect uncontaminated pollen, (2) covering the ear shoot before silks appear, (3) collecting pollen after silks appear, and (4) dusting pollen onto the silk of the plant selected as female and immediately covering the ear shoot (*54*). In the collection of pollen, the tassels are covered with large paper bags of heavy paper and waterproof glue. A paper clip or staple is used to secure the bag tightly around the neck of the tassel. This is commonly achieved

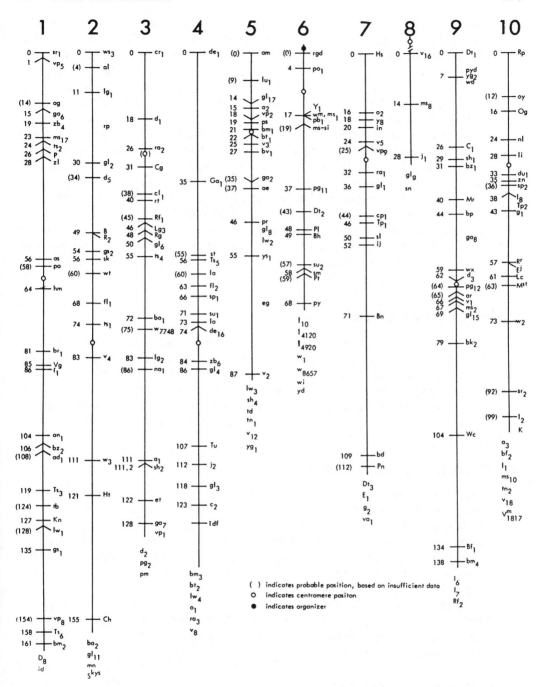

FIGURE 13.1. Abbreviated linkage map of maize.
Adapted from Neuffer *et al.* (*114a*).

by placing the edges together and folding corners before securing with staples or paper clips. When the tassel bag is removed care must be taken to avoid spilling the pollen. Pollination should be done in a manner that will not expose silk to pollen from other plants: the tassel bag is folded at midpoint and slipped over the ear shoot; simultaneously the ear bag is removed, the tassel bag straightened, and pollen dispersed over the silk by a quick accordionlike shaking of the tassel bag so that as the pollen falls it becomes airborne. The tassel bag is then pulled down carefully with one edge of the bag wedged firmly between the shoot and the stalk. The bag is folded around the stalk, stapled, and left in place until harvest. Commonly, date pollinated and pollen source are recorded on the tassel bag. It is necessary to inspect bags occasionally to ensure they are in place and that elongating ear shoots have not pushed through the bags to expose new silks to foreign pollen. Normally, 1 or 2 days before pollination, a knife is slipped under the ear shoot bag to cut the tip of the ear shoot so that the elongating silks emerge all at the same length. The resulting shoot brush of silk tips facilitates pollination. Exposure to foreign pollen is less in the pollen exchange process and the breeder can more easily direct pollen from the tassel bag toward a maximum number of silk tips. Contamination is reduced, and seed set improved.

ENDOSPERM COMPOSITION

The product for consumption in vegetable corns, as well as in most other forms of *Z. mays,* is primarily endosperm tissue. For more than half a century plant geneticists have been interested in the inheritance of variations in corn endosperm. Although gene action in controlling carbohydrate synthesis and deposition in endosperm tissue is not yet fully understood, genetic variation in endosperm carbohydrate composition has been greatly exploited by sweet corn breeders, and the potential for further emphasis in this area is high (*16,68*).

Relationship of Sweet Corn to Field Corn

The primary genetic difference between standard sweet corn and field corn involves the *sugary* (su) locus on chromosome 4, which affects endosperm chemistry. Sweet corn is homozygous for the recessive allele at this locus, while field or dent corn and other non-sweet types are homozygous for the dominant starchy allele (*Su*). Gametophyte factors, if present in an appropriate allelic configuration, can distort the segregation ratio (*114*); however, this distortion has no major breeding significance as *su* kernels can still be selected from segregating ears.

In addition to the major gene difference at the *su* locus, sweet corn and field corn also differ in many minor genetic factors that affect kernel quality. As a result, considerable effort is often needed to recover the original kernel quality present in the *su* line following a cross with non-sweet maize.

Endosperm Carbohydrate Synthesis

The biochemical effects associated with the *su* locus involve alterations in kernel carbohy-drate composition. Sweet corn accumulates more sugar than *normal (Su)* maize (Table 13.3). The primary sugar is sucrose, with lesser amounts of glucose, fructose, and maltose present (*30*). The main effect associated with *su* is the accumulation of phy-toglycogen to 25% or more of the kernel dry weight. Phytoglycogen, a polysaccharide

TABLE 13.3. **Carbohydrate Composition and Total Dry Matter in Whole Kernels of** *normal* **(***Su/Su***) Maize and Sweet Corn (***su/su***) cv. Golden Cross Bantam at Four Stages of Development**[a]

Genotype	Kernel age (days)	Reducing sugars[b]	Sucrose[b]	Total sugar[b]	WSP[b,c]	Starch[b]	Total carbohydrate[b,d]	Dry matter[e]
Normal	16	9.4	8.2	17.6	3.7	39.2	60.5	15.7
	20	2.4	3.5	5.9	2.8	66.2	74.9	27.1
	24	1.6	2.6	4.8	2.8	69.2	76.1	37.2
	28	0.8	2.2	3.0	2.2	73.4	78.6	43.8
Golden Cross	16	8.1	15.4	23.6	7.8	28.7	60.1	16.1
Bantam	20	3.2	5.5	8.7	27.0	35.5	71.3	26.8
	24	1.9	3.9	5.9	33.3	38.5	77.7	33.5
	28	1.6	1.9	3.6	34.8	33.9	72.3	37.0

[a]Adapted from Creech (*37*).
[b]Percentage dry matter.
[c]WSP, water-soluble polysaccharide.
[d]Sum of weight of reducing sugar, sucrose, WSP, and starch/dry weight.
[e]Percentage fresh weight.

consisting of glucose molecules linked by α-D-(1,4) bonds with α-D-(1,6) branch points, is the main component extracted in the water-soluble polysaccharide (WSP) fraction in sweet corn, though essentially none is present in *normal* maize (*134*).

Starch content in sweet corn is much lower than that in *normal*. Starch is composed primarily of the two glucose polysaccharides amylose and amylopectin. Amylose consists of glucose molecules linked by α-D-(1,4) bonds, while amylopectin structure is similar to that of phytoglycogen, except that branch points are not as frequent (*134*). Starch concentration in sweet corn, expressed as a percentage of dry weight, increases until around 20 days after pollination and then remains constant. However, due to further increases in phytoglycogen concentration, total polysaccharide concentration increases through 30–40 days postpollination, with the total carbohydrate concentration approaching that in *normal* kernels (*134*).

Cellular Development of Endosperm and Plastids

Previously we have reviewed the gross changes that occur in kernel chemical composition during kernel development; to appreciate fully the changes that occur, it is necessary to examine the changes in the endosperm on a cellular basis. The value of this approach can be seen when one considers that the values in Table 13.3, for example, represent the sampling of a population of cells with different physiological ages (*134*).

The morphology and development of *su* maize plastids and kernels have been described in numerous studies at the light (*17*) and electron (*8,167*) microscopic levels. This work has shown that starch granule initiation is similar to that in normal or nonmutant kernels; however, the one to several starch granules that form in each amyloplast (*167*) enlarge only slightly (*134*). Within the most mature endosperm cells, which are located in the central crown region, the initially formed starch granules are mobilized and are replaced with phytoglycogen (*17*). As a result, within developing *su* kernels, plastid types range from amyloplasts with compound starch granules, to amyloplasts containing phytoglycogen and a few small starch granules, to amyloplasts containing phytoglycogen plus many very small starch granules and/or granule fragments, and to plastids containing only phytoglycogen (*17*). The cells with the different plastid types are located in specific regions of the endosperm and are apparently related to the physiological age of the cells, with phytoglycogen plastids found in the most mature cells (*17,105*). The *su* kernels go through the developmental sequence characteristic of *normal* except that the older cells fill with phytoglycogen rather than starch. Later in development, phytoglycogen plastids in some cells appear to rupture (*17*), and kernel polysaccharide synthesis sometimes appears to be inhibited (*134*).

Differences in the extent, rate, and/or beginning cell age for starch granule mobilization and for phytoglycogen have been described in *su* material by Garwood and Vanderslice (*68*), and these authors suggest that the *su* gene may be a regulatory gene consisting of a multiple allelic series. Selection for improved sweet corn hybrids may be altering the timing of these developmental events, though further studies of different types of hybrids are needed to test this hypothesis. Ideally one would like to develop a sweet corn hybrid that has this process altered so that more sugar is retained at the market stage of development, but after that time increased amounts of polysaccharide would be formed so that mature kernel quality, i.e., seed quality, would be improved. The composition for cv. Golden Cross Bantam in Table 13.3 is representative of that occurring in sweet corn; however, it should be emphasized that considerable variation does occur among *su* sweet corn lines and cultivars for sugar, starch, and phytoglycogen, permitting selection for improved composition.

Sweet corn kernel carbohydrate composition also changes during postharvest storage (Table 13.4). These changes are detrimental to sweet corn quality; however, they can be minimized with proper postharvest storage and through development of hybrids with high-sugar genotypes as discussed in the next section.

Sweet corn hybrids with increased kernel sugar have been developed by using several different genetic approaches. The first approach involves selection within material homozygous for the standard *su* gene. Sweet corn hybrids developed in this way include Silver Queen, Mainliner, and Tendertreat (Table 13.5), which, when compared to such standards as Iochief (IA5125 × IA453) have higher sugar content (Table 13.4). The sugar levels in the best standard (*su*) hybrids are not as high as when *su* is replaced or augmented with other endosperm mutant genes.

Alternative approaches involve (a) replacing the *su* allele with its *Su* counterpart, which conditions starchy endosperm, and using a mutant gene at one or more other loci, which also alter endosperm carbohydrate composition; (b) combining the mutant form of such a gene(s) with the *su* gene; and (c) using *se* as a recessive modifier of *su*. The use of various genetic backgrounds provides further opportunity for the breeder.

Additional Endosperm Mutants and Their Usefulness

Beginning with early work by Laughnan (*106*), Kramer (*104*), Creech (*37*), and others (*24,96*), some 12 mutants had received considerable attention by the mid-1970s (*67*). Of these, *sh2, ae, du, wx,* and *bt2* have been intensively used in cultivar development, although considerable research has been conducted with various others (Table 13.6). More recently an additional mutant, *se,* has been characterized and has shown considerable potential for commercial use (*16,56*).

In their comprehensive review of the use of the endosperm mutants in sweet corn improvement, Boyer and Shannon provide the biochemical considerations as well as the

TABLE 13.4. Carbohydrate Content of IA5125 × IA453 (*su*), Illini Xtra Sweet (*sh2*), and Pennfresh ADX (*ae du wx*) after Various Intervals of Postharvest Storage at 4° or 27°C[a]

Storage treatment		Carbohydrate (% dry wt)[b]			
Time (hr)	Temp (°C)	Sucrose	WSP[c]	Starch	Total carbohydrate[b]
sugary					
0	—	14.4a	32.6d	12.1a	65.2a
24	4	10.5b	38.2bcd	7.9bc	61.7abc
48	4	12.9a	37.0cd	6.9bc	61.6abc
96	4	9.9b	42.7abc	6.0c	63.6ab
24	27	5.7c	45.2a	7.3bc	61.8abc
48	27	4.6c	41.4abc	7.4bc	57.1c
96	27	2.4d	43.3ab	9.2b	58.0bc
shrunken-2					
0	—	36.5a[d]	0.9a	2.9b[d]	45.5ab
24	4	32.7b	0.7a	5.0ab	45.5ab
48	4	30.6b	0.7a	5.9a	45.0ab
96	4	33.9ab	0.8a	5.6a	47.8a
24	27	29.0b	0.8a	5.2a	40.7b
48	27	28.2b	0.8a	8.3a	43.3ab
96	27	13.5c	0.6a	9.7a	28.1c
amylose-extender dull waxy					
0	—	24.8a	7.0a[d]	23.3b	62.0ab[d]
24	4	22.0a	1.2c	25.9ab	56.8bc
48	4	26.0a	1.3bcd	21.7b	58.1b
96	4	22.1a	1.7bd	30.0ab	61.2a
24	27	16.7b	1.4cd	30.7ab	55.2bc
48	27	12.9bc	1.6cd	34.6a	54.8bc
96	27	11.1c	2.5b	31.4ab	50.4c

[a]Adapted from Garwood *et al.* (*69*).

[b]Separation of means within a column for a given genotype using the procedure of Waller and Duncan (*159a*). $k = 100$, except as designated.

[c]Total carbohydrate is the sum of reducing sugar, sucrose, WSP, and starch.

[d]Mean separation for five designated carbohydrate component–genotype combinations based on unpaired *t* tests, 5% level.

procedures in current use for developing high-sugar hybrids, and the difficulties encountered (*16*).

Except for the utilization of *bt* and *bt2* by Brewbaker (*19,21*), the breeding strategy principally has been one of the backcross conversion of proven *su* inbreds. The converted inbreds generally are *SuSu,* but are always recessive at a second locus.

Historically, the three most successful examples of single-mutant exploitation other than *su* resulted from the use of *sh2* and the *brittle* genes *bt* and *bt2*. Illini Xtra Sweet, based on *sh2* backcrossed into the parents of Iochief, was developed more than 25 years ago and was the first high-sugar hybrid released. Since then several other *sh2* hybrids have been developed, notably Florida Sweet, based on the converted *su* inbred parents of Iobelle (*169*), and Florida Staysweet. All of these hybrids are of the genetic constitution *sh2 Su*. The Florida releases have shown improvement in the poor germination that was a deterrent to widespread acceptance of Illini Xtra Sweet. The *sh2* hybrids have established a niche in the home and market garden trade. This is particularly true in the Far East, where they dominate the market.

The *brittle* (*bt* and *bt2*) genes have been used by Brewbaker to develop the open-pollinated, tropically adapted cultivars Hawaiian Super-sweet #6 and Hawaiian Super-Sweet #9, respectively (*19,21*). These have replaced *su* cultivars in Hawaii and show promise of wider usage in tropical regions. Their high sugar content permits longer retention of good quality at high temperatures in the tropics.

Various double endosperm mutant combinations also have been exploited. Seed quality problems have limited the use of double and multiple combinations for the fresh market. In commercial breeding programs, an increasingly popular approach has been to develop hybrids in which one or both of the two genes segregate on the ear of the hybrid. Seed that gives rise to the hybrid plant is heterozygous for one or both genes and consequently germination is improved. Two examples can be cited. First, there has been the use of *sh2* or *se* segregating in an *su* background to provide a 25% (3 : 1 segregation) modification for higher sugar on the ear of the hybrid. Many hybrids released recently use one or the other of these two genes, and the success of these combinations is rather well established. The *Se/se, su/su* combination may have been available for some time in hybrids marketed as Everlasting Heritage (E.H.) types but of unknown genetic constitution (*79*). Second, the recent use of *sh2* combined with *se* to modify an *su* background in a 9 : 7 ratio giving a 44% modification can be cited as an example that, according to Courter and Rhodes, is found in the bicolor hybrid Symphony (*33*).

Hybrids said to be homozygous for *se* and *su* also have been developed (*33*). There has been much recent emphasis on exploiting *se,* a recessive modifier of *su* that was found originally in the *su se* inbred, Illinois 677a (*55*). The *su se* genotypes have a higher sugar content comparable to that of the *sh2* types, but the WSP (mainly phytoglycogen) content is similar to that of standard (*su/su, Se/Se*) sweet corn. Hence, appearance and mouth-feel are closer to that of the standard. Analysis of the sugars in *su se* endosperm has shown the presence of high levels of maltose (*55*).

Because *se* genotypes are also *su* in their genetic constitution, the transfer of the *se* modifier gene into *su* inbreds has been somewhat difficult. Selection of *se su* based on slight differences in kernel color, or dry-down rate at various stages of maturity, on ears segregating for *su Se* and *su se* is not reliable and the *se* gene may easily be lost. The need

TABLE 13.5. Total Sugar Content (% dry wt) of Sweet Corns at Various Stages of Development[a]

Days after pollination[b]	Florida Sweet	Silver Queen	H74-205	H74-204	Tendertreat	Mainliner
16	43	35	22	18	36	33
20	35	23	18	17	45	22
22	37	19	10	21	20	21
24	27	16	13	15	25	16
28	20	8	10	12	11	15
32	14	11	11	11	11	11

[a]After Hannah and Cantliffe (*79*).
[b]Sweet corn normally is harvested 18–24 days after pollination, depending on rate of development.

TABLE 13.6. Maize Endosperm Genes in Use or of Potential Use for Sweet Corn[a]

A. Individual Mutant Genes

Gene[b]	Gene symbol	Chromosome	Kernel phenotype[c]
amylose extender	*ae*	5	Tarnished, translucent, or opaque; sometimes semifull
brittle	*bt*	5	Shrunken, opaque to tarnished
brittle-2	*bt2*	4	Shrunken, opaque to tarnished
dull	*du*	10	Opaque to tarnished; S.C.[d] semi-collapsed translucent with some opaque sectors
floury	*fl*	2	Opaque
floury-2	*fl2*	4	Opaque
opaque-2	*o2*	7	Opaque
shrunken	*sh*	9	Collapsed, opaque
shrunken-2	*sh2*	3	Shrunken, opaque to translucent
shrunken-4	*sh4*	5	Shrunken, opaque
sugary	*su*	4	Wrinkled, glassy; S.C. not as extreme
sugary-2	*su2*	6	Slightly tarnished to tarnished
sugary enhancer	*se*	Unknown	Light-colored, slow-drying kernels observed in homozygous *su* lines only[e]
waxy	*wx*	9	Opaque

B. Examples of Gene Interaction in Double and Triple Mutant Gene Combinations[c]

Genotype	Interaction[f]	Phenotype
ae bt	Epistasis (*bt*)	Shrunken, opaque to tarnished
bt2 su	Epistasis (*bt2*)	Shrunken, translucent to tarnished
sh su	Complementary	Extremely wrinkled, glassy with opaque sectors
o2 sh	Epistasis (*sh*)	Collapsed, opaque
ae sh2 wx	Epistasis (*sh2*)	Shrunken, opaque
su wx	Epistasis (*su*)	Wrinkled, glassy to opaque
ae su	Complementary	Not as full as *ae*, translucent (tarnished in sweet corn), may have opaque caps
ae su su2	Complementary	Partially wrinkled, translucent to tarnished
ae du wx	Complementary	Shrunken, opaque to tarnished; S.C.[d] semicollapsed opaque

[a] Adapted from Boyer and Shannon (*16*).
[b] All gene loci are named and symbolized using the revised rules for genetic nomenclature.
[c] Adapted from Garwood and Creech (*67*).
[d] S.C.-Sweet corn background differs from dent background.
[e] Ferguson *et al.* (*56*).
[f] Gene symbol is given in parentheses.

for laboratory testing as a routine part of the early phase of the breeding program may be necessary, at least until we learn more about the genetics of *se*.

Finally, some use has been made of the triple mutant combination *ae du wx*. Culminating in the hallmark hybrid Pennfresh ADX (*66*), this combination has improved eating quality and has a reduced rate of postharvest carbohydrate transformation compared with *ae wx* used alone (*69*).

Today more than 60 hybrids employing one or more mutant endosperm genes other than, or in addition to, the *su* of traditional sweet corn are on the market and are advertised under various trade names and descriptions (*33*). This number is increasing rapidly. The need for a generic classification of these new mutant hybrids has become apparent in advising growers and seed retailers on isolation requirements to safeguard against contaminant pollination (*33*). The shrunken, brittle, and ADX hybrids must be protected from pollen produced by standard (*su*) sweet corn hybrids as such cross-pollination will make the product starchy.

Use of the various endosperm mutant genes for the development of hybrids with higher sugar levels has illustrated the difficulty that plant breeders often encounter when phenotype is rather dramatically altered, even by a single gene. Several problems have tempered the use of many of the mutant endosperm genes. Stand establishment has been more difficult than with *su* types, especially in cool soils (*4,121,128*). The somewhat poorer germination of *su* compared with *Su* (*normal* starchy) has been corrected gradually through the years, not only through improved seed production and handling processes and seed treatments, but also by continued genetic gain. Similarly, some opportunity for improvement of germination with other mutant endosperm types undoubtedly exists. One must remember that the endosperm biochemistry and the difficulties of producing, drying, and processing commercial seed of these mutants will continue to challenge the seedsman (*9*). In terms of survival capacity, the endosperm mutants are biochemical defectives. This is particularly the case with *sh2* and *bt* and with combinations including these alleles.

Pericarp thickness can increase inadvertently during development of high-sugar types as selection is practiced to overcome associated germination problems (*168*). As sugar levels increase, less storage polysaccharide is formed in mature kernels. Thus, less energy is available during germination and problems can develop. With a thicker pericarp, the kernels are less susceptible to damage and subsequent disease.

Another problem with the endosperm mutants has been their acceptance by the public. Some consumers dislike high sugar levels and object to small but significant quality differences: lighter or darker kernel color, small changes in general appearance of the processed product, and a slightly harsher texture or mouth-feel (*136*). However, these objections may be overcome by proper marketing strategy, processing technology, exploitation of genetic background, and choosing certain endosperm mutant combinations over others.

GERMPLASM SOURCES AND DEVELOPMENT

Development of Open-Pollinated Cultivars

Development of cultivars progressed at a rapid rate beginning in the late nineteenth century. Quoting Huelsen (*87*):

> Stowells Evergreen, a hybrid between Menomy soft corn and Northern Sugar, developed by Nathan Stowell of Burlington, New Jersey, is mentioned in Hovey's Magazine of Horticulture in 1851. The United States Patent Office Report of 1853 mentions two cultivars of sweet corn, Mammoth Sweet and Stowell Late Evergreen. Thereafter, the multiplication of cultivars was rapid: 6 in 1858, 12 in 1866, 33 in 1884 and 63 in 1899. Obviously, these included a number of duplications. The 1884 list contains Stowells Evergreen, Crosby's Early and Ne Plus Ultra (the forerunner of Country Gentleman), all but the last are still being grown. The 1899 list contains, in addition, Country Gentleman.

Golden Bantam, introduced in 1902, is viewed as a source from which most of our modern sweet corn releases originated. More extensive and detailed description of open-pollinated sweet corn cultivars is available from Galinat's extensive work on the evolution of sweet corn (59).

As a result of breeding efforts at the state agricultural experiment stations (AES) two fairly distinct germplasm pools arose. The first, developed in the 1920s and 1940s by W. R. Singleton and D. F. Jones, and some others in the region, could be called the Eastern pool. The second, or Midwestern, pool arose in the 1930s to 1950s from the work of G. Smith, W. Huelsen, and E. Haber at the Indiana, Illinois, and Iowa AES, respectively. In the East, emphasis was placed on earliness and on ear type for the fresh market. Earliness was also of great importance in the Midwest, but in that region greater emphasis was placed on attributes needed for processing, such as type, uniformity, and cut-corn yield. Since the 1950s, important additional contributions have been made by the private sector, with breeders utilizing both germplasm pools.

Population Development

To gain time and reduce risk, the common practice is to use germplasm best adapted to the particular geographic area. Such an approach has been successfully practiced over a long period of time. Long-range considerations, however, indicate that a broader-based gene pool may be a more promising source for extracting lines in the next step of sweet corn improvement (126,145).

Experience and theory also indicate that when several cultivars with similar yields are composited into a breeding population, the average yields of the new population will be higher than the average of the parental cultivars (146). Obviously, a number of sweet corn germplasm pools can be synthesized. Experience with field corn should be readily transferable to sweet corn source development (46,144,159).

Exotic Synthetics

Introgression of exotic germplasm has been practiced in field corn breeding (22,32,78). This experience suggests that at least several generations are required to allow the emergence of recombinations that incorporate the desired characteristic into adapted genetic systems. Currently, several public sweet corn breeding programs in the United States are introgressing field corn and selected Latin American races and cultivars into adapted sweet corn backgrounds.

Developing Breeding Populations

The theoretical basis for population development has been given (77,109,113,146), and a variety of procedures is available for the improvement of maize (50,130). This work is applicable to both field corn and sweet corn (80). Synthetic populations in sweet corn have been built around (a) elite inbreds of related origin, (b) adapted hybrids, (c) incorporation of disease and insect resistance, and (d) emphasis on quality, earliness, yield potential, row number, color enhancement, mutant endosperms, and other characters. With recurrent selection, the next cycle of selection may be initiated immediately after the first intercrossing among selected parents, but the choice of an appropriate tester and procedures will vary, depending upon objectives associated with each population. In every case, however, simultaneous selection must be exercised for yield and desired horticultural characteristics, as well as for quality attributes.

Coordinated efforts by publicly funded institutions, as in the United States by a cooperative state and federal regional project, may be the most efficient route to provide improved germplasm to the sweet corn industry for further upgrading and evaluation. Included in this effort is the development of composites designed to pool germplasm contributing to (1) high yield, (2) high quality, (3) resistance to classic diseases, (4) resistance to exotic diseases, and (5) resistance to insects. These five composites have recently been initiated and may not be directly useful for a number of years.

Various populations, composites, and synthetics have been developed by a number of state AES and by the USDA for release to commercial breeders in recent years, e.g.:

> *Hawaii AES*
> Supersweet composites based on *bt* or *bt2* or *sh2*
> High-lysine supersweet composites
> Composites having polygenic resistance to tropical diseases
> *New York AES*
> Composites with European corn borer resistance
> Composites with maize dwarf mosaic resistance
> *Minnesota AES*
> Composites with European corn borer resistance
> *USDA*
> Composites with corn earworm resistance

Line Conversions and Special Breeding Lines

To date, much of the more dramatic improvement of inbreds and hybrids, particularly for such traits as disease resistance and endosperm quality, has been attained through backcrossing. The inbreds thus converted can be readily assimilated into the commercial breeding program.

A detailed descriptive listing of public releases can be obtained from individual state AES.

MAJOR ACHIEVEMENTS

Almost all sweet corn cultivars currently available are single crosses. Hybrid usage is not new, however, since it dates from the introduction of Redgreen in 1924 (*94*). Widespread adoption of hybrids followed the introduction of Golden Cross Bantam in the early 1930s, and by 1947 hybrids accounted for more than 75% of the total crop grown (*139*). Currently, only a small amount of open-pollinated sweet corn is produced as a novelty item for home gardeners, although some recently improved types are being used in tropical regions (*18,21*). Following the success of Golden Cross Bantam, hybrids were developed in the 1930s and 1940s with varying maturities and/or with resistance to bacterial (Stewart's) wilt. Since then, new hybrids have continued to exhibit important improvements in one or more traits.

Until the 1970s perhaps the most economically significant achievement since the 1930s was the gradual improvement of standard *su* sweet corn hybrids for fresh market, canning, and freezing. Significant improvements have been made in yield, quality, uniformity of seedling vigor, and adaptation to mechanization. In the last decade more emphasis has been given to the development of bicolor, straight row white, and mutant endosperm hybrids (Table 13.7).

TABLE 13.7. Number of Sweet Corn Cultivars Released in the United States from 1924 to 1980[a]

	Yellow	Bicolor	White Country Gentleman	White Straight row	Other	Total
1924–1949	71	—	4	12	1	88
1950–1954	43	—	2	3	4	52
1955–1959	72	—	2	6	1	81
1960–1964	61	—	1	4	2	68
1965–1970	60	—	1	3	5	69
1971–1980	78	13	1	14	17	123

[a]Data prior to 1968 are from New Vegetable Varieties List XIII, as prepared by the National Sweet Corn Breeders Association and approved by the Garden Seed Research Committee, American Seed Trade Association; data from 1968–1980 courtesy of A. D. Taylor.

Evidence is accumulating that endosperm mutants will be successfully put to practical use in the near future. Although this work began long ago and the first high-sugar hybrid, Illini Supersweet (*sh2*) has been available for 25 years, the other high-sugar mutant endosperms are only now coming into prominence. Additional breeding work has been stimulated by developments such as Pennfresh ADX (*ae du wx*) (*66*), Florida Staysweet (*sh2*) (*168*), Hawaiian Super-sweet #6 (*bt*) (*21*), Hawaiian Super-sweet #9 (*bt2*) (*18*), the release of various private sector endosperm mutant hybrids, improvement of marketing strategies, and careful attention to cultural practices.

GOALS AND TRENDS IN BREEDING PROGRAMS

Current Efforts

The improvement of sweet corn in the United States emanates from programs associated with nine commercial seed firms, two processors, and 10 publicly funded institutions. According to our survey, the total sweet corn breeding effort in the United States is conducted by 102 full-time-equivalent (FTE) employees, with 26 of these at the professional level (Table 13.8).

The same survey revealed that the major breeding effort, utilizing approximately 60% of allocated resources, currently is focused on improving yellow, straight-rowed corn. A significant part of the total breeding work (10%) is directed toward utilization of endosperm mutants (Table 13.9). Breeding of straight-rowed white and bicolor hybrids receives approximately equal attention (5%) in improvement, while low priority is given to

TABLE 13.8. Number of FTE Employees Engaged in Sweet Corn Improvement, United States, 1982

	Seed industry	Processing industry	Public institutions	Total
Professionals	17.2	3.8	5.0	26.0
Technical help	32.1	5.5	8.5	46.1
Seasonal help	20.9	5.8	3.7	30.4
Total	70.2	15.1	17.2	102.5

TABLE 13.9. Evaluation of a Typical Breeding Project in Relation to Its Competitive Position, 1982

	Estimated resource allocation (%)	Seed sales volume (%)	Estimated sales potential of material in testing stage[a]	Market saturation[a]	Current market share by the company[a]
A. Hybrid development					
Yellow					
Early	15	30	5[b]	3	2
Full season	30	60	3	5	3
Country Gentleman	0	0	5[b]	0	0
White					
Country Gentleman	10	0	2	4	0
Straight-rowed	5	0	3	5	0
Bicolor	5	10	3	4	3
Endosperm mutants	10	0	4 (sh2)	2	0
se, sh2, ae, du, wx					
B. Supporting activity					
Male-sterile conversion	10				
Ga^s incorporation	5				
Breeding for specific disease tolerance	10				
Population development	3				
Other	2				

[a]Rating scale: 1, very low; 5, very high.
[b]Review objectives because of high sales potential.

the development of irregularly rowed white and yellow corns, known in the trade as Country Gentleman types.

Hybrid development has disappeared from the objectives in publicly funded programs. At the same time there is little evidence of high-risk projects or long-term objectives being pursued by the private sector. Apparently, in the public sector, other considerations have overshadowed the importance of early work, such as that done at Connecticut (*139*), Purdue (*141*), Iowa, and Illinois (*87*). This trend of divorcing academic pursuits from applied aspects is likely to be reexamined and reversed, benefiting the industry, academic training programs, and ultimately the consumer of the product.

Sweet corn is consumed as a fresh-market and processed vegetable. Most seed producers conduct programs addressing both markets, with emphasis on the processing hybrids. The programs conducted by processors focus entirely on the needs of that industry. The majority of the 10 publicly supported programs view the fresh market as their primary responsibility.

Specific Goals

Ideally, the breeder does not release a new hybrid unless it is superior in at least one respect to the hybrid it replaces and equal in all other respects. However, a number of compromises frequently will be accepted after weighing individual traits according to their perceived economic importance. For example, high yield may persuade the breeder to compromise on certain other traits, e.g., huskability. The trait compromised will vary with the intended use and marketing strategies. By rating the relative importance of commonly measured traits in hybrids targeted for fresh market and different processing styles, an illustration can be developed as shown in Table 13.10.

Expression of uniqueness in any one of the traits listed in Table 13.10 obviously enhances the potential of a hybrid. This is particularly true in respect to quality attributes, stress tolerance, and disease resistance, provided that yield and adaptation requirements have not been sacrificed.

It must be noted, however, that no concessions will be accepted in traits of hybrids intended for multiple processing styles. Ear appearance, relatively unimportant for cream-style use, will not be sacrificed for frozen corn on the cob; conversely, a low recovery, insignificant for frozen corn on the cob, will not be tolerated in whole-kernel and cream-style processes. Hybrid development for a specific processing style limits the marketing potential for the hybrid.

Regional Adaptation

Adaptation and intended use influence the choice of hybrid for a production region. For the fresh market, the straight-rowed white and bicolor types prevail in the eastern United States, while yellow hybrids dominate the Florida winter crop and the home and market

TABLE 13.10.   Relative Importance[a] of Traits of Sweet Corn Hybrids in the Fresh-Market and Processing Trades

		Processing		
Trait[b]	Fresh market	Whole-kernel	Cream-style	Frozen corn on cob
Yield				
Green weight	1	3	3	2
Weight of cut corn	1	3	3	1
Usable ears/acre	3	2	2	3
Ear characteristics				
Husk cover	3	2	2	2
Flag leaves	3	2	2	2
Ear length	2	2	2	3
Light silk color	3	2	3	3
Tip fill	3	2	2	3
Ease of husk removal	1	3	3	3
Appearance of husked ear	3	2	2	3
Color of cob (light)	3	3	3	3
Kernel characteristics				
Size	2	3	1	2
Depth	1	3	2	2
Color	2	3	3	3
Tenderness	2	3	2	2
Flavor	3	3	2	3
Silk attachment color	2	2	2	2
Black layer	1	3	2	1

[a]Rated 1, relatively unimportant; to 3, very important.
[b]Seedling vigor, uniformity, resistance to insects and diseases, lodging resistance, and stress tolerance are examples of traits that are equally important in fresh-market and processing sweet corn.

gardens in all areas except in the East. The white Country Gentleman hybrids are grown in the upper Midwest, but the processed product is marketed primarily in the South. The yellow Country Gentleman type is an insignificant component in the processing market.

Good sweet corn hybrids are not area specific. Within a wide belt across the continental United States and in corresponding latitudes in Europe, the established cultivars have performed well under widely different climatic and soil conditions (129). An extensive discussion on regionally adapted materials derived from open-pollinated cultivars as genetic sources of earliness and other characteristics was made by Huelsen (87).

Future Considerations

A breeder should maintain a proper balance between the demands of present-day markets and technology, and those changes he anticipates in both areas in the future.

Success of mutant endosperms will depend on changes in marketing demands. Work on vestigial glume anticipates technological changes in the removal of kernels from the cob (60,97). Deliberate introduction of a more tapered ear may be required to facilitate orientation of an ear in automated processing lines. The possibility of a sweet corn combine may place a premium on high-yielding, prolific sweet corn hybrids with a concomitant disregard for ear appearance, row symmetry, taper, and other currently important attributes in hybrid development. Additional quality traits, such as corn aroma, may be exploitable (57).

SELECTION FOR SOME SPECIFIC CHARACTERS

No attempt will be made in this section to enumerate all techniques and procedures the breeder can assimilate from supporting disciplines. Research done on field corn is readily transferable to sweet corn. Only a brief and incomplete listing of references can be provided in conjunction with topics addressed here.

Pest Resistance

On a world basis, maize, like other crops, has a large number of diseases (157) and insect pests (26,115). In sweet corn, losses due to insects have been established at 14% in the United States and even higher in tropical regions (20). Except for several serious problems such as corn earworm and European corn borer, sweet corn production when compared to that of other vegetable crops has not been highly restricted by major pest problems in the main temperate growing region of the United States. Consequently, much of the sweet corn germplasm currently used in commercial production, while more resistant than cultivars used in earlier decades, has only a low level of resistance to many of the pests that attack maize (137). Recently, increased emphasis has been placed on breeding for pest resistance. There are many reasons for this. Certain diseases, such as some of the viruses, have increased their range into the main production areas. In addition, the expansion of commercial sweet corn into other growing regions, particularly European, Asian, and tropical countries, exposes North American germplasm to potentially new pests. Furthermore, a gradual increase in the use of endosperm mutant types, which in general have greater stand establishment difficulty, requires greater attention to breeding for resistance to soilborne pathogens. High economic yield (as opposed to biological yield) is more difficult to achieve in sweet corn because the product is consumed directly and there are rigid consumer and/or industry specifications. In addition, stand establishment is more difficult in sweet corn than in field corn. In most of the main production regions, planting of sweet corn is scheduled over a long period of time to permit harvest extension. Consequently, emergence and development almost always coincide for a part of the season with a point in time when each pest is likely to have an advantage.

Host resistance is the most feasible long-term method for controlling sweet corn disease and insect pests. It is unlikely that a hybrid will be developed with resistance to all pests, but it is sound, routine practice to retain pest-free segregates from among horticulturally acceptable phenotypes. Specific breeding efforts are underway to incorporate resistance to one or more pests. In the United States, a coordinated effort among breeders is underway via a cooperative state and federal regional project to develop composites that carry multiple-pest resistance. Reports of pest resistance in sweet corn germplasm have increased in recent years (38,39,149).

Insect resistance, in particular, has stimulated research on genetic parameters associated with resistance to attack of the primary ear. These include number and tightness of husks, silk channel length and tightness, silk longevity, and suitablility of these tissues as substrates, not only to earworms, borers, army worms, nitidulid beetles, and other insects, but also to bacteria and fungi that may follow insect damage. At the same time, it is recognized that these characters, if modified, may undesirably influence seed set, ease of machine harvest, husk removal, and other factors that are of critical importance to the sweet corn industry.

Corn Earworm

Because of its wide geographical occurrence, the corn earworm (*Heliothis zea* Boddie) probably causes more loss than any other insect, in current principal production regions. Widstrom and colleagues identified traits associated with earworm resistance (165). Despite low heritability, low variability of temperate-zone genetic material for earworm resistance, and genotype—environmental interaction (164,165), initial gain for reduced ear damage was good in recurrent selection programs using S_1 progeny selection (165) and mass selection (161,171).

Husk length and tightness have long been thought to be associated with resistance to earworm (142). Greater husk number has been shown to result in greater resistance if husk length is sufficient to cover the ear trip (120). Selection for greater husk number has been suggested as a practical route to greater resistance to earworm and to other ear-attacking insects (20). It is an easily evaluated trait and is less influenced by environment or by ear length, taper, and tip fill than are husk length, tip cover, and husk tightness. Tandem recurrent mass selection for corn earworm resistance has resulted in good progress while maintaining resistance to northern and southern corn leaf blights, high row number, light silks, few suckers, cylindrical ears, and yellow kernels (161). This work has culminated in the release of composite 9E-79, which has both antibiotic and physical (husk length and tightness) resistances (160).

European Corn Borer

Until the past 10–15 years, the European corn borer (*Ostrinia nubilalis* Huebner), primarily a problem at latitudes above 35°, did not receive much attention in sweet corn breeding even though in the main production areas of the central and northern United States control of this insect has been a significant cost item in production. As in the case of corn earworm, these losses reflect chemical control costs, by-passed fields, more need for increased attention to ear culling and trimming in processing or in the marketing of fresh

corn, and possibly lower grades of processed product. In field corn considerable information is available on ecological aspects (*26*), insect rearing, host resistance (*73,150,152*),
and breeding aspects (*117*).

The insect in its larval stage attacks many or all above-ground parts of the maize plant.
The ultimate concern in field corn and in sweet corn seed production is the lodging and ear
drop that occur as a result of stalk tunneling followed by damage from fungi and bacteria.
In commercial sweet corn production, stalk tunneling is much less important, since the
crop is harvested while the stalks are still green and sturdy and generally before severe
stalk rot occurs. Direct damage to the ear by kernel feeding or by cob tunneling is of much
greater concern.

Low but possibly useful levels of resistance to ear feeding have been found in adapted
sweet corn (*5,149*). Breeding work has emphasized simple recurrent mass selection with
ear evaluation following artificial infestation. Evaluation methods have been developed
(*73,76*). Higher levels of resistance have been found in Zapalote Chico (*40*) and in
Antigua (*152*). Moderately high resistance to leaf feeding has been found and further
improved within sweet corn germplasm (*39*).

Seed Rots and Seedling Blights

A number of species of fungi are capable of infecting corn seed and seedlings. Most of
these pathogens become aggressive at low soil temperatures. A noteworthy exception is
Penicillium oxalicum, which can cause disease at temperatures optimal for seed germination and seedling growth (*42*). Little is known of the basis for resistance to seed rot and
seedling diseases (*157*). Research suggests that many genes are involved (*86*). Some
progress has been made in selecting genotypes more tolerant to fungal organisms. Early
nursery plantings and laboratory cold tests facilitate screening for tolerance to seedling
diseases. Seed treatment remains the most effective control method under adverse growing conditions (*119*).

Stalk Rots

Stalk rot of corn is a complex disease, which may involve several pathogenic fungi and
bacteria. Two treatises dealing with several aspects of stalk rots of corn have been
published (*29,102*). The several stalk rots are usually named after the particular incitant
organism, such as *Gibberella, Diplodia, Pythium,* and *Fusarium*. Ullstrup (*157*) indicates
that field corn lines B14, B37, Oh43, and H60 may be considered as source material for
improved stalk strength. Resistance to stalk rot apparently is inherited in a quantitative
manner (*93*). Huelsen (*87*) observed that the most susceptible inbreds are usually eliminated as a part of breeding routine in all programs.

The standability of a line is particularly important in seed production. Senescence of
stalk tissue occurs at different rates among different lines and apparently is correlated with
stalk sugar content (*35*). Methodology for inoculation in breeding for improved stalk rot
resistance is available (*170*).

Ear Rots

A number of organisms can incite ear rot in sweet corn. Most common among them are
Diplodia (*155*) and *Gibberella zeae* (*153,154*). For each, marked differences in resistance
have been observed, but we are aware of no effort to develop inbred lines with high
resistance.

Fusarium kernel rot is an important disease for sweet corn producers. In varying degrees it can be found every year as a whitish-pink to lavender mold at the ear tip. Infection often is associated with incomplete husk cover or with earworm damage. An adequate tip cover and selection for disease-free ears appear at present to be the only control measures available. There is no evidence to support use of chemicals either at edible or dry-seed stages. No specific resistance source has been reported, and quantitative gene action has been postulated (*13*).

Kernel crown spot, a bacterial disease that may attack before harvest maturity, has occasionally caused losses in sweet corn for processing (*123*). Losses are caused primarily by lower quality due to discoloration. Inheritance and genotypic sources of resistance are unknown.

Leaf Diseases

The prevalence of various leaf diseases fluctuates from year to year. Important to the sweet corn industry are the leaf blights. Northern corn leaf blight has received considerable attention and research effort. Two types of resistance have been established to the causal fungus, *Helminthosporium turcicum* Pass: (a) monogenic (*Ht*) resistance present in the field corn inbred GE440, Ladyfinger popcorn (*83*), and other sources (*85,156*), and (b) polygenic resistance. The latter appears to control the number rather than the size of lesions (*88*). The best control has been obtained by combining in the hybrid both types of resistance. The disease is easy to induce in the nursery by using ground leaf tissue collected from diseased plants in the previous year (*82*).

Southern corn leaf blight emerged in 1970 when virulent race T of *Helminthosporium maydis* Misik & Mry attacked hybrids with Texas male-sterile cytoplasm (*151*). Resistance to race T is primarily cytoplasmically controlled and the threat of disease was greatly minimized by abandoning Texas cytoplasm. Resistance to a second race O has been reported to be polygenic (*34*) and as a monogenic recessive in some germplasm (*140*).

Bacterial Wilt

Bacterial wilt or Stewart's wilt, caused by the bacterium *Erwinia stewartii*, depends upon the presence of insect vectors and their overwintering (*118*). Resistance to bacterial wilt appears to be determined by two major and one minor dominant, independently inherited genes (*163*). The release of Golden Cross Bantam, a resistant hybrid, minimized its economic importance. Disease severity is affected by imbalance in the potassium/nitrogen ratio (*143*), and winter temperatures have been used to predict disease prevalence in the following season (*148*).

Leaf Rust

Leaf rust, caused by *Puccinia sorghi* Schw., was long considered a minor disease in sweet corn. However, the disease is present every year in some production regions. More severe outbreaks in 1977 and 1978 and recent research have shown that the disease can cause serious economic damage (*75*). A hypersensitive form of resistance found in field corn has been reported as governed by a single dominant gene *Rp*, localized on chromosome 10 (*124*). Later work identified several related factors closely linked to *Rp* and two independent loci with several alleles *Rp*³, *Rp*⁴ (*84,166*). In addition to this hypersensitive or "strong" form of resistance, various sweet corn inbreds and hybrids have low but detectable levels of partial resistance (*74*). Methodology for artificially producing epiphytotics has been developed (*82*). Inheritance studies of partial resistance to leaf rust in field corn have been reported (*101*). Both types of resistance are being exploited, namely, conversion of sweet corn breeding material to the hypersensitive type of resistance and improvement in levels of partial resistance.

Smut

Sweet corn is susceptible to two smut diseases: common smut and head smut. The galls of common smut, caused by *Ustilago maydis* (DC) Cda, can appear on various parts of the plant. Reaction to ear infection seems to be inherited in the same way as total plant infection, and the F_1 is intermediate (*90*). Inheritance apparently is controlled by a large number of genes (*132*). Selection of smut-free segregates is simple because symptoms are easy to recognize. Heavy applications of barnyard manure or nitrogen, low stand densities, early plantings, and use of inoculum have been utilized to induce smut in the nursery. A monograph on the disease and its causal organism has been published (*28*).

Head smut, caused by *Sphacelotheca reiliana* (Kühn) Clint, is a relatively minor sweet corn disease, but can be seedborne. It is more frequently observed in Idaho seed production areas, and has occurred as an economic problem in California. Some striking differences have been recorded in disease resistance among sweet corn inbreds (*138*) and among hybrids (*116*). Little is known about inheritance of head smut, but seed treatment can protect seed from infection. Long-term head smut nurseries are being maintained.

Virus Diseases

The recent, explosive appearance of maize dwarf mosaic (MDM) strains A and B in major sweet corn production regions has revived interest in breeding for resistance. Resistance sources in field corn include inbreds Pa405, B68, and GA209. The procedures for developing inoculum, strain differentiation, and inoculation techniques have been described (*122*).

Resistance to MDM is conditioned primarily by a small number of genes (*43,108,112*). Commercial sweet corn hybrids with resistance to strains A and B and adapted to the main production regions are now available (*38*).

In addition to MDM, maize chlorotic dwarf and corn stunt are two more prominent representatives among nine other virus and mycoplasma diseases reported on field corn (*72,110*). Relatively little is known about the mode of inheritance of resistance to these diseases, and their economic importance is believed to be for the most part regional and relatively minor.

Stress Tolerance and Barrenness

The ideal environment for the breeding nursery should enhance the expression of characters being selected. The performance of lines per se in the selection environment should be correlated with the performance of their hybrids across the use environments; there should be good yield stability. Since production environments commonly are suboptimal for many factors affecting yield and its stability, the stress on genetic systems is more a rule than an exception. Incorporating stress tolerance is therefore an important aspect of the breeding effort.

Russell and Teich (*131*) concluded that at high populations, inbred line performance per se was at least as effective as selection by extensive testcrossing, and far more efficient. These workers also verified the important observation that under drought conditions there is a strong association between barrenness (absence of the ear shoot) and delay of silk emergence relative to pollen shedding. This and other reports suggest that one should select at higher stand densities plants that have synchronized ear shoot and tassel development (92). In some programs this has been done by leaving the first 10 hills at two plants/hill for evaluation and then selecting plants for selfing in the thinned portion of the row at stands approximating 21,000 plants/acre.

Row Number

Kernel row number influences kernel shape. Higher row number is associated with a narrower kernel, preferred in both processing and fresh-market trades. Row number is a typical quantitatively inherited character and its inheritance is governed by multiple factors (107). In a series of crosses in which parents had 8–20 rows, the F_1 was intermediate and the F_2 segregated for both parental types (48). Linkage relations have been reported between red cob and high row number and between dense, condensed tassel and the tendency for high row number (2). Selection in the F_2 and subsequent generations is effective in recovering the desired type (49).

Pericarp Tenderness

Pericarp tenderness is one of the primary quality factors in sweet corn (98). The structural features have been described and can be related to the organoleptic sensation perceived as toughness. This parameter should not be confounded with endosperm texture and kernel size. In the process of maturation, endosperm density increases and kernel size affects the total amount of pericarp tissue encountered in a subjective taste test.

Pericarp is a maternal tissue and differs widely in the number of cell layers covering the integuments. The thickness of pericarp can be measured in the dry-seed stage and measurements interpreted as indicative of a line's potential for transmitting tenderness. The procedure is relatively simple. A clinical microtome is helpful to obtain sections from pericarp tissue at a plane intersecting the tip of the germ and dorsal to it. Four or five sections can be used to obtain as many measurements per kernel. Trimming the kernel to minimize cross-secting the endosperm and clamping the kernel firmly in a specially designed clamp permits measurements in a routine manner. Crystal violet stain is recommended for a better resolution under high–dry mangification.

Several other devices have been used to measure tenderness but no method of measurement has proved to be entirely reliable (135). The mode of pericarp thickness inheritance has not been determined, but genetic advance through mass selection for tenderness has been reported (91). Commonly a tender pericarp line is used as the pollen parent in planning hybrid combinations, where a good combining inbred has the disadvantage of transmitting toughness to the hybrid.

Cytoplasmic Male Sterility (*cms*)

A combination of manual and mechnical detasseling is currently practiced on an estimated 95% of the sweet corn seed production fields and constitutes a major item in seed costs. By using *cms* the need for such detasseling may be circumvented. In the period from 1950 to 1970, *cms* in Texas-type cytoplasm (*cms*-T) came into increasing use. In 1970, a nationwide epidemic of *H. maydis* race *T* spread across host fields, causing severe losses in field corn on a national scale. Current seed production developments concentrate on creating diversity of cytoplasmic and nuclear sources (*31,45*). Extensive reviews on the subject are available (*12,44,146*).

Restoration of fertility is under the control of conventional nuclear factors. Restoration of full fertility to *cms-T* is controlled by two complementary dominant factors *Rf* and *Rf2*. Beckett reported that some additional, ill-defined factors confer partial restoration to *cms-T* (*12*). Restoration of fertility to *cms-S* is conferred by *Rf3* in a homozygous state (*23*).

A procedure for developing both male-sterile and fertility-restoring strains of an inbred line was proposed by Eckhardt (*47*). The procedure eliminates testcrosses that are required if normal cytoplasm is used in the transfer of the fertility-restoring gene. The basic starting material is the F_1 cross between the cytoplasmic male sterile source and the fertility restorer source. This fertile F_1 carrying male sterile cytoplasm is used as a female parent in the cross with an inbred that has the fertility restorer factors added. Backcrossing is now carried out using the fertile male-sterile-cytoplasm segregates as female and the inbred as recurrent parent. After adequate backcrossing the following occurs:

1. Fertile segregates can be selfed and the homozygous fertile line selected. This is the inbred line with restorer factors added.
2. Sterile segregates can again be crossed with the inbred. This is the male sterile inbred line.

Complications that may develop in this scheme are discussed by Duvick (*44*).

Genetic Male Sterility

A genetic system to produce hybrid corn seed without detasseling by utilizing the close linkage of the genes *y* (white endosperm), and *ms* (male sterile) was suggested but was never used because of contamination from 5% recombination. Galinat, however, described a two-step seed production system that resolved this problem by using electronic color sorters to separate yellow kernels from white (*63*). This approach has not been utilized commercially.

DESIGNING THE BREEDING PROGRAM

It has been said that plant breeding encompasses elements of both science and art. A third aspect may be added, one that deals with efficiency in execution, vision about future needs, and a prudent dissemination of achievements. By its nature any breeding program requires long-term commitment and the return on investment has a long lead time. The primary goal of the breeding program is improvement of performance in relation to economic use. Recognition of these facts is necessary for all concerned, beginning with the project leader and ending with those who administer the resources. These realities also mandate development of a well-prepared plan. It must be acknowledged that the plan has some flexibility as parameters change.

The Planning Phase: A Case Study

Time devoted to program planning may be one of the most efficient ways to utilize professional skills. The topics listed at the top of the next page need to be considered sequentially, and to some extent, in tandem.

1. Development of objectives
2. Estimation of resources
3. Outlining of procedures
4. Anticipation of problems
5. Selection of best alternatives
6. Design of the total program

The steps in program planning will be illustrated by using a hypothetical program supported by a seed company in which the sweet corn breeder is the project leader, whose assignment, according to the job description, is to develop a program that will provide hybrids for national and international marketing needs. The project is an umbrella for breeding different corn types, including endosperm mutant types, and is responsive to both processing and fresh-market requirements. The breeder is also responsible for the design and the supervision of a testing program. The breeder prepares and submits budgets, compiles and interprets trial data, and participates in introducing new releases into marketing channels. The program, in progress for several years, is funded independently of sweet corn seed sales. As a part of cyclic corporate review, the breeder is required periodically to reevaluate the project, state objectives of different activities, and provide rationale for allocating resources within the project. It is apparent that we have a busy project leader who, though concerned with current activities, must also be ready to project into the future. The breeder must thoroughly understand the industry and the public it serves.

Objectives

The general scope of the project must be summarized as a brief, all-encompassing statement, e.g.: "The goals of the sweet corn breeding project are to develop hybrids that will (1) maximize profitability, (2) stabilize production by reducing vulnerability to diseases, insects and climatic changes, and (3) provide a product with some distinctive marketing attributes."

The next step in the analytical process is to assess project goals in relation to the current and desired competitive position in the marketplace. Although aversion to structured analysis is the common reaction from most people, judgments such as those represented in Table 13.11 have to be made at some point in an integrated decision-making process. A cursory view of this table points to two easy targets for review. First, early hybrids are a highly marketable commodity in a relatively saturated overall market; and second, a yellow Country Gentleman type of hybrid can be a spin-off from breeding for white Country Gentleman types. The yield potential and high quality of Country Gentleman types would place such a hybrid in a favorable marketing position. Consequently, a new program designed to develop yellow Country Gentleman corn will need a definition of objectives.

A detailed and specific statement is prepared for each subproject. It may be advisable to remember that most goals will be met if the new hybrid equals in all respects the one currently used and exceeds it in at least one measurable parameter. Success is most likely if a step-by-step process is used, where each new hybrid represents some improvement. As an example, the new objective for breeding an early hybrid is restated as: "To develop a corn hybrid equal to the standard (Hybrid 207) in yield and quality but which reaches 74 percent moisture 2 days or 40 heat units earlier and which has better seedling vigor. The hybrid must be acceptable to the canner/freezer trade. Schedule completion in 5 years."

TABLE 13.11. Summary of the Scope of a Comprehensive Sweet Corn Breeding Program

	Activity	Estimated resource allocation (%)		Reasons for change
		Current	Future	
A.	Hybrid development			
	Yellow, early	15	20	Anticipated marketing of hybrid 207; new program for potential replacement
	Full season	30	35	Extensive sampling of new High-Yield synthetic; increased testcross program
	Country Gentleman	0	2	Outstanding yellow Country Gentleman for market development
	White Country Gentleman	10	10	Infuse new germplasm to create a more distinctive hybrid; a narrower kernel is required
	Straight row	5	5	Maintain program with emphasis on a better quality early hybrid
	Bicolor	5	10	Marketing pressure
	Endosperm mutants	10	5	Discontinue work on $su1$, $su2$, and su du; concentrate on ae du wx for fresh market, and on su se, $sh2$
B.	Supporting activity			
	Male sterility	5	5	Eliminate conversion of obsolete lines and C cytoplasm
	Ga^S	5	0	Program has not succeeded in Ga^S transfer to major lines; discontinue work
	Disease tolerance	10	5	Reduce MDM program; eliminate insect nursery; continue test for $H.$ $turcicum$
	Population development	3	1	Support work in university programs; maintain early synthetic
	Other	2	2	Evaluation of cold-tolerant material and USDA Plant Introductions

Specific goals must be realistic. They are not public relations statements, and should not frustrate a profession that has made its contribution but is not immune to wishful demands. Administration or management must be made aware of what progress is probable or possible within alternative time and resource frameworks.

It is advisable to develop a goal statement for each activity. Expectation of how much time a professional can allocate to freelance exploration should be considered in advance. Such research may be the most rewarding aspect for both the project leader and the funding organization.

Estimation of Resources

We have assumed an atypical project, one not restricted by funding. At what point the project rests on the curve of diminishing returns is for the most part a matter of judgment. The obvious question concerning what may happen if additional resources are applied has to be viewed in the light of cost–benefit considerations.

Outlining Procedures

Developing and outlining procedures depends on objectives and resources. For example, a project to convert an inbred to a male-sterile restorer genotype must be viewed from the perspective of the promise and utility of the inbred, the time conversion will be completed, impact on seed production costs, and how close the inbred may be to obsolescence. There is need to consider a winter program and adherence to a strict timetable. As a second example, in breeding for disease and insect resistance the project leader may have to consider equipment, inoculation procedures, choices of location, and—not least—the leader's own skills before embarking on a full-scale program.

Anticipation of Problems

Anticipating potential problems releases the pessimistic component in a breeder's personality. The many reasons as to why the program may be impeded does not affect the one positive statement embodied in the objective: "it will be accomplished." The anticipation of potential hurdles, problems, and detours is a useful exercise as long as it does not develop an attitude of defeat. Breeding as a profession thrives on optimism.

The hurdles may reside in budgetary underestimates, improper choice of parental material, unforeseen changes in technology, changes in market demand, and lack of familiarity with the problem at the planning stage. The recognition of potential difficulties may lead to the best choices for reaching the stated objective.

Selection of Best Alternatives

Experience and intuition are commonly applied to evaluate various alternatives. Elements embodied in a more formalized evaluation scheme are given in the following simplified example:

Alternative	Time required	Cost estimate	Likelihood of success
1	Short	High	High
2	>2 years	Low	Uncertain
3	>3 years	Low	Likely

The obvious trade-offs in such an exercise force analyses of additional inputs: Can the alternative be "sold"? Does it have desirable side effects? Does it meet resource constraints?

Design of the Total Program

The development of the program approach can be illustrated by using as a breeding objective the development of an early hybrid, as stated earlier. Before developing procedures the breeder catalogs the genetic resources available. Let us assume that Hybrid 207 is a cross between a midseason inbred A used as a seed parent, and an early inbred B. The other inbreds available in the program have not indicated yielding potential in combination within the targeted maturity range. As source material, the program has maintained an early, high-row synthetic; but selection pressure for greater yield and higher ear placement has lengthened maturity. The decision is made to modify both parents by crossing the seed parent inbred to two or three earlier inbred lines known to combine well with the pollen parent inbred, and to modify the latter by introducing improved early vigor via the backcross procedure. The key consideration is whether this modification will achieve the desired improvement in earliness and retain good combining ability for yield and other characters.

Considering the importance of the project, the breeder decides to start the work with three different populations and to concentrate on the more promising one or two after comparing and evaluating the source material.

Based on these considerations, the breeder outlines the following procedures:

Current Situation

Hybrid 207

Inbred A (seed parent): high yield; long ear; late; good kernel type; tough pericarp; good plant type   ×   Inbred B (pollen parent): early; starchy; tender pericarp; low ear placement; wide kernels; poor seedling vigor

Potential Sources for Improving Inbreds A and B

Inbred 101, 102, 103: expected to combine well with line B; 3 days earlier than A

Inbreds 201, 202, 203: expected to combine well with line A; 1 day later than B; vigorous in cold soils

The relative value of above sources will be established from testcross performance as outlined below. The results will determine which line will be emphasized in further improvement work.

Breeding Outline

Year 1—Summer

(a)   Initial improvement crosses:
A × 101, 102, 103: To develop an improved "yield" line (A).
B × 201, 202, 203: To develop a new early line with improved seedling vigor in cold soil and higher ear placement.

(b)   Testcrosses (make two crosses in each combination):
101, 102, 103 × B
201, 202, 203 × A
Purpose: To grow F_2 yield test along with F_2 populations in Year 2. It is assumed that yield test will indicate the best combining source with the "new" improved line.
Further breeding efforts will be emphasized in the corresponding improvement cross.

Winter Program—F_1

Grow 20 plants from each improvement cross; self five and harvest all; grow 10 F_1 plants from each testcross; harvest three ears from sibbed plants; bulk seed for planting replicated test in year 2.

Year 2—Summer—F_2

(a)   Improvement crosses: Plant early. Grow maximum plant population, i.e., 600–1000 plants/cross. Self and date 100 early segregates. At harvest visually select among selfed plants for ear type, ear height, standability, freedom of disease, and plant characteristics. Grow frequent parent checks. In F_2 populations involving line B compare populations for seedling vigor and cold tolerance. Backcross five or six earliest segregates from each F_2 population to the recurrent parent.

(b) Grow testcrosses in a replicated yield test. Harvest at an estimated 73% moisture. Obtain data on earliness, seedling vigor, plant and ear type. On basis of testcross performance reduce F_2 breeding load to 1 or 2 populations in each A and B group. Include Hybrid 207 as a check.

Winter Program—F_3

Self and testcross F_3 plants to the appropriate A or B tester as a silk parent. Identify each self with its testcross. Attempt to make 20 such crosses advanced to F_3. Make five selfs in each F_3 progeny row. Harvest all selfs and crosses.

Year 3—Summer—F_4

(a) Plant early. Grow F_4 progeny rows. Evaluate on an among-F_3 family basis and particularly on testcross performance. Make decision about bulking material and move toward a seed increase program. Consider bulking the selfs within an F_4 progeny row if F_3 family shows phenotypic uniformity.

(b) Plant 200–400 plants of each BCF_2 population. Judge F_4 lines for phenotypic uniformity. Handle as the original F_2.

(c) Place testcrosses of $F_3 \times$ tester in a single-plot trial. Obtain data on yield and quality, earliness, and agronomic type. Use frequent checks.

(d) Reduce the nursery to a manageable size. Review objectives and discard any material outside the tolerance limits for the objective.

(e) Emphasize the BC_1 population. Use BC_2 if some critical characteristic has not been adequately transferred into recovered material.

Winter program—F_4

Increase seed and make crosses as needed to meet testing requirements. Intercross the most promising selected improved A and B lines (now F_4) for a performance trial in year 4. Use results of that trial to make major decisions about advancing a small number of test hybrids. Oustanding hybrids may merit advanced testing but a conservative judgment often will prove to be prudent. Review objectives to meet the next challenge.

Typical Impediments

In program development the primary concerns deal with the reliability of measurements used in decision making and the frustrating suspicion that some goals may be mutually exclusive in their realization.

To illustrate the latter, one may reflect on pericarp tenderness and wonder to what extent it affects quality of the seed and seedling vigor. Is the undesirable, starchy *Su* endosperm related to the early seedling vigor? In selecting for one trait we may be counterproductive by shifting some other trait in an undesirable direction.

Addressing the problem of measurements, how reliable are our yield estimates in single-plot evaluation? Should an extensive and expensive testcross program be replaced by evaluation of S_1 lines per se?

The Comprehensive Program

It is doubtful that the formalized structure as outlined in Table 13.11 for illustration is in common use. For the most part it exists as a project leader's mental concept translatable into a nursery plan. Obviously it can be reconstructed at the end of the season by adding the number of selfs, crosses, and plots utilized in any particular project. Although the project leader cannot be expected to develop a lengthy dissertation on all facets of the project, every new and changed activity should be supported by a statement of objectives and procedures.

Breeding work can be conducted in many different ways. The best results will be attained by a project leader who is disciplined to analyze all aspects of the project in an organized manner that will lead to an efficiently designed breeding program.

HYBRID EVALUATION

Efficient evaluation of hybrids is vital in any sweet corn breeding program. Newly developed hybrids must be carefully compared against standards for yield, quality, adaptation to processing or fresh-market needs, agronomic type, environmental adaptation, and seed production requirements. The testing phase can be the most expensive part of the program. An important issue that will be forced quickly upon the breeder is the degree of testing sophistication appropriate to each stage of hybrid development. Each stage consists of a sample from which a subsample is elevated to the next stage. A cursory evaluation at earlier stages permits the screening of larger numbers of combinations. Finally, in the evaluation of advanced hybrids the cost per entry reaches its peak.

Throughout this scenario, a decision must also be made regarding appropriate number of locations, population densities, and crop cultural regimes. Application of modern field plot techniques is required for the efficient conduct of performance trials and intelligent interpretation of results. A body of information is available on this subject (6,125,127,147). There is, however, little published data on the efficiency and precision with which sweet corn hybrids can be evaluated. Information available from other crops is not transferable to sweet corn, because of the nonuniformity of maturity and quality at harvest, as will be shown later in this section. There are, however, numerous reports that describe the response of sweet corn hybrids to variations in plant arrangement and population (3,15,111); and such variables as irrigation and/or plant nutrition (27,52,58,158). Methods have been developed for timing of harvest (41,71), and for evaluating product quality, particularly for the processing sector (70,103).

Locations and Replications

The soil, climatic, and growing conditions in the test field should be similar to those under which the crop will be grown commercially. It is important that within individual replications the soil is uniform and that variations in soil structure and fertility, previous cropping, and topography do not affect yields.

Increasing the number of replications usually reduces experimental error much more effectively than increasing the plot size. The mean of several replications provides a more accurate measure of the performance of an entry than does one plot. In general, the accuracy is increased in proportion to the square root of the number of replications. Uniform replicates differing in fertility or rate of planting may add to the information obtained from the test. Different planting dates of one or two replicates per date may be used to substitute replicates at one date.

Size and Shape of Plot

Size and shape of plot affect test accuracy. Increasing plot size within certain limits usually results in decreasing error. Increasing the plot size also increases within-replication land area, resulting in greater heterogeneity within the replication. Actual plot size usually is a compromise between these two opposing tendencies. The single-hill method, based on the principle that F_1s from homozygous parents are uniform, was explored in testing field corn hybrids. Intriguing from the standpoint of efficiency, the method requires perfect uniformity of stand, and of soil and seed quality. It has been tested by some sweet corn workers but has not been used for critical measurements in commercial programs.

Replicated testing of advanced hybrids usually involves, as a minimum, replicated single-row plots, but more likely a three to five-row plot, with the inner rows used for data on such characters as yield, quality, lodging resistance, plant and ear height, and general response to environment. Multiple-row, replicated testing on a split-plot basis across population and/or crop cultural regimes has been practiced in the breeding programs of the processing industry, but probably has been less common in the seed and fresh-market sectors.

Controlling Competition

Border effect is important in sweet corn performance trials. Competition between hybrids differing in stature and maturity can be controlled by subgrouping hybrids similar for such characters and by planting appropriate border rows. Traditional (*su*) programs have ignored the xenia effect, which has been considered significant by some workers in both sweet corn (*7*) and field corn (*100*) evaluation. Current broadening of many programs to include additional mutant endosperm genes will demand more attention to controlling pollen contamination.

Factors Affecting Testing

We have noted that because the two crops are similar with respect to general breeding strategy, sweet corn workers have drawn heavily on the work done by field corn breeders and agronomists. The problems encountered in testing are, however, entirely different and more difficult to resolve in sweet corn than in field corn. These difficulties arise from numerous differences between the two types (Table 13.12). Sweet corn breeding and testing must address these differences.

Nonuniformity of Sampling

The number or proportion of ears at the same maturity varies considerably from sample to sample at harvest. This is a consequence of lack of uniformity of silking time as

TABLE 13.12. Relative Importance of Characteristics in Comparisons among Sweet Corn and Field Corn Hybrids

Characteristic	Sweet corn	Field corn
Uneven germination or plant spacing	Critical influence on uniformity of maturity	Some importance
Maturity at harvest (% moisture)	68–78	13–18
Perishability	Extremely rapid	Stable
Second ears contribute to yield	Partially	Fully
Shape of cob	Critical	Unimportant
Appearance of ear	Critical	Unimportant
Utilization of growing season	Partial	Full
Quality evaluation	Difficult to measure	Easy to evaluate
Adaptation to processing requirements	Many considerations	Few critical
Seed production	Difficult	Relatively routine

TABLE 13.13. Silk Emergence (% of Plants) of Three Sweet Corn Hybrids in Two Testing Environments

Hybrid	Test	Day of silk emergence						Not silked
		1	2	3	4	5	6	
A	1	3	12	23	17	20	16	9
	2	2	18	31	17	17	8	7
B	1	1	6	20	32	18	11	12
	2	2	5	26	20	27	13	7
C	1	1	2	12	30	14	14	27
	2	1	2	18	45	22	7	4

illustrated in Table 13.13 by silk counts on 200 plants grown under uniform conditions in a test thinned to obtain reasonable uniformity of emergence and spacing. Table 13.13 shows that from 4 to 27% of the plants failed to silk after a 6-day period of silking, and the modal silk day was either the third or fourth day after the first silks emerged. The variability is also reflected in the harvested sample.

Seasonal Effect

The effect of planting date on hybrid yields can be seen from Fig. 13.2. Because of adverse climatic conditions, the recovery of cut corn usually is reduced in late plantings in the Midwest. In this particular test it was accentuated by an outbreak of maize dwarf mosaic prevalent in plantings after June 3. Such sequential planting also reveals hybrid A as better adapted to late plantings or, possibly, as possessing better tolerance to MDM than hybrid B.

Important Characters

Performance trials must address the needs of different segments in the industry (Tables 13.10 and 13.14). These needs may differ considerably between the fresh-market sector and the different segments of the processing industry (Table 13.10). The usefulness of the characteristics listed in Table 13.14 depends on the segment of the industry for which the evaluation is being conducted.

Designing the Testing Program

Let us assume that we are designing a testing program to identify superior hybrids for a canning–freezing operation in a relatively narrow latitude belt. The program over the years has undergone several changes reflecting uncertainty about which procedures furnish reliable results efficiently. The dispersal of tests is guided by the perceived magnitude of genotype–environment interaction in relation to the experimental error.

Further assume that most of the hybrids for advanced testing originate from the breeding project. Outside entries are periodically compared in a special trial, and those approximating program needs are subsequently tested in the advanced trials. A suggested testing sequence by years is given on the next page.

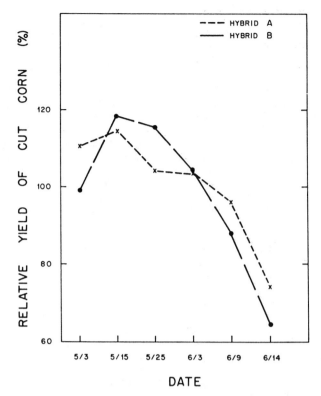

FIGURE 13.2. Cut corn yield of two hybrids planted at six planting dates at Le Sueur, Minnesota, in 1978, and harvested at the same maturity. Yield is expressed as percentage of average yield for the year.

Year 1: Single-plot tests are grown at one or two locations. Hybrids for further testing are retained after selection for yield, quality, and adaptation to process and market requirements.

Year 2: A preliminary trial of 30–50 entries is conducted at one location for yield and quality evaluation. The test consists of three replications and two harvests in each plot. A single-row, nonreplicated observation planting may be made for observing hybrids under irrigation or a particular growing condition.

Seed for the next testing stage is obtained either by making crosses concurrently with the test or by utilizing a winter program to make crosses for those hybrids to be advanced or retested.

Year 3: Advanced testing includes yield trials at two locations on soils believed representative of the general growing area. Two plantings, one early and one mid season, are made in three replications at each location. Harvests are made at 74% moisture; a quality test is conducted at one location; and a single-plot planting from which six harvests are made beginning with 78% moisture and reaching 68% at 2% intervals.

Year 4: The harvest range and testing scope may be expanded for hybrids in the fourth year of testing and targeted for production use. Special tests are made, such as testing for cultural practices, stress tolerance, disease resistance, and observation plant-

TABLE 13.14. Measurements Used to Compare Sweet Corn Hybrids for Fresh-Market and for Processing Trades

	Unit	Description
Measured yield values		
Green weight of sample	lb	Green-harvested sample wt
Cut corn weight	lb	Cut, but unwashed
Calculated yield values		
Green weight		Expressed as mwt/A or tons/acre
Cut corn		Expressed as cases/acre or cwt/acre
Cases/mwt		Based on pounds of cut corn per standard case
Recovery		% of cut corn of harvested green wt
Ear values		
Weight of usable ear	lb	
Length of usable ear	in.	Av. estimate for sample
Diameter of usable ear	in.	Av. of 10 ears
Usable ears/plant, av.	no.	May express per 1000 plants
Av. wt of harvested ear	lb	
Av. wt of usable ear	lb	
Modal row number	no.	
Silk color (rating)	no.	Rate, dark vs. light
Appearance (rating)	no.	Poor vs. good
Uniformity (rating)	no.	High vs. low
Tip fill (rating)	no.	Poor vs. good
Worm damage	%	Of usable
Huskabiity	%	Of ears requiring rehusking
Broken in husker	%	Of usable
Nonusable	%	Of total harvested
Smutted	%	Of total harvested
Shape (rating)	no.	Tapered vs. cylindrical
Curvature (rating)	no.	
Plant type		
Seedling vigor (rating)	no.	
Maturity to 50% silk	no.	Heat units or days
Plant height	ft	Av. for plot
Ear height	in.	Av. height
Lodging	%	
Tillering (rating)	no.	None vs. much
Husk cover (rating)	no.	Indicates short husk
Shank length	in.	
Flag leaves (rating)	no.	Long vs. short
Branch ears (rating)	no.	None vs. much
Seed production		
Adaptation of seed parent	no.	Integrated rating
Expected seed yields	%	Estimated yield in % of check
Seed parent tillering rate	no.	None vs. much
Stalk strength (rating)	no.	At harvest
Required spread planting	HU	Or expressed as stage of growth
Quality		
Kernel type (rating)	no.	
Color (rating)	no.	
Tenderness (rating)	no.	
Flavor/taste (rating)	no.	
Quality grade (rating)		Integrates all quality attributes; usually related to maturity.
Maturity at harvest		
Moisture	%	Determined on cut corn

ings in some appropriate production location. These trials are focused on material in a prerelease stage and designed to collect specific information. Evaluation for disease reaction usually is done in a special test after artificial inoculation with disease organisms. Frequent checks of known susceptible and resistant hybrids facilitate evaluation of material. The reading of disease prevalence and severity is made at a time appropriate to discern differences among entries on the test. Occasionally disease and insect resistance may be evaluated in an environment where a particular pest is endemic and artificial inoculation is not required.

In the processing sector, yield tests typically are planted in 30-in. rows. Stands might be thinned to a spacing of 10 in. to give a plant population of 21,000 plants/acre. Row length is 30 ft and harvest is made from 30 consecutive plants. Plants at the end or surrounding a missing hill are excluded from the harvest, but may be used for moisture determination. Second ears judged as likely to be picked by the mechnical harvester are included in the sample. Ears from broken plants or plants lodged on the ground are excluded from harvest, but all plants are counted for yield computation.

The sample is delivered to an experimental processing location in which determinations are made as follows:

1. Green weight of unhusked ears is recorded.
2. Husking is done in a non-butting husker. Incompletely husked ears are thrown back on the husking bed and the number of such throws is recorded.
3. Inspection of the sample consists of arranging ears on the table and removing nonusable, immature, and broken ears. The usable ears are counted, and average length and diameter of 10 ears are recorded. Broken ears are counted but only usable pieces are processed through the cutter with the rest of the sample.
4. Cut corn weight is recorded and a 100-g sample removed for moisture determination.
5. Other determinations are made as required. Recording is done in a manner that does not require further manual transcription prior to analysis.

Fresh-market hybrids are evaluated by emphasizing aspects critical for the particular market, such as husk cover, tip fill, ear appearance, row symmetry, type uniformity, kernel size, color, and taste properties (Table 13.10).

It is important that results be summarized quickly. Timely decisions minimize work in harvest and processing, and contribute to efficiency and cost control.

The evaluation of trial data is the culminating activity for the testing season. Generally it includes various compromises and trade-offs. In the processing industry, high yield may be attended by either single or combined negative characteristics such as lowered recovery, reduced quality, and/or poor huskability. An intimate knowledge of the economic impact that each unit change in measurement produces on profitability is required if decision making includes intuitive judgment. A formalized profitability analysis usually will be a more precise and better documented way to support decisions when significant trade-offs are involved. Appearance of critical weaknesses in a new hybrid under test conditions signals a need for an early decision.

Hybrids commonly are evaluated in relation to the performance of a long-term check or a hybrid having wide adaptation. The choice of the latter is common in commercial programs. The combination of both is advised for studies with long-term implications and hybrid comparison for disease and insect tolerances.

Performance comparisons commonly are based on tolerance limits for any particular character. There is a zero tolerance for some characteristics, such as fasciated tips, because they may be even more likely to appear in production. Such hybrids must be judged unacceptable regardless of other positive attributes. The appearance of several negatives is more important than one favorable character. It must not be assumed that shortcomings will become minimized or disappear entirely with continued testing, or that a hybrid will gain trade acceptance because of a single outstanding performance measurement.

Repeated evaluation of a hybrid at a given stage or level of testing is a widely used practice, but hybrids should not be returned to a lower-level test or recycled in the testing sequence.

Conclusions and Recommendations

The final steps of a testing program classify hybrids in one of the following categories:

1. Recommend for production
2. Advance to the next testing stage
3. Retest at the same level
4. Discard

Recommendations for production require a carefully prepared hybrid description in which all known comparisons are made available to the production or marketing areas within the organization.

Repeated testing provides broader exposure to environments and gives more reliable comparisons. It must be remembered that selection for yield is based on data affected by genetic and error components. Selection for yield caused by upward bias in error will be minimized in extended testing, and conversely an average performance may show improvement for the same reason when the original bias was in a negative direction.

Assignment to special tests is required when a character requires a more thorough evaluation. Such testing may characterize more detailed aspects of quality or may reveal sensitivity to a particular pesticide, disease reaction, or environmental stress. This usually is practiced near the end of the testing cycle and before the entry of a hybrid into the market. Such tests minimize surprises, which for the most part are negative in their nature.

Discarding of hybrids becomes more difficult as the testing advances. Excessive testing is unproductive. The objective of the program—to identify an elite hybrid—can best be attained when the breeder is not overwhelmed by a mass of mediocracy in the trials. The required ruthlessness to discard material may be evoked by remembering that an element of art is involved in the pursuit of breeding as a profession. The use of some artistry permits the breeder to concentrate on material that is inherently superior.

Recording of Data

Data should be taken at appropriate times on all characteristics important to each experiment. Both measurements and ratings, on a predetermined scale, can be used. Transcribing of notes should be minimized. The precision with which measurements are taken has to be guided by the use of the data as well as by the ease with which measurements can be obtained. Notebooks should be provided with headings and data arrangement in chronological order.

Breeding programs with access to computer facilities must present data in a predetermined manner as indicated by the computer program. Yield–quality regression analysis provides meaningful information on hybrid comparisons. It may be substituted by yield–maturity analyses when maturity is measured in terms of moisture percentage. This, however, can be done only in trials that provide a sufficiently wide range of harvest maturities and adequate number of observations, as would be available in the quality test from six harvests at six planting dates, with harvest date determined by time or by heat unit accumulation (*71*).

In trials designed for the sweet corn processor, the green yield is expressed in mwt/acre or tons/acre. The yield of cut corn commonly is reduced to yield of cases per acre. This value is obtained by dividing the cut corn yield in pounds/acre by a standard weight of cut corn used to fill one case of processed corn, commonly from 16–20 lb per case. The ratio of cases/mwt is indicative of profitability. The more cases that are obtained from 1 ton of green produce, the more profitable the hybrid will be from the processors' standpoint.

Testing for Hybrid Seed Yield

There is little published information about evaluation of sweet corn inbreds for yield and for adaptation during hybrid seed production. The work done by individual seed producers for the most part is of a proprietary nature. For field corn, detailed reviews have been provided by Craig (*36*), Jugenheimer (*95*), and others (*53*). Sweet corn inbreds are as a rule less vigorous than field corn inbreds. Furthermore, selection of tenderness has resulted in thinner pericarp, which renders the seed more susceptible to mechanical damage and seedborne microorganisms. Inbreds selected for earliness commonly are of short stature and have low ear placement on the plant. These and other characteristics make sweet corn seed production difficult and mandate that hybrid combinations be planned so that at least one parent, preferably the seed parent, is acceptable to seed production needs.

Well in advance of hybrid release it is advisable to introduce the inbreds into the seed production environment to determine the following characteristics:

1. Yield potential, percentage barrenness, extent of second ears
2. Stalk strength, lodging resistance, husk cover, ease of silking, tillering, adaptation to picker–husker harvest
3. Disease reaction
4. Tolerance to pesticides, heat, and drought
5. Pattern of silk emergence and anthesis, maturity
6. Expected mill-out by seed size, seed germination, and cold test vigor

This can be done by planting a single observation row of 25 plants in two replications or at two planting dates. In summarizing results, an integrated rating is given for each inbred relative to its value as a seed parent or pollen parent. Such information is a valuable supplement to the description of the hybrid, which must be available before planning the proper marketing strategy.

MAINTENANCE OF INBRED LINES

The maintenance of established lines is done in a manner to prevent changes in their breeding behavior due to delayed segregation, mutations, and out-crossing. Maintenance may involve self-pollination, sib pollination, or a combination of the two. Procedures may vary, but the following illustrates an approach which distinguishes among the types of seed:

Breeder's seed: As an aggregate of 100–150 selfs from plants of identical genetic make-up and identified by a line designation. Breeder's seed is usually maintained by the breeder and the selfing process is repeated after supply is depleted to provide foundation seed stock.

Foundation seed stock: An ear-to-row planting of selfs from the breeder's seed. The planting requires adequate isolation, particularly from other sweet corn. It must be rogued before anthesis and the entire row eliminated if an off-type segregate is detected. Bulk harvest in such open-pollinated plantings constitutes foundation seed stock. Responsibility for maintaining foundation seed is assigned to the breeder or a designated person on the production staff.

Stock seed: An increase from the foundation seed source and done in volume to satisfy the hybrid seed production plan. Type purity is maintained by observing isolation requirements and by roguing individual off-type plants. The responsibility for such seed increase commonly is assigned to a seed production representative.

All seed increase blocks must be well isolated from foreign pollen sources that do not produce a visible xenia effect and preclude removal of contaminants. Breeder's seed should be renewed in the same year the existing stock is used as a seed source for foundation seed. Particular care must be given to prevent mechanical mixtures. Equipment used for planting, harvesting, and seed processing has to be carefully cleaned and all contaminations removed at all steps of seed handling.

ACKNOWLEDGMENTS

The authors wish to acknowledge the substantial assistance of Dr. D. L. Garwood, Garwood Seed Company, Stonington, Illinois, in the preparation and review of the manuscript.

REFERENCES

1. Anderson, E. 1946. Maize in Mexico: A preliminary survey. Ann. Mo. Bot. Gard. *33*, 147–247.
2. Anderson, E., and Brown, W. L. 1948. A morphological analysis of row number in maize. Ann. Mo. Bot. Gard. *35*, 323–336.
3. Andrew, R. H. 1967. Influence of season, population, and spacing on axillary bud development of sweet corn. Agron. J. *59*, 355–358.
4. Andrew, R. H. 1982. Factors influencing early seedling vigor of shrunken-2 maize. Crop Sci. *22*, 263–266.
5. Andrew, R. H., and Carlson, J. R. 1976. Evaluation of sweet corn inbreds for resistance for European corn borer. J. Am. Soc. Hortic. Sci. *101*, 97–99.
6. Andrew, R. H., and Weckel, K. G. 1965. Yield and quality of sweet corn for canning as influenced by season and maturity. Univ. Wis. Res. Rep. *19*.
7. Andrews, F. S. 1963. Xenia effects, soil influence, and ear evaluation as an index of progeny performance in sweet corn quality evaluation. Proc. Am. Soc. Hortic. Sci. *83*, 522–530.
8. Badenhuizen, N. P. 1969. The Biogenesis of Starch Granules in Higher Plants. Appleton-Century-Crofts, New York.
9. Baenziger, P. S., and Glover, D. V. 1980. Effect of reducing plant population on yield and kernel characteristics of sugary-2 and normal maize. Crop Sci. *20*, 444–447.
10. Beadle, G. W. 1980. The ancestry of corn. Sci. Am. *242*, 112–119.

11. Beadle, G. W. 1981. Origin of corn: Pollen evidence. Science *213*, 890–892.

12. Beckett, J. B. 1971. Classification of male sterile cytoplasms in maize in the presence of Texas sterile cytoplasm. Crop. Sci. *61*, 183–184.

13. Boling, M. B., and Grogan, C. O. 1965. Gene action affecting host resistance to fusarium ear rot of maize. Crop Sci. *5*, 305–307.

14. Bonnett, O. T. 1953. Developmental morphology of the vegetative and floral shoots of maize. Res. Bull—Ill., Agric. Exp. Stn. *568*.

15. Bowers, J. L. 1943. The effect of spacing and number of plants per hill on the yield of eleven sweet corn hybrids. Proc. Am. Soc. Hortic. Sci. *43*, 275–277.

16. Boyer, C. D., and Shannon, J. C. 1982. The use of endosperm genes in sweet corn improvement. Plant Breed. Rev. *1*, 139–161.

17. Boyer, C. D., Daniels, R. R., and Shannon, J. C. 1977. Starch granule (amyloplast) development in endosperm of several *Zea mays* L. genotypes affecting kernel polysaccharides. Am. J. Bot. *64*, 50–56.

18. Brewbaker, J. L. 1971. Breeding tropical supersweet corn. Hawaii Farm Sci. *20* (1), 7–10.

19. Brewbaker, J. L. 1977. 'Hawaiian Super-sweet #9' corn. HortScience *12*, 355–356.

20. Brewbaker, J. L., and Kim, S. K. 1979. Inheritance of husk numbers and ear insect damage in maize. Crop. Sci. *19*, 32–36.

21. Brewbaker, J. L., and Banafunzi, N. 1975. 'Hawaiian Super-sweet #6' corn. HortScience *10*, 427–428.

22. Brown, W. L., and Goodman, M. M. 1976. Races of corn. *In* Corn and Corn Improvement. G. F. Sprague (Editor), pp. 49–88. Am. Soc. Agron., Madison, WI.

23. Buchert, J. G. 1961. The stage of the genome–plasmon interaction in the restoration of fertility to cytoplasmically pollen-sterile maize. Proc. Natl. Acad. Sci. U.S.A. *47*, 1436–1440.

24. Cameron, J. W., and Teas, H. J. 1956. Carbohydrate relationships in developing and mature endosperms of brittle and related maize genotypes. Am. J. Bot. *41*, 50–55.

25. Carter, G. F. 1948. Sweet corn among the Indians. Geogr. Rev. *28*, 206–221.

26. Chiang, H. C. 1978. Pest management in corn. Annu. Rev. Entomol. *23*, 101–123.

27. Chipman, E. W., and MacKay, D. C. 1960. The interactions of plant populations and nutritional levels on the production of sweet corn. Proc. Am. Soc. Hortic. Sci. *76*, 442–447.

28. Christensen, J. J. 1963. Corn Smut Caused by *Ustilago maydis*, Monogr. No. 2. Am. Phytopathol. Soc., Madison, WI.

29. Christensen, J. J., and Wilcoxson, R. D. 1966. Stalk Rot of Corn, Monogr. No. 3. Am. Phytopathol. Soc., Madison, WI.

30. Cobb, B. G., and Hannah, L. C. 1981. The metabolism of sugars in maize endosperms. Plant Physiol. (London) *67*, 107 (abstr.).

31. Coe, E. H., Jr., and Neuffer, M. G. 1976. The genetics of corn. *In* Corn and Corn Improvement. G. F. Sprague (Editor), pp. 111–223. Am. Soc. Agron., Madison WI.

32. Committee On Genetic Vulnerability Of Major Crops 1972. Genetic Vulnerability of Major Crops. National Academy of Sciences, Washington, DC.

33. Courter, J. W., and Rhodes, A. M. 1982. A classification of vegetable corns and new cultivars for 1983. Ill., Agric. Exp. Stn., Mimeo.

34. Craig, J., and Fajemisin, J. M. 1969. Inheritance of chlorotic lesion resistance to *Helminthosporium maydis* in maize. Plant Dis. Rep. *53*, 742–743.

35. Craig, J., and Hooker, A. L. 1961. Relation of sugar trends and pith density to diplodia stalk rot in dent corn. Phytopathology *51*, 376–382.

36. Craig, W. F. 1977. Production of hybrid corn seed. *In* Corn and Corn Improvement. G. F. Sprague (Editor), pp. 671–719. Am. Soc. Agron., Madison, WI.

37. Creech, R. G. 1965. Genetic control of carbohydrate synthesis in maize endosperm. Genetics *52*, 1175–1186.

38. Dale, J. L., McFerran, J., Wann, E. V., and Bove, R. L. 1982. Evaluation of sweet corn hybrids for virus resistance, yield, and ear quality. Rep. Ser.—Arkansas, Agric. Exp. Stn. *267*.

39. Davis, D. W., and Grier, S. L. 1980. AS9 sweet corn population. HortScience *15*, 666–667.

40. Davis, D. W., and Grier, S. L. 1979. Resistance of some sweet corn germplasm to second brood European corn borer. Annu. Plant Resistance Insects Newsl. *4*, 23–25. Purdue University.

41. Davis, R. M., Jr., Fildes, R., Baker, G. A., Zahara, M., May, D. M., and Tyler, K. B. 1970. A method of estimating the potential yield of a vegetable crop obtainable in a single harvest. J. Am. Soc. Hortic. Sci. *95*, 475–480.

42. Diachun, S. 1939. The effect of some soil factors on penicillium injury of corn seedlings. Phytopathology *29*, 231–241.

43. Dollinger, E. J., Findley, W. R., and Williams, L. E. 1970. Inheritance of resistance to maize dwarf mosaic virus in maize (*Zea mays* L.). Crop Sci. *10*, 412–415.

44. Duvick, D. N. 1959. Genetic and environmental interactions with cytoplasmic pollen sterility of corn. Proc. 14th Annu. Hybrid Corn Ind.—Res. Conf. pp. 42–52.

45. Duvick, D. N. 1972. Potential usefulness of new cytoplasmic male steriles and sterility systems. Proc. 27th Annu. Corn Sorghum Res. Conf. pp. 192–201.

46. Duvick, D. N. 1977. Genetic rates of gain in hybrid maize yields during the past 40 years. Maydica *22*, 187–196.

47. Eckhardt, R. C. 1953. A technique to simultaneously incorporate male sterile cytoplasm and restorer factor(s) into inbred lines of maize. Proc. 50th Annu. Conv. Assoc. South. Agric. Workers (abstr.), p. 42.

48. Emerson, R. A., and East, E. M. 1913. The inheritance of quantitative characters in maize. Res. Bull.—Nebr., Agric. Exp. Stn. *2*.

49. Emerson, R. A., and Smith, H. H. 1950. Inheritance of number of kernel rows in maize. Mem.—N.Y., Agric. Exp. Stn. (Ithaca) *296*.

50. Empig, L. T., Gardner, C. O., and Compton, W. A. 1972. Theoretical gains for different population improvement procedures. Nebr., Agric. Exp. Stn., Bull. *M26* (rev.), 1–22.

51. Ervin, A. T. 1951. Sweet corn, mutant or historic species? Econ. Bot. *5*, 302–306.

52. Evans, D. D., Mack, H. J., Stevenson, D. S., and Wolfe, J. W. 1960. Soil moisture, nitrogen and stand density effects on growth and yield of sweet corn. Oreg., Agric. Exp. Stn., Tech. Bull. *53*.

53. Everson, L. E. 1972. The Use of Test Information in Quality Control and Sales, Mississippi Short Course for Seedsmen, No. *15* (mimeo.). Mississippi State University, State College.

54. Fehr, W. R., and Hadley, R. (Editors) 1980. Hybridization of Crop Plants. Am. Soc. Agron., Madison, WI.

55. Ferguson, J. E., Dickinson, D. B., and Rhodes, A. M. 1979. Analysis of endosperm sugars in a sweet corn inbred (Illinois 677a), which contains the sugary enhancer (*se*) gene, and comparison of *se* with other corn genotypes. Plant Physiol. *63*, 416–420.

56. Ferguson, J. E., Rhodes, A. M., and Dickinson, D. B. 1978. The genetics of sugary enhancer (*se*), an independent modifier of sweet corn (*su*). J. Hered. *69*, 377–380.

57. Flora, L. F., and Wiley, R. C. 1978. Classification of sweet corn aromas by stepwise discriminant analysis. J. Food Qual. *1*, 341–348.

58. Fulton, J. M. 1970. Relationships among soil moisture stress, plant populations, row spacing and yield of corn. Can. J. Plant Sci. *50*, 31–38.

59. Galinat, W. C. 1971. The evolution of sweet corn. Res. Bull.—Mass. Agric. Exp. Stn. *591*.

60. Galinat, W. C. 1966. The evolution of glumeless sweet corn. Econ. Bot. *20*, 441–445.

61. Galinat, W. C. 1969. The evolution under domestication of the maize ear: String cob maize. Res. Bull.—Mass. Agric. Exp. Stn. *577*.

62. Galinat, W. C. 1973. Preserve Guatemalan teosinte, a relic link in corn's evolution. Science *180*, 323.

63. Galinat, W. C. 1975. Use of male-sterile 1 gene to eliminate detasseling in production of hybrid seed of bi-color sweet corn. J. Hered. *66*, 387–388.

64. Galinat, W. C. 1976. The origin of corn. *In* Corn and Corn Improvement. G. F. Sprague (Editor), pp. 1–47. Am. Soc. Agron., Madison, WI.

65. Galinat, W. C. 1971. The origin of maize. Annu. Rev. Genet. *5*, 47–478.

66. Garwood, D. L., and Creech, R. G. 1979. 'Pennfresh ADX' hybrid sweet corn. HortScience *14*, 645.

67. Garwood, D. L., and Creech, R. G. 1972. Kernel phenotypes of *Zea mays* L. genotypes possessing one to four mutated genes. Crop Sci. *12*, 119–121.

68. Garwood, D. L., and Vanderslice, S. F. 1982. Carbohydrate composition of alleles at the sugary locus in maize. Crop Sci. *22*, 367–371.

69. Garwood, D. L., McArdle, F. J., Vanderslice, S. F., and Shannon, J. C. 1976. Postharvest carbohydrate transformations and processed quality of high sugar maize genotypes. J. Am. Soc. Hortic. Sci. *101*, 400–404.

70. Gather, L. A., and Calvin, L. D. 1963. Relation between preference scores and objective and subjective quality measurements of canned corn and pears. Food Technol. *17*, 97–100.

71. Gilmore, E. C., Jr., and Rogers, J. S. 1958. Heat units as a method of measuring maturity in corn. Agron. J. *50*, 611–615.

71. Gordon, D. T. 1974. Distinguishing symptoms and latest research findings on corn virus diseases in the United States. Proc. 29th Annu. Corn Sorghum Res. Conf. pp. 153–173.

73. Grier, S. L., and Davis, D. W. 1980. Infestation procedures and heritability of characters used to estimate ear damage caused by second-brood European corn borer (*Ostrinia nubilalis* Hübner) on corn. J. Am. Soc. Hortic. Soc. *105*, 3–8.

74. Groth, J. V., Davis, D. W., Zeyen, R. J., and Mogen, B. D. 1983. Ranking of partial resistance to common rust in 30 sweet corn hybrids. Crop Prot. *2*, 219–223.

75. Groth, J. V., Zeyen, R. J., Davis, D. W., and Christ, B. J. 1983. Yield and quality losses caused by common rust (*Puccinia sorghi* Schw.) in sweet corn (*Zea mays*) hybrids. Crop Prot. *2*, 105–111.

76. Guthrie, W. D. 1971. Resistance of maize to second-brood European corn borers. Proc. 26th Annu. Corn Sorghum Res. Conf. pp. 165–179.

77. Hallauer, A. R. 1981. Quantitative Genetics in Maize Breeding. Iowa State University Press, Ames.

78. Hallauer, A. R., and Sears, J. H. 1972. Integrating exotic germplasm into corn belt maize breeding programs. Crop Sci. *12*, 203–206.

79. Hannah, L. C., and Cantliffe, D. J. 1976. Levels of various carbohydrate constituents and percentage germination of four Everlasting Heritage sweet corns. Proc. Fla. State Hortic. Soc. *89*, 80–82.

80. Hansen, L. A., Baggett, J. R., and Rowe, K. E. 1977. Quantitative genetic analysis of ten characteristics in sweet corn (*Zea mays* L.). J. Am. Soc. Hortic. Sci. *102*, 158–162.

81. Hitchcock, A. S. 1951. Manual of the grasses of the United States. Misc. Publ.—U.S., Dep. Agric. *200*.

82. Hooker, A. L. 1954. Relative efficiency of various methods of inducing field infections with *Helminthosporium turcicum* and *Puccinia sorghi*. Plant Dis. Rep. *38*, 173–177.

83. Hooker, A. L. 1963. Inheritance of chlorotic-lesion resistance to *Helminthosporium turcicum* in seedling corn. Phytopathology *53*, 660–662.

84. Hooker, A. L. 1969. Widely based resistance to rust in corn. *In* Disease Consequences of Intensive and Extensive Culture of Field Crops. J. A. Browning (Editor), Spec. Rep. No. 64. Iowa Satate University, Ames.

85. Hooker, A. L., Hilu, H. M., Wilkinson, D. R., and Van Dyke, C. G. 1964. Additional sources of chlorotic-lesion resistance to *Helminthosporium turcicum* in corn. Plant Dis. Rep. *48*, 777–780.

86. Hoppe, P. E. 1929. Inheritance of resistance to seedling blight of corn caused by *Gibberella saubinetii*. Phytopathology *19*, 79–80.

87. Huelsen, W. A. 1954. Sweet Corn. Interscience Publishers, New York.

88. Hughes, G. R., and Hooker, A. L. 1971. Gene action conditioning resistance to northern leaf blight in maize. Crop Sci. *11*, 180–184.

89. Iltis, H. H. 1972. The taxonomy of *Zea mays* (Gramineae). Phytologia *23*, 248–249.

90. Immer, F. R. 1927. The inheritance of reaction to *Ustilago zeae* in maize. Minn., Agric. Exp. Stn., Tech. Bull. *51*.

91. Ito, G. M., and Brewbaker, J. L. 1981. Genetic advance through mass selection for tenderness in sweet corn. J. Am. Soc. Hortic. Sci. *106*, 496–499.

92. Jensen, S. D. 1971. Breeding for drought and heat tolerance in corn. Proc. 26th Annu. Corn Sorghum Res. Conf. pp. 198–207.

93. Jinahyon, S., and Russell, W. A. 1969. Evaluation of recurrent selection for stalk rot resistance in an open-pollinated variety of maize. Iowa State J. Sci. *43*, 229–251.

94. Jones, D. F., and Singleton, W. R. 1934. Crossed sweet corn. Bull.—Conn. Agric. Exp. Stn., New Haven *361*.

95. Jugenheimer, R. W. 1958. Hybrid Maize Breeding and Seed Production, F.A.O. Agric. Dev. Pap. No. 62. Food Agric. Organ., Rome.

96. Kaukis, K., and Haunold, A. 1966. Effect of maturity on amylose content in the sweet corn endosperm. Crop Sci. *6*, 577–579.

97. Keys, N. A., and Andrew, R. H. 1977. Expression of the vestigial glume character in adapted sweet corn inbreds. Crop Sci. *17*, 662–664.

98. Khalil, T., and Kramer, A. 1971. Histological and histochemical studies of sweet corn (*Zea mays* L.) pericarp as influenced by maturity and processing. J. Food Sci. *36*, 1064–1069.

99. Kiesselbach, T. A. 1949. The structure and reproduction of corn. Res. Bull.—Nebr., Agric. Exp. Stn. *161*.

100. Kiesselbach, T. A. 1961. Significance of xenia effect on kernel weight of corn. Res. Bull.—Nebr., Agric. Exp. Stn. *191*.

101. Kim, S. K., and Brewbaker, J. L. 1977. Inheritance of general resistance in maize to *Puccinia sorghi* Schw. Crop Sci. *17*, 456–461.

102. Koehler, B. 1960. Corn stalk rots in Illinois. Res. Bull.—Ill., Agric. Exp. Stn. *658*.

103. Kramer, A. 1952. A tri-metric test for sweet corn quality. Proc. Am. Soc. Hortic. Sci. *59*, 405–413.

104. Kramer, H. H., Pfahler, P. L., and Whistler, R. L. 1958. Gene interactions in maize affecting endosperm properties. Agron. J. *50*, 207–210.

105. Lampe, L. 1931. A microchemical and morphological study of the developing endosperm of maize. Bot. Gaz. (Chicago) *91*, 337–376.

106. Laughnan, J. R. 1953. The effect of the *sh2* factor on carbohydrate reserves in the mature endosperm of maize. Genetics *38*, 485–499.

107. Lindstrom, E. W. 1931. Genetic tests for linkage between row number genes and certain qualitative genes in maize. Res. Bull.—Iowa, Agric. Exp. Stn. *142*.

108. Loesch, P. J., Jr., and Zuber, M. S. 1967. An inheritance study of resistance to maize dwarf mosaic in corn (*Zea mays* L.). Agron. J. *59*, 423–426.

109. Lonnquist, J. H., and McGill, D. P. 1956. Performance of corn synthetics in advanced generations of synthesis and after two cycles of recurrent selection. Agron. J. *48*, 249–253.

110. Louie, R., Gordon, D. T., Knoke, J. K., Gingery, R. E., Bradfute, O. E., and Lipps, P. E. 1982. White-line mosaic in Ohio. Plant Dis. *66*, 167–170.

111. Mack, H. J. 1972. Effects of population density, plant arrangement, and fertilizers on yield of sweet corn. J. Am. Soc. Hortic. Sci. *97*, 757–760.

112. Mikel, M. A. 1983. Maize dwarf mosaic virus in *zea mays* L: Inheritance of resistance, yield loss, serology and seed transmission. Ph.D. Dissertation. University of Illinois, Champaign-Urbana.

113. Moll, R. H., and Stuber, C. W. 1974. Quantitative genetics—empirical results relevant to plant breeding. Adv. Agron. *26*, 277–313.
114. Nelson, O. E., Jr. 1952. Non-reciprocal cross-sterility in maize. Genetics *37*, 101–124.
114a. Neuffer, M. G., Jones, L., and Zuber, M. S. 1968. The Mutants of Maize. Crop Sci. Soc. Am., Madison, WI.
115. Ortega, A., and De Leon, C. 1974. Insects of maize. *In* World-Wide Maize Improvement, pp. 1–36. CIMMYT, El Batan, Mex.
116. Paulus, A. O., Otto, H. W., Nelson, J., and Shibuya, F. 1975. Controlling sweet corn smut. Calif. Agric. (Feb), 12–13.
117. Penny, L. H., Scott, G. E., and Guthrie, W. D. 1967. Recurrent selection for European corn borer resistance in maize. Crop Sci. *7*, 407–409.
118. Pepper, E. H. 1967. Stewart's Bacterial Wilt of Corn, Monogr. No. 4. Am. Phytopathol. Soc., Madison, WI.
119. Pieczarka, D. J., and Wolf, E. A. 1978. Increased stand of 'Florida Staysweet' corn by seed treatment with fungicides. Proc. Fla. State Hortic. Soc. *91*, 290–291.
120. Poole, C. F. 1940. Corn earworm resistance and plant characters. Proc. Am. Soc. Hortic. Sci. *38*, 605–609.
121. Pucaric, A., Crane, C. L., and Glover, D. V. 1975. Stand, early growth and plant and ear height of endosperm mutants and normal maize hybrids under field conditions. Z. Acker-Pflanzenbau *141*, 317–325.
122. Reeves, J. T., Jackson, A. O., Paschke, J. D., and Lister, R. M. 1978. Use of enzyme-linked immunosorbent assay (ELISA) for serodiagnosis of two maize viruses. Plant Dis. Rep. *62*, 667–671.
123. Reiling, T. P., Kommedahl, T., and Vidaver, A. 1977. Kernel crown spot of sweet corn. Plant Dis. Rep. *61*, 827–829.
124. Rhoades, V. H. 1935. The location of a gene for disease resistance in maize. Proc. Natl. Acad. Sci. U.S.A. *21*, 243–246.
125. Robinson, H. F., and Moll, R. H. 1959. Implications of environmental effects on genotypes in relation to breeding. Proc. 14th Annu. Hybrid Corn Ind.—Res. Conf. pp. 24–31.
126. Robinson, H. F., Comstock, R. E., and Harvey, P. H. 1955. Genetic variances in open pollinated varieties of corn. Genetics *40*, 45–60.
127. Rojas, B. A., and Sprague, G. F. 1952. A comparison of variance components in corn yield trials. III. General and specific combining ability and their interaction with locations and years. Agron. J. *44*, 462–466.
128. Rowe, D. E., and Garwood, D. L. 1978. Effects of four maize endosperm mutants on kernel vigor. Crop Sci. *18*, 709–712.
129. Russell, W. A. 1974. Comparative performance for maize hybrids representing different eras of maize breeding. Proc. 29th Annu. Corn Sorghum Res. Conf. pp. 81–101.
130. Russell, W. A., and Eberhart, S. A. 1975. Hybrid performance of selected maize lines from reciprocal recurrent selection and testcross selection programs. Crop Sci. *15*, 1–4.
131. Russell, W. A., and Teich, A. H. 1967. Selection in *Zea mays* L. by inbred line appearance and testcross performance in low and high plant densities. Res. Bull.—Iowa Agric. Exp. Stn. *552*.
132. Saboe, L. C., and Hayes, H. K. 1941. Genetic studies of reactions to smut and of firing in maize by means of chromosomal translocations. J. Am. Soc. Agron. *33*, 463–470.
133. Sass, J. E. 1976. Morphology. *In* Corn and Corn Improvement. G. F. Sprague (Editor), pp. 89–110. Am. Soc. Agron., Madison, WI.
134. Shannon, J. C., and Garwood, D. L. 1983. Genetics and physiology of starch development. *In* Starch: Chemistry and Technology. R. L. Whistler, J. N. BeMiller, and E. F. Paschall (Editors), 2nd Edition, pp. 25–86. Academic Press, New York.
135. Showalter, R. K. 1960. Measuring pericarp content of sweet corn. Proc. 57th Annu. Conv. Assoc. South. Agric. Workers pp. 208–209.

136. Showalter, R. W., and Miller, L. W. 1962. Consumer preference for high-sugar sweet corn varieties. Proc. Fla. State Hortic. Soc. *75*, 278–280.

137. Sim, L. E., and Garwood, D. L. 1978. Sweet Corn Disease Evaluation Summary, Annu. Rep. CSRS Reg. Proj. NE-66 (mimeo.). Agric. Exp. Stn., Pennsylvania State University, University Park.

138. Simpson, W. R., and Fenwick, H. S. 1968. Chemical control of corn head smut. Plant Dis. Rep. *52*, 726–727.

139. Singleton, W. R. 1948. Hybrid sweet corn. Bull.—Conn. Agric. Exp. Stn., New Haven *518*.

140. Smith, D. R., and Hooker, A. L. 1973. Monogenic chlorotic-lesion resistance in corn to *Helminthosporium maydis*. Crop. Sci. *13*, 330–331.

141. Smith, G. M. 1933. Golden Cross Bantam sweet corn. U.S., Dep. Agric., Circ. *268*.

142. Snyder, R. J. 1967. The relationship of silk balling, husk length, hust tightness and blank tip to earworm and sap beetle resistance in maize. Proc. Am. Soc. Hortic. Sci. *91*, 454–461.

143. Spencer, E. L., and McNew, G. L. 1938. The influence of mineral nutrition on the reaction of sweet-corn seedlings to *Phytomonas stewarti*. Phytopathology *28*, 213–223.

144. Sprague, G. F. (Editor) 1976. Corn and Corn Improvement. Am. Soc. Agron., Madison, WI.

145. Sprague, G. F. 1971. Genetic vulnerability in corn and sorghum. Proc. 26th Annu. Corn Sorghum Res. Conf. pp. 96–104.

146. Sprague, G. F., and Eberhart, S. A. 1976. Corn breeding. *In* Corn and Corn Improvement. G. F. Sprague (Editor), pp. 305–362. Am. Soc. Agron., Madison, WI.

147. Sprague, G. F., and Federer, W. T. 1951. A comparison of variance components in corn yield trials. II. Error, year × variety, location × variety and variety components. Agron. J. *43*, 535–541.

148. Stevens, N. E., and Haenseler, C. M. 1941. Incidence of bacterial wilt of sweet corn 1935–1940: Forecasts and performance. Plant Dis. Rep. *25*, 152–157.

149. Straub, R. W. 1977. European corn borer control in early sweet corn: Role of pre-silk applications and leaf feeding resistance. J. Econ. Entomol. *70*, 524–526.

150. Sullivan, S. L., Gracen, V. E., and Ortega, A. 1974. Resistance of exotic maize varieties to the European corn borer *Ostrinia nubilalis* (Hübner). Environ. Entomol. *3*, 718–720.

151. Tatum, L. A. 1971. The southern corn leaf blight epidemic. Science *171*, 1113–1116.

152. Tingey, W. M., Gracen, V. E., and Scriber, J. M. 1975. European corn borer resistant maize genotypes. N.Y. Food Life Sci. Q. *8*.

153. Tuite, F., Shanor, G., Rambo, G., Foster, F., and Caldwell, R. W. 1974. The gibberella ear rot epidemics of corn in Indiana in 1965 and 1972. Cereal Sci. Today *19*, 238–241.

154. Ullstrup, A. J. 1966. A Widespread Epiphytotic of Gibberella Ear Rot in the U.S.A., p. 46. Int. Symp. Plant Pathol., New Delhi, India.

155. Ullstrup, A. J. 1949. A method for producing artificial epidemics of diplodia ear rot. Phytopathology *39*, 93–101.

156. Ullstrup, A. J. 1963. Sources of resistance to northern corn leaf blight. Plant Dis. Rep. *47*, 107–108.

157. Ullstrup, A. J. 1976. Diseases of corn. *In* Corn and Corn Improvement. G. F. Sprague (Editor), pp. 391–500. Am. Soc. Agron, Madison, WI.

157a. U.S. Department of Agriculture (1981). "Agricultural Statistics." U.S. Government Printing Office, Washington, DC.

158. Vittum, M. T., Peck, N. H., and Carruth, A. F. 1959. Response of sweet corn to irrigation, fertility level and spacing. Bull.—N.Y., Agric. Exp. Stn. (Ithaca) *786*.

159. Walden, D. B. (Editor) 1978. Maize Breeding and Genetics. Wiley-Interscience, New York.

159a. Waller, R. A., and Duncan, D. B. 1969. A Bayes rule for the symmetric multiple comparisons problem. J. Am. Stat. Assoc. *64*, 1484–1503.

160. Wann, E. V. 1980. Earworm resistant sweet corn composite 9E–79. HortScience *15* (3), 317.

161. Wann, E. V., and Hills, W. A. 1975. Tandem mass selection in a sweet corn composite for earworm resistance and agronomic characters. HortScience *10* (2), 168–170.

162. Weatherwax, P. 1935. The phylogeny of *Zea mays*. Am. Midl. Nat. *16*, 1–71.
163. Wellhausen, E. J. 1937. Genetics of resistance to bacterial wilt in maize. Res. Bull.—Iowa Agric. Exp. Stn. *224*.
164. Widstrom, N. W., and McMillian, W. W. 1973. Genetic effects conditioning resistance to earworm in maize. Crop Sci. *13*, 459–461.
165. Widstrom, N. W., McMillian, W. W., and Wiseman, B. R. 1970. Resistance in corn to the earworm and the fall armyworm. IV. Earworm injury to corn inbreds related to climatic conditions and plant characteristics. J. Econ. Entomol. *63*, 803–808.
166. Wilkinson, D. R., and Hooker, A. L. 1968. Genetics of reaction to *Puccinia sorghi* in ten corn inbred lines from Africa and Europe. Phytopathology *58*, 605–608.
167. Williams, B. R. 1971. The ultrastructure of plastid differentiation in endosperm cells of *normal* and *sugary-1 Zea mays* L. Ph.D. Dissertation. Pennsylvania State University, University Park.
168. Wolf, E. A. 1978. Florida Staysweet. Circ.—Fla., Agric. Exp. Stn. *S-259*.
169. Wolf, E. A., and Showalter, R. K. 1974. Florida-Sweet. A high quality sh2 sweet corn hybrid for fresh market. Circ.—Fla. Agric. Exp. Stn. *S-226*, 1–13.
170. Zuber, M. S. 1962. A mechanical method for evaluating stalk lodging. Proc. 17th Annu. Hybrid Corn Ind.—Res. Conf. pp. 15–23.
171. Zuber, M. S., Fairchild, M. L., Keaster, A. J., Fergason, V. L., Krause, G. F., Hildebrand, E., and Loesch, J. P., Jr. 1971. Evaluation of 10 generations of mass selection for corn earworm resistance. Crop Sci. *11*, 16–18.

14
Asparagus Breeding

J. HOWARD ELLISON

Asparagus is grown commercially only in certain regions in the United States. This is controlled largely by climatic conditions. The highest yields per acre historically have been produced in the Sacramento and San Joaquin valleys of central California and the Yakima Valley of south–central Washington. These areas both have bright, warm days and cool nights during the summer, which favor photosynthesis by the green brush (fern) and storage of carbohydrates in the massive asparagus root system.

Less ideal, but satisfactory growing conditions are found in western and southwestern Michigan, southern New Jersey, the Delmarva Peninsula (Delaware and the eastern shores of Maryland and Virginia), and the Connecticut River valley of Massachusetts. The southeastern and Gulf Coast states are not suited to asparagus culture. This is due primarily to the mild winters, when asparagus sends up spears at the expense of stored reserves, thus reducing the yield during the spring harvest. Long, hot summers are

521

TABLE 14.1.   U.S. per Capita Consumption (lb) of Asparagus in All Forms Expressed in Farm Weight and on a Fresh Equivalent Basis[a]

	1959–1964 mean	1968	1969	1970	1971	1972	1973	1974	1975	1976	1977	1978	1979
Fresh	0.62	0.50	0.40	0.50	0.41	0.50	0.40	0.40	0.40	0.40	0.30	0.30	0.30
Canned	0.98	0.87	0.83	0.81	0.73	0.70	0.84	0.62	0.64	0.67	0.48	0.46	0.33
Frozen	0.35	0.30	0.28	0.28	0.24	0.19	0.21	0.19	0.17	0.21	0.17	0.17	—

[a]1959–1964 mean data after Ehlert and Seeling (8). 1968–1979 data supplied courtesy of United Fresh Fruit and Vegetable Association, Alexandria, Virginia. 1979 data are preliminary.

unfavorable for carbohydrate accumulation due to the high rate of respiration. Mild winter temperatures in the Imperial Valley of southern California encourage unwanted winter spears, as in the Gulf Coast states. However, little or no rain falls in the Imperial Valley, and growers merely withhold irrigation during winter and thus eliminate new spears. Summer temperatures are extremely high in the Imperial Valley and yield is reduced by excessive respiration. However, the harvest season is the earliest in the country and reduced yield is compensated by very high prices received for the first fresh asparagus of the year. Harvesting usually begins in January, after growers remove the brush and apply irrigation.

The gradual but steady decline in per capita consumption of asparagus (Table 14.1) undoubtedly is a result of the decline in production and the resulting scarcity of the product (8). Mean farm price for asparagus in the United States, calculated from production and crop value data in Table 14.2, for 1959, 1969, and 1979 are 0.11, 0.19, and 0.51 $/lb, respectively (39). Although inflation accounts for some of the price rise, the small 1979 crop is the biggest factor involved in the sharp price rise that year. Factors responsible for declining acreage since 1959 are economic: a shortage of "stoop-labor" willing to harvest the asparagus by hand, a lack of efficient mechanical harvesters, and the loss of many acres of asparagus to fusarium stem and crown rot in California, Illinois, New Jersey, Massachusetts, Delaware, and Maryland.

Major changes have taken place in asparagus production in the United States since the late 1950s (Table 14.2) (39). Acreage in production in 1979 was a scant one-half of the acreage in 1959, with the biggest reductions occurring in California, New Jersey, and Illinois. Acreage in Washington and Michigan increased somewhat. The main cause of the loss of production was an epiphytotic of fusarium stem and crown rot in California and New Jersey. Yield per acre in California actually increased during the acreage decline,

TABLE 14.2.   Asparagus for Fresh Market and Processing[a]

	Acreage			Yield per acre (cwt)			Value ($1000)		
	1959	1969	1979	1959	1969	1979	1959	1969	1979
California	77,800	44,700	26,400	24	29	35	21,004	26,438	48,556
Illinois	8,600	9,000	3,100	17	16	11	1,518	2,534	1,594
Michigan	11,200	13,900	17,000	15	15	13	1,858	4,368	12,951
New Jersey	31,500	22,700	1,600	25	21	17	8,833	9,254	2,203
Washington	15,800	17,400	21,000	23	29	28	3,670	9,090	26,371
U.S. total	160,000	123,830	78,160	23	24	24	40,544	57,465	95,779

[a]After U.S. Department of Agriculture (39).

TABLE 14.3. World Production of Canned Asparagus by Country[a]

	Production by year (millions of pounds)		
	1964	**1969**	**1974**
United States	192.0	159.3	132.1
Taiwan	2.7	153.2	199.5
Spain	22.5	44.2	30.4[b]
Japan	14.0	—[c]	25.6[b]
France	7.1	9.1	12.1
Australia[d]	9.1	9.0	9.6[b]
Canada	6.5	9.3	13.0
West Germany	8.8	—[c]	1.6
Mexico[e]	—[f]	1.2	9.5

[a]Adapted from U.S. International Trade Comission (40).
[b]1973 data substituted.
[c]Not available.
[d]Year beginning July 1.
[e]Data shown represent exports only.
[f]Less than 50,000 lb.

due to a shift of the industry from the old disease-ridden Delta area near Stockton to a new area in the Salinas Valley. Improved strains of asparagus from the University of California also have increased yield per acre. In New Jersey, there has been no shift to "new land," and growers still are plagued by the *Fusarium*-contaminated land, which has been used for asparagus for many years.

The United States led all nations in the production of canned asparagus in 1964, when Taiwan production was negligible (Table 14.3) (40). By 1969 Taiwan production nearly equaled that of the United States, which had declined approximately 17% during those 5 years. By 1974 Taiwan had 46% of the world production. It captured most of the European (largely German) market for canned white asparagus, by developing a new, efficient cultural method for growing asparagus and by having a large, inexpensive labor force for growing and processing the crop. Other countries that increased canned asparagus production somewhat from 1964 to 1974 were Spain, Japan, Australia, Canada, and Mexico. The Mexican production is mostly for export to the United States.

ORIGIN AND GENERAL BOTANY

Boswell, Sturtevant, and Vavilov place the origin of asparagus in the eastern Mediterranean and Asia Minor, according to Ehlert and Seeling (8). Before it was used for food, asparagus had quite a reputation as a medicine for almost anything from the treatment of bee strings to heart trouble, dropsy, and toothache. The Greeks, who gave asparagus its name, cultivated asparagus as a luxury vegetable as early as 200 B.C. The Romans grew the plant in trenches, a method generally adopted in other countries to which asparagus spread, according to Kidner (23). An asparagus species (thought to be *Asparagus officinalis altilis,* but more probably *Asparagus martimus*) was grown in gardens, particularly by the Romans, who prided themselves on growing large spears (23). Of the wild species with which the author is familiar, *A. martimus* (referring to the seashore) is the most similar in appearance to *A. officinalis.* Although *A. officinalis* is the only asparagus species cultivated as a vegetable, even today rural people from Spain to Greece delight in eating tender, young spears of the spiny species *Asparagus acutifolius,* which grows wild in the coastal areas of the Mediterranean. Needless to say, the young spears have no spines.

Most of the cultivated "varieties" of asparagus today are merely strains of *A. officinalis,* with little to separate them morphologically one from another. The present strains or selections may vary greatly in local adaptation, yield and size of spears, disease resistance, longevity, etc., but not much in general appearance, growth habit, etc.

Bailey (*1*) lists *Asparagus* as a large genus (about 150 species) of herbaceous perennials and tender woody shrubs and vines, grown mostly for ornamental habit and foliage, but one of them (*A. officinalis* Linn.) for food. Of the latter species, Bailey writes, "*officinalis,* Linn. Asparagus. An erect herb from a woody crown with long fleshy roots:

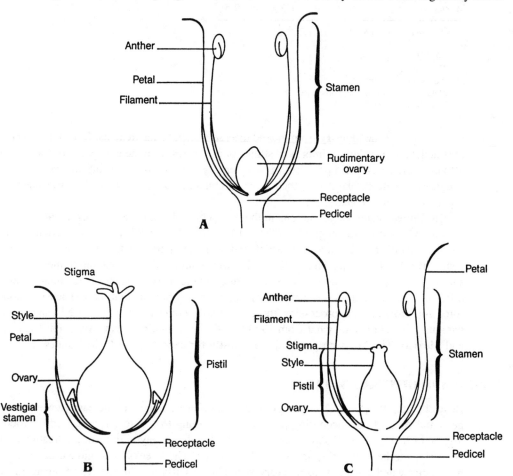

FIGURE 14.1. (A) Normal flower of male asparagus. Note rudimentary sterile ovary. Anthers are yellow and produce much pollen. (B) Normal flower of female asparagus. Note vestigial stamens. Anthers are white and do not produce pollen. (C) Androgenous flower of andromonoecious asparagus. Such flowers are rare, are self-pollinated, and can set one or more viable seeds. Functional androgenous flowers are limited to 1–10 per andromonoecious plant.
Drawing by J. J. Kinelski.

stalks smooth, much branched above, 4–12 feet high; cladodes 3–8 in a fascicle, ½–1 inch long, terete: leaf-scale with a short soft spur at base: flowers 1–4, in axils with cladodes or branches, campanulate, yellowish green: berries red ¼–⅜ inch, 1–9 seeded; seed germinate in 12–14 days in warm house, often taking a month when planted outdoors in spring'' (*1*, p. 406). See Fig. 14.1 for sketches of flowers. A well-developed 1-year-old crown of *A. officinalis* is shown in Fig. 14.2, and a bud cluster from a mature crown is seen in Fig. 14.3. Three views of a bud cluster during the cutting season are shown in Fig. 14.4. Note the sequence in which the spears grew and were cut. Notice also that new buds were laid down *during* the cutting season.

The climate along the northern coast of the Mediterranean Sea varies considerably with differences in elevation. Most of the coast near sea level is free from freezing temperatures, but mountainous regions are subjected to hard freezes. This may account for the adaptation of asparagus to areas where it never freezes (Imperial Valley of California) as well as to areas in Minnesota, where the frost penetrates the soil several feet.

No closely related species to *A. officinalis,* to the author's knowledge, have been useful horticulturally in crossing with *A. officinalis. Asparagus springeri* is highly resistant to *Fusarium* spp., but will not cross with *A. officinalis.* The author collected numerous seed samples of *A. acutifolius* in the wild in Crete, mainland Greece, Italy, and Spain, and received seed of wild *A. martimus* from Yugoslavia for possible use in breeding for resistance to *Fusarium.* Unfortunately, all of the specimen plants were susceptible to the *Fusarium* isolates common to old asparagus fields in New Jersey. Neither did any of the exotic plants flower under greenhouse conditions at New Brunswick, New Jersey, during a period of 5 years.

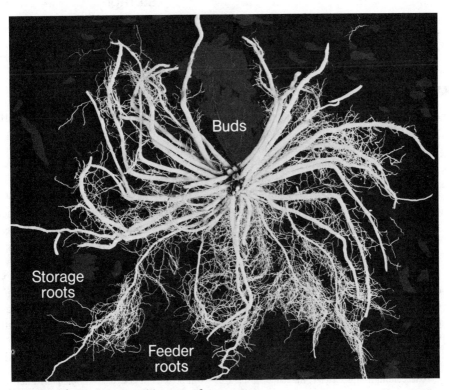

FIGURE 14.2. A 1-year-old crown of asparagus.

FIGURE 14.3. A bud cluster from a mature crown of asparagus showing developing spears.

FLORAL BIOLOGY AND CONTROLLED POLLINATION

Asparagus officinalis is dioecious, with male and female flowers borne on different plants.[1] Male flowers have rudimentary pistils, a few of which are capable of setting fruit with one or more seed. Female flowers have vestigal stamens, none of which have been known to produce pollen.

Asparagus flowers are borne on new shoots and reach anthesis before maturity of the shoots and the cladodes. Male plants flower 210–235 days and female plants 269–295 days after sowing seed in the greenhouse (J. H. Ellison, unpublished data). The fact that females start to flower only when the plants are older and larger than when males first flower seems to be the result of natural selection. Considerable energy goes into the seed, and female plants that set fruit before having the vigor to mature seed apparently have been eliminated from the population. Even mature females, when in a state of low vigor, often abort flowers instead of setting fruit. It seems logical also to assume that natural selection would favor early and prolific flowering with males, which would increase the probability of fertilizing mature females in the area. Males do flower more prolifically than females.

The fact that asparagus is dioecious, and that male and female plants differ signifi-

[1]Henceforth in this chapter, the name ''asparagus'' will be used to designate *A. officinalis* Linn.

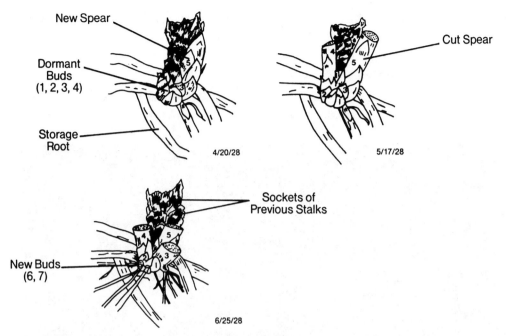

FIGURE 14.4. Three views of a bud cluster of a mature crown of asparagus showing the sequence of spear production and new bud development during the cutting season.
Drawing by G. T. Nightingale.

cantly in several characters of horticultural interest, makes the inheritance of sex expression very important to the breeder. Male and female plants occur in about equal numbers according to Flory (*14*), Robbins and Jones (*31*), and unpublished data of C. M. Rick and G. C. Hanna. Rick and Hanna (*30*) and Flory (*14*) attempted to distinguish the sex-determining part of chromosomes cytologically, but were not able to observe a heteromorphic pair. Subsequently, another authority, A. Thuesen (*38*), described a pair of sex chromosomes. Asparagus has 10 chromosomes ($2n = 20$), and Thuesen (*38*) describes no. V as the heteromorphic pair. The X chromosome has a median constriction, whereas the Y chromosome has a submedian constriction. Both chromosomes distinguish themselves by the greater distance at which their satellites are situated compared to the other five chromosomes, which have closely attached satellites. The Y chromosome is further conspicuous by having its satellite attached to the short arm, in contrast to other chromosomes, all of which have the satellites attached to the long arm. Both Rick and Hanna (*30*) and Thuesen (*38*) agree that sex expression in asparagus is inherited as though controlled by a single-gene factor dominant for maleness. Although Norton (*27*) did not study the inheritance of sex expression with asparagus, he did report finding "hermaphrodite" plants now and then, which had a preponderance of normal male flowers plus a few androgenous flowers with small but functional pistils. Such plants today are known as andromonoecious, since they are strongly male. *Andro,* from the Greek, refers to maleness; and monoecious (*mon* + Greek *oikos,* meaning "one house") refers to having both sexes in the same individual.

Pollination of the hermaphrodite (androgenous) flowers on male plants is most likely by self-pollination. The stigma of the male pistil is below and completely surrounded by the anthers of the narrow, campanulate flower (Fig. 14.1c). When the anthers dehisce, they form a nearly solid mass of pollen above and around the small stigma, thus making the probability of cross-pollination practically nil.

Pollination Techniques

Controlled pollination with asparagus is achieved either by hand or by the use of honey bees. Asparagus is dioecious, with female and male flowers on separate plants (see Fig.

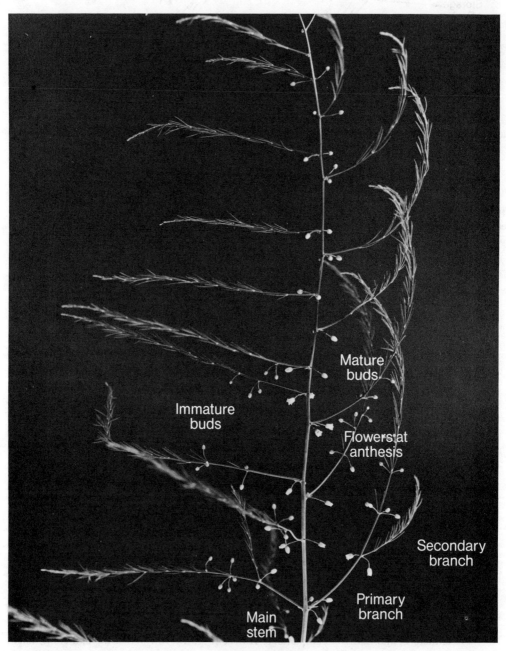

FIGURE 14.5A.   Flowering stalk of female asparagus plant.

14.5). Crossing can be achieved in an insect-proof greenhouse merely by using a fresh male flower at anthesis as a hand-held "brush," in order to deposit pollen on the stigmas of female flowers. This, of course, presupposes that both sexes are in bloom at the same time, which is not always the case. The author has found it quite convenient to collect pollen when available and store the pollen in tightly closed vials in an ordinary domestic freezer. The pollen remains viable for 6 months and longer, but should be replaced every 6 months for security. Anthers may be collected before they dehisce by scraping them

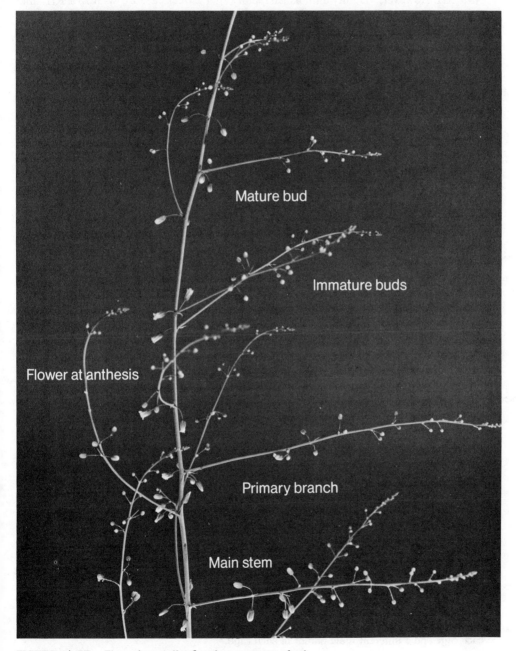

FIGURE 14.5B. Flowering stalk of male asparagus plant.

from the flowers, or pollen can be collected from intact flowers by suction after the anthers dehisce. A clean toothpick or a sterilized camel's hair brush is convenient for transferring the pollen to stigmas. Ample seed for greenhouse testing can be produced by hand pollination, in that 100 fruit generally yield 300 seed.

For long-term storage of asparagus pollen, Snope and Ellison (*37*) found that lowering temperature and relative humidity (RH) increased pollen life considerably. Viability was undiminished after 60 weeks at $-20°C$ with no control of RH and at $1°C$ with 15% and 45% RH. However, viability diminished rapidly after 2 weeks at $20°C$ with 75% RH, after 8 weeks at $1°C$ with 75% RH, and after 8 weeks at $20°C$ with 15, 45, and 75% RH.

For the production of sufficient seed for field experiments and for grower's farm demonstrations, one should use honey bees for controlled pollination. When many crosses are desired, as in diallel crossing of a number of females with a number of males, caged plants in the field are convenient. For example, in order to make 100 diallel crosses between 10 parents of each sex, one would have to produce 10 divisions of each female genotype (clone). One division of each female clone would be planted in the field in each of 10 cages (approx. $10 \times 6 \times 6$ ft high). With the 10 such cages established (see Fig. 14.6) one male genotype can be introduced into each of the 10 cages to complete the 10×10 diallel combinations.

Galvanized pipe is good for the framing of the cage, which can be covered with a plastic screen tent with zippered openings. Small hives of bees are put in the screened cage at the time of flowering, and can be maintained for 3 or 4 months with proper care and feeding. A breeder wishing to use different males in different years should not plant the male in the cage, but should use potted males from the greenhouse. One wishing to use pollen from a plant in the field may use "cut flowers" in a bucket of water. The stalk can be changed once or twice a week during the bloom period if the male plant is vigorous. Care must be taken to cut the stalks before the flower buds are open or to pick off any open flowers in case they are contaminated by foreign pollen.

Some situations require the pollination of a large number of female parents (50–100) with one male parent in a topcross. A seed garden isolated by a minimum of 300 m from other asparagus can be used for such crossing. The male parent, however, must be propagated vegetatively to produce sufficient pollen.

Seed Production

There are four simple but important rules to keep in mind when producing asparagus seed in the field:

1. *Do not harvest the spears:* Harvesting spears reduces stored reserves for maximum stalk growth and maximum seed production.

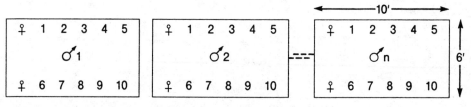

FIGURE 14.6. Asparagus pollination cages with 10 female parents (clones) repeated in each cage and a different male parent in each cage.

FIGURE 14.7. Asparagus berries, mean diameter 9.0 mm.

2. *Keep honey bees in the field:* A large population of bees, two to four hives per acre, will ensure good pollination.

3. *Keep asparagus beetles out of the field:* Asparagus beetles (*Crioceris duodecim-punctato* L.) will ruin the seed if not carefully controlled.

4. *Harvest only red-ripe berries:* Low germination is a common problem with asparagus seed. Scheer *et al.* (*33*) found that machine grading was effective in removing light, immature seeds of low germination when there was a large proportion of this type of seed in the sample. Seed from red-ripe fruit was superior in germination to seed from less mature bronze-colored fruit. Likewise, germination was poor in seed harvested from late brush vs. early brush.

A convenient spacing of plants in the seed block is 5 ft between rows and 2 ft between plants in the row. One row of males is needed for every set of four female rows. It is good to plant males in the outside row, then four female rows, then one male row, followed by four female rows, and so on across the field. This arrangement provides one male plant for every four female plants.

Aspargus seed can be harvested with a grain combine properly adjusted for asparagus, so that the berries (Fig. 14.7) are crushed and the seed (Fig. 14.8) are not. The crushed berries must not be allowed to sit overnight because heating can occur, reducing seed germination. The day's harvest should be washed with water to separate the berry skins and pulp from the seed, which should be spread out on screens to dry overnight.

Fusarium spores are common on asparagus seed, and the clean, dry seed should be treated with a mixture of benomyl (50% neat) at 5 g/kg of seed and Thiram (50% neat) at 8 g/kg of seed. This treatment is very effective in disinfecting the seed.

MAJOR ACHIEVEMENTS OF THE RECENT PAST

The year 1913 is not recent, but no review of achievements in asparagus breeding would be complete without reference to J. B. Norton's classic 1913 publication, ''Methods Used

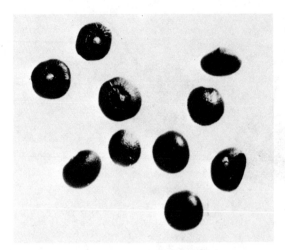

FIGURE 14.8. Asparagus seed, mean diameter
4.0 mm.

in Breeding Asparagus for Rust Resistance'' (27). Asparagus rust was first discovered in
America in 1896 by B. D. Halstead (17) of the New Jersey Agricultural Experiment
Station (AES). The same year brought reports of its occurrence on Long Island (New
York), Connecticut, and Massachusetts. The disease was thought to have been introduced
from Europe, where asparagus and probably also rust originated. By 1902 rust had spread
to the large asparagus fields of California, causing serious losses in all parts of the country
where asparagus was grown.

    Norton (27) reported that ''all cultivated varieties of asparagus are readily affected by
the rust, although it has been found that some varieties, notably Palmetto, are less
susceptible to its attack than others.'' French Argenteuil was found to be very similar to
Palmetto.

    Fungicide spray programs were developed by the various experiment stations, but for
reasons of cost and trouble, most growers did not take up the practice of spraying to
control the rust.

    It soon became apparent that breeding for rust resistance was a logical and necessary
approach to the problem (27). Commercial asparagus growers in the Concord, Massachusetts, area suffered such serious losses due to rust that they persuaded the Massachusetts AES to request the cooperation of the USDA in establishing such a breeding
program. Norton, a USDA employee, was put in charge of the project. He collected seed
samples of more than 100 asparagus ''varieties'' from nearly every state in the East, from
Illinois, from Michigan, and from England, France, and Germany. Norton grew seedlings
in the greenhouse during the winter in Washington, D.C., and transported them to eastern
Massachusetts for field evaluation during the summer season. Rust was so prevalent in
Massachusetts that he had little trouble evaluating the genetic lines for resistance.

    Norton made many plant selections and crosses in the field, after which he would plant
the seed in the Washington, D.C., greenhouse and evaluate the seedlings in Massachusetts the following year.

    The 1913 publication was ahead of its time. Norton studied his progenies by using
advanced statistical methods to correlate mature plant yield and vigor in the field as
related to mean height of seedlings in the greenhouse. The positive correlations enabled

him to do preliminary evaluations of progeny yield and vigor in the seedling stage, thus saving much time and money. He also found that the plant no. A7-83, selected for vigor and rust-resistant phenotype, transmitted larger seed size and taller seedlings than other selected males did when crossed on the same females. Also, at the same seed weight, A7-83 transmitted greater seedling height than other males, showing that the transmitted vigor of the seedlings was genetic and did not depend merely on more stored energy available in heavier seeds. Progenies of A7-83 were very rust resistant and had more vigor than open-pollinated progenies no matter what female was used. Norton gave A7-83 the name Washington and used it as the pollen parent for his two famous cultivars. A7-83 was selected from New American stock grown by Anson Wheeler of Concord, Massachusetts. Female B32-39, selected from Reading Giant stock grown by Sutton and Sons, Reading, England, transmitted better rust resistance than any other female tested. It was a rather small plant, but when crossed with Washington, it produced a progeny "best for resistance and type of any seedlings grown," to quote from Norton (27, p. 55). Martha was the name given to female B32-39, and Martha Washington the name given to the first generation progeny of B32-39 × A7-83.

In 1919 Norton (28) published a paper in which he described a second female plant, A5-11, which he named Mary. It was a giant plant, which when crossed with Washington produced the largest seed and largest seedlings of any combination he tested. Mary Washington was the name for the first generation progeny of A5-11 × A7-83. Mary Washington Stock was the designation for the F_2 seed of Mary Washington, and the same system of nomenclature was used for Martha Washington and Martha Washington Stock.

In this publication, Norton (28) pointed out the variability from plant to plant even in the F_1 generation of both Martha and Mary Washington and gave explicit instructions for proper seed production of the two cultivars. He mentioned the isolation necessary from other asparagus, the roguing of off-type plants from the seed block, and even the importance of proper sorting of crowns to be used for seed production. What a pity neither the USDA nor the Massachusetts (nor any other) AES heeded his words. Not even an industrial asparagus association made any effort to maintain pure stocks of Martha or Mary Washington asparagus. As a consequence, many growers saved seed from their production fields, usually without any knowledge of proper methods for maintaining the genetic qualities of the original cultivars. The result was the loss over the years of perhaps the best asparagus strains ever produced in the United States.

In 1919, J. B. Norton retired from the USDA and took a position with the Coker Pedigreed Seed Company of Hartsville, South Carolina, a company well known because of its excellent research and breeding programs in corn, cotton, and other important crops grown in the South. Norton used his Washington parents and foundation stock plants to put the Coker Company in the forefront of the asparagus seed business. Norton remained with the company until his death in 1936.

Yield data presented by Jones and Hanna (21) in Davis, California, indicated (Table 14.4) that Mary Washington from the Coker Pedigreed Seed Company yielded more than six other strains of asparagus (Palmetto plus five types of Washington), and produced larger-sized spears than five of the other strains (Table 14.5). Jones and Hanna (21) summarized as follows: Mary Washington yielded more than the other cultivars and also was superior in earliness and size of spear. Mary Washington outyielded Palmetto by a total of 3444 lb/acre for the 12 years and the lowest yielding strain of the Martha Washington by 20,868 lb/acre. Highly significant differences occurred between strains of the same cultivar: one strain of Martha Washington outyielded the other by a mean of 657

TABLE 14.4.   Yield of Spears (lb/acre) from Different Varieties and Strains of Asparagus, Davis, California[a]

Variety	1925	1926	1927	1928	1929	1930
Palmetto[c]	590	2,482	4,195	5,024	6,653	6,756
Mary Washington[d]	712	2,670	4,446	5,352	7,298	7,269
Washington A[e]	396	2,448	3,978	4,766	5,956	6,880
Washington B[d]	344	1,829	3,109	3,842	4,862	5,561
Martha Washington A[f]	426	2,146	3,583	4,262	5,582	6,088
Martha Washington B[c]	436	2,204	3,604	4,496	5,613	6,342
Reselected Washington[d]	477	2,408	4,026	4,586	5,666	6,734

[a]After Jones and Hanna (*21*).
[b]The smallest difference between varieties and strains that can be considered significant under the conditions of this test is 285 lb.
[c]Seed produced in the Delta region of California.

lb/acre, or a total of 7884 lb for the 12 years. Mary Washington produced significantly larger spears than Palmetto or Martha Washington.

Essentially every strain of asparagus that has been developed in the United States and Canada since 1930 has been a selection out of Martha or Mary Washington.

One of the most successful selections out of Mary Washington stock was California 500, produced by G. C. Hanna, a Vegetable Specialist at the University of California, Davis. In the early 1950s, California 500 was thought to have been grown on more than half of the then 78,000 acres in the state. This cultivar was known for high yield, good uniformity, and especially for the attractive appearance of the spears for fresh market. Hanna selected for tight tips, a full sized (vs. narrow) "neck," and smooth, tightly clasped bud scales on the stem of the spear. (The spear on the left of Fig. 14.9 has the

TABLE 14.5.   Mean Weight (g) of Spears of Different Varieties and Strains of Asparagus, Davis, California[a]

Variety	1925	1926	1927	1928	1929
Palmetto[c]	17.8	25.5	29.7	29.0	32.6
Mary Washington[d]	23.6	32.1	34.8	35.4	39.0
Washington A[e]	22.7	31.2	37.5	37.2	37.6
Washington B[d]	19.8	26.0	30.0	30.6	32.2
Martha Washington A[f]	21.5	28.2	31.2	31.4	33.6
Martha Washington B[c]	22.2	27.0	30.9	32.2	34.2
Reselected Washington[d]	21.8	29.2	33.3	32.2	33.4

[a]After Jones and Hanna (*21*).
[b]The smallest difference between varieties and strains that can be considered significant under the conditions of this test is 1.21 g.
[c]Seed produced in the Delta region of California.

Table 14.4. (*cont.*)

1931	1932	1933	1934	1935	1936	Mean[b]
10,436	9,117	8,205	7,918	7,635	9,362	6,531
10,638	9,266	8,520	8,097	7,994	9,553	6,818
9,061	8,444	7,606	6,844	7,388	8,434	6,017
8,264	7,628	6,902	6,398	7,117	8,244	5,342
8,274	6,890	6,290	5,356	5,622	6,433	5,079
8,792	7,788	7,497	6,625	7,029	8,400	5,736
9,298	8,474	7,816	7,386	7,740	8,879	6,124

[d]Seed from the Coker Pedigreed Seed Company of Hartsville, South Carolina.
[e]Seed from the USDA.
[f]Seed from the Market Garden Field Station, Waltham, Massachusetts.

desirable conformation.) The spear colors, both purple tip and green background, were not as dark as in the original Mary Washington, nor did California 500 carry the rust resistance of the Mary Washington stock. For this reason, California 500 was poorly adapted in the humid eastern states, where rust is more prevalent than in the desert climate of California.

J. Sneep (*35*), working in the Netherlands, stated that O. Banga (*2*), director of the Institute for Plant Breeding, Wageningen, toured the United States in 1945–1946, right after World War II, in search of recent advances in plant breeding. Banga met G. C. Hanna in California and brought to the Netherlands the theoretical scheme for breeding all-male asparagus. Sneep (*35*) surveyed commercial asparagus fields searching for andromonoecious plants. He cautioned against selecting ''double'' plants, namely, a male

Table 14.5. (*cont.*)

1930	1931	1932	1933	1934	1935	1936	Mean[b]
27.1	28.8	26.2	25.1	23.2	25.8	22.4	26.1
32.7	33.2	29.3	28.4	26.4	28.3	24.7	30.7
34.2	31.9	29.6	28.6	26.2	28.2	24.6	30.8
29.8	29.2	27.5	25.8	24.0	26.4	23.2	27.0
30.2	28.8	26.2	26.8	23.8	26.2	22.4	27.5
31.2	30.0	27.4	26.5	25.0	26.8	24.6	28.2
31.0	30.0	27.8	27.2	25.1	27.8	24.7	28.6

[d]Seed from the Coker Pedigreed Seed Company of Hartsville, South Carolina.
[e]Seed from the USDA.
[f]Seed from the Market Garden Field Station, Waltham, Massachusetts.

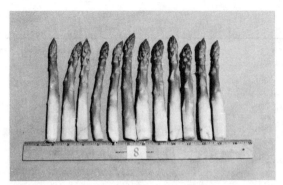

FIGURE 14.9. Asparagus spears cut for market.
Spear on left has ideal conformation.

plant and a female plant that grew as one plant in the nursery and were transplanted as such in the field. Thus, the male and female shoots growing intermixed from the same clump could incorrectly appear to be an andromonoecious (berry-bearing male) plant. Sneep reported the number of andromonoecious plants found in the fields fluctuated between 0 and 2%. He also reported that of 234 descendants of 10 andromonoecious plants, 56 were female, 160 were male, and 18 were andromonoecious.

Homogametic (MM) supermales can be distinguished from heterogametic (Mm) siblings by a sex genotype progeny test. Progeny of the supermale (MM) will be all male (Mm) and progeny of the heterogametic (Mm) male will segregate male (Mm) and female (mm) in a 1:1 ratio. Female sex genotype is homogametic recessive (mm).

In the same paper Sneep (*35*) pointed out the possibility of using andromonoecism in the breeding of uniform male hybrids by crossing female inbreds with supermale (MM) inbreds. To produce the female inbreds he proposed selfing an andromonoecious plant (Mm), selecting an Mm in S_1 for selfing, and continuing the selfing series to homozygosity (Fig. 14.10). To produce inbred supermales (MM) he proposed a similar selfing program beginning with an andromonoecious (MM) plant.

In a subsequent paper, Sneep (*36*) elaborated on the inheritance of andromonoecism. He pointed out that a certain homogametic (MM) F_3 andromonoecious plant produced 55

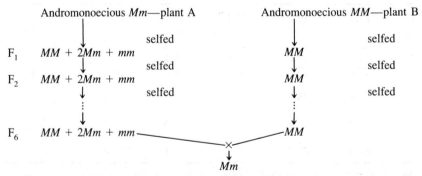

FIGURE 14.10. Scheme for inbreeding andromonoecious asparagus plants for the production of uniform male hybrids.
After Sneep (*35*).

seedlings through selfing, and that all were andromonoecious—none female. He concluded that all such offspring also would be homogametic (*MM*) and could be used for the production of male asparagus hybrids. He also stated that the inheritance of andromonoecism involved dominant factors and could lead to the production of all andromonoecious [or *hermaphroditic* (italics mine)] hybrids instead of all male hybrids. Sneep (*36*, pp. 227–228) stated that this problem might occur and referred to a *preliminary* scheme for preventing andromonoecism in commercial seeds by means of lines that are heterozygous for this condition. He suggested that "in addition to plants homozygous for andromonoecism, these heterozygous lines would produce heterozygous plants and plants of which the genetic factors determining andromonoecism are recessive. The heterozygous plants can be used to continue the process of inbreeding. The non-andromonoecious (*MM*) plants, which have only recessive factors for andromonoecism, can be used as father plants in producing commercial seed of the male variety, without the fear that this seed will yield andromonoecious plants."

For reasons unknown to the writer, the advanced-generation hybrids that came out of the male line breeding program initiated by Sneep did not prove to be popular in the Netherlands. I suspect that it was due to the problem of producing andromonoecious and/or hermaphroditic hybrids due to the selection pressure for andromonoecism that resulted from inbreeding successive generations of andromonoecious plants. The writer observed one of Sneep's F_4 generation asparagus inbred lines that was very uniform, quite low in vigor, and produced more berries and seed than an ordinary dioecious asparagus line. This F_4 line was completely hermaphroditic.

Another worker in the Netherlands, A. A. Franken, assisted by C. Backus, a technician, produced a number of inbred lines by sibbing for several generations (A. A. Franken, personal communication). By using females from certain inbreds crossed with males from other inbreds, Franken and Backus produced 125 hybrids for field testing. The writer saw the hybrids at the end of the first growing season, and several of the hybrids looked outstanding for vigor and yield potential. After a second year of observation, the best hybrids were to be selected for a replicated yield trial. Commercial seed of the chosen hybrids will be easy to make because of the inbred nature of the parent dioecious lines. The seedsman would need only to grow the required number of females from the seed parent line and the males from the pollen parent line to establish the permanent seed block. This program led to the development of the Limbras Hybrids, numbered 10, 18, 22, and 26, which make up about 95% of the newly set fields in the Netherlands. Reports from Europe and New Zealand indicate that the Limbras Hybrids produce high yields, but disease resistance of these hybrids is unknown to the author.

Because of the excellent paper on breeding methods by Lucette Corriols-Thevenin (*6*), the author presents some details of her scheme below. The first commercial hybrids released in France by Corriols-Thevenin were double-cross hybrids. Figure 14.11 shows how this program was carried out and is self-explanatory. The double-cross hybrid approach was used because it permitted large-scale seed production in relatively few years. This was done before large-scale tissue culture had been developed. From 1974 to 1977 four double-cross hybrids were released (Diane, Junon, Minerve, and Larac), which increased total yield over the commercial standard more than 30% and early yield more than 60% (*6*).

After commercial tissue culture was perfected, Corriols-Thevenin released four clonal hybrids (Aneto, Lesto, Bruneto, and Steline). These clonal hybrids increased total yield over the standard by 75% and early yield at a similar 76% (*6*). The scheme used to

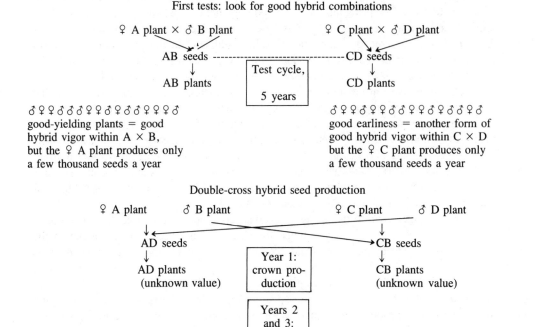

First tests: look for good hybrid combinations

♀ A plant × ♂ B plant   ♀ C plant × ♂ D plant

AB seeds -CD seeds

AB plants | Test cycle, | CD plants

                 | 5 years |

♂♀♀♂♂♂♀♀♂♀♂♂♀♀♂   ♂♀♀♂♀♀♂♂♀♀♂♀♂♂♀♂
good-yielding plants = good   good earliness = another form of
hybrid vigor within A × B,   good hybrid vigor within C × D
but the ♀ A plant produces only   but the ♀ C plant produces only
a few thousand seeds a year   a few thousand seeds a year

Double-cross hybrid seed production

♀ A plant   ♂ B plant   ♀ C plant   ♂ D plant

AD seeds   →CB seeds

AD plants | Year 1: | CB plants
(unknown value) | crown pro- | (unknown value)
                 | duction |

                 | Years 2 |
                 | and 3: |
♀♀♀♀♀♀♀♀♀♀♀♀♀♀♀ | planting | ♂♂♂♂♂♂♂♂♂♂♂♂♂♂♂
                 | flowering |
All the male plants are discarded |←sexing→| All the female plants are discarded

                 | Year 4: |
                 | commercial seed production |

Set the AD females and the CB males in a field far from any other asparagus culture. Then, insects will
transfer pollen from male to female plants.
The AD female plants will bear ADCB seeds (double hybrids); these seeds will have again the hybrid
vigor previously detected in both AB and CD initial couples.

FIGURE 14.11. Scheme for the production of double-cross hybrid asparagus seed.
Adapted from Corriols-Thevenin (6).

develop the clonal hybrids and the seed production is presented in Fig. 14.12. The clonal
hybrids are more homogeneous than the double-cross hybrids.

Corriols-Thevenin (6) is now developing uniform F_1 hybrids from homozygous parents, which are propagated by tissue culture for seed production. The seed parents are
obtained by starting with haploids from polyembryonic seeds, which are doubled by the
use of colchicine. Both pollen parents and seed parents may be obtained from anther
culture, in which haploid embryoids are automatically doubled during the anther culture
process. The male plants that originate in anther culture will sire only male hybrids and
are called supermales, because they do not carry the gene for femaleness. One such
homogeneous F_1 hybrid, called no. 54, is very promising and comparable to Cito and
Steline for total and early yield (6).

An efficient method to detect monoploid seedlings was developed by Bassett et al. (4),
who used a recessive seedling marker gene (basal stem color) to detect parthenogenesis in
female plants. Red-stalk is the marker. When red-stalk female plants were crossed with a

homozygous dominant green-stalk male, any seedlings in the progeny that showed the recessive red-stalk character were very likely to be monoploid. This method was much less time consuming than examining twin seedlings cytologically.

Also included in the Corriols-Thevenin paper (6) is a detailed scheme (Fig. 14.13) for producing a dioecious pure line by backcrossing five generations to a homozygous female. The male and female offspring (except for the sex chromosomes of the male) will have the same genotype as the homozygous recurrent female parent. This pure line can then be propagated by seed. Uniform F_1 hybrid seed can be produced by crossing female plants from one such pure line with male offspring from a genetically unrelated pure line that is produced and maintained in the same way.

It is obvious from the yield data presented by Corriols-Thevenin (6) that her hybrids are uniform and high yielding. The concern of the author is that the French hybrids have not been selected for disease resistance, and certain ones displayed considerably more suscep-

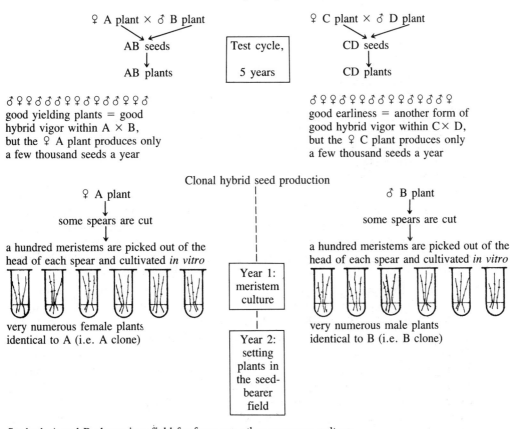

Set both A and B clones in a field far from any other asparagus culture.
Then insects will transfer pollen from male to female plants. As soon as planted, the A females will be able to bear AB seeds (= clonal hybrid): these seeds are the same as those previously found valuable in the first tests.

FIGURE 14.12. Scheme for producing hybrid seed by clonal propagation of the inbreds.
Adapted from Corriols-Thevenin (6).

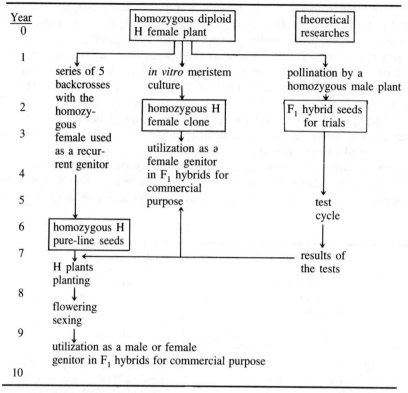

FIGURE 14.13. Scheme for producing pure-line asparagus pollen parents by backcrossing males to a homozygous female plant and the time required to develop uniform hybrids.
Adapted from Corriols-Thevenin (6).

tibility to *Fusarium* spp. in 1978 than did Rutgers experimental hybrids growing side by side on *Fusarium* infested soil in Rio Grande do Sul, Brazil. With a perennial crop such as asparagus, which cannot be rotated like an annual crop, there is the danger that a strain of *Fusarium* will develop that will be virulent on a particular hybrid. When this happens, every plant in the field will be equally susceptible because of the uniformity of the hybrid.

The most impressive asparagus breeding program in Europe with which the writer is familiar is conducted by Dr. Hans Rolf Spath, Director, Sudwestdeutsche Saatzucht GMBH, Oberwald 2, Rastatt/Baden, West Germany, F.R. This private seed company had the services of a father and son team of asparagus breeders, which probably spanned 75 years of continuous selecting, crossing, and testing of lines. They had developed asparagus hybrids highly resistant to rust and botrytis. This was the only botrytis resistance known to the author at that time, even though the disease was serious in France, Germany, and Spain, and perhaps elsewhere. In 1972, the son produced experimental dioecious F_1 hybrids through eight generations of sib mating to produce inbreds (Fig. 14.14). He also had produced experimental male hybrids, using S_1 generation supermales (Fig. 14.15). The seed parents of the male hybrids probably were the same seed parents developed through inbreeding, as shown in the scheme in Fig. 14.14. The best dioecious hybrids yielded 50% more than their own standard commercial strain, and the best male

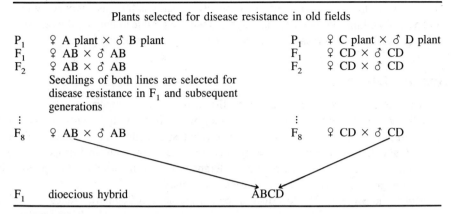

Plants selected for disease resistance in old fields

P₁	♀ A plant × ♂ B plant		P₁	♀ C plant × ♂ D plant
F₁	♀ AB × ♂ AB		F₁	♀ CD × ♂ CD
F₂	♀ AB × ♂ AB		F₂	♀ CD × ♂ CD

Seedlings of both lines are selected for disease resistance in F₁ and subsequent generations

F₈   ♀ AB × ♂ AB   F₈   ♀ CD × ♂ CD

F₁   dioecious hybrid   ABCD

FIGURE 14.14. Scheme for sib mating asparagus to produce inbred parents and hybrid lines.
After F. Bohne, personal communication.

hybrid yielded 100% more than the same standard. The commercial hybrids now are available as Lucullus numbered hybrids.

CURRENT GOALS OF BREEDING PROGRAMS

Male Hybrids

Studies by Robbins and Jones (*31*), Currence and Richardson (*7*), Yeager and Scott (*44*), and Ellison and Scheer (*10*) indicate that male asparagus plants yield more than female plants. Yeager and Scott (*44*) also found that males live longer than females, a very important characteristic in a perennial crop. Robbins and Jones (*31*) and Ellison and Scheer (*10*) also pointed out that males produced yields earlier in the spring than females. Although females generally yield larger spears than males (*11*), certain male hybrids have larger spears than dioecious lines (J. H. Ellison, unpublished data).

Further advantages of male hybrids are that they produce no seedling weeds, which are a problem in dioecious asparagus fields. Much more important, growers cannot save their

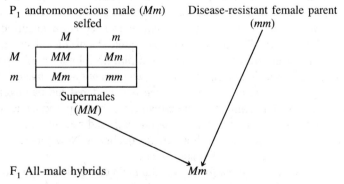

FIGURE 14.15. Scheme for breeding all-male hybrids.
After F. Bohne, personal communication.

own seed, resulting in the loss of a new genetic asparagus line, as happened with Martha and Mary Washington. Seed of male hybrids will be produced only by the seedsman from selected seed parents crossed with the selected supermale pollen parent.

Fusarium Resistance

Resistance to *Fusarium* spp. has to be listed as the most important goal in asparagus breeding today. Fusarium is recognized as a limiting factor to asparagus production in all of the important growing areas of the United States, Canada, Brazil, Japan, Taiwan, and New Zealand. Fusarium is now being recognized as a problem in Europe also. A new asparagus cultivar without at least field tolerance to fusarium would not be popular in the asparagus-growing areas of the countries listed above.

Molot and Simone (25) described a replant problem in Southern France, in which the plant symptoms are similar to those of fusarium, as being caused by *Rhizoctonia violacea* Tul. The replanting problem in Germany and the Netherlands was associated with symptoms similar to those of *Fusarium* spp. in the United States.

Fusarium oxysporum Schlecht f. sp. *asparagi* Cohen has been recognized as a serious pathogen to asparagus in New Jersey for many years, but it had been considered mainly a debilitating pathogen, since it was found generally to rot only the small fibrous roots. It was of economic importance since it reduced plant vigor and thus yield and size of spears, but it was thought not to kill plants outright. The author observed during the 1950s and 1960s that in the poorly drained parts of any field, plants did die outright, and survivors were markedly stunted.

In the early 1970s, four consecutive rainy seasons made most New Jersey asparagus fields appear to be poorly drained. In August, 1972, asparagus stalks turned yellow and many lodged. A survey of problem fields revealed that the crowns were severely rotted. Up to 50% of the plants were dead, and the survivors were severely stunted. The fusarium was no respecter of age of bed, and many acres were put to the plow. Most of the fields had been planted previously to asparagus, and so the soil was thoroughly infested with *Fusarium* spp.

Stephen A. Johnston made a detailed study of the asparagus fusarium problem in New Jersey. In studying 12-year-old plants in a declining field, Johnston *et al.* (20) isolated *Fusarium moniliforme* Sheldon from stem and crown lesions. *F. moniliforme* was isolated infrequently from lesions on roots, stems, and crowns of plants in a 2-year-old field. Typical reddish lesions and vascular discoloration developed on roots of seedlings inoculated with *F. oxysporum* f. sp. *asparagi,* but complete root collapse occurred after seedlings were inoculated with *F. moniliforme.* To quote from the abstract, "*F. moniliforme* caused more extensive crown rot and stem pith discoloration of asparagus plants in crown and stem pathogenicity tests than did *F. oxysporum* f. sp. *asparagi.* Based on isolation and pathogenicity studies, *F. moniliforme* is considered the pathogen of a separate disease of asparagus. Fusarium stem and crown rot is the proposed name for this disease. Both fusarium stem and crown rot and fusarium wilt and root rot caused by *F. oxysporum* f. sp. *asparagi* are associated with asparagus decline in New Jersey" (20, p. 778). Symptoms of the two diseases overlap somewhat.

It is the author's opinion that *F. moniliforme* was the main cause of the dying out and dwarfing of asparagus plants observed for years in poorly drained areas of asparagus fields mentioned above. Since it was common knowledge that asparagus could not tolerate "wet feet'' no one really investigated the cause of asparagus decline in isolated low areas of the fields.

In California the fusarium problem is ascribed to *F. oxysporum* f. sp. *asparagi* and little evidence points to problems with *F. moniliforme*. Fusarium has been a major problem in the Delta region of California for many years and was responsible for much of the asparagus acreage being shifted to the Salinas Valley to new land not infested with *Fusarium*.

In New Zealand and Japan, fusarium is a serious problem, and as in California, the cause is related to *F. oxysporum* f. sp. *asparagi* and little blame is put on *F. moniliforme*. The situation in Rio Grande do Sul, Brazil, is more like that in New Jersey, in that both *F. oxysporum* and *F. moniliforme* cause serious damage.

Regardless of the area in the world or the organism associated with the problem, poor drainage always seems to produce the same conditions—dying plants and stunted survivors.

Rust Resistance

Resistance to rust (*Puccinia asparagi* D.C.) is an important asparagus breeding objective in the humid eastern and midwestern United States and corresponding areas in Canada, but not so much in the arid western states of California and Washington. Asparagus rust is an obligate parasite, and does not have an alternate host as do many other species of *Puccinia*. Rust defoliates the plants prematurely, thus cutting off the source of photosynthate for storage in the extensive asparagus root system. The asparagus is more vulnerable to rust in the nursery and the first 2 years in the field, when generally it is not harvested. Removing the spears during a 2-month harvest interrupts the spring aeciospore stage of the rust, which prevents or delays the summer (urediospore) stage, which is so destructive to the asparagus fern. Kahn *et al.* (*22*) gave a detailed description and life cycle of *P. asparagi* D.C.

Rust generally is less of a problem in the northern European countries (England, Germany, The Netherlands) than in the eastern United States. Many asparagus cultivars from these countries proved to be quite susceptible to rust under New Jersey conditions (J. H. Ellison, unpublished data).

It is not known whether races of *P. asparagi* exist or not. There are no cultivars of asparagus immune to rust that might show a differential response to races of *Puccinia* if they do exist.

Yield and Adaptation

Good productivity is essential in any new asparagus cultivar because the grower must have a good return on investment. Local adaptation is very important in the performance of asparagus cultivars and has a great effect on yield. An example of the importance of local adaptation is illustrated by the performance of two Rutgers F_1 hybrids in New Jersey and Michigan. Both hybrids yielded well in New Jersey, and although one yielded the best of nine lines in a trial in Michigan, the other yielded only seventh from the top, and both of these hybrids had the same pollen parent! A similar, though reverse, situation occurred with two Rutgers asparagus F_1 hybrids that were evaluated in New Zealand and New Jersey. Both hybrids yielded at the top of 11 lines at each of three locations in New Zealand, but only one of the hybrids performed well in New Jersey, where they both originated. These two hybrids also had a common pllen parent. Adaptation to different soil types and climates is an important characteristic of an asparagus hybrid; and needless to say, it is difficult to predict.

Spear Quality

The price for large spears usually is greater than the price received for small or medium-sized spears. Good conformation of the spear is preferred to a narrow neck below the tip. The tip should be composed of tightly closed, smooth buds, and lateral buds should be smooth and covered by tightly formed bracts. A round cross section of the spear is preferred to oval, and flat cross section spear shape should be avoided. A dark-green background color with dark-purple bracts and tip is the original Mary- and Martha Washington-type spear color, which is popular with the trade. White asparagus, etiolated by being cut underground in a high ridge, is preferred without a purple spear tip, which is a telltale sign that the tip was exposed to light at the time of harvest. The telltale purple tip is associated with an older emerged spear, which is thus likely not to be as tender as a spear with a white or pale-green tip. Since it is impossible to harvest all spears before the tip is exposed to light, which causes spears to color somewhat, an all-green mutant is desired for white-spear production. This mutant would be incapable of developing anthocyanin pigments; thus, spears that are cut after having emerged slightly more than ideal still would not have the telltale purple tip.

Long, fiber-free spears are desirable and this may be an objective in some asparagus breeding projects. Although there are differences in fiber content of spears of different genetic lines, the greatest difference in spear fiber content is most apt to be due to the postharvest method of handling. Fiber starts to develop the moment the spear is cut, and the rate of fiber development is positively correlated with the temperature in which the spear is kept. Prompt removal of field heat by hydrocooling and storage of the spear at 2°–3°C will minimize fiber development, especially if the butt of the spear is in water or resting on a wet pad.

Genetic Variation Available

The potential for improvement in asparagus is great due to the vast genetic variability associated with dioecism. This is true if the breeder has access to large commercial acreage for plant selection, and especially if the region has been in production for many years. In replanted fields and in old fields, where fusarium pressure is high, and in fields suffering epiphytotics of rust, natural selection offers the astute breeder an abundance of highly valuable germplasm from which to select disease-resistant breeding material. This type of breeding material should have good local adaptation.

Local adaptation seems to be more important with asparagus than with most annual vegetable crops because of the prolonged exposure to the pathogens (season after season) of the same plants in the same site. Annual crops usually are rotated and not subjected to the accumulative effects of one particular site.

Marker Genes

Few marker genes have been found in asparagus (Table 14.6). Irizarry *et al.* (*19*) reported one such distinctive character as persistent green, inherited as a single recessive gene (g/g). In the homozygous state, this gene maintains dark-green foliage and the foliar pigments, chlorophyll and carotene, at maximum levels in the late fall when normal

**TABLE 14.6.   Simply Inherited Characters of Asparagus and
                      Their Gene Symbols**

Symbol	Character description	Reference
G	Normal green	19
g	Persistent green	19
p	Anthocyanin pigmentation (red stalk)	4
P	All green (no anthocyanin)	4
M	Male sex expression	35
m	Female sex expression	35

asparagus has lost most of the chlorophyll and the brush has turned yellow. Persistent-green plants never turn yellow, but turn from green to brown when the tissue dies. Although this writer thought that the persistent-green gene would increase yield by maintaining green brush later in the fall than normal brush and extending the photosynthetic period, increased yield is not the case in New Jersey (J. H. Ellison, unpublished data). It may be that increased metabolic activity occurs in persistent-green plants in late fall in New Jersey, and this may predispose the plants to low-temperature injury in early winter. Yield of a persistent-green experimental cultivar is outstanding in Mexico, where the growing season is long and hard freezes do not occur. However, there are no critical comparative yield data between persistent-green cultivars and yellow controls in Mexico as in New Jersey.

Red stalk (recessive to green stalk) is a marker that was found to be useful by Bassett *et al.* (*4*) in the identification of haploid seedlings (see section on Major Achievements of the Recent Past).

SELECTION TECHNIQUES FOR SPECIFIC CHARACTERS

Rust Resistance

Norton (*27*) selected phenotypes resistant to rust (*P. asparagi* D.C.) and made numerous crosses among the selected parents. The progenies were field tested for rust resistance, and parents were selected on the basis of low rust and high vigor of the progeny.

The author selected 50 plants for rust resistance and vigorous phenotype among an estimated 250,000 plants from 30 or more acres of cultivated asparagus in 1960 (J. H. Ellison, unpublished data). The farms were widely scattered in three counties of southern New Jersey. The severity of the rust epiphytotic varied from field to field where the plants were selected, and this showed up when all the plants were inoculated under uniform conditions in the greenhouse. Although there was in general a correlation between severity of symptoms of the plants in the field and in the greenhouse, a few plants with little rust in the field developed severe symptoms in the greenhouse and were discarded. As with Norton's work (*27*), the plants selected by the author for rust-resistant phenotype produced rust-resistant progenies. Figure 14.16 shows such a progeny versus the more susceptible Roberts Super strain of Mary Washington.

Fusarium Resistance

Little is published regarding techniques for selecting asparagus for resistance to *Fusarium* spp., but the method described on the next page is fairly typical of those used by asparagus breeders in the United States and overseas.

FIGURE 14.16. Left: Roberts Super strain showing severe rust damage.
Right: A resistant experimental cross showing less damage.

In 1961 the author and his assistant collected 200 specimens (isolates) of *Fusarium* spp., from brown vascular bundles of yellowing asparagus stalks 5–6 cm above ground. These were collected in 20 or so asparagus fields in southern New Jersey. Each isolate came from a different stalk, was given an accession number, and was catalogued as to name and address of the owner of the farm from which the specimen came. Each isolate was cultured separately and a drench of mycelia and spores was prepared and applied to asparagus seeds sown in pasteurized soil. The pathogenicity of the isolates ranged from nil to highly virulent, in which 90–100% of the asparagus seedlings were killed, and the distribution pattern of all isolates fit a bell-shaped curve. Twelve of the most virulent isolates were put in two systems of long-term storage: (1) in soil, which was allowed to dry, and (2) in agar, under sterile mineral oil. Four of the 12 selected for widely differing geographic locations were cultured separately and then mixed before being used to inoculate soil. The soil, about 100 gal. in volume, had been taken from two old asparagus fields, one in central and the other in southern New Jersey. Both were highly infested with *Fusarium* spp., and after being inoculated at the rate of about 3 gal. of the mixed *Fusarium* inoculum per week for 6 months, truly became a deathbed for asparagus seedlings.

The author had assumed that the isolates of *Fusarium* collected in 1961 and used in the inoculation work during the 1960s and early 1970s were all specimens of *F. oxysporum* f. sp. *asparagi*. This was the species of *Fusarium* associated with asparagus root rot, which had been recognized for some years. When Johnston *et al.* (*20*) ascribed asparagus stem and crown rot as being caused by *F. moniliforme,* the writer was afraid he had not been selecting against the more pathogenic of the two fusaria involved. In the mid 1970s Johnston identified the isolates that had been put in long-term storage in the 1960s. Actually, 8 of the 12 most virulent isolates saved for long-term storage were *F. moniliforme,* and three of the four isolates used to create the deathbed inoculum were *F. moniliforme.* This should have been expected, because *F. moniliforme* is generally more pathogenic than *F. oxysporum,* and only the most pathogenic isolates were saved.

An estimated 100,000 seeds of F_1 asparagus progenies selected for rust resistance were put through the fusarium deathbed. Most of the seedlings were killed within 6 weeks. The strongest were selected among the survivors, and after several transplantations in the inoculated soil, requiring up to 24 months in the greenhouse, approximately 0.64% (144 female and 500 male seedlings) were selected for planting in the field. Other workers also report a higher proportion of male than female seedling survivors under high-titer *Fusarium* inoculation even before the seedlings mature (G. C. Hanna and J. A. Huyskes, personal correspondence). These 644 seedlings were the most vigorous and had the least symptoms of *Fusarium* infection of the estimated 50,000 seedlings studied. Approximately 50% of the 100,000 seeds never emerged or died in the first few weeks after emergence. After 15 years in the field under conditions favoring fusarium stem and crown rot (heavy soil and wet seasons in the 1970s) only about 20% of the fusarium-selected plants survived, and of these only 14% of the females and 1% of the males were outstanding in fusarium-resistant phenotype. Progeny testing of these and other parents for fusarium resistance is underway at the present writing. Preliminary results indicate that of these F_1 plants, selected for resistance to fusarium, only 1 seed parent is better than the best P_1 plants selected originally for rust resistance and vigor in old cultivated fields where natural selection had prevailed. Perhaps the above results should not be surprising, considering the heterogeneity of asparagus populations and the fact that only one generation of selection pressure was imposed, rigorous though it was. Perhaps it is put in better perspective to say that the best field-selected parents, after natural selection had operated, were as resistant genotypically as the F_1 plants which were selected rigorously for fusarium resistance in the greenhouse. Figure 14.17 shows plants selected for resistance to fusarium by Ellison: plant (A) has phenotypic but not genotypic resistance; plant (B) shows little resistance; and the Mary Washington plant (C) shows none.

In one asparagus breeding program in which a plant pathologist selected several generations for resistance to fusarium, experimental hybrids with resistance to fusarium were developed, but yield was very low. This resulted because of not selecting for vigor and yield while selecting for resistance (R. Gilmer, personal conversation).

Sex Expression

The author (J. H. Ellison, unpublished data) has observed a positive correlation between the level of fruitfulness of the andromonoecious father of supermales and the tendency to obtain berry-bearing males in the male hybrids sired by these supermales. For this reason, it is wise to select supermales only from andromonoecious plants that set relatively few (1–10) fruits per year. The tendency for berry-bearing males to occur in male hybrids is influenced also by the seed parent of the hybrids. Apparently there are modifying genes carried by seed parents that govern the level of andromonoecism expressed in the male hybrids. It behooves the breeder to test experimental male hybrids thoroughly for maleness before releasing them.

[See the reference to sex expression in the section on Floral Biology and Controlled

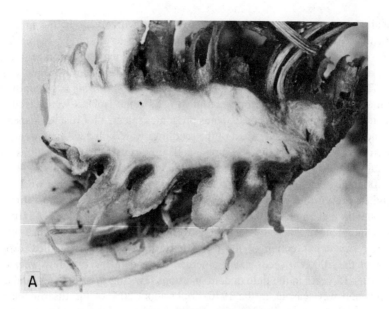

FIGURE 14.17. (A) Selected plant showing phenotypic resistance to fusarium crown rot. (B) Although selected for resistance, plant shows little resistance. (C) Mary Washington plants killed by fusarium crown rot. Note the removal of parenchyma tissue by the rot.

Pollination, as well as the reference to Sneep (*35,36*) in the section on Major Breeding Achievements of the Recent Past for selection techniques regarding sex expression.]

Yielding Ability and Spear Quality

There are two theories concerning yielding ability with asparagus. One is based on the premise that there are major genes for vigor and high yield in the parents, and that a favorable combination of these genes produces a vigorous, high-yielding progeny. The other theory is based on heterosis. Since asparagus naturally is cross pollinated, inbreeding brings about a depression of vigor and yield, which is more than restored when unrelated inbreds are crossed in favorable combination. Evidence to support the "gene for yield" theory is seen in Mary Washington, the very high-yielding progeny of two giant plants. The author recently released Jersey Centennial, a vigorous, high-yielding F_1 hybrid, the parents of which also are very vigorous. There is ample evidence also to support the heterosis theory. One example comes from The Netherlands where low-vigor sibling inbreds produce vigorous, high-yielding hybrids; another comes from France, where doubled monoploids (haploids), crossed with inbred males, produce hybrids with high yield and vigor (*6*).

Large-sized spears are desirable in asparagus. As most plants age, they produce smaller-sized spears than in their prime. Some few plants maintain large spears into old age, and by selecting for this character, the breeder can produce cultivars that maintain good spear size as the plants age (G. C. Hanna, personal conversation; J. H. Ellison, personal observation).

Over the years, efforts have been made to select high-yielding plants in the seedling or the 1-year-old crown stage. Most have failed. Although Haber (*16*) found positive correlation with male plants between crown weight and number and total weight of spears during the first three cutting seasons, this was not the case with female plants. They showed a negative correlation between crown weight and number and total weight of spears in one year, and no correlation in the other two years. When data from both sexes were pooled, no significant correlations were found among these factors. Wellensiek (*41*) cut the tops off of asparagus seedlings 4–6 weeks after emergence and selected seedlings that produced new shoots within 2–3 weeks. The selected plants were more vigorous than the unselected ones 3 months later. Yield data reported by Wellensiek and Jonkers (*42*), however, showed that selection for vigorous regeneration of growth had no relationship with mature plant yield. Scheer and Ellison (*32*) studied the relationships between low-, medium-, and high-vigor asparagus seedlings, selected 2½ months after sowing in the greenhouse, as related to field performance. The selection technique was based on an index of number of stalks multiplied by stalk diameter. Survival in the third year in the field of the medium- and high-vigor seedlings (97 and 94%, respectively) was superior to that of the low-vigor seedlings (64%). Although yield of female plants the first season was positively correlated with seedling stalk number, the correlation coefficient was .39, too low for prediction purposes. With male plants there were no correlations between greenhouse and field performance data.

In order to select high-yielding plants (phenotypes), Ellison and Schermerhorn (*12*) studied yield as related to earliness with mature male and female plants. Several criteria were used to study earliness: (1) number of spears emerged per plant 1 day prior to the beginning of harvest; (2) number of days to first harvest of each plant after regular harvesting began; and (3) full-season as related to early yield (2 weeks). Figure 14.18

FIGURE 14.18. Early asparagus plant 1-34 with seven spears 1 day
prior to first harvest. Plant with white stake had only one spear, and
several plants in background had not emerged.

depicts an early plant with seven spears emerged one day prior to the first harvest. The
plant with the white stake produced only one spear and several plants in the background
had produced none.

Full-season yield had a strong, positive relationship to earliness as measured by each
criterion. Male plants emerged earlier and yielded more than female plants. Thirteen male
and two female plants (out of 217 males and 205 females) with five or more spears
emerged 1 day prior to harvest, and yielded 78 and 94% more during two consecutive
harvest seasons than control plants selected at random on the basis of proximity to early
plants. There was no difference in spear diameter between early and control plants.

Yield of individual plants as related to brush vigor the preceding autumn was studied
by Ellison and Scheer, (10). Brush vigor was defined as (a) number and (b) diameter of
stalks, each of which had a specific influence on yield. Although stalk diameter per se was
not correlated with yield of spears, it was correlated positively with diameter of spears.
Stalk diameters also influenced the relationship between stalk number and total weight of
spears. Maximum yields came from plants having many large stalks, whereas plants with
many small stalks did not produce high yields due to small spear diameter. Also, plants
with few large stalks failed to produce high yields because they made few, though large,
spears. The above relationships were consistent with the same plants for three consecutive
seasons, indicating that brush vigor one season is a good criterion for predicting long-term
yielding ability. This enables the breeder to observe a very large population of plants and
select the high-yielding ones, and it permits the seedsman to rogue weak plants from seed
fields on the basis of weak brush.

A study by Ellison *et al.* (11) showed that yield of individual asparagus plants was
related to number and size of stalks (brush) as well as to earliness and sex expression.
Full-season spear diameter is positively correlated with early (2 weeks) spear diameter and

with stalk diameter and is negatively correlated with number of stalks per plant. Important multiple correlations indicated that the plants with the highest yield had relatively high levels of earliness, high brush vigor, and early (2 weeks) spear diameter. Selecting high-yielding plants on the basis of earliness and brush vigor can save the breeder much time and labor compared to obtaining costly full-season yield data.

The system used to breed asparagus in California is provided below through the courtesy of Frank H. Takatori (personal correspondence), Asparagus Specialist at the University of California, Riverside.

In California, where growers produce green asparagus for the fresh market, earliness of production is of prime importance. The experimental progenies are screened for earliness the first two seasons after planting in the field. By this process, from 60 to 70% of the new crosses can be eliminated.

The second most important factor is large spear size. Under desert conditions, where temperatures are high and the plantings are made on raised beds, the crowns are only 3–4 in. deep and spear size is small. The progenies are checked for spear size by harvesting 2 or 3 weeks in the third and fourth season, and more data are developed on early production. Harvesting in the fifth season is made for the full season, but by this time the number of lines is reduced to a manageable size.

The third factor in the selection process is compact or tight spear tips. High desert temperatures cause the tips to open prematurely. The parent material is allowed to go to fern (brush) in the spring under mild conditions. At this time measurements are taken on height to the first lateral branch, number of nodes to the first branch, diameter of spear, etc. The fern is removed in June or July when air temperatures are very high, and the same measurements are repeated on the summer fern. Generally, the second measurements show a drastic change, e.g., lower branching, associated with smaller spears. Takatori then selects parent plants that show the least change in fern characteristics due to increased temperature. He calls these "temperature-insensitive" plants.

Takatori has found that male plants in the progeny resemble the pollen parent in spear size, earliness, etc., and the female progeny tend to resemble the seed parent. To obtain the most uniformity, he crosses parents of similar morphological and performance characteristics. To facilitate the evaluation of new progenies, the male and female plants are identified in the greenhouse and planted separately in the field. This is helpful in the evaluation and selection of the seed and pollen parents.

To breed for resistance to fusarium, Takatori inoculates seedlings in the greenhouse, where most are killed. The survivors are planted in the field and intercrossed. Although many progenies do not have true resistance, certain ones have good field tolerance. These grow well in a field that has been replanted to asparagus continuously for over 20 years and is highly infested with fusarium. Progenies that grow well in this experimental field show no fusarium problems in growers' fields, indicating Takatori is picking up strong field tolerance to fusarium (F. H. Takatori, personal correspondence).

Adaptation

Without previous experience it would seem to be impossible to predict whether a new asparagus cultivar would have good local adaptation or not. However, when the breeding program is designed mainly for solving production problems for a specific state or province, then it behooves the breeder to select breeding stocks from that local production area. In this way, the breeder increases the probability that the new cultivars will have local adaptation compared to exotic breeding stock or cultivars. Natural selection undoubtedly enables the breeder to select superior plants that are superior simply because they are well adapted to local conditions.

In order to test for wide geographic adaptation, the breeder should arrange for hybrids to be evaluated under as many diverse conditions in as many different geographic locations as possible.

High-Branching Spears

Spears that branch low often have a loose or open tip before they can be harvested, thus rendering them unmarketable. Spears that branch high maintain a tightly closed tip for a longer time after emergence, and thus have a better chance of being marketable when cut. Harvesting is done generally once a day, and many low-branching spears are lost because of premature opening of the heads.

There is a good correlation (J. H. Ellison, unpublished data) between height of branching of summer stalks and tightness of spear tips in the spring. In addition to selecting for tall spears with tight tips in the spring, the breeder can select for high-branching stalks when selecting for other desirable brush characters in the summer and fall.

FIGURE 14.19. The selection apparatus is placed behind the spear in such a way that the spear top coincides with a point on the oblique upper side.
After Huyskes (18).

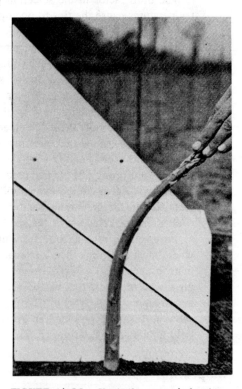

FIGURE 14.20. Next, the spear is bent over until it breaks at the junction of the fibrous and fiberless region. The snapped-off top is entirely free from fiber.
After Huyskes (18).

FIGURE 14.21. The situation after breaking off the fiberless top. The break lies below the so-called selection line, and so the examined spear is free from fiber over a great length. The stumps of these long tender spears are cut off at an oblique angle in order to indicate they have long fiberless tops.
After Huyskes (*18*).

FIGURE 14.22. The situation after harvest. From the stumps cut off at an angle it appears that this plant has only produced spears with long fiberless tops. For, if the fiberless top is too short, the stump is not cut off at an angle. The number and the thickness of the spears, which are usually slightly shriveled, are decisive for the yield in this season.
After Huyskes (*18*).

Long Tender Spear Tips

Huyskes (*18*) described a method for selecting asparagus plants for long, fiberless spears. It involved snapping spears at the point where the fibrous stub will be left on the plant and the tender spear will be held in the hand, as described by Barrons (*3*), Carolus (*5*), and Schermerhorn (*34*). By using a specially constructed board, with a sloping line to separate the long tender spear tips from the short tender spear tips, Huyskes snapped and classified spears in one operation. He marked the stubs of long tender spear tips by cutting the tops of the stubs at an oblique angle. The stubs of the short tender spear tips were left uncut. Thus at the end of the harvest season, the yield, the approximate spear diameter, and the proportion of long to short tender spear tips could be ascertained merely by examining the stubs on each plant. The selection apparatus and the way to use it are depicted in Figs. 14.19–14.22.

DESIGN OF THE COMPLETE BREEDING PROGRAM

The author has chosen to describe his own asparagus breeding program because of the diversity of the problems presented and familiarity with the details of the techniques employed in the effort to solve the problems. This does not imply the best breeding program, but only the one most familiar to the writer.

Estimating Yield without Complete Harvest Records

Obtaining complete yield records with asparagus involves up to 50 daily harvests per year, requiring much labor and cost. Yield data with many individual male and female plants from three different populations indicated a very strong positive correlation between full-season yield and earliness (12). Thus, earliness can be substituted for costly harvest records.

Yield also was found to be highly correlated with brush vigor the previous season (7,10,27,45), which means that plants with high yielding ability can be selected in autumn on the basis of vigor. This is an especially valuable technique when selecting plants for resistance to foliar diseases, such as rust, botrytis, stemphylium, or cercospera.

The best estimates of all in predicting individual asparagus plant yield are based on multiple correlations that involve both earliness and brush vigor (11). This method is useful in evaluating individual plant phenotypes when selecting parent material, and also for roguing late and non-vigorous plants from fields in the seed production of synthetic asparagus cultivars.

Evaluating Cultivars from Other Areas

Before starting an asparagus breeding program, which is long term and expensive, existing cultivars from other states and countries should be evaluated. In some cases none will have good, long-term, local adaptation, but some may provide genetic material useful for breeding. Although not probable, it is always possible that one of the cultivars may prove to be excellent, in which case a breeding program is not needed. In this case, a reliable source of seed should be verified. More likely, though, is that one of the cultivars may be superior in some respects to the local commercial asparagus strain, and might serve in the interim until further improvements can be made through breeding.

The most expensive aspect of conducting a yield trial of asparagus cultivars is the cost of harvesting. Most of this can be saved by substituting early-yield (2 weeks) or brush vigor data. Ellison et al. (9) found full-season yield of 20 cultivars to be highly correlated with early yield or brush vigor, but the best estimates of full-season yield were based on a combination of both.

One important precaution must be kept in mind when evaluating asparagus cultivars from other states or countries. A thorough evaluation, preferably in several sites, should be made of cultivars from other areas before they are recommended to local growers. An example of the need of this precaution is illustrated below. The author found two out of 17 cultivars from other areas to be superior to local commercial strains after 5 years of evaluation in the field, and recommended the two cultivars to local growers on a trial (limited-acreage) basis. Unfortunately, the two lines started to deteriorate compared to local strains in the sixth year of the test and continued to decline thereafter. In this case, 5 years was not long enough to determine that the two cultivars did not have good local adaptation.

Another finding from the evaluation trial of the 17 cultivars above was that two asparagus male hybrids from The Netherlands yielded 50% more than local asparagus strains for 8 years at a site where rust was not prevalent. At two other sites, where rust was prevalent, the two male hybrids proved to be very susceptible, but still yielded about the same as the more rust-tolerant local lines. The high-yielding ability of the male hybrids convinced the writer that male line breeding should be one of his objectives.

Pathological Problems

Identification of the major diseases of asparagus, and the evaluation of the relative importance of each disease to the industry, should be considered early in the planning of the breeding program. Either the breeder should be well versed in plant pathology or should have the full cooperation of the resident plant pathologist.

In New Jersey in the 1950s there were 32,000 acres of asparagus concentrated in two main growing areas of three southern counties. The density of asparagus culture was great and rust was a major problem. The crop had been grown for many years and there were many volunteer asparagus plants along roadsides, hedge rows, and railroads, which served as a source of rust inoculum. The volunteers produced abundant spring aeciospores because of not being harvested, and thus became the source of inoculum for the damaging summer urediospore stage. Another source of aeciospores in the spring was the many 1- and 2-year-old asparagus fields, which were not cut at all, or not cut long enough to break the spring aecial spore stage of the fungus, and thus interrupt the life cycle of the rust. The density of asparagus culture also increased the severity of the rust problem by providing a vast area for the production of countless billions of urediospores once the summer stage of the rust began.

Asparagus yield was declining in New Jersey in the 1950s; and because rust was severe and the symptoms so dramatic (entire fields turning from green to brown during August), the main cause of yield decline was thought to be due to rust.

Although rust was a significant problem, a second, more serious problem was the replant problem, the cause of which was not yet understood. Asparagus, replanted in the same fields where it had been grown previously, never did as well as the first bed. This was true even if many years intervened between crops. The second crop had less yield and less vigor, and developed poor stand earlier than the first crop. Possible causes of the replant problem were thought to be (1) a soil deficiency brought about by years of growing the first crop, (2) a possible toxic residue left by the first crop (allelopathtic substance), or (3) a soilborne disease.

Research in California by Grogan and Kimble (*15*) and in New Jersey (J. H. Ellison, unpublished data) during the late 1950s indicated that a root rot was the main cause of the asparagus decline and replant problem, and that *F. oxysporum* f. sp. *asparagi* was considered the most important pathogen associated with it. The symptoms of F. oxysporum were mainly confined to the feeder rootlets, but the effects on asparagus production were more damaging than the effects of rust. The crop losses due to fusarium were less well recognized because most of the symptoms were below the soil, whereas rust symptoms were spectacular on the brush. Fusarium was more harmful than rust because the root rot took a toll every year, and the rust was severe only about 1 or 2 years out of 7, causing relatively little damage most years.

In the summer of 1972, in southern New Jersey, entire asparagus fields of all ages were stricken by a sudden and severe stalk collapse and crown rot. This occurred during the third consecutive year having excessive rainfall. The 1973 season also was a wet year, and the crown rot continued in fields not devastated in 1972. Acreage in New Jersey declined from about 20,000 acres in 1972 to less than 5000 acres in 1974. Johnston *et al.* (*20*) named the new disease fusarium stem and crown rot of asparagus and ascribed the disease to *F. moniliforme*. Both fusarium stem and crown rot and fusarium root rot, caused by *F. oxysporum* f. sp. *asparagi,* are associated with asparagus decline in New Jersey, but the crown rot is much more devasting than the root rot.

Different breeders may use different systems to identify parent plants and progeny resulting from selfing, crossing, etc. The system presented in Fig. 14.23 was given to the author many years ago by a colleague, Oved Shifriss. It is simple, systematic, and provides for the identification of all parents and progeny from crossing, selfing, and sib mating.

Breeding for Rust Resistance

In selecting asparagus for rust resistance, one must judge each selected plant in comparison to its neighbors, because rust infection can be influenced by microclimate. Wind direction, factors affecting dew formation and persistence, density of foliage, etc. can influence the severity of rust in different parts of the same field. Selected plants must be above average in vigor, both in number and size of stalks. This is related to yield

---

A. Each plant will be identified by a capital letter (or letters) followed by a number. For example, C40, R1, Md10, G101, RR27, in which C is Coker; R, Raritan; Md, Maryland; G, Greenwich; and RR, Rust Resistant.

B. Selfing: Parent number (R1) will be followed by a hyphen and an S_1 plant selection number. For example, R1-1, C32-8. Each successive generation will be separated by hyphens:

Generation:	P_1	S_1	S_2	S_3	S_4	S_5
Pedigree:	R1	-1	-6	-14	-5	-13
Pedigree:	C32	-8	-1	-3	-23	-16

C. Crossing
 1. Each cross or sibbing will be given a special serial number. For example, 001, 056, 137.
 2. Selected F_1 plants will be given the serial number, followed by a capital letter. For example, 001A, 056D, 137H.
 3. If selfing is made in (2), the pedigree system will be as in (B):

Generation:	F_1	S_1	S_2	S_3	S_4
Pedigree:	001A	-3	-21	-6	-13
Pedigree:	056D	-1	-7	-33	-17

 4. If crossing (or sibbing) is done with F_1 plants, each cross will be given a special serial number [as in (1)]:

Number	Pedigree
059	001A × 056D
333	137H × 012C
456	123E × R1-1

FIGURE 14.23. Plant identification system used by author.

potential. If possible, select plants from old fields and from replanted fields, where natural selection has eliminated weak and short-lived plants, and where plants with outstanding vigor may have fusarium tolerance in addition to rust resistance. In very old fields it is important to search for plants with large stalks, which may transmit this desirable character to their progenies.

Rust-resistant plants were selected in late summer and fall, and were marked by driving tall stakes into the soil adjacent to the selected plants, so that they could be relocated, dug, and moved to the greenhouse at the end of the growing season. The plants were propagated by crown division, and five divisions of each plant were placed in a high-temperature greenhouse (27°C day and night), with high humidity. Newly divided asparagus crowns establish new roots and shoots better under high temperatures than low temperatures. High relative humidity helps to prevent desiccation of the tender shoots. Of 50 plants selected for rust resistance, only three died in the greenhouse.

The surviving 47 plants were inoculated with rust in the greenhouse, using the method published by Kahn *et al.* (22). Subsequently, 23 of these plants were discarded because of a lack of sufficient rust resistance. The remaining plants were used in crossing and progeny testing.

The progeny tests were carried out on a commercial asparagus farm in southern New Jersey rather than on the Vegetable Research Farm at New Brunswick, where soil and growing conditions were atypical of the asparagus-growing area. For instance, rust was rarely seen in New Brunswick and then only at low levels, whereas rust was more prevalent in the asparagus-growing area.

Fortunately, the field used for the progeny test had been planted to asparagus twice before, providing a good test site for evaluating the progenies for fusarium tolerance as well as rust resistance and yield. Sixteen F_1 progenies were studied for 10 years; and although several showed good rust resistance, only two were outstanding in yield and vigor. These two had a common pollen parent, which had been selected for great vigor, large stalks, and rust resistance in a 15-year-old field in southern New Jersey. Both lines showed evidence of good tolerance to fusarium, and both produced high yields in New Jersey. They were included in a cultivar test in Michigan, conducted by personnel of Michigan State University, and one proved to be the top yielding line, whereas the "sister" line was below average. The seed parent of the superior progeny was selected for great vigor and large stalks in a 35-year-old field in southern New Jersey. Not all selections for vigorous plants in old fields will prove to be good parents. However, the 35-year-old female parent cited above is a perfect example of the result of natural selection during many years of harvesting pressure, drought years, wet years, and exposure to rust and fusarium.

Breeding for Resistance to Fusarium

Since asparagus is composed of a highly heterogeneous population of plants, natural selection has a great impact on plants in an old field where *Fusarium* titer builds up to a high level. This effect is even more important after several generations have been exposed to natural selection. This was the case in southern New Jersey, where asparagus was first cultivated about the turn of the century, and growers saved their own seed.

Genes for tolerance to fusarium can be found in parents selected for vigor in old fields. One female, no. 27 in our records, was given the notation "healthy roots" the day it was dug in a 12-year-old replanted field. This plant, with the healthy-roots phenotype also has a good genotype, in that it transmits good fusarium field tolerance to its F_1 progeny.

Number 27 was the only plant among 50 selected for rust resistance and vigor that was given the notation "healthy roots," although several others transmit the same level of fusarium tolerance.

In order to combine fusarium resistance with rust resistance, seed of the best rust-resistant F_1 progenies were subjected to very high levels of *Fusarium* inoculation pressure in hopes of finding rust-resistant seedlings with fusarium resistance. As stated in the section above, inoculum from highly pathogenic isolates of *F. oxysporum* f. sp. *asparagi* and *F. moniliforme* were combined in preparing the deathbed inoculated soil to be used in screening for resistant seedlings.

The survivors of the *Fusarium* greenhouse inoculations that had the least symptoms and the most vigor were transplanted to the field for holding and for evaluation for earliness and brush vigor. The inoculated soil in the greenhouse pots was shifted with the plants, so that fusarium pressure would continue in the field. After several seasons, plants of both sexes with the highest levels of earliness and brush vigor were selected for crossing and progeny testing.

Diallel crosses were made between 10 females and 10 males in pollination cages using bees. The 100 progenies are being evaluated at the present time on high-titer *Fusarium*-infested soil at the Rutgers Research and Development Center, Bridgeton, New Jersey. So far, the results indicate that the best progenies from fusarium-selected parents have about the same fusarium tolerance as the best progenies from the plants selected in the field for rust resistance and vigor.

It is obvious that the fusarium breeding program is a long one, and will necessitate selecting, crossing, and testing for a number of generations. Inbreeding through sib pollination to concentrate genes for resistance in certain families, followed by outcrossing among families to obtain hybrid vigor, should be a promising method.

Breeding Male Asparagus Hybrids

The high-yielding ability of the male asparagus hybrids from The Netherlands (50% more than local commercial cultivars) in our evaluation program in the late 1950s and early 1960s excited the author's interest in breeding all-male asparagus. There are several advantages to male compared to dioecious asparagus: (1) no seed to make seedling weeds in the field, (2) no seed to compete with the storage roots for photosynthate, and (3) no seed for growers to save and thus ruin the genetic quality of the cultivar by using F_2, F_3, F_4, etc., seed, which are inferior to F_1 seed.

In working with three different populations of asparagus, the author observed 20% of the male plants to produce one or more viable seeds within 3 years. The plants were maintained in the greenhouse, where conditions seem to favor the expression of andromonoecism more than conditions in the field.

The best supermales have come from two andromonoecious plants selected in the field for rust resistance and vigor. One plant is no. 22 in our records and is the pollen parent of Jersey Centennial. The other selection is no. 14 male. They both transmit fusarium field tolerance to their progeny. Since supermales are the result of self-pollination, and since these two andromonoecious male parents are genotypically fusarium tolerant, their first generation selfed (S_1) offspring may segregate for good fusarium tolerance or resistance in the supermale (*MM*) genotype. This indeed seems to be the case. The first supermale from andromonoecious plant no. 22 produced outstanding F_1 progenies, with high yield and good field tolerance to fusarium. This resulted without any selection pressure on the supermale (*MM*) parent. Andromonoecious plant no. 14 has produced only four super-males, one of which sires fusarium-tolerant, high-yielding F_1 progenies. Several hetero-gametic (*Mm*) male plants, which were identified after rigorous greenhouse and field selection for fusarium resistance, show phenotypic resistance to the disease. However, their progenies do not have as much field tolerance to fusarium as do the progenies of the two good supermales mentioned above.[2]

We now have several new supermales from andromonoecious plant no. 22, and other supermales derived from a number of andromonoecious males selected for resistance to fusarium. The fusarium selected andromonoecious plants have fusarium-resistant phe-notypes, and when self-pollinated, it is hoped, will produce some supermales with fusarium-resistant genotypes.

There are some precautions to observe in breeding all-male hybrids. These do not involve the seed parents, which would be selected and tested in the same manner as when breeding dioecious lines. The results of selfing dioecious plants are varied—good and bad. We dealt above with the segregation for desirable genotypes in S_1. Likewise, there can be segregation for undesirable gene combinations as a result of selfing. Some charac-ters, such as very low vigor of the supermale, can be recognized easily, but others, such as low viability of pollen, might not be recognized early and could result in the loss of valuable time and money spent in testing a worthless supermale. Also, some supermales have anthers that may not dehisce normally, and others produce a low yield of pollen from otherwise normal flowers.

Even though S_1 supermales are inbred only one generation, most have low vigor compared to noninbred males and may require special cultural practices for good survival and growth in the seed production field. During the wet 1970s in New Jersey, the low vigor 14-4 supermale survived poorly when planted in the conventional 6-in-deep furrow. When planted on a raised bed, however, survival and growth were improved but not satisfactory. Another character to check when selecting a supermale is its response to tissue culture. A supermale that will not respond to tissue culture would not be usable. The progenies of each supermale must be observed carefully for sex expression before release, because male lines can have plants that produce berries and seed at an objectional level. Some seed parents carry modifying genes that contribute to the problem of berry-bearing plants in male hybrids, and therefore must be screened for this character.

The same kind of segregation applies to female as well as male plants in the S_1. The author has not attempted to use S_1 females as seed parents because of (1) low vigor that would limit seed production, and (2) the probability that S_1 seed parents could contribute to the berry-bearing problem in all male hybrids. The selection pressure that results from using seed parents derived by selfing andromonoecious males could result in significant levels of andromonoecism in the F_1 all-male hybrid. However, there is the possibility that a desirable S_1 seed parent could be found by persistent research. Figure 14.24 illustrates how one might select a seed parent among female plants derived from a self-pollinated andromonoecious male plant. Desirable characters that might be discovered are (1) in-creased disease resistance and (2) increased uniformity in the F_1 progeny due to the inbred nature of the seed parent. Low-vigor S_1 females could be discarded to avoid low seed yield, and careful progeny testing of selected S_1 seed parents would be required to avoid the berry-bearing character in the F_1 male progeny. This would be an absolute require-

---

[2]Two male hybrid cvs., Jersey Giant and Greenwich, have been patented and released. Both have improved yield and field tolerance to rust and fusarium.

P$_1$ Andromonoecious male (M/m) selfed.

S$_1$ 1 *MM*   :   2 *Mm*   :   1 *mm*   **FIGURE 14.24.** Selection of seed parent
          Supermale   Normal male   Female   among S$_1$ female plants.

ment in the case of F$_1$ male hybrids, but also would be important even in dioecious F$_1$ hybrids. Seed parents should not be selected in S$_2$ and subsequent selfed generations because of poor vigor of the seed parent and the higher probability of berry-bearing F$_1$ males than expected from S$_1$ seed parents. Another reason the author has not attempted to use S$_1$ seed parents is due to the low frequency of such females. Although equal numbers of S$_1$ supermales and S$_1$ females are expected, the actual number of female plants found at Rutgers is approximately one-half the expected number. This may be due to the fact that we inoculate the S$_1$ seedlings with *Fusarium*, and as mentioned above, fewer females than males survive such inoculation. The low frequency of females may be due to sex-linked lethal factors.

Diallel crossing between several selected female and supermale plants is the best way to study general and specific combining ability. The tendency for each parent to transmit andromonoecism or hermaphroditism may be observed in the hybrids along with all other important characters. It is good from a strategic standpoint to use more female than male parents, because males need to be isolated during seed production whereas females do not.

Tissue Culture

Tissue culture (cloning) is an efficient way to propagate asparagus plants vegetatively (asexually). The only means of asexual propagation available before the advent of tissue culture was crown division, which is very slow and inefficient in producing large numbers of plants from one plant. Crown division still is useful as a simple means of producing a few (5–10) plants from a single individual. Tissue culture, on the other hand, has opened avenues of modern asparagus breeding that otherwise would be closed. Most of the new improved asparagus cultivars from the United States and Europe actually are clonal hybrids, which are F$_1$ progenies from single crosses. These would not be possible without tissue culture. In order to produce commercial seed, the two parents must be propagated vegetatively, and the only efficient way to do this currently is by tissue culture.

Murashige *et al.* (*26*) have led in asparagus tissue culture research, and Yang (*43*) has made significant contributions in developing useful modifications of the Murashige method. One must be careful that genetic changes do not occur in the tissue culture process. Malnassy and Ellison (*24*) showed that an early method based on callus tissue culture led to ploidy changes. Currently used systems depend on meristem or bud culture, and they are known to yield genetically stable explants. Callus that may develop during the culturing of meristems or buds is carefully trimmed away.

Figure 14.25 shows four stages of the asparagus meristem tissue culture process. Since the original meristem section is approximately 0.5 mm in diameter and nearly microscopic, that first step is not shown. When the meristems form roots and reach a size approximately that of the explants shown in Fig. 14.25A, they are transferred from the petri dish to the growth tube. Two weeks later the explant appears as the one in Fig. 14.25B. Within another 8 weeks, the explant grows to the stage shown in Fig. 14.25C and is ready to be transplanted to soil in the greenhouse. After 3–4 months, the explant reaches the size shown in Fig. 14.25D and is ready for field planting. For successful

FIGURE 14.25. Asparagus tissue culture. (A) Rooted meristems 4 weeks after sectioning. (B) Explant 2 weeks after transfer from petri dish. (C) Explant 10 weeks after transfer from petri dish, ready to be potted. (D) Explant 3–4 months after potting in the greenhouse.

transplanting to the field, the root system should be fully developed in a 3- to 4-in.-diameter pot.

Parent material must be maintained while progenies are being evaluated, which may require from 5 to 10 years or more. The parent plant is multiplied by cloning, and the new explants are field planted in a holding block, where they can be kept indefinitely. An efficient way to do this is to establish an experimental seed block, isolated by 1000 ft or more from other asparagus, in which the female holding block can be used to produce topcross seed. This is done by introducing one male clone one season, and another the next. The male clone can be managed best in pots rather than transplanting the males in the soil. Male plants should be spaced no more than 10 ft apart in either direction.

In the initial stage of progeny testing, a single female plant per genotype produces sufficient seed. In the final testing of reselected seed parents, 10–20 explants of each clone will be needed for sufficient seed for large-scale tests on commercial farms. Figure 14.26 shows such a seed block of reselected female parents. In the foreground are spears,

FIGURE 14.26. A holding block for reselected seed parents. Clone no. 32 on right shows spears for harvest records and young brush beyond. Clone no. 51 (seed parent of Jersey Centennial) on left shows desirable high-branching habit.

where harvest records are being taken. In the background is young brush about to bloom. Notice the low-branching habit of the clone on the right versus the desirable high-branching habit of the clone on the left. Notice also the uniformity of growth within each clone. Out of sight on the left is the first row of the supermale pollen parent.

Experimental pollen parents can be planted in a male block, spaced 3 ft apart in rows 5 ft apart, for long-term holding.

A TYPICAL FULL-YEAR FIELD AND GREENHOUSE OPERATION

Field Research

By its very nature, field work is seasonal. Seed is planted in the spring, and so are asparagus crowns and transplants. Asparagus yield data also are recorded in the spring, but rust observations, brush vigor ratings, and stand counts are made in early fall. The summer months are spent in plot maintenance and data analysis. Reports concerning field research generally are written in the winter. The following is an account of field operations for a typical year.

Yield Records

Asparagus in New Jersey generally is ready to harvest in late April. In late March or early April, field experiments must be fertilized and limed as needed and ridged as soon as the spears start to emerge. In early April, plot labels must be made for each experiment and set out to mark the plots as soon as the tillage is completed. Cutting boxes 9 in. long are used for fresh-market spears so they may be trimmed to a uniform length. Gram balances are used for weighing, and special data sheets are used for each experiment. A spear diameter gauge of ⅜ in. is used to judge culls less than ⅜ in. in diameter, and a ⅝-in. gauge is used to judge jumbo spears greater than ⅝ in. in diameter. Medium-sized spears are ⅜ to ⅝ in. inclusive. A normal harvest season is 6 weeks, but in some experiments we harvest only 2 weeks in order to reduce labor costs. Most demonstrations harvested by cooperating county agricultural agents are cut only 2 weeks, due to time limitations of the agents. Yields for 2 weeks are highly correlated with yields for 6 weeks.

Nursery

A progeny nursery was seeded about May 1, consisting of a number of experimental supermales topcrossed with a common seed parent. Two standard supermales were included as controls. Mary Washington was used as a commercial control. Approximately 300 seed of each progeny were sown in a single row 20 ft long. Rows were 30 in. apart. The soil reaction (pH) should be adjusted to 6.5–6.8 by the use of dolomitic limestone, and 500 lb/acre of 5–10–15 fertilizer should be disked into the soil prior to planting. Another 500 lb/acre of the same fertilizer should be side dressed during late July. Great care must be taken to control thrip and asparagus beetle because these pests can devastate young seedlings. A few herbicides are available for weed control, but extreme care must be taken to follow directions on the package. Some hand weeding likely will be needed.

New Yield Test

A small replicated yield test of new cultivars was planted in early May. The soil pH was adjusted to 6.5–6.8, and 1000 lb/acre of a 5–10–15 fertilizer were plowed down with the cover crop. No side dressing will be needed. Since the cultivars are being tested for resistance to fusarium, the experimental site is located on land previously used for asparagus and known to be heavily infested with fusarium. Plot size should be 20–25 plants set 12 in. apart in rows 5 ft apart. Four to six replications are needed. Weed control will be achieved by smothering weeds when they are small as the furrows are gradually filled in by cultivation. Some hand hoeing will be required. Experimental results this fall will consist of stand counts, vigor ratings, and observations on symptoms of rust or fusarium if present.

New Observational Progeny Test

Twenty experimental male hybrids plus a standard male hybrid and Mary Washington were planted in May on a commercial asparagus farm with the cooperation of a County Agricultural Agent. The plot size in this case was a single row several hundred feet long, reaching across the field. Only one replication was planted. Observations on plant stand, vigor, and disease will be made in the early fall. Observational yield data on 100 ft per row may be taken after 2 years.

Holding Breeding Material

Experimental supermales were transplanted from the greenhouse to a permanent holding block in the field while they are being progeny tested. Observations will be made on vigor of the supermales in the field to determine their suitability as pollen parents, if needed, in future seed blocks.

Seed Blocks

A new seed block was planted to produce seed for our growers. It consisted of two seed parents and a common supermale. In order to eliminate any errors that may have occurred in the cloning process, all the plants in seed parent rows will be examined during bloom to remove any male plants, which would be rogues.

Two previously established seed blocks required special maintenance. Seedling volunteers were removed to avoid pollen contamination, and volunteer asparagus within 1000 ft of the seed block were removed for the same reason. Morning glory is an especially bad weed in asparagus seed blocks because the seeds of these two species are so similar in size, shape, and color. It is nearly impossible to remove this weed seed from the asparagus seed. Domestic bees at the rate of one hive per acre of seed block should be introduced at the onset of bloom. They may be removed on August 15 in New Jersey, insofar as seed set after that date does not mature well.

Summer is the time to see that good culture is provided for all the asparagus experiments and seed blocks. This involves weed control, irrigation, insect control, and other maintenance. For example, in some sites damage by rabbits is severe, and fencing must be provided to protect the seed block. In late summer, berries ripen on mature stalks and many of these lodge. The fallen stalks should be gathered up and put under cover so as not to lose the seed. These stalks may be kept in dry storage 1 month or more until the remainder of the crop is ready to be harvested.

Autumn Field Work

Early fall is the time to study experimental hybrids in observational trials on commercial farms. Ratings on brush vigor, rust, and fusarium symptoms should be made, and the percentage plant stand should be recorded. This is a good time to acquaint the county agent and the grower with the differences observed among the hybrids.

Autumn is a good time to move plants selected in the field from the field to the greenhouse. They are most apt to survive the move at this time of year. Also, there is sufficient time to tissue culture these plants over winter for planting in the field the following spring in a seed block or a pollination cage.

Greenhouse Research

Because we can work the year round in the greenhouse, this research is not considered seasonal. The objective of some greenhouse research is to confirm results found in the field and/or to explain them. Another greenhouse objective is to simulate conditions in the field and try to duplicate field results. This has been done in the present program concerning rust and fusarium screening of asparagus parents and progenies.

Rust

Rust inoculation under uniform conditions in the greenhouse was useful when comparing the relative rust resistance of parent plants selected under different levels of rust pressure in different fields. Screening of asparagus progenies for rust resistance by greenhouse inoculation also was valuable.

Fusarium

All of the initial selecting for resistance to fusarium was done in the greenhouse. Seedlings were inoculated with high titer inoculum of *F. oxysporum* and *F. moniliforme* in order to kill susceptible individuals. Survivors were reinoculated repeatedly and only the most tolerant were selected for field observation.

For many years hybrids have been screened in the greenhouse for tolerance to fusarium. The objective was to correlate greenhouse results with field results of good and poor cultivars in order to screen new experimental hybrids in the greenhouse before testing them in the field. If susceptible hybrids could be eliminated through greenhouse screening, much time, space and labor could be saved in the field. Results have been inconclusive and research continues.

Hand Pollination

Some hand pollination seems to be inevitable. Whereas it is easy to topcross several or many females with one male with bees in a cage or in isolation in the field, the reciprocal crosses cannot be done in such a manner. Pollen parents must be isolated, and topcrossing or diallel crossing with several males is best done by hand in the greenhouse unless relatively large quantities of seed are needed. In this case, cages must be used. Hand pollination is ideal for crossing numerous S_1 males on a cloned female, as we do in sex genotype testing. We set a minimum of 100 berries (about 300 seeds), from which we use 15 seeds for the sex genotype test. The remainder of the seed of those which are all male is used for greenhouse or field evaluation. In male hybrid breeding, the dioecious seed generally is not used. Another task performed in the greenhouse, as implied above, is growing to maturity (flowering) the progeny of S_1 males to determine the sex genotype. The all-male progenies naturally identify the respective pollen parents as supermales.

PHENOTYPES SELECTED FOR CLONING

Certain plants selected for resistance to fusarium show strong phenotypic but not genotypic resistance (Fig. 14.17A). This means that the plant does not transmit resistance to its progeny, but does exhibit good resistance itself. An asparagus plant with phenotypic resistance to rust and fusarium, high yield and quality of spears, and wide geographic adaptation could become the basis of a cloned cultivar. Asparagus tissue culture has the potential to become efficient enough to be used commercially to propagate a cloned cultivar for home gardeners, market gardeners, and even large-scale asparagus growers. Such a cultivar, vegetatively propagated from one original phenotype (plant), would have a high probability of being superior to a hybrid because of the extremely large population from which such a plant could be selected. It is inconceivable that any breeder, or indeed all breeders, could produce a population of hybrids as large as the population of individual plants found in any major asparagus production area from which desirable phenotypes might be selected.

Research on cloned cultivars is beginning at Rutgers and several other AES in several different countries. Success along these lines could lead to the selection of phenotypes for asexual propagation instead of genotypes for sexual reproduction. In addition to this large population of commercial plants, breeders may select for disease-resistant phenotypes among large populations of inoculated plants in the greenhouse and/or field.

Plant Patenting

Asparagus parents and clonal hybrids qualify to be patented (29). Patenting the parents and hybrids protects them from being used without permission of the inventor and also generates royalty revenue, which usually can be used to support the breeding program.

The first three patents of this writer's program are (1) female asparagus clone no. 51, (2) male asparagus clone no. 22, and (3) Jersey Centennial, the clonal hybrid of 51 × 22. Jersey Centennial is being released jointly by the New Jersey AES at Rutgers University, and Michigan State University, where much of the early testing was done. Data presented by Ellison et al. (13) showed Jersey Centennial to yield 80% more large spears early in the season and 38% more marketable yield than Mary Washington on *Fusarium*-infested land in New Jersey. Jersey Centennial also displayed superior stand and vigor as compared to Mary Washington for 8 years on *Fusarium*-infested land in two New Jersey sites in addition to the one where yield was measured, demonstrating field tolerance to fusarium. In Michigan, Jersey Centennial yielded substantially more than eight other asparagus cultivars when grown on land virgin to asparagus.

TRIALS OF ADVANCED LINES

It is important to test asparagus hybrids for wide geographic adaptation before releasing them. In addition to the two dioecious hybrids (mentioned above) that both performed equally well in New Jersey, but only one of which performed well in Michigan, there is a reverse case. The author sent several hybrids to New Zealand, where two performed equally well, surpassing all other hybrids in each of three widely separated locations. In New Jersey, where both hybrids originated and had the same pollen parent, only one of these excelled. The better hybrid was outstanding also in Delaware, Virginia, and Oklahoma. Since adaptation cannot be predicted, it must be tested.

Since fusarium resistance is such an important character in any new asparagus release, the field trials should be situated on replanted old asparagus land, which is infested with *Fusarium*.

A more adequate screening technique in the greenhouse to test experimental progenies for resistance to fusarium is much needed before taking the progenies to the field. Field testing requires so much time and money that only progenies that pass the greenhouse screening test should go to the field. Current research in the United States and New Zealand on such a test is based on *Fusarium* inoculation of asparagus cultivars in the greenhouse known to be tolerant or susceptible to fusarium in the field. Progress on the technique has been slow so far, and it may be because the problem in the field is largely due to crown rot, and most greenhouse data have been based on root rot. Recent greenhouse research at Rutgers University indicated that certain cultivars did not differ markedly in resistance to root rot, but were quite different concerning crown rot resistance. The greenhouse crown rot results were well correlated with crown rot responses of the same cultivars in the field. In order to evaluate cultivars for crown rot in the greenhouse, it seems that plants with well-developed crowns must be used. Most of the research in the past is based on direct seeding into *Fusarium*-contaminated soil in the greenhouse, or on seedling transplants only 12 weeks old, which have very little crown tissue. More experiments are currently being planned with 1-year-old crowns, which we hope will simulate field crowns better than younger seedlings do.

Observational trials of a single replication can be very useful in evaluations of large numbers of progenies on commercial farms in the asparagus-growing areas of the state. County agricultural agents usually can be counted on to cooperate in finding good growers who will be pleased to see the new progenies on their farms. Single rows 100 m in length are sufficient for trial. Three county agricultural agents in New Jersey cooperate by obtaining early yield records (2 weeks).

Replicated trials of the more promising hybrids should be conducted at the AES nearest the asparagus-growing area. Little is published concerning plot size and number of replications needed to give various levels of precision. Unpublished data from New Zealand indicated that single-row plots 3 m in length with 14 replications would be the best design for testing approximately 20 unrelated cultivars. A trial consisting of diallel combinations would not need so many replications to measure main effects because of the repetitions of each parent of one sex being crossed with all the parents of the opposite sex. Most workers use four to six replications of single-row plots approximately 15 m in length.

The author has received requests for seeds of advanced experimental hybrids from workers in several states as well as in many foreign countries. The cooperation provided by these workers in evaluating the Rutgers hybrids has been invaluable and mutually beneficial. There are plans to make several of the hybrids available in the regions where they show high yield and local adaptation.

SEED PRODUCTION

There are commercial tissue culture laboratories that can clone the parents for commercial hybrid seed production. Seed companies should produce certified seed, inspected by the state, to ensure future availability and genetic integrity of the new hybrids. Care must be taken that only the parent clones are included in the seed block. Seedling volunteers must be removed annually before they can flower and contaminate the source of pollen. Isolation of at least 300 m from other asparagus fields and volunteers must be maintained. Seed harvesting, cleaning, packaging, and labeling usually is supervised by state inspectors so that the term Certified Seed has real meaning to the buyer.

ACKNOWLEDGMENTS

The photographs for many of the illustrations (Figures 2, 3, 5A, 5B, 7, 8, 9, 16, 17, 18, 25, and 26) were supplied by courtesy of the New Jersey Agricultural Experiment Station, New Brunswick, New Jersey.

REFERENCES

1. Bailey, Liberty Hyde 1942. The Standard Cyclopedia of Horticulture, Vol. 1, pp. 406–407. Macmillan Publishing Co., New York.
2. Banga, O. 1949. Veredeling van de asperge in Californie. Meded. Dir. Tuinbouw (Neth.) *12*, 264–270.
3. Barrons, K. C. 1945. The field snapping method of harvesting asparagus. Mich., Agric. Exp. Stn., Q. Bull. *28*, 111–114.
4. Bassett, M. J., Synder, R. J., and Angell, F. F. 1971. Efficient detection of asparagus monoploids for the production of colchiploid inbreds. Euphytica *20*, 299–301.
5. Carolus, R. L. 1949. Yield and quality of asparagus harvested by the field snapping method. Mich., Agric. Exp. Stn., Q. Bull. *31*, 370–377.
6. Corriols-Thevenin, L. 1979. Different methods in breeding asparagus. *In* Eucarpia, Section Vegetables, Proceedings of the Fifth International Asparagus Symposium. G. Reuther (Editor), pp. 8–20. Forschungsanstalt Geisenheim, F. R. Germany.
7. Currence, T. M., and Richardson, A. L. 1937. Asparagus breeding studies. Proc. Am. Soc. Hortic. Sci. *35*, 554–557.
8. Ehlert, G. R., and Seeling, R. A. 1966. Fruit and Vegetable Facts and Pointers (3rd Rev.), pp. 1–16. United Fresh Fruit and Vegetable Association, North Washington at Madison, Alexandria, VA.
9. Ellison, J. H., Reynard, G. B., Scheer, D. F., and Wagner, J. J. 1960. Estimating comparative yields of asparagus strains without full season harvest records. Proc. Am. Soc. Hortic. Sci. *76*, 376–381.
10. Ellison, J. H., and Scheer, D. F. 1959. Yield related to brush vigor in asparagus. Proc. Am. Soc. Hortic. Sci. *73*, 339–344.
11. Ellison, J. H., Scheer, D. F., and Wagner, J. J. 1960. Asparagus yield as related to plant vigor, earliness and sex. Proc. Am. Soc. Hortic. Sci. *75*, 411–415.
12. Ellison, J. H., and Schermerhorn, L. G. 1958. Selecting superior asparagus plants on basis of earliness. Proc. Am. Soc. Hortic. Sci. *72*, 353–359.
13. Ellison, J. H., Vest, G., and Langlois, R. W. 1981. 'Jersey Centennial' asparagus. HortScience *16*, 349.
14. Flory, W. S., Jr. 1932. Genetic and cytological investigations on *Asparagus officinalis* L. Genetics *17*, 432–467.
15. Grogan, R. G., and Kimble, K. A. 1959. The association of *Fusarium* wilt with the asparagus decline and replant problem in California. Phytopathology *49*, 122–125.
16. Haber, E. S. 1932. Effect of size of crown and length of cutting season on yield of asparagus. J. Agric. Res. (Washington, D.C.) *45*, 101–109.
17. Halstead, B. D. 1896. Asparagus rust reported in New Jersey. Gard. For. *9*, 394.
18. Huyskes, J. A. 1959. A method for selecting asparagus varieties with a long fiberless top. Euphytica *8*, 21–28.
19. Irizarry, H., Ellison, J. H., and Orton, P. 1965. Inheritance of persistent green color in *Asparagus officinalis* L. Proc. Am. Soc. Hortic. Sci. *87*, 274–278.
20. Johnston, S. A., Springer, J. K., and Lewis, G. D. 1979. *Fusarium moniliforme* as a cause of stem and crown rot of asparagus and its association with asparagus decline. Phytopathology *69*, 778–780.
21. Jones, H. A., and Hanna, G. C. 1939. A comparison of some asparagus varieties in California. Canner (April 15).

22. Kahn, R. P., Anderson, H. W., Hepler, P. R., and Linn, M. B. 1952. An investigation of asparagus rust in Illinois. Ill., Agric. Exp. Stn., Bull. *559.*
23. Kidner, A. W. 1959. Asparagus. Faber and Faber, London.
24. Malnassy, P., and Ellison, J. H. 1970. Asparagus tetraploids from callus tissue. HortScience *5,* 444–445.
25. Molot, P. M., and Simone, J. 1969. Mise au point des connaissances sur les maladies de l'Asperge. Pepinieristes Horticulteurs Maraichers No. 100, pp. 6041–6045.
26. Murashige, T., Shabde, M. N., Hasegawa, P. M., Takatori, F. H., and Jones, J. B. 1972. Propagation of asparagus through shoot apex culture. I. Nutrient medium for formation of plantlets. J. Am. Soc. Hortic. Sci. *97,* 158–161.
27. Norton, J. B. 1913. Methods used in breeding asparagus for rust resistance. U.S. Dep. Agric. Bur. Plant Ind. Bull. *263.*
28. Norton, J. B. 1919. Washington Asparagus: Information and suggestions for growers of new pedigreed rust-resistant strains. U.S. Dep. Agr. Bur. Plant Ind., C., T., & F. C. D. Circ. 7.
29. Q and A about Plant Patents 1970. A U.S. Dept. of Commerce Patent Office Publication, 0-381-362. U.S. Government Printing Office, Washington, D.C.
30. Rick, C. M., and Hanna, G. C. 1943. Determination of sex in *Asparagus officinalis* L. Am. J. Bot. *30,* 711–714.
31. Robbins, W. W., and Jones, H. A. 1925. Secondary sex characters in *Asparagus officinalis* L. Hilgardia *1,*183–202.
32. Scheer, D. F., and Ellison, J. H. 1960. Asparagus performance as related to seedling vigor. Proc. Am. Soc. Hortic. Sci. *76,* 370–374.
33. Scheer, D. F., Ellison, J. H., and Johnson, M. W. 1960. Effect of fruit maturity on asparagus seed germination. Proc. Am. Soc. Hortic. Sci. *75,* 407–410.
34. Schermerhorn, L. G. 1947. Is field snapping of asparagus practical in New Jersey? Hortic. News, Rutgers Univ. *29,* 1956–1957.
35. Sneep, J. 1953. The significance of andromonoecy for the breeding of *Asparagus officinalis* L. Euphytica *2,* 89–95.
36. Sneep, J. 1953. The significance of andromonoecy for the breeding of *Asparagus officinalis* L. II. Euphytica *2,* 224–228.
37. Snope, A. J., and Ellison, J. H. 1963. Storage of asparagus pollen under various conditions of temperature, humidity, and pressure. Proc. Am. Soc. Hortic. Sci. *83,* 447–452.
38. Thuesen, A. 1960. Cytogenetical studies in *Asparagus officinalis* L. R. Vet. Agric. Univ., Yearb., Copenhagen, *47,* 47–71.
39. U.S. Department of Agriculture. 1960, 1970, 1980. Agricultural Statistics. U.S. Government Printing Office, Washington, DC.
40. U.S. International Trade Commission 1976. Asparagus, Publ. No. 755. U.S. Int. Trade Comm., Washington, DC.
41. Wellensiek, S. J. 1949. The selection of one year old male asparagus plants. Meded. Dir. Tuinbouw (Neth.) *12,* 876–889.
42. Wellensiek, S. J., and Jonkers, H. 1957. The effect of selecting very young asparagus plants. Meded. Dir. Tuinbouw (Neth.) *20,* 506–511.
43. Yang, H.-J. 1977. Tissue culture technique developed for asparagus propagation. HortScience *12,* 140–141.
44. Yeager, A. F., and Scott, H. 1938. Studies of mature asparagus plantings with special reference to sex, survival and rooting habits. Proc. Am. Soc. Hortic. Sci. *36,* 513–514.
45. Young, R. E. 1937. Yield-growth relations in asparagus. Proc. Am. Soc. Hortic. Sci. *35,* 576–577.

Index of Scientific
and Common Names

A

A